Alistair Barros • Daniel Oberle
Editors

Handbook of Service Description

USDL and Its Methods

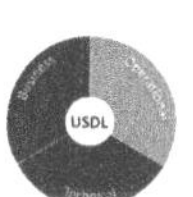

Editors
Alistair Barros
Queensland University of Technology
Brisbane, Australia
alistair.barros@qut.edu.au

Daniel Oberle
SAP Research
Karlsruhe, Germany
d.oberle@sap.com

ISBN 978-1-4899-9897-2 ISBN 978-1-4899-9897-2 (eBook)
DOI 10.1007/978-1-4614-1864-1
Springer New York Heidelberg Dordrecht London

Softcover reprint of the hardcover 1st edition 2012

Printed on acid-free paper

Springer is part of Springer Science+Business Media (www.springer.com)

To my wife Kylie, and our beautiful babies, Marion, Thomas, Emily and Veronica — with love from Alistair

Preface

The Positioning of Services

We are at the dawn of the long anticipated services revolution. To be sure, the notion of a service is hardly new, for services have been an ostensible feature of the way labor is organized to deliver consumer value since at least the shift to the post-industrial age. Indeed, services, together with goods, characterize the outputs of human organized systems, as understood by macroeconomics, no less. They have increased in prominence, under globalization and deregulation, as units of functionality that influence organizational restructures and outsourcing on a global scale. Look into most company operational plans, and services are among the key reference points for how work is coordinated both internally and externally, and against that which productivity is measured. Technologically, it is now some 10 years since Web services were proposed as the mechanism for unlocking valuable, often stove-piped logic, from software applications, and interoperating these across heterogeneous stacks, applications, business units and company boundaries. In the intervening years, the maturity towards the Service-oriented Architecture (SOA) has followed, through considerable investments and efforts by business, technology and research sectors.

So, why the excitement about services now? The reason is simple: the Internet and mobile communications, coupled with new and disruptive business models, are lifting up the conventional barriers to service access in an unprecedented way. Beyond familiar Web consumer services such as Facebook, Twitter, eBay, Amazon, iTunes, Google Maps, PayPal, FlickR, technological breakthroughs, especially around smart devices and cloud computing, are ushering in a dramatic growth of services. Mobile "Apps" and software-as-a-service are growing by the day. Business process outsourcing on-demand, multi-enterprise business process in the cloud, service marketplaces, service-centric business networks, and platform and infrastructure-as-a-service, are also on the rise and remonetizing services beyond their original settings. Moreover, mainstream segments such as transportation and logistics, banking and finances, public sector and manufacturing, and enterprise software providers, are slowly but surely following suit, when one considers the following Web-based services entering the "long tail:" carrier bookings and track-and-trace of shipments, world-wide tariff look-ups, news events, loan originations and servicing, healthcare (the American Health Level 7 standard), business formation, enterprise software services, water/energy utility monitoring, platform services such as business process management and enterprise services bus, and virtualized IT infrastructure services.

With the growth in number and sophistication of services widely available, the question turns to how effectively consumers can discover, understand and access services — with relative independence and without full reliance on providers. Experience has so far shown that any attempt to describe services faces a common stumbling block: *what is a service*? Despite the widespread phenomenon of "service" in economic, political, business, communal and individual walks of life — and, undoubtedly, because of that diversity — there are still many uncertainties and tensions in arriving at a general conception of services.

Some services concern human endeavor or largely human interactions, such as project management, sales, consultancy, therapy of different sorts, church worship, and bus, train and other transportation services. These clearly don't fit the motif of services understood as Web services. At the other extreme are technical services providing platform and infrastructure functionality whose complexities and resource dependencies challenge the consumer "Apps" motif of services. In between are business applications delivered through business units, actors and designated work centers and channels. A mix of human and automated tasks are involved, where service knowledge is dispersed across operational procedures, transactions and the data inside databases. Even in this regular form encountered in businesses, the notion of service strikes ambiguities. In the example of a travel agency offering flight bookings, what is the service? Is it flight bookings as a business function, the flight booking application, or the flight services involving different airplanes that are sought?

The Need for Explicit Service Descriptions

In the IT community, SOA languages such as the Web Services Definition Language (WSDL) [4], the Web Application Description Language (WADL) [10], or the Web Services Business Process Execution Language (WS-BPEL) [1], have focused on describing services and their interactions in a uniform way, for leveraging heterogeneous technologies. By consequence, other Web services specifications such as WS-Policy [2] and Web Service Level Agreements [17], even if they concern operational issues, are fixed to the particular view of a service as software. The Universal Description and Directory Service (UDDI) [5] specification was defined for a standard naming and directory service, as part of a Web services architecture, so that consumers can discover and interact with services. UDDI, too, has a software focus while allowing arbitrary non-technical and non-functional attributes (e.g., pricing) to be supported through a service description scheme which has to be defined by the user, though.

Approaches for service description in the realm of Semantic Web Services, e.g., OWL-S [13], Web Services Modeling Ontology (WSMO) [16], Semantic Annotations for WSDL (SAWSDL) [8] or Semantic Annotations for REST (SA-REST) [9], have anchored programmatic descriptions of services with conceptual meaning through ontologies in order to automate *Web* services discovery, composition, access and interoperability. The prospect of improving automatic discovery, interaction and composition of services has additionally spurred efforts to conceptualize the wider context in which services are accessed, seen through the SOA Reference Model (SOA-RM) [7] and its semantic form [14]. Through these efforts, concepts such as *capability*, to define the exposed functionality of a service, *policy*, to constrain how a service can operate, and *service provider*, to capture the agent responsible for delivering a service, *interaction protocol* etc. — have gained consensus in the community. As with the work on Semantic Web Services, the target of these languages is on service interoperability through architectural frameworks. Thus, the form of service under consideration remains with software.

With the emergence of on-demand applications, the notion of software-as-a-service has arisen, covering software applications (e.g., customer relationship management on-demand) and business process outsourcing (e.g., gross-to-payroll processing, insurance claims processing) to cloud and platform services. The emphasis of service here implies that the consumer gets the designated functionality he/she requested together with hosting through a consumption-based model (such as pay-per-use). Thus, software-as-a-service is not synonymous with Web services and the service providers need to carefully disclose non-functional aspects such as pricing and availability and to factor these into the overall service they deliver. Services, in other words, are more than core functions that are accessed by consumers. They are delivered by a provider to a consumer possibly over a specified period of time, in a particular geographic context, with a pricing model and payment structure, monitored with a service level agreement, and related legal obligations of the consumer and the provider [6]. The functionality together with constraints, rights, obligations and penalties understood between providers and consumers for delivery, moves toward an understanding of service encountered in commercial practice.

A further dichotomy in the understanding of services is the distinction between business and software (or technical) services. Ironically enough, software practitioners appeal to the notion of business or enterprise services to emphasize the business relatedness of software solutions, while business practitioners take it for granted that their software applications are used in commercial operations. More specifically, different parts of commercial organizations catalogue their services assets to different ends. The focus of governance portfolios tends to focus on business services (and business processes) as integral to business operations — meaning their cost centers, organizational objectives, customer segments and volumes, operating margins, profits, revenue targets etc. An explicit alignment of business and IT services through formal mechanisms such as governance and enterprise architectures is rare in practice. Instead, IT services are separately managed through software registries, which describe software services, their technical dependencies and supportive platforms. The separation of concerns across business and technical portfolios largely explains the conceptual impediment for a holistic cognizance of services; indeed, one that is still prevalent today.

One of the first attempts at comprehensively describing services was the work of O'Sullivan [15]. This work drew from practical insights into how everyday services such as hotel accommodation, hair-dressing, house building and insurance, are advertised and offered, to a scheme for describing services and a variety of delivery aspects including locative, temporal, pricing, payment, security, trust and rewards. As O'Sullivan presciently observed: "The everyday services that surround us, and the ways in which we engage with them, are the result of social and economic interaction that has taken place over a long period of time. Any attempt to provide automated electronic services that ignores this history will deny consumers the opportunity to negotiate and refine over a large range of issues, the specific details of the actual service to be provided."

About USDL

The need for a new stage of maturity for service conceptualization across all key aspects, and shaping the standardization of a next generation service description language, has paralleled a wider development.

As the different research and development efforts in SOA, software-as-a-service, cloud computing, service management methodologies and governance, a dedicated intellectual foundation for the services — as a field in its own right — has been sought. In 2006, Henry Chesbrough and Jim Spohrer published *A Research Manifesto for Services Science* [3] that argues for a new multi-disciplinary academic to integrate across academic silos and advance service innovation more rapidly. Accordingly, several strategic research initiatives and flagships were established, notably the following:

- The EU Framework Programme 7 has had as a key strategic theme, viz., the *Internet of Services*, leading to several millions of Euros in research investments across at least 20 projects. These concerned different aspects of business and IT service management, beyond single organizations, out to service-based hubs, communities and business networks, and ultimately out to the Internet;
- *Future Internet Public Private Partnership (FI-PPP)* [11] with a budget of 90 million Euros aiming to advance Europe's competitiveness in future internet technologies and systems and to support the emergence of future internet-enhanced applications of public and social relevance. It addresses the need to make public service infrastructures and business services/processes significantly smarter (i.e., more intelligent, more efficient, more sustainable) through tighter integration with Internet networking and computing capabilities. The frameworks of the Internet of Services and Internet of Things underpin the Future Internet vision;

- THESEUS/TEXO [18], one of Germany's largest publicly funded IT research projects in recent years, addressing the fact that it has become commonplace to sell content such as music and videos on the Internet, yet Web-based services are not as widely used. The goal of the project is develop an infrastructure that will make it easier to combine and utilize the electronic services in wide service-based hubs, communities and networks, as an important step towards fostering an Internet of Services;
- In the US, an industry consortium led by IBM sponsors the *Service Research and Innovation Institute (SRII)*,[1] a non-profit organization aimed at improving the productivity and transformation for the technology industry, organizations and society at large — around services. It brings together industry, technology, solutions, research and academic organizations to share their work and experiences on all the key areas of services. It strives for shaping the science and engineering for service delivery in healthcare, financial, telecom, retail, education, government, and energy, to name just a few verticals.
- In the APJ region, a similar endeavor has been initiated called the *Smart Services CRC*.[2] The CRC is a commercially focused collaborative research initiative, developing innovation and productivity improvements for the services sector, especially for small-to-medium businesses. It has drawn representations from the enterprise service and solution specialists (Infosys), media (Fairfax Digital), government (the state governments of Queensland and New South Wales), health (Austin Hospital) to collaborate with a number of Australian universities in yearly projects over a seven year horizon. It aims at innovative (smart) services, agile tools for aggregation and next-generation service delivery platforms.

Out of a number of these mega-investments, the *Unified Service Description Language (USDL)* was born. It has been developed across several research institutes and publicly funded projects across Europe and Australia, and this now extends to the Americas as part of a standardization push through W3C.[3] The overarching philosophy of development has been inspired from the design science approach [12]. In addition, USDL required a highly collaborative and interdisciplinary approach. Previously developed service description concepts, languages and experiences were harnessed, and USDL, at the outset, was situated at the conceptual level so that a variety of aspects could be analyzed without constraint of any one implementation language or technology.

Clearly, the key challenge for USDL is scope: what sorts of services can be uniformly described? That is to say, is a uniform conception of services across political, economic, business, entertainment, technological, individual and other spheres, in the first place, possible or desirable? In the case of USDL, the scope of service has been on services as understood for business and supportive IT provisioning, i.e., the socio-technical sense of services. In this respect, purely human, purely automated and mixed human/automated services were considered, that have a boundary

[1] `http://www.thesrii.org/`

[2] `http://www.smartservicescrc.com.au/`

[3] `http://www.w3.org/2005/Incubator/usdl/`

of cognizance that is available through the tasks of service provisioning, discovery, access and delivery. Services from various domains including cloud computing, service marketplaces and business networks, were investigated. Noteworthy were use cases involving service marketplaces procuring services as complex as those from SAP's portfolio and ecosystem, advancing previous insights into aspects such as service bundling and business contexts in which services are requested and consumed. Further use cases from the corporate world shed insights into commercial management of services, such as cost center ownership, releasing, and service granularity validated against enterprise-grade software portfolios.

Taken together, key dichotomies, encountered in the current state-of-the art service languages and techniques, that were addressed in USDL included: technical and business; structure and behavior; intra- and inter-organizational; single- to third-party provisioned; single-flavored composition (process-based) to multi-flavored composition (process, data dependency, functional inheritance/import and bundling); singular to plural pricing models; functional to non-functional service delivery (service level agreements).

About the Book

This book provides a state-of-the-art insight into previous developments, the design and specific proposals of USDL, and different methods that use USDL for service design, engineering and management.

Part I of the book provides an in-depth overview of existing state-of-the-art service description approaches. The plethora of approaches is grouped into different strands each devoted a separate chapter. Experts were invited to survey *Product-Service System Approaches*, *Service Network Approaches*, *Service System Approaches*, *Service-oriented Architecture (SOA) Approaches*, as well as *Semantic Web Services*, respectively.

The remaining parts of the book deal exclusively with USDL and can be read independently from the state of the art. Part II is concerned with the actual meta-model of USDL providing both an overview chapter on the design rationale as well as several chapters that highlight specific aspects of the language. Contributions to this part came from researchers in the different projects that developed the different parts of USDL.

In light of the efforts and insights of multiple institutes and disciplines — comprising business management, information systems, IT and computer science (incl. SOA, security and cloud), and law — the design rationale set the ground for a consensual design of USDL across diverse and distributed teams that contributed. The design proceeded through a constructivist synthesis, whereby a services discourse and requirements illuminating on key challenges for USDL concept formation ("signposts" model development), were established. The result is that USDL contains:

- concepts (either well-established or new and agreed upon) that are essential to service descriptions, and self-contained in USDL;
- concepts that are part of USDL but serve to align service descriptions with other artifacts, e.g., USDL should relate to, not overlap with, languages dedicated to other organizational phenomena (e.g., business processes, organizational resources, WSDL and other SOA aspects);
- concepts that can support domain-specific (e.g., industry specific) specializations, since a "silver bullet" language for all service domains and industries is infeasible.

Given the complexity of the service domain, USDL has been designed to be conceptual and modular. Specifically, USDL's structure can be seen from the following, broad logical arc:

- The essential descriptors of a service that are central to understanding it and that tie together other parts (*Service*);
- The structural aspect of functionality (*Functional*);
- The behavioral aspect of functionality (*Interaction*);
- How a service is interfaced with for delivery (*Technical*);
- The participants involved in the provisioning, delivery an consumption of a service (*Participants*);
- The non-functional aspects of pricing (*Pricing*), legal constraints (*Legal*), and service level agreements (*Service Level*).

Knowledge about the USDL meta-model is the prerequisite for Part III whose chapters deal with methodological aspects of USDL, e.g., basic tooling or variant management. Finally, Part IV documents different evaluation endeavors of USDL such as case studies.

Acknowledgements

The people who were instrumental in development of USDL and this book are too numerous to mention. Nevertheless, we would like to draw attention to the following for special mention.

To Lutz Heuser and Joachim Schaper, who as (former) leaders of SAP Research installed "Internet of Services" as a flagship of its research and development, and who ensured a proper dissemination and opening up of USDL for wider consolidation through the community and for standardization, once it arrived at a sufficient level of maturity. This book owes foremostly to the priority on dissemination that they placed.

Secondly, the authors are grateful to Hervé Couturier (Executive Vice President of SAP Business Technology and Research) and to Warren Brady (CEO of the Smart Services CRC) for supporting the ongoing development of USDL and the development of this book.

Last but by no means least, the development of USDL involved many more people from research and industry than is indicated through the authorship of the book's USDL chapters and in the publicly released specification of USDL (available at www.internet-of-services.com). USDL, as it stands today, and its intellectual baseline available through this book for on-going refinement and maturity, rests of the shoulders of many "giants."

Brisbane, Australia & Karlsruhe, Germany,
September 2011

Alistair Barros
Daniel Oberle

References

1. A. Alves, A. Arkin, S. Askary, C. Barreto, B. Bloch, F. Curbera, M. Ford, Y. Goland, A. Guizar, N. Kartha, C.K. Liu, R. Khalaf, D. König, M. Marin, V. Mehta, S. Thatte, D. van der Rijn, P. Yendluri, and A. Yiu. Web Services Business Process Execution Language Version 2.0. OASIS Standard 11 April 2007. http://docs.oasis-open.org/wsbpel/2.0/OS/wsbpel-v2.0-OS.html.
2. S. Bajaj, D. Box, D. Chappell, F. Curbera, G. Daniels, P. Hallam-Baker, M. Hondo, C. Kaler, D. Langworthy, A. Nadalin, N. Nagaratnam, H. Prafullchandra, C. von Riegen, D. Roth, J. Schlimmer, C. Sharp, J. Shewchuk, A. Veadmuthu, Ü. Yalcinalp, and D. Orchard. Web Services Policy 1.2 - Framework (WS-Policy). W3C Member Submission 25 April 2006. http://www.w3.org/Submission/WS-Policy/.
3. H. Chesbrough and J. Spohrer. A Research Manifesto for Services Science. *Commun. ACM*, 49(7):35–40, 2006.
4. E. Christensen, F. Curbera, G. Meredith, and S. Weerawarana. Web Services Description Language (WSDL) 1.1. W3C Note 15 March 2001. http://www.w3.org/TR/wsdl.
5. L. Clement, A. Hatley, C. von Riegen, and T. Rogers. UDDI Version 3.0.2 UDDI Spec Technical Committee Draft, Dated 20041019. http://www.uddi.org/pubs/uddi_v3.htm.
6. B. Dietrich. Resource planning for business services. *Communications of the ACM*, 49(7):62–64, 2006.
7. J. A. Estefan, K. Laskey, F. G. McCabe, and D. Thornton. Reference Architecture Foundation for Service Oriented Architecture Version 1.0. OASIS Committee Draft 02, 14 October 2009. http://docs.oasis-open.org/soa-rm/soa-ra/v1.0/soa-ra-cd-02.pdf.
8. J. Farell and H. Lausen. Semantic Annotations for WSDL and XML Schema. W3C Recommendation 28 August 2007. http://www.w3.org/TR/sawsdl/.
9. K. Gomadam, A. Ranabahu, and A. Sheth. SA-REST: Semantic Annotation of Web Resources. W3C Member Submission 05 April 2010. http://www.w3.org/Submission/SA-REST/.
10. M. Hadley. Web Application Description Language. W3C Member Submission 31 August 2009. http://www.w3.org/Submission/wadl/.
11. L. Heuser and D. Woods. Is Europe Leading the Way to the Future Internet? *IEEE Internet Computing*, 14:91–94, 2010.
12. A. R. Hevner, S. T. March, J. Park, and S. Ram. Design science in information systems research. *MIS Quarterly*, 28(1):75–105, 2004.

13. D. L. Martin, M. Paolucci, S. A. McIlraith, M. H. Burstein, D. V. McDermott, D. L. McGuinness, B. Parsia, T. R. Payne, M. Sabou, M. Solanki, N. Srinivasan, and K. P. Sycara. Bringing Semantics to Web Services: The OWL-S Approach. In J. Cardoso and A. P. Sheth, editors, *Semantic Web Services and Web Process Composition, First International Workshop, SWSWPC 2004, San Diego, CA, USA, July 6, 2004, Revised Selected Papers*, volume 3387 of *Lecture Notes in Computer Science*, pages 26–42. Springer, 2004.
14. B. Norton, M. Kerrigan, A. Mocan, A. Carenini, E. Cimpian, M. Haines, J. Scicluna, and M. Zaremba. Reference Ontology for Semantic Service Oriented Architectures Version 1.0. OASIS Public Review Draft 01, 5 November 2008. `http://docs.oasis-open.org/semantic-ex/ro-soa/v1.0/see-rosoa-v1.0.html`.
15. J. J. O'Sullivan. *Towards a Precise Understanding of Service Properties*. PhD thesis, Queensland University of Technology, 2006.
16. D. Roman, J. de Bruijn, A. Mocan, H. Lausen, J. Domingue, C. Bussler, and D. Fensel. WWW: WSMO, WSML, and WSMX in a Nutshell. In R. Mizoguchi, Z. Shi, and F. Giunchiglia, *The Semantic Web - ASWC 2006, First Asian Semantic Web Conference, Beijing, China, September 3-7, 2006, Proceedings*, volume 4185 of *Lecture Notes in Computer Science*, pages 516–522. Springer, 2004.
17. W. Sun, J. Zhang, and F. Liu. WS-SLA: A Framework for Web Services Oriented Service Level Agreements. In *Proceedings of the 10th International Conference on CSCW in Design, CSCWD 2006, May 3-5, 2006, Southeast University, Nanjing, China*, pages 714–717. IEEE, 2006.
18. O. Terzidis, A. Fasse, B. Flügge, M. Heller, K. Kadner, D. Oberle, and T. Sandfuchs. Texo: Wie THESEUS das Internet der Dienste gestaltet — Perspektiven der Verwertung. In L. Heuser and W. Wahlster, editors, *Internet der Dienste*, acatech diskutiert, pages 141–161. Springer, 2011.

Contents

Part II USDL — Meta-Model

Part III USDL — Methods

Part IV USDL — Evaluation

List of Contributors

Hans Akkermans
VU University, Amsterdam, The Netherlands

Matthias Allgaier
SAP Research Karlsruhe, Germany

Alistair Barros
Queensland University of Technology, Australia

Christian Baumann
Berkeley Center for Law & Technology, University of California, Berkeley, School of Law, USA
Center for Applied Legal Studies (ZAR), Karlsruhe Institute of Technology, Germany

Jörg Becker
University of Münster, European Research Center for Information Systems, Germany

Daniel Beverungen
University of Münster, European Research Center for Information Systems, Germany

Michele Bezzi
SAP Research Sophia-Antipolis, France

Dominik Q. Birkmeier
University of Augsburg, Germany

Achim D. Brucker
SAP Research Karlsruhe, Germany

Stephen Dawson
SAP Research Belfast, UK

Michael Dietrich
SAP Research Karlsruhe, Germany

Keith Duddy
Queensland University of Technology, Australia

Roberta Ferrario
ISTC-CNR, Laboratory for Applied Ontology, Trento, Italy

Matthias Flügge
Fraunhofer FOKUS, Berlin, Germany

Andreas Friesen
SAP Research Karlsruhe, Germany

G. R. Gangadharan
IBM India Research, India

Jaap Gordijn
VU University, Amsterdam, The Netherlands

Nicola Guarino
ISTC-CNR, Laboratory for Applied Ontology, Trento, Italy

Alan Hartman
IBM Research Lab, Haifa, Israel

Matthias Heinrich
SAP Research Dresden, Germany

Steffen Heinzl
SAP Research Darmstadt, Germany

Markus Heller
SAP Research Karlsruhe, Germany

Stijn Heymans
SemanticBits, Herndon, VA, USA

Jörg Hoffmann
INRIA, Nancy, France

Edzard Höfig
Fraunhofer FOKUS, Berlin, Germany

Andrea Horch
University of Stuttgart, Institute IAT, Germany

Christian Janiesch
Institute of Applied Informatics and Formal Description Methods (AIFB), Karlsruhe Institute of Technology, Karlsruhe, Germany

Kay Kadner
SAP Research Dresden, Germany

Tom Kiemes
SAP Research Karlsruhe, Germany

Maximilien Kintz
University of Stuttgart, Institute IAT, Germany

Andreas Klein
SAP Research Karlsruhe, Germany

Martin Knechtel
SAP Research Dresden, Germany

Thomas Kohlborn
Queensland University of Technology, Australia

Ronny Kursawe
Technische Universität Dresden, Faculty of Computer Science, Germany

Uwe Kylau
SAP Research Brisbane, Australia

Ken Laskey
The MITRE Corporation, McLean, VA, USA

Pieter De Leenheer
VU University, Amsterdam, The Netherlands
Collibra nv, Brussels, Belgium

Bastian Leferink
B2M Software AG, Karlsruhe, Germany

Torsten Leidig
SAP Research Karlsruhe, Germany

Jens Lemcke
SAP Research Karlsruhe, Germany

Maria Maleshkova
Knowledge Media Institute (KMi), The Open University, Milton Keynes, UK

Annapaola Marconi
Fondazione Bruno Kessler, Trento, Italy

Florian Marienfeld
Fraunhofer FOKUS, Berlin, Germany

Martin Matzner
University of Münster, European Research Center for Information Systems, Germany

Norman May
SAP AG, Walldorf, Germany

Oliver Müller
University of Liechtenstein, Martin Hilti Chair of Business Process Management, Liechtenstein

Maria Niedziella
SAP Research Karlsruhe, Germany

Michael Niemann
KOM - Multimedia Communications Lab, Technische Universität Darmstadt, Darmstadt, Germany

Francesco Novelli
SAP Research Darmstadt, Germany

Daniel Oberle
SAP Research Karlsruhe, Germany

Sven Overhage
University of Augsburg, Germany

Maryam Panahiazar
Kno.e.sis Center, Wright State University, Dayton, OH, USA

Jonas Pattberg
Fraunhofer FOKUS, Berlin, Germany

Carlos Pedrinaci
Knowledge Media Institute (KMi), The Open University, Milton Keynes, UK

Joshua Phillips
SemanticBits, Herndon, VA, USA

Ivan S. Razo-Zapata
VU University, Amsterdam, The Netherlands

Jörg Rech
SAP Research Karlsruhe, Germany

Philip Robinson
SAP Research Belfast, UK

Marcello La Rosa
Queensland University of Technology, Australia

Thorsten Sandfuchs
SAP Research Karlsruhe, Germany

Martin Schäffler
Siemens AG, Munich, Germany

Alexander Schill
Technische Universität Dresden, Faculty of Computer Science, Germany

Sebastian Schlauderer
University of Augsburg, Germany

Benjamin Schmeling
SAP Research Darmstadt, Germany

Wolfgang Karl Rainer Schwach
SAP Research Karlsruhe, Germany

Gabriel Serme
SAP Research Sophia-Antipolis, France

Virginia Smith
Hewlett-Packard Company, Roseville, CA, USA

Josef Spillner
Technische Universität Dresden, Faculty of Computer Science, Germany

Michael Stollberg
SAP Research Dresden, Germany

Gunther Stuhec
SAP Global Partner and Ecosystems Group, Walldorf, Germany

Orestis Terzidis
SAP Research Karlsruhe, Germany

Anke Thede
B2M Software AG, Karlsruhe, Germany

Wolfgang Theilmann
SAP Research Karlsruhe, Germany

Romano Trampus
University of Trieste, Italy

Klaus Turowski
Otto-von-Guericke University Magdeburg, Germany

Ingo Weber
The University of New South Wales, School of Computer Science & Engineering, Sydney, Australia

Monika Weidmann
Fraunhofer Institute for Industrial Engineering IAO, Stuttgart, Germany

Moritz Weiten
SEEBURGER AG, Bretten, Germany

Maciej Zaremba
Digital Enterprise Research Institute (DERI), National University of Ireland, Galway, Ireland

Chapter 1
The Internet of Services and USDL

Orestis Terzidis, Daniel Oberle, Andreas Friesen, Christian Janiesch, and Alistair Barros

Abstract A prominent research focus, especially in the context of EU public funding, has been the systematic use of the Internet for new ways of value creation in the services sector. This idea of service networks in the Internet, frequently dubbed the *Internet of Services* or Web service ecosystems, wants to make services tradable in digital media. In order to enable communication and trade between providers and consumers of services, the Internet of Services requires a standard that creates a "commercial envelope" around a service. This is where the *Unified Service Description Language (USDL)* comes into play as a normative and balanced unification of service information. The unified description established by USDL is machine-processable, considers technical and business aspects of a service as well as functional and non-functional attributes.

1.1 Services Sector: Key Driver of Developed Economies

The services sector is an economic growth driver in most developed economies. As an example, consider the Federal Republic of Germany, where the largest part of the macroeconomic value of 2009 is generated by the service industry [15, page 637].

Orestis Terzidis, Daniel Oberle, Andreas Friesen
SAP Research Karlsruhe, Vincenz-Priessnitz-Str. 1, 76131 Karlsruhe, Germany,
e-mail: orestis.terzidis@sap.com, e-mail: d.oberle@sap.com,
e-mail: andreas.friesen@sap.com

Christian Janiesch
Institute of Applied Informatics and Formal Description Methods (AIFB), Karlsruhe Institute of Technology, Englerstr. 11, Geb 11.40, 76131 Karlsruhe, Germany,
e-mail: christian.janiesch@kit.edu

Alistair Barros
Queensland University of Technology, GPO Box 2434, Brisbane, QLD 4001, Australia,
e-mail: alistair.barros@qut.edu.au

As can be seen in Fig. 1.1, jobs are only created in the services sector in Germany during recent years.

Many manufacturing firms transformed their orientation from products to hybrid value creation through complementary services during the last two decades [5]. Furthermore, a new breed of service businesses entered the stage, growing at a speed that is unprecedented in the history, changing the rules of the existing markets, and creating new ones. Powered by globalization, competition, and the Internet, this process happens globally and at accelerating speed [14]. It breaks existing product supply chains and transforms them into more volatile networks of collaborating businesses which are called *business value networks* or short *business networks* [28]. Business networks increasingly take the form of *service networks* where a business network forms around service value propositions of the participants in order to achieve win-win situations through joint value creation [8]. A service network is a logical collection of services whose exposure and access are subject to constraints characteristic of business service delivery. In these networks, service consumers procure services through different distribution and delivery channels, outsourcing service delivery functions such as payment, authentication, and mediation to specialist intermediaries. Service networks make explicit the notion of service procurement, separating it from that of conventional service supply. [7]

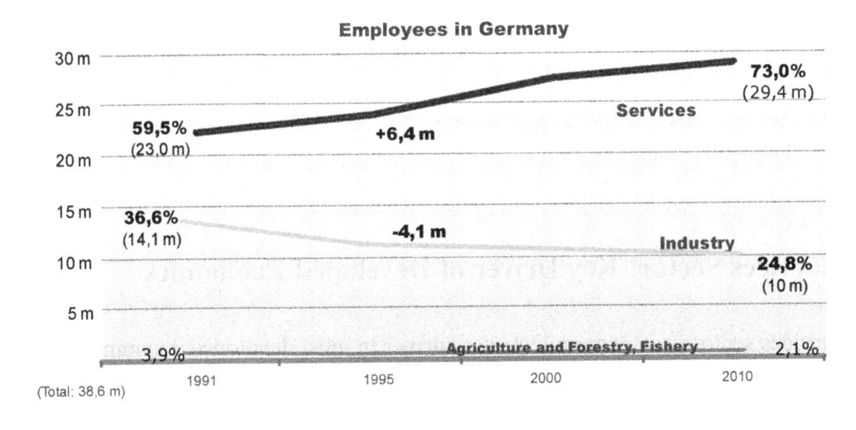

Fig. 1.1: Growth through services: Jobs created in Germany in the services sector [15, page 87].

This trend coincides with the ongoing industrialization of the services sector in developed economies. The industrialization, i.e., the commoditization of a good, can be traced into the steps of innovation, bespoke, product, and service [26]. In general, the three driving forces of every industrialization process have been automation, standardization, and specialization.[1] The following paragraphs will ex-

[1] As an example, consider the automotive industry. In this particular case, automation started by introducing production lines as early as Henry Ford up to the increasing application of robots. In

plain what the three drivers mean with respect to the transition from product to a services-focused sector. First, the Information and Communications Technology (ICT) serves as an *Automation* and transformation factor. Automation is the replacement of human activity by machine activities [21] in order to relieve humans from heavy, dangerous, complex, boring and time-consuming tasks [19]. At the low end, automation in the services sector implies, e.g., the digital brokering of physical services such as car repair. At the high end, service automation considers services that can be electronically consumed in wide settings, e.g., on the Web, which have grown over the last few years. Beyond Web services and services available through the Web, new and disruptive models have emerged that are accelerating the ubiquity of services. Software-as-a-service, business process outsourcing, cloud computing and infrastructure-as-a-service, platform-as-a-service, service marketplaces, and service-centric business networks are a growing list of examples where services are commoditized, exposed and accessed beyond conventional boundaries. In addition to Web consumers, the reach extends to mainstream industries like transportation and logistics, banking and finances, public sector, and manufacturing, when one considers the following sorts of Web services now available: track-and-trace of shipments, tariff look-ups, health insurance comparisons, medical assistance, business formation and ERP hosting.

Second, and similar to the area of physical products, *Standardization* [11] is the basis and prerequisite of every further development of an industrial sector. Therefore, standards will play a significant role also in the services industry. Standards are expected to drive the professionalization and industrialization of the service industry, to increase the transparency, and to lead to higher value services, and, thus, to contribute to the overall development of the service economy. [1] The need for standardization becomes apparent when one considers the uptake of service networks as explained above. When participants in a service network specialize to play a specific role in the provisioning and delivery of services, they act as intermediaries between other participants in the network. Such new specialized roles need to disclose and exchange, as well as comprehend business information about services (pricing, general terms and conditions, service-level agreements, etc.) in a standardized way. A standardized and machine readable description of such information will facilitate interoperability between such roles on the business level.

Besides the emergence of specialized intermediaries, there is another aspect of *Specialization* [20]: services once targeted at a specific market, are rebranded and repurposed to fit new consumer needs or other markets in order to extend the reach. Repurposing is facilitated by the automation, since even physical services can be brokered digitally and might be accessed by additional markets and regions.

The ongoing industrialization of the services sector spawned many research activities and even the call for a new discipline, viz., service science, surveyed in Section 1.2. A particular research focus is the systematic use of the Internet for new

---

the beginning, each car manufacturer developed its own tires, radios, or electronic components. By standardization of such components and their interfaces, specialization became possible. That means, component suppliers emerged that specialize in the production of tires, radios, or electronic components sold to a multitude of car manufacturers.

ways of value creation in the services sector — frequently called the *Internet of Services*. The basic concepts of the Internet of Services are explained in Section 1.3 where the need for a standardized service description is elaborated. The latter is represented by the *Unified Service Description Language (USDL)* (cf. Section 1.4). Finally, strategic implications are given in Section 1.5.

1.2 Intense Research During the Past Years

Today, the dominating understanding of economic activities is still centered on the product as the main focus of value exchange. A new foundation based on service-driven principles and establishing a service-dominant mindset is required to understand and exploit the innovative value proposition offered by the Service-orientation and the On Demand paradigms to its full potential in the context of software-based complex service systems. In 2006, Henry Chesbrough and Jim Spohrer published *A Research Manifesto for Services Science* [10] that argues for a new multi-disciplinary academic approach in order to integrate academic silos and advance service innovation more rapidly. As we have learned in the previous section, the services sector has grown over the last 50 years to dominate economic activity in most advanced industrial economies. Yet, the scientific understanding of modern services is rudimentary [10]. In 2008, a white paper proposing a framework to progress in service science and innovation has been published as a result of the Cambridge Service Science, Management and Engineering Symposium and a subsequent consultation process with the academic community [2]. According to [23], the notion of a service system serves as the proper basic category for systematic development of theory and practice around service innovation. In addition, [22] identifies the notion of service system as the basic abstraction of service science. [3] and [12] (cf. also Chapter 4) further refined and formalized the notion of service system, respectively.

Following the research manifesto and the seminal scientific groundwork, several initiatives, institutions, and research projects have been established in recent years that revolve around the subject of service science. In the US, an industry consortium led by IBM currently sponsors the *Service Research and Innovation Institute (SRII)*[2] which is a non-profit organization aiming at the improvement of productivity and quality for the technology industry, organizations and society at large. The SRII is accompanied by events such as the SRII Global Conference as a forum for industry, research/professional organizations, and academia to share their research work on all the key areas of services and especially connecting science and engineering to services delivered through major verticals such as health care, financial, telecom, retail, education, government, and energy, to name a few. In the APJ region, a similar endeavor has been initiated called the *Smart Services CRC*.[3] The CRC is a commer-

[2] http://www.thesrii.org/

[3] http://www.smartservicescrc.com.au/

cially focused collaborative research initiative, developing innovation, foresight and productivity improvements for the services sector.

In Europe, the European Union is currently spending a significant amount of money for service research. An example is the *Future Internet Public Private Partnership (FI-PPP)* with significant budget aiming to advance Europe's competitiveness in future internet technologies and systems and to support the emergence of future internet-enhanced applications of public and social relevance. It addresses the need to make public service infrastructures and business processes significantly smarter (i.e., more intelligent, more efficient, more sustainable) through tighter integration with Internet networking and computing capabilities.

The term *Internet of Services* has been coined in the context of EU public funding as a technology strategy for explicitly supporting service networks in the Internet.[4] The observation was that the Internet provides the access mechanisms but not the required service supply technology. Service supply comes from companies and communities using dedicated platforms for SOA and service delivery, and the Internet will not magically replace that. However, Internet supportive protocols and technologies will need to be extended so that services can be accessed more seamlessly than is currently the case. For example, when a service is being interacted with, it should be possible through Internet supportive technologies for other potentially relevant services and resources to be sensed through the Internet. Service discovery should not be stove-piped through keyword search on particular resources, but should be as seamless as accessing Web pages. The current barriers to matchmaking of service needs and capabilities — whether it be for finding the cloud services that can support the hosting needs of an application, a transportation service that works best in a geographic locality for a certain line of goods, or a B2B gateway that has capabilities for enabling trading partners to talk should operate on an Internet, not an individual repository, level.

The situation throughout Europe is comparable to national spendings for research projects. As an example, consider Germany's largest publicly funded ICT research programme in recent years called THESEUS [25]. The project addresses the fact that it has become commonplace to sell content such as music and videos on the Internet, yet Web-based services are not as widely used. The goal of the project is develop an infrastructure that will make it easier to combine and utilize Internet services, an important step toward creating an Internet of Services. An accompanying event has been the International Research Forum 2008 [27] inviting top researchers from around the world. Participants from business, academia and government examined the topic of the Internet of Services. The discussion illuminated how the Internet of Services will change enterprise computing. It will help businesses leverage core strengths, find partners in new business networks, collaborate and tap new global markets. Software as a service, cloud computing and other trends are democratizing innovation as never before. The infrastructure barriers to doing business in the Internet of Services are falling away.

---

[4] A synonymous term is "Web service ecosystems," i.e., service ecosystems in the Internet introduced in [7].

1.3 The Internet of Services — Basic Concepts

The basic idea of the Internet of Services is to systematically use the Internet for new ways of value creation in the services sector. There are different angles from which one may look at this approach. From an IT perspective, service-oriented architectures [9], software as a service [13], as well as business process outsourcing [24] and business process out-tasking are related trends. In this context, the concept of *service* is referring to a technical understanding of software functions provided as Web services.

But *services* in a broader sense are more than technical capabilities that can be invoked by computer program interfaces. When referring to the importance of the services economy in Section 1.1, we clearly went beyond the purely technical perspective. It is true that the modern service economy is already making an intense use of information technologies, but it is also true that other factors are relevant as well. Therefore, it is important to clarify what we mean by the term *service* Here is our proposal for a definition:[5]

Definition 1.1. A service is a commercial transaction where one party grants temporary access to the resources of another party in order to perform a prescribed function and a related benefit. Resources may be human workforce and skills, technical systems, information, consumables, land and others.

Let us illustrate the definition with some examples. First, think about a traditional service such as a taxi. It is a commercial transaction, as you pay for it. The taxi company grants temporary access to a socio-technical system including the driver, the car, the navigation system (or maps), the billing system and the call centre. The function and the related benefit are the transportation from one point to another.

Another example may be a rating service which defines the creditworthiness of companies. Before selling something to a company, one may want to know how creditworthy it is, so payment conditions can be arranged accordingly. For companies with poor ratings, payment in advance will be asked for while companies with good rating may pay on delivery. The rating service, by nature, is information. As a consequence, the main resources involved are information systems and, most importantly, their content. The temporary access to this information is identical to the service delivery. There is some price scheme to grant the access to this information; examples could be a flat rate or a pay-per-use model. The function is to receive a rating and the benefit to reduce the transaction risk.

Yet other examples include business services such as event management, transportation, insurances, attorney services, public services or medical services. It is also worth noting that not all services are commercial transactions, as some may have a social character such as nursing services that integrate family members.

[5] Other definitions of the term have surfaced in diverse disciplines (a nice discussion is given in [12]). This particular definition serves the needs to explain the basic concepts of the Internet of Services.

1.3.1 The Digital Footprint of Services

The service definition above covers a broad range of services. It is therefore useful to find criteria to differentiate the various kinds of services. One obvious criterion relates to the digital footprint of a service: how and to what degree are information technologies used to instantiate a service.

As an example, think about logistic services. The traditional service is about granting access to some socio-technical resources to ship something from one point to another. With digital media, the information flow associated to the material flow became more and more important. As an example, information about delays in *just-in-time* delivery is almost as important as the material flow itself. Even without introducing a strict metric, it is intuitive to say that the digital footprint of *just-in-time* logistics is higher than in traditional transportation.

In the example above, information technologies contribute to the coordination of the core process. But digital media can also add to customer experience. Take the example of a haircut. It is in itself a manual service. But the service may be *enriched* by digital media: the appointment may be done over the web, the hair cut may be chosen in advance at home out of a photo gallery or the hair colour may be presented as an overlay on the portrait photo of the customer.

Another dimension of digital footprint comes from new ways of customer *co-creation*. E-banking is a good example for an efficient co-creation. Here, the customer has web access to the bank's information systems directly over digital media. This *digital self service* makes the overall process more efficient and provides more transparency to the customer. Part of the service delivery is shifted to the service consumer, but creates value for both the service consumer and the service provider.

In general, a simple thought experiment can help to illustrate the digital footprint of a service. Just imagine all computer system would shut down for a given time. To what degree would it be possible to deliver the service? If you apply this thought experiment to the services of our daily life, it becomes quite clear that the *digital footprint* in services is already significant.

Yet, as of today, there is no normative way of describing the services in a unified and machine-readable way. Such a description would *wrap* a service and would expose it in a novel way. This kind of digital footprint is what we aiming for with the Unified Service Description Language.

1.3.2 Complementing the Service-Oriented Architecture paradigm

For Web services, the SOA paradigm including the Web Service Description Language (WSDL) did provide standards for technical service ecosystems. Based on these standards, the promise of SOA was to lower effort of integration of services coming from different information systems. Even if in many practical situations there are still organizational and technical obstacles to fully leverage the potential of SOA, it is clear that the paradigm does indeed offer possibilities to combine and integrate technical services in a faster, more flexible and more consistent way.

The Internet of Services and the Unified Service Description Language take this approach to the next level. They complement the SOA approach by adding the operational and business aspects to it. To illustrate this, consider the following. In definition 1.1 above, we made clear that a service should not only be considered as the invocation of a technical interface, but rather as an economic or social transaction with a broader context. A rating service may indeed be implemented as the invocation of a technical Web service of a given information system. For the context of this service, it is not only important to define how to technically invoke it (give an address and the interfaces), but it is essential to define the price scheme, the service level agreement, and the terms and conditions when consuming the service and paying for it.

There is another extension of the SOA paradigm. It relates to the roles that come into play in service networks. In SOA, three basic roles have been defined: the service provider, the service consumer and the service mediator or broker. In addition to these roles, we follow the proposal of [6] to also consider the service hoster, gateway, aggregator, and channel maker (cf. Fig. 1.2). The *service hoster* is an example for an intermediary that catalogues special types of services, namely infrastructure-as-a-service and platform-as-a-service offerings (commonly termed cloud-computing services). It also provides means to interface uniformly with the providers of these services, i.e., re-hosts services through cloud computing environments. Likewise, the *service gateway* is a specific intermediary that provides interoperability through cataloguing and interfacing with a choice of a 3rd Party B2B gateway, which provides services such as message translation and store-forward processing. The *service aggregator* provides additional value by packaging and combining services. Finally, the *service channel maker* is positioned at the consumer end of the service provisioning chain where services are channeled into user environments and consumed. Other roles may emerge when service networks establish, e.g., a clearinghouse role.

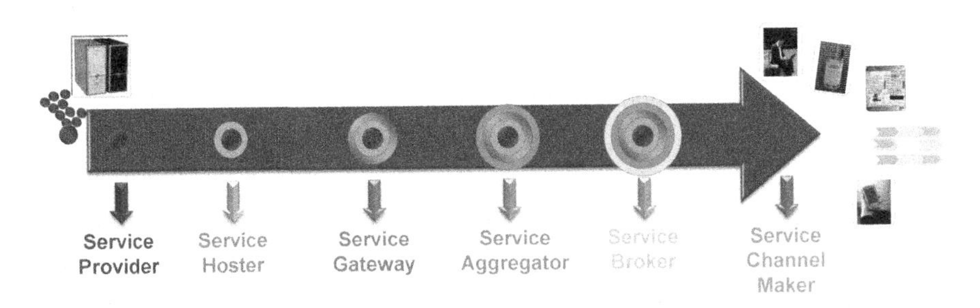

Fig. 1.2: Original SOA roles and additional roles identified by [6].

What is important in our context is the fact that with the many roles involved in the Internet of Services, the need for a common standard in the description of operational and commercial metadata becomes important. Just think about the role of a service aggregator. This role will aggregate atomic services to service bundles or aggregates. As an example for aggregates, you may think about packaged tours where airport transfers, flights, hotel booking and other services are bundled in one package. In a similar way you could think about an export service as a bundle of logistic services, insurances, services of the customs authority, technical certifications for the target country, export credit guarantees and the like.

The economics of bundling is defined by the cost of aggregation in relation to the perceived benefit. Obviously, the cost is dominated by the integration cost. It is exactly this integration cost that can be lowered by a common standard such as USDL. Ideally, a composition of USDL-described services can be based on a standardized tool chain and therefore be performed efficiently and at low cost.

Summarizing, the Internet of Services wants to make services tradable in digital media. In particular, with USDL, we want to offer a standard that creates a *commercial envelope* around the services. Technical services may be lifted to business services, but the same standard should also be able to describe more manual or physical services. As many services have a hybrid character with both, a digital and physical or manual footprint, a unified service description language can facilitate the combination and aggregation of such services.

1.3.3 Software Applications and Services

One question both for SOA and for the Internet of Services is the relationship between services and traditional applications such as Enterprise Resource Planning Systems. Here, we suggest the following two-dimensional approach.

In one dimension, we distinguish according to the *frequency of use* of a certain function. Consider an entry in an accounting system. This is something that happens daily in a business. The frequency of use for such a function is therefore very high. On the other side, you may have a rare business situation such as an export of your product in some unusual country. In order to deliver to this country, you are looking for special export services. This happens rarely, maybe only once in the history of your company. Between these two extremes, you may have situations where your ratings for the creditworthiness of new customers are required, which you need, e.g., 10 times a year.

As a second dimension, we suggest the *degree of integration* into the existing core processes of a business. The degree of integration concerns the amount of data that must be exchanged and the number of interaction points within a process. As an example, the creditworthiness check is needed at exactly one interaction point (when defining the payment conditions); the input to the service is the customer name and registration, while the output is a rating on some scale. This could be considered a shallow integration, while the export example needs much more interaction points and more data. On the scale, it can be considered to be a service with a high degree of integration.

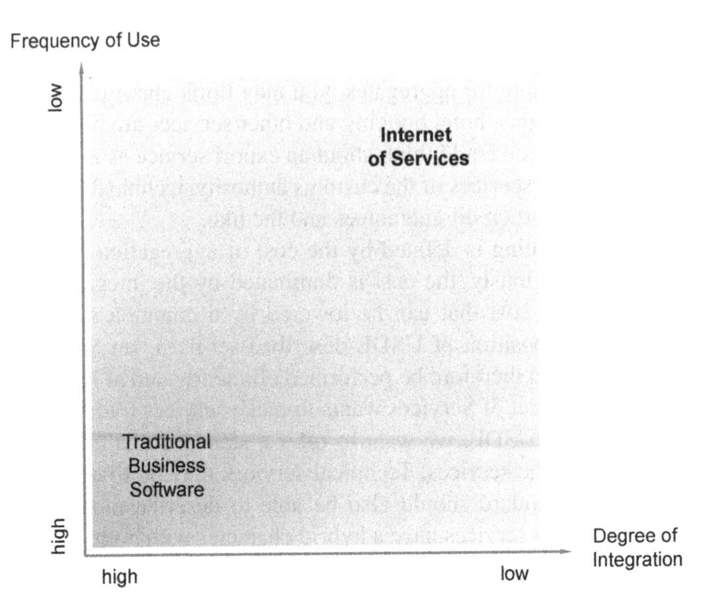

Fig. 1.3: Classification of services according to frequency of use and degree of integration.

The two dimensions span a function matrix as depicted in Fig. 1.3. Traditional application suites, such as SAP's Business Suite, with a high degree of integration are in the lower left corner, while we anticipate a whole range of services in the other sectors of the matrix. All these functions are provided by some sort of socio-technical system: for traditional software, the application is complemented by consulting, maintenance, call centres and training services, while other parts of the matrix may have an even stronger emphasis on the involvement of human resources.

Let us discuss two simple implications for this landscape. First, look at pricing. The traditional model for software is that of a flat rate, either licensing and maintenance fees, or Software-as-a-Service monthly flat rates. In any case, you can use the system as much as you like, while if you do not use it, you still pay the costs. For services, that you rarely use, you would expect a completely different pricing scheme. You may accept a moderate fixed rate to have access to a service ecosystem, but you will be willing to pay a higher price for a special service that fulfils your needs.

A second consequence is related to the degree of automation for exception handling of a service. In highly integrated functions with frequent use, automated exception handling usually covers more than 99% of all conceivable states of a service. This is different for services that are in other parts of the service matrix. Here, it may be a better choice to always keep some significant social interaction (by phone, mail or in person) deliver value. This allows for a higher flexibility and makes it possible to also respond to less standardized requests.

According to the functional matrix, we anticipate that standard software in the future will be complemented by some form of service delivery framework, e.g., [6], which enables a broad ecosystem, i.e., the Internet of Services, to deliver services on a common technological basis. We consider USDL to be one of the foundational technologies to set up such an Internet of Services around core enterprise systems as we know them today.

1.3.4 Applying the Internet Economy to the Services Sector

From an economic perspective, one of the basic features of the Internet is that it connects communities at marginal communication costs. Markets that were fragmented without the Internet suddenly are connected. Sometimes, this phenomenon is referred to as the long-tail opportunity [4].

A well-known example is books with a limited audience. There may be some hundreds amateurs around the globe that appreciate a book such as "Butterflies of the Caribbean" or similar titles. Without the Internet, it would be very hard for an editor or a bookshop to find the people that show an interest and a willingness to pay for such a book. The barrier to produce the book would be very high and it would be most likely that the book never appears on the market. With the Internet, the limited community can be reached and the long-tail books suddenly can be marketed profitably. Products for smaller groups with special interests or needs suddenly can be marketed profitably over Internet media.

The same situation also holds for service ecosystems. Apart from providing highly standardized and low-priced services with economies of scale, or highly individualized premium services, we anticipate mass-customized services to play a prominent role in future service ecosystems. As an intermediate between an economy of scale [16] and an economy of scope [18], we speak about an economy of micro-scale. We expect new business opportunities for such services to materialize systematically in the future service ecosystems. The economic set-up and an efficient channel for such services depend on service delivery platforms and a corresponding service ecosystem design including the right rules and business models for such ecosystems.

1.4 The Unified Service Description Language

The previous section made clear that an Internet of Services requires a way of describing services to wrap a service and expose it in a novel way. We have learned that a service is not only considered as the invocation of a technical interface, but rather as an economic or social transaction with a broader context. Therefore, it is essential to describe the price scheme, the service level agreement, or the terms and conditions when consuming the service and paying for it. With the many roles involved in service networks, the need for a common standard in the description of operational and commercial metadata becomes important.

None of the existing approaches responds to the needs of having a comprehensive service description in an Internet of Services setting. In particular, SOA description efforts provided means for formally describing IT services but do not allow describing the commercial conditions under which the service is consumed or under what operational conditions a service can be invoked. Further, SOA description approaches do not consider physical or hybrid services.

Therefore, we contribute *USDL, the Unified Service Description Language*, as a standard that creates a commercial envelope around a service. Technical services may be lifted to business services, but the same standard should also be able to describe more manual or physical services. As many services have a hybrid character with both, a digital and physical or manual footprint, a unified service description language can facilitate the combination and aggregation of such services. We consider USDL to be one of the foundational technologies to set up such an Internet of Services around today's core enterprise systems.

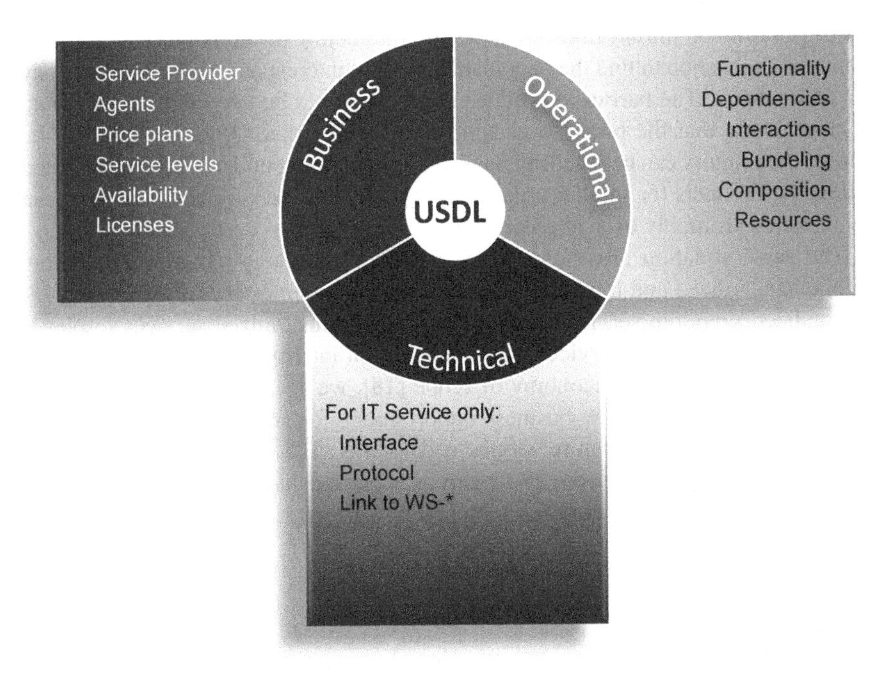

Fig. 1.4: USDL unifies business, operational, and technical master data of a service.

USDL comes into play for the aspect of standardization (cf. Section 1.1) as a normative and balanced unification of service information, which is required to further capitalize on the growth potential of the services industry. The unified description established by USDL is machine-processable, considers technical and business aspects of a service as well as functional and non-functional attributes, and provides both a blueprint and means for extensibility.

The Unified Service Description Language (USDL) is proposed as a normative and comprehensive master data model for the commercial meta data of IT, physical, or hybrid services. More specifically, USDL allows a unified description of business, operational and technical aspects of services as depicted in Fig. 1.4. USDL aims at a holistic service description putting a special focus on business aspects such as ownership and provisioning, release stages in a service network, composition and bundling, pricing and legal aspects among others, in addition to technical aspects. It proposes a consolidated foundation for service-based systems enabling different roles to participate in diverse aspects of provisioning in service networks.

1.5 Strategic Implications

In the previous sections, we have presented the approach of the Internet of Services, its relevance, the state of the discussion, the basic concepts, and the Unified Service Description Language as a standard to facilitate interoperability and fluidity of novel service networks. What are the strategic implications of the approach?

First let us talk about the role of platforms as *trust gates*. Service networks will imply many smaller players, often specialized in a small range of specific services (e.g., export services). In most cases, a potential service consumer will not know the party that is offering such a service. From a buying psychology point of view, one main consideration of the service consumer will be: *Can I trust this service provider?*

Bigger platforms can be trust gates in this situation. Combined with community ranking, well known from places such as eBay or Amazon, the platform can act as a business mediator in a double sense: it can match request and offering, but it can also act as a trust gate for the integrity of the offerings. This can happen by some form of checking the offerings before they are published on the platform (such as certifications in app stores), it can make use of community based reputation and rating systems, but it can also be done by sanctions against business partners that do not comply with certain standards.

These measures can lead to a reasonable level of trust on the consumer's side to also acquire new or unknown offerings. Of course, this trust federation — from a known platform brand to an unknown service provider — requires a strong brand of the platform operator. This brand may build on companies, cooperatives, a regional organization (such as a chamber of commerce) or even the public sector. In any case, strong related brands — software companies, internet companies, telecom operators or strong local brands — have a significant opportunity to extend their business in their installed base.

What we have seen in the Internet Economy, is an immense concentration of the market power of the big platforms. Some have been talking about *Internet Islands*

[17], where non-interoperable segments in the web are dominated by some platform operator. We think that a similar approach in the service sector would not be sustainable and are looking for a more pluralistic approach. A right balance between the service providers and the bigger platforms can lead to a powerful and sustainable growth in the service sector, leveraging the possibilities of Internet technologies for new forms of value creation.

In this light, the Unified Service Description Language can be seen as a conscious choice to shift market power to the service provider. Once a provider has described his or her service in USDL, he or she can publish it multiple times and on different platforms. This is in contrast to proprietary platform services descriptions that would create barriers between the platform islands imposing additional effort on the providers if they wanted to use more than one platform as mediation and delivery channels.

In return, we think that this will also be beneficial for the platforms, as they will be able to expose and mediate a much wider range of services and service bundles quickly. This is also better for the customers of the ecosystem, as there are no artificial technical barriers to profit from a broad spectrum of service providers. We anticipate that bigger platforms with powerful brands, combined with agile and innovative service providers can create ecosystems that are as open as possible, and as controlled as necessary to guarantee quality and reliability. With USDL, we want to create the right technical basis for a win-win interplay between internet-scale service mediation and delivery platforms and small and mid-sized service providers.

We anticipate that the basic set-up described above will develop new processes that we have not seen in the service economy so far. An example may be the innovation process around services. Most services can indeed not be protected by patents or similar forms of IP protection. A platform with a certain market power could of course introduce something similar — just based on the contracts with service providers in the ecosystem. They could set-up a registration instance, where a novel service could be registered with a time stamp. If some sort of fair decision body acknowledges the innovative character of the service, the platform could guarantee for a certain period (probably rather months than years) that no similar service may be offered in the marketplace of the platform. This would create a strong incentive for the service providers to create unique selling propositions through service innovations.

We anticipate that the overall approach for the Internet of Services and USDL will help to unlock the long tail in service networks. We think that both for the business area and consumers this will lead to a wide range of relevant new services. They will create tangible benefits for their users and economic growth with new value propositions.

References

1. Grundstruktur für die Beschreibung von Dienstleistungen in der Ausschreibungsphase. PAS 1018:2002-12, Deutsches Institut für Normung (DIN), Beuth Verlag, Berlin, 2002.
2. Succeeding through service innovation: A service perspective for education, research, business and government. A white paper based on cambridge service science, management and engineering symposium, IfM and IBM, University of Cambridge Institute for Manufacturing (IfM) and International Business Machines Corporation (IBM), 2008. `www.ifm.eng.cam.ac.uk/ssme`.
3. S. Alter. Service system fundamentals: work system, value chain, and life cycle. *IBM Systems Journal*, 47(1):71–85, 2008.
4. C. Anderson. *The Long Tail: Why the Future of Business is Selling Less of More*. Hyperion, 2006.
5. K. Backhaus, J. Becker, D. Beverungen, M. Frohs, R. Knackstedt, O. Müller, M. Steiner, and M. Weddeling. *Vermarktung hybrider Leistungsbündel*. Springer, Heidelberg, 2010.
6. A. Barros and U. Kylau. Service delivery framework — an architectural strategy for next-generation service delivery in business network. In P. Kellenberger, editor, *Proceedings 2011 Annual SRII Global Conference SRII 2011, 30 March - 2 April 2011, San Jose, California, USA*, pages 47–58. IEEE Computer Society Conference Publishing Services (CPS), 2011.
7. A. P. Barros and M. Dumas. The rise of web service ecosystems. *IT Professional*, 8(5):31–37, 2006.
8. B. Blau, C. van Dinther, T. Conte, Y. Xu, and C. Weinhardt. How to coordinate value generation in service networks. *Business & Information Systems Engineering*, 1(5):343–356, 2009.
9. B. Bygstad and T.-M. Grønli. Service oriented architecture and business innovation. In *44th Hawaii International International Conference on Systems Science (HICSS-44 2011), Proceedings, 4-7 January 2011, Koloa, Kauai, HI, USA*, pages 1–10. IEEE Computer Society, 2011.
10. H. Chesbrough and J. Spohrer. A research manifesto for services science. *Commun. ACM*, 49(7):35–40, 2006.
11. P. A. David and W. E. Steinmueller. Economics of compatibility standards and competition in telecommunication networks. *Information Economics and Policy*, 6:217–241, 1994.
12. R. Ferrario and N. Guarino. Towards an ontological foundation for services science. In J. Domingue, D. Fensel, and P. Traverso, editors, *Future Internet - FIS 2008, First Future Internet Symposium, FIS 2008, Vienna, Austria, September 29-30, 2008, Revised Selected Papers*, volume 5468 of *Lecture Notes in Computer Science*, pages 152–169. Springer, 2008.
13. D. Hilkert, C. M. Wolf, A. Benlian, and T. Hess. The 'as-a-service'-paradigm and its implications for the software industry - insights from a comparative case study in crm software ecosystems. In P. Tyrväinen, S. Jansen, and M. A. Cusumano, editors, *Software Business - First International Conference, ICSOB 2010, Jyväskylä, Finland, June 21-23, 2010. Proceedings*, volume 51 of *Lecture Notes in Business Information Processing*, pages 125–137. Springer, 2010.
14. H. Kagermann and H. Österle. *Geschäftsmodelle 2010. Wie CEOs Unternehmen transformieren*. Frankfurter Allgemeine Buch, Frankfurt am Main, 2nd edition, 2006.
15. S. Krings, editor. *Statistisches Jahrbuch 2010 für die Bundesrepublik Deutschland mit Internationalen Übersichten*. Statistisches Bundesamt, Wiesbaden, Germany, 2010.
16. P. Krugman. Scale economies, product differentiation, and the pattern of trade. *The American Economic Review*, 70(5):950–959, DEC 1980.
17. K. Lischka. Internet der Zukunft — Im Netz der Giganten. Spiegel Online, JAN 2010.
18. J. C. Panzar and R. D. Willig. Economies of scope. *The American Economic Review*, 71(2):268–272, May 1981.
19. R. Parasuraman and M. Mouloua. *Automation and Human Performance: Theory and Applications*. Lawrence Erlbaum Associates, Mahwah, NJ, USA, 1996.
20. G. Richardson. The organisation of industry. *Economic Journal*, 82(327):883–896, 1972.
21. P. Satchell. *Innovation and Automation*. Ashgate Publishing, Aldershot, England, 1998.

22. J. Spohrer, S. L. Vargo, N. Caswell, and P. P. Maglio. The service system is the basic abstraction of service science. In *41st Hawaii International International Conference on Systems Science (HICSS-41 2008), Proceedings, 7-10 January 2008, Waikoloa, Big Island, HI, USA*, page 104. IEEE Computer Society, 2008.
23. J. C. Spohrer, P. P. Maglio, J. H. Bailey, and D. Gruhl. Steps toward a science of service systems. *IEEE Computer*, 40(1):71–77, 2007.
24. J. Tas and S. Sunder. Financial services business process outsourcing. *Commun. ACM*, 47(5):50–52, 2004.
25. O. Terzidis, A. Fasse, B. Flügge, M. Heller, K. Kadner, D. Oberle, and T. Sandfuchs. Texo: Wie THESEUS das Internet der Dienste gestaltet — Perspektiven der Verwertung. In L. Heuser and W. Wahlster, editors, *Internet der Dienste*, acatech diskutiert, pages 141–161. Springer, 2011.
26. S. Wardley. Innovation, the future and why nothing is ever simple. In *FOWA Expo. Future of Web Apps. London.*, 2008.
27. D. Woods, L. Heuser, and C. Alsdorf, editors. *International Research Forum 2008*. Evolved Technologist Press, 2009.
28. J. Word. *Business Network Transformation: Strategies to Reconfigure Your Business Relationships for Competitive Advantage*. Jossey-Bass, San Francisco, CA, 2009.

Part I
State of the Art

The first part of the book provides an overview of existing service description approaches. The plethora of approaches is grouped in different strands each devoted a separate chapter. Chapter 2 starts off by focussing on *Product-Service Systems (PSS)*, i.e., a conceptual framework for the cooperative design and delivery of customer solutions. Customer solutions can be designed and delivered by the cooperation of manufacturing companies with external service providers or by manufacturing companies themselves. As the integration of customers is a constitutive characteristic of service processes, customers are to be acknowledged as co-creators of value that provide a variety of inputs. The chapter extracts central concepts from several disciplines that are engaged in researching business aspects of PSS to develop a catalogue of modeling requirements to be accounted for in service description. Consecutively, these requirements are utilized to assess the current state of conceptual modeling languages for (product-related) service description.

Chapter 3 continues by discussing several approaches to design, analyze, describe and compose *Service Networks*. The chapter analyzes the technical and business-related aspects of such approaches, their evolution, and the trends they will be likely to follow. Further, Chapter 4 draws attention to the newly emerged discipline of *Services Science* as an effective means to understand services and the socio-technical systems in which they are deployed. This systemic view requires a genuinely interdisciplinary approach to the study of services. Consequently, the chapter reviews a number of significant approaches to analyze, understand and model service systems, with an emphasis on showing similarities and differences that highlight the many aspects of a rich service ecosystem.

With regards to purely technical services, different standardization efforts have been proposed by various consortia to enable interaction among heterogeneous environments in the paradigm of *Service-oriented Architecture (SOA)*. Chapter 5 overviews the most prevalent of such SOA approaches and shows how technical services can be described, how they can interact with each other and be discovered by users.

Finally, the research area of *Semantic Web Services* investigates the annotation of services, typically in a SOA, with a precise mathematical meaning in a formal ontology. These annotations allow a higher degree of automation. The last decade has seen a wide proliferation of such approaches, proposing different ontology languages, and paradigms for employing these in practice. Chapters 6 and 7 provide an understanding of the fundamental techniques, from Artificial Intelligence and Databases, on which they are built as well as an overview of such approaches, respectively.

Chapter 2
Product-Service System Approaches
A Business Perspective on Service Modeling

Daniel Beverungen, Martin Matzner, Oliver Müller, and Jörg Becker

Abstract For some time, increasing importance is attached to services, both from an economical and a managerial perspective. First, the notion of "service as basic unit of exchange" emphasizes the application of specialized competencies for the benefit of someone else, while disregarding if a physical good or any other resource is used for exchanging value. Second, service-orientation allows enterprises to enter new markets by extending their existing portfolio of products by related services or realizing entire new offerings that are enabled by recent advances in information technology. Service description is a key challenge in developing and providing services to and with customers. Further it is a premise for coordinating several providers of an integrated customer solution. This chapter is an effort to explain how conceptual modeling can facilitate service description. We use Product-Service Systems (PSS) as an exemplary domain. We extract central concepts from several disciplines that are engaged in researching business aspects of PSS to develop a catalogue of modeling requirements to be accounted for in service description. Consecutively, these requirements are utilized to assess the current state of conceptual modeling languages for (product-related) service description. The review leads to the identification of further prospects to be accounted for by service description.

Jörg Becker, Martin Matzner, Daniel Beverungen
University of Münster, European Research Center for Information Systems, Leonardo-Campus 3, 48149 Münster, Germany, e-mail: `firstname.lastname@ercis.uni-muenster.de`

Oliver Müller
University of Liechtenstein, Martin Hilti Chair of Business Process Management, Fürst-Franz-Josef-Strasse, 9490 Vaduz, Liechtenstein, e-mail: `oliver.mueller@uni.li`

2.1 The Need for Conceptual Modeling of Services

2.1.1 The Trend Towards Service

Services have had a lasting effect on last decade's business development. From an *economical perspective*, we have been witnessing a transition from a primarily goods based to a more and more service-based economy in most developed countries [53]. Today, services are ubiquitous and they account, for example, for more than 80% of the gross domestic product (GDP) and total employment of the United States and about 70% of the GDP in Germany [15, 38, 74]. These figures reflect a development that has been prescribed by Vandermerwe and Rada under the term "Servitization of Business." In their understanding "servitization" refers to the increasing offer "of 'bundles' of customer-focused combinations of goods, services, support, self-service, and knowledge" [83, p. 314] — with services dominating the bundles. As an umbrella term the notion of 'service' has been coined as "the application of specialized competencies, through deeds, processes, and performances for the benefit of another entity or the entity itself." [84, p. 2] They coined 'service' to be the new basic unit of exchange in economies [84, 85, 86].

From a *managerial perspective*, enterprises at the same time struggle to efficiently provide adequate services to their customers. In order to generate superior business returns and as a result of ever increasing competition, companies face the need to specialize on their core resources and core competencies. However, at the same time they need to address complex business needs of their customers. Previous research [44, 69] has stated that the formation of relational ties and networks with further suppliers is a viable means for addressing both needs at the same time.

Though the service sector now accounts "for most of the world's economic activity, [...] it's the least-studied part of the economy" [73, p. 71]. Consistent with this is the observation that still no general consensus of the structure and nature of services has been reached (cf. [26] and also Chapter 4). In the recent past, researchers from different disciplines have so far investigated the phenomenon from rather distinct angles, e.g., from an economic, business, or technical perspective [4]. Only recently, an interdisciplinary research effort under the headwords of Service Science or Service Science Management and Engineering (SSME) has been emerging.

2.1.2 Beyond Goods and Services: Customer Solutions as Value Offers

Accordingly, many industries are today experiencing a transition from a goods-based to a service-based focus: Traditional manufacturing companies strive to provide physical goods and services as integrated customer solutions [23], which are delivered in relational processes with customers [82].

Physical goods and services are no longer perceived to be dichotomous [25]. Instead they are rather seen as complementary vehicles to offer value propositions to customers [84, 85]. This trend is especially recognized in the German Mechanical Engineering and Electrical Engineering industries. Evaluating results from two broad empirical studies in both sectors, Stille concludes that turnover related to services has doubled in the Electrical Engineering sector from 9.6% (1997) to 18.5% (2000), while significant gains from 16.8% (1997) to 22.5% (2000) could be identified in the Mechanical Engineering Sector [76]. Mercer Management Consulting points out, that half of the growth in German Mechanical Engineering in the years 1998–2003 can be allocated to exploiting the potential of the service business. Likewise, the margin realized in the service business (10%) was estimated to be significantly higher than the margin realized in the physical goods business (2.3%). Furthermore, Mercer states that margins gained from services could be even higher when looking at some leading edge services only, which constantly catch margins of up to 18% [50]. Additional empirical research shows, that companies attribute a high (38.1%) or very high (59.8%) impact on their revenues to their service business. Services are also seen as a good means for differentiation from competitors as well as for customer retention [23]. Consistent with these findings, 94.9% of the examined companies plan to expand their business by offering customer solutions [77].

Unless the matter still is heavily debated, the following characteristics have been proposed for customer solutions:

- Customer solutions comprise separately marketable tangible goods and intangible service which are purposefully combined to solve a problem for a customer or for a group of customers [23].
- Physical goods components and service components might (but need not) be substitutable with other components without changing the solution provided [23].
- For customers, outcomes of the solution can have tangible and intangible effects. One goal of providing customer solutions is to create an outcome for customers and/or providers which is superior to the simple sum of outcomes of the components [23]. In the end, the value of a solution is co-created by the supplier and the customer by integrating resources with other resources; the value realized from this relational process is, therefore, determined by each of the parties [86].
- Customer solutions are value propositions offered by the supplier to a customer. If a value proposition is accepted by a customer, customer solutions are co-created in service processes that are closely integrated with the customer's business processes and, therefore, require customer input (such as information, objects, personnel or other resources) [85].

Services which are often offered as part of customer solutions may correspond to different stages of the traditional product lifecycle, such as a start-up, operation or disposal stage (cf. Fig. 2.1). Services in the start-up stage may constitute pre-sales services such as engineering, consulting or technical assembly. During the operation stage, service activities such as preventive maintenance, corrective maintenance or spare parts logistics are mainly conducted to uphold the operability of the physical

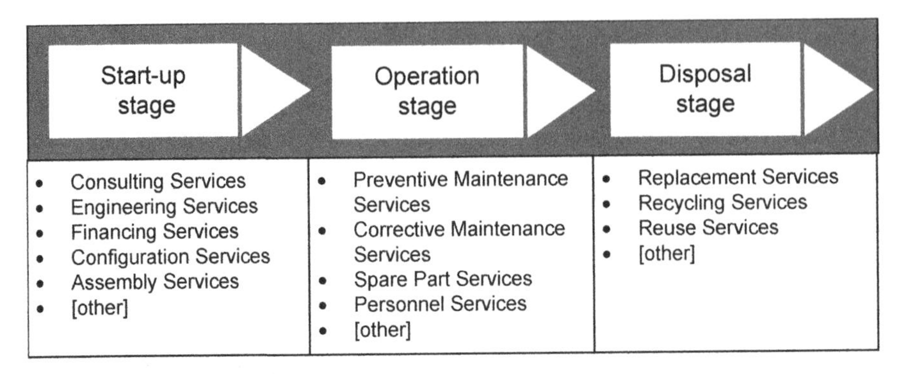

Start-up stage
Operation stage
Disposal stage
Consulting Services
Engineering Services
Financing Services
Configuration Services
Assembly Services
[other]
Preventive Maintenance Services
Corrective Maintenance Services
Spare Part Services
Personnel Services
[other]
Replacement Services
Recycling Services
Reuse Services
[other]

Fig. 2.1: Exemplary services, arranged with respect to a traditional lifecycle of physical goods.

good. Referring to the disposal stage, any of the physical goods components might be replaced, refurbished and reused, or recycled.

In practice, the most widely offered services in industrial settings are spare part logistics, preventive maintenance/fault repair, consulting, and assembly. It is striking, that most of these services are in fact not new and have a strong physical goods focus. In contrast to this, highly innovative services such as capacity management in value networks, performance contracting business models, or on-demand personnel provision are as yet seldom offered.

Some topologies have been proposed to grasp the characteristics of integrated physical goods and services, most notably by Engelhardt, Kleinaltenkamp and Reckenfelderbäumer [25] (cf. Fig. 2.2). The authors take a marketing dominant perspective and systematize outputs for customers in two dimensions, each employing two parameters: On the one hand, the output perceived by the customer may be rather immaterial (such as additional knowledge as the result of training to operate a vertical lath) or rather material (such as a vertical lath, that has been delivered and assembled). On the other hand, processes involved to deliver customer solutions might have to be tightly integrated into the processes of customers (for example processes to design and deliver an engineered-to-order vertical lath) or can be handled rather autonomously (for example producing spare-parts and inventory management).

In this topology, customer solutions can be systematized as being co-created in relational processes of suppliers and customers and can provide tangible as well as intangible results.

2.1.3 Product-Service Systems

It has been argued, that customer solutions can be designed and delivered by the cooperation of manufacturing companies with external service providers [59] or

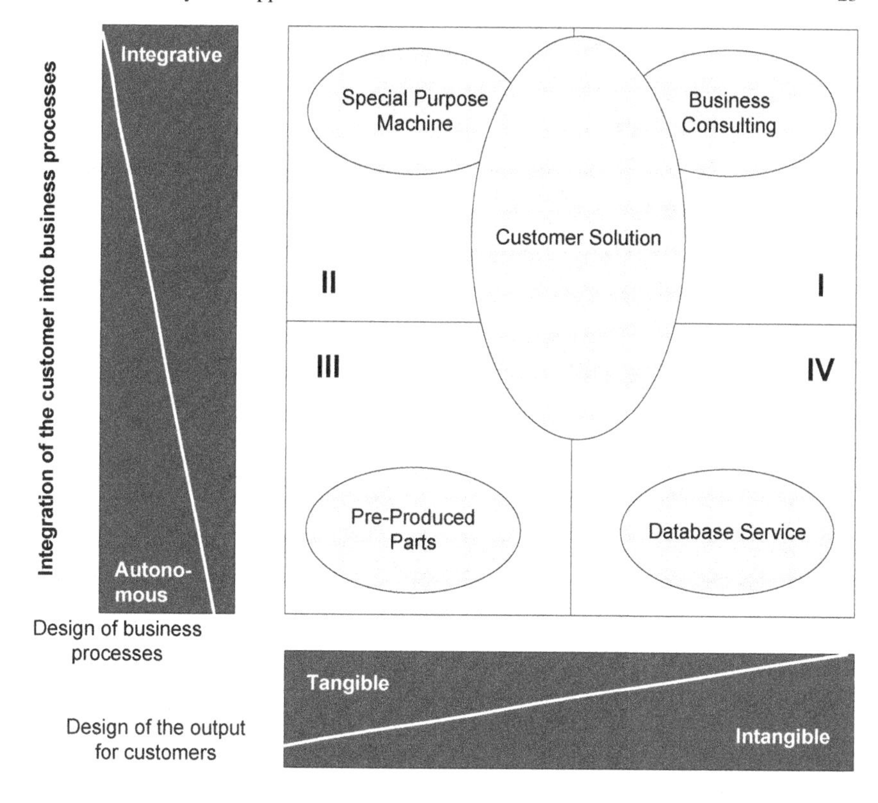

Fig. 2.2: Customer solutions as sales objects (cf. [25]).

by manufacturing companies themselves. As the integration of customers is a constitutive characteristic of service processes, customers are to be acknowledged as co-creators of value that provide a variety of inputs [63]. Providers may not offer value but only value proposition, while the creation of value is performed cooperatively with customers [85]. During this cooperation, value propositions are applied to generate value for customers (i.e., the customer solution). Customers might be consumers (B2C market), other companies (B2B market) or the public sector.

Drawing from existing definitions of Service Systems [46], we use the term Product-Service System (PSS) [5, 52, 79] as a conceptual framework for the co-operative design and delivery of customer solutions (cf. Fig. 2.3). Information flows to integrate business processes to design and deliver customer solutions (represented by arrows in Fig. 2.3) are of special interest to foster an efficient and effective design of the cooperation process.

Here, service description is a means to facilitate the integration between the several providers of a customer solution as well as to integrate the customer as co-

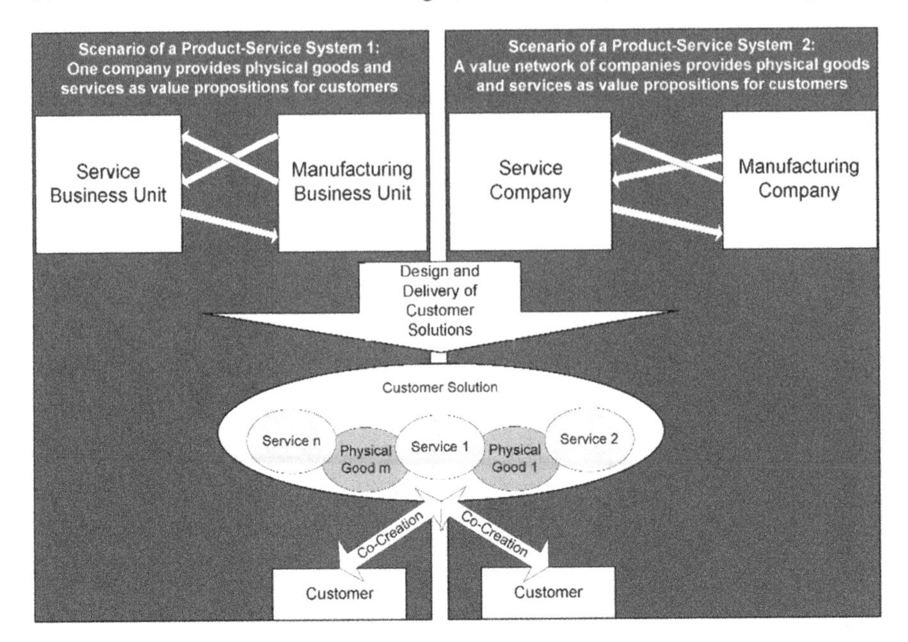

Fig. 2.3: Cooperative design, proposition and delivery of customer solutions in Product-Service Systems (PSS) (see also [5]).

creator of value into processes. Further, a purposeful digital representation of a service might allow providers for exploiting new sales channels (e.g., digital service marketplaces), and might even lead to creating and providing entirely new offers that arise from dynamically integrating value offers. Customers can be provided with enhanced functionalities in searching and finding services and composing purposeful solutions.

2.1.4 Analysis of Conceptual Modeling Approaches for Services

This work is an effort to explore the status-quo and prospects for further research of service description approaches using PSS as an example. The remainder of the chapter is organized as follows:

In the consecutive section, concepts that need to be represented in conceptual models of services from a business perspective on customer solutions are identified from four sub-disciplines of service science. These perspectives are ENGINEERING, SUPPLY CHAIN MANAGEMENT, SERVICE MARKETING, and ENVIRONMENTAL RESEARCH. We therefore review the particular needs and contributions. The identified key concepts are combined in an "evaluation sheet."

The evaluation sheet is applied in the subsequent section to discuss the current state of conceptual modeling approaches for service description in the context of

PSS. The analysis leads to the identification of further prospects for using conceptual models to describe services. The conclusion and outlook section summarizes the results and postulates directions for further research.

2.2 The Interdisciplinary Study of PSS Conceptual Modeling: Extracting Concepts

2.2.1 Conceptual Modeling and Languages

In the field of Information Systems (IS) Research, conceptual models are often used to describe, abstract from, emphasize and explain information concepts. On the one hand, conceptual models are designed with respect to unambiguously defined (i.e., specified by means of a meta-model) modeling languages. On the other hand, they should convey a degree of intuitive understanding for their users. Well-designed conceptual models enable members of an interdisciplinary project team to communicate with each other more effectively, regarding concepts such as the structural organization of a company or its business processes [19, 89].

Conceptual modeling has been argued to hold great business potential, for instance to grasp and redesign business processes in the field of business process modeling. Conceptual models used for the development of information systems may explicitly aim at addressing targeted users, senior executives, application designers, and programmers in software development processes. Thus, conceptual models can simultaneously address management issues as well as aid software and business engineering projects on an operational level.

Generally, a modeling language comprises a conceptual aspect and a representational aspect [35]. The conceptual language aspect (ortho-language) defines the meaning of the modeling constructs and relationships among them and constitutes the expressiveness of conceptual models designed with this modeling language. The representational aspect (notation) assigns representation formalisms to the specified constructs to make them easier to grasp and use for stakeholders by reducing the cognitive load imposed on human interpreters. Modeling languages determine the rules according to which conceptual models (or even reference models) can be designed. Modeling languages are usually formally described by meta-models, which represent the language concepts and their (mostly graphical) representation and can also enable advanced model operations such as specifying a dynamic semantics of models [34].

Using well-established modeling languages can accelerate the process of conceptual modeling, since modelers and users may already be familiar with the modeling language's constructs (think of modeling languages such as UML) and therefore using models might facilitate more effective communication processes. The application of well-established modeling languages may guide the modeling process and thus decreases the risk of wrong methodological decisions. Therefore, in this chapter we choose the support of current modeling languages for generating conceptual models in a PSS context as the focus of our exploration.

Conceptual models can be used to support designers in dealing with the specific requirements in a PSS context, such as (see also [7]):

- What are business processes in a PSS context like? How need the front-stage and back-stage of service systems be integrated with each other to provide customer solutions in a consistent and efficient way? How and to what extent might business processes in PSS be improved?
- Which organizational units are involved in the process of value creation of customer solutions? What is their role in the process of value creation and which components of customer solutions do they provide?
- What is the overall productivity or efficiency of a service process? Which key performance indicators should be selected to assess the performance of a service process? Against which other processes shall a service process be benchmarked in order to reason about its performance?
- How is a customer solution or an entire portfolio of customer solutions structured? Which components do these offerings comprise? What resources will be necessary to create the customer solutions? What costs will be associated with providing the customer solutions along their entire lifecycle, even if this lifecycle spans several years or even decades?
- How might individual value propositions for customers be derived from the portfolio by combining previously defined physical goods and service modules into customer solutions (i.e., bundling)? Which configuration rules do exist and how are they specified?
- How much money is a customer willing to pay for his or her configured solution? What preferences, needs, wants, and demands does a customer have? Which solutions shall providers offer to a customer from an economic point of view? Which solutions shall not be offered to a customer, because their creation is undesirable (e.g., non-profitable for the provider)?
- How much negative impact do alternative customer solutions impose on the environment? Which customer solution should be selected to minimize the ecological footprint of the value creation process?

2.2.2 Extracting Concepts from PSS Research Disciplines

The design and delivery of customer solutions is currently addressed by research in several academic disciplines [4, 86], each of which imposes its own point of view on the subject. We now provide a brief introduction of the main issues emphasized by four specific disciplines and deduce criteria as requirements to be addressed by modeling languages in a PSS context. It should be noted that these disciplines nevertheless also overlap to some extent.

ENGINEERING disciplines, such as Mechanical Engineering or Electrical Engineering, traditionally focus on engineering, constructing, and operating physical goods. In this discipline, it is often argued that a shift towards a new service economy is taking place, since manufacturing companies have strived to professionalize their service business in the recent years. To account for this shift, engineering disciplines strive to apply the common techniques of product development to the development of services also. In this respect, they deal services as units of outputs that have other characteristics than physical goods have. This point of view has been criticized as being rooted in a Goods-Dominant (G-D) logic mindset that is based on the assumption that value is created in the form of units of output rather than focusing on the relational creation of value that is favored by the Service-Dominant logic view [86].

Modeling languages for specifying physical goods and production processes have long been established in research and practice. The representation of bills of materials is one common and widespread manifestation. A bill of materials represents the model of the physical good and may break down its physical structure into components, parts or even raw materials. Each component or part is created in a definable manufacturing process, whose steps can be represented by work plans and other process models.

Creating physical goods according to formalized specifications has thus long been the focus of engineering disciplines, which has led to a considerable degree of standardization concerning ways to formally describe manufacturing processes. STEP (STandard for the Exchange of Product model data, ISO 10303-41: Fundaments of Product Description and Support; ISO 10303-42: Geometric and Topological Representation; ISO 10303-46: Visual Presentation) for example has gained particular importance in the product engineering domain [2, 57].

Additionally, a 'Service Engineering' research movement has emerged in Germany [29] in which engineering disciplines strive to apply engineering methods to the design of business services [14, 67].

Drawing from modeling bills of materials in the engineering disciplines, the service engineering discipline attempts to decompose services into sub-modules. These components can then be described by process models which closely resemble the work plans used in manufacturing. Fig. 2.4 depicts a bill of materials of a physical product (left) as well as a bill of materials of a service (right). The structural analogy of both models is striking since they both display the structure of sales objects that can be sold to customers by utilizing different hierarchical levels.

As has been shown, from an engineering point of view, representing the structure of physical goods (product engineering) and services (service engineering) and their components (*customer solutions subdivided into components*) is crucial. Based on this specification, work plans comprising activities in production processes, sequence planning, and machine capacity can be designed. Work plans are one common feature of current ERP systems. Since we will deal with processes and work plans from a Supply Chain Management (SCM) perspective in more detail, we identify three characteristics *control flow*, *capacity*, and *activities* arising from the engineering perspective.

Choosing a mass-customization approach as the underlying business strategy of the firm may enable companies to exploit economies of substitution by *reusing*

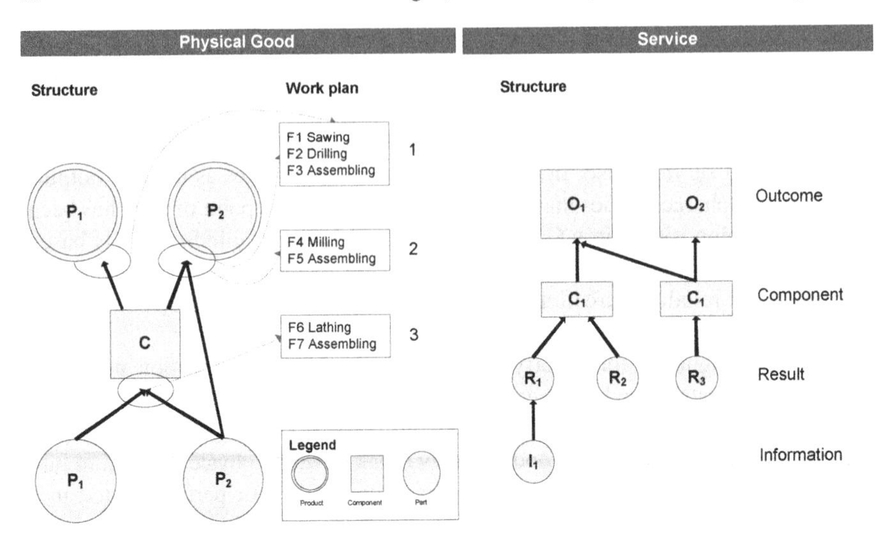

Fig. 2.4: Comparing a bill of materials for a physical good with a bill of materials for a service (cf. [42, 64]).

modules (reuse of the components as modules) [30]. Benefits to be gained from modularization include re-using existing knowledge associated with physical good and service modules, reducing performance slippage when incorporating additional modules into the bundle, reducing incorporation costs for suppliers and customers and, perhaps most importantly, making customer solutions modularly upgradeable to cope with changing customer demand (*substitution of modules*) [3, 30]. A prerequisite to assemble customer solutions from modules is to describe them in taxonomies (i.e., with is-part-of relationships) of modules as well as establishing non-hierarchical relationships (i.e., configuration rules) between modules. These might be inclusive (*configuration rules to specify inclusion (may)*) or exclusive (*configuration rules to specify exclusion (must not)*).

Components might be described by a variety of attributes (*attributes of components*), a particularly important one of which is information about the longevity of physical good or service modules (*longevity*). With respect to these attributes, services such as consulting and maintenance might differ significantly due to the size, configuration and longevity of a physical good or its components. To provide information in sufficient detail, product models have to account for a variety of lifecycle phases of the traditional product lifecycle, ranging from the start-up stage to disposal. Quality assurance is especially challenging for service processes, because inspections at the end of the service creation process can only occur after delivery to customers. Therefore, quality standards are important to be followed during any production or service process (*quality standards*). The quality of service as perceived by customers might be explained by a gap-model, as the deviation between expected and perceived service quality [54].

The discipline of SUPPLY CHAIN MANAGEMENT as an integrative discipline drawing from business, engineering, and computer science / IS points-of-view [60], emphasizes the need for managing business processes based on information modeling. The focus of study here is the actual business process that is carried out to deliver the customer solution (G-D logic point of view), or the outline how the relational process of value creation between suppliers and customers (S-D logic point of view) is or needs to be performed.

In order to document and improve the effectiveness and efficiency of the business processes in PSS, the information flows that are used to cooperatively create value are analyzed and designed. This analysis is often focused on the touch points in a service system, where value is co-created between different actors in a service system. After analyzing the interacting at these points, business processes might be redesigned and new IT artifacts might be designed to increase the overall quality or efficiency of the cooperation.

In addition, setting up service processes might be aided by drawing from past successes in disciplines such as Supply Chain Management, Materials Requirement Planning (MRP) and Enterprise Resource Planning (ERP) [21].

The discipline of Operations Management (OM) emphasizes the need for multi-disciplinary cooperation across several functional areas, such as Human Resources, Marketing or Accounting, to maximize the efficiency in providing customer solutions for customers. In OM, steps from customer analysis to product/service engineering, delivery, and disposal are not viewed in detail, but rather seen as one output of the cooperation [32].

Modeling languages for PSS have to account for these characteristics. Business processes necessary to design and deliver customer solutions include the activities to be carried out in the process (*activities*), the order in which they have to be carried out (*control flow*), materials to be procured and transported (*flow of materials*), information to be utilized (*flow of information*), and money for physical goods bought or sold to customers (*cash flow*). For services, the most important aspect to model is the work steps of the corresponding service process [65]. It has been stated, that manufacturing processes and service processes differ significantly and managerially from each other [63]. Therefore, a suitable modeling language for PSS must not model service processes in the same way as manufacturing processes, but should take distinctive characteristics of service processes such as customer integration into consideration. Because resources and inputs provided by customers are individual in each service process, modeling languages must be able to represent various customer inputs and various sequences of service processes (*resources to be introduced by customers*). Also production processes for different physical goods or their variants might be different and affect the control flow in manufacturing.

As each process uses or consumes (operand) resources in order to be carried out, process cycle times are important attributes for managing different types of manufacturing and service processes (*process cycle time*). These resources need to be reserved in order to be utilized by the process. This process can be carried out by applying techniques proposed by the discipline of production planning and control. Manufacturing as well as service processes might be subject to failure. Consequently, the failure rate should be estimated and considered for planning and scheduling the resource allocation for processes (*failure rate*).

Services processes per se are intangible, but might involve a variety of operant and operand resources during their execution. Operant resources (such as personnel, knowledge, business processes, culture, business relationships) can be seen as the fundamental source of competitive advantage [45, 85]. Operand resources are the resources acted upon when conducting service processes, such as machinery, components, parts, materials and other objects (*consumption of operand resources (economic point of view)*).

In the light of services being non-storable, suppliers have to make sure they have sufficient resources at their disposal to carry out the service process when they are requested at the 'moment of truth.' Yet on the other hand, they may want to minimize the time their resources remain idle, waiting for customer input. This optimization problem motivates a resource planning for service processes, as has been applied in manufacturing processes for years. Even if techniques such as yield management and queuing strategies have been successfully applied to manage the critical resources in service processes, a resource planning for services is still argued to lag behind resource planning in manufacturing [21].

As resources for service processes are perishable, conceptual models for business services must be able to represent resources and their capacity. Resources should be displayed in process models, such that for each function to be carried out the resources to be consumed are depicted and scheduled. During the service process, some organizational units and IT systems and applications are likely to be involved.

Because PSS might comprise different actors, such as manufacturers, service providers, and customers (as stated in Section 2.1), it is important to carry out business processes smoothly even across business units and organizations. Integrating information and processes found in the front-end and back-end of service systems has already been identified to be a considerable challenge [58, 80]. For example, to offer and deliver a managed truck fleet as a value proposition for customers, a truck manufacturer and a consulting agency have to synchronize their businesses by exchanging documents such as order and bidding documents, schedule dates, or product master data. Therefore, business processes in PSS have to be able to display, which sub-processes must be carried out by manufacturers, service providers, or customers and which activities are to be made visible to others stakeholders (see also the consecutive section).

Due to differing needs of customers and due to different resources and inputs to be introduced into the service process by customers, service processes might be carried out differently each time. Standardizing services can help to provide them more consistently across these instances.

Main points of interest from a SERVICE MARKETING PERSPECTIVE on customer solutions (notably presented by [71][84]; as well as a Journal of the Academy of Marketing Science special issue in 2008) comprise a relational view on how value is created [84, 85], determining adequate prices and business models to successfully market these value propositions [78], and integrating the customer into service processes as a co-creator of value [55, 71, 70].

The emerging research disciplines of Service Science and Service Science, Management and Engineering (SSME) respectively, focus on the design and delivery of services in Service Systems, comprising providers (or even value networks of providers) as well as customers as co-creators of value. This point of view is based on the philosophical foundation of the Service-Dominant logic (of Marketing) [86, 84]. The S-D logic view posits an alternative view on the creation of value that is based on the application of operant resources (i.e., knowledge and skills) for the benefit of another entity. Consequently, all value is created in relational processes by combining operant resources with each other. Value is assumed to be determined by the beneficiary, while a supplier cannot offer value but only value propositions. Physical goods that are exchanged between suppliers and customers are perceived to be vehicles for the application of operant resources. Therefore, all economies are perceived to be service economies, whereas the emergence of a new service economy is denied. In essence, S-D logic moves the understanding of how value is created from a focus on what is exchanged to a focus on relational value creation between suppliers and customers. Since Service Science is emerging as in interdisciplinary research discipline, it seems to go beyond the traditional boundaries of service marketing. However, we will still focus on the 'inner core' of service marketing here in order to derive criteria on the co-creation of value and the marketing on services as 'offerings' that can be advertised in the marketplace.

From a "traditional" service marketing perspective, a modeling language for services must take distinctive characteristics of services into account. The distinctive feature of service processes is the integration of customers as co-creators of value into service processes (*resources to be introduced by customers*). Therefore, it is crucial to account for the *line of visibility* and *line of interaction* towards customers [70]. Sampson and Froehle [63] emphasize, that other often cited characteristics of services (however critized by [85]) including perishability, simultaneity, intangibility, heterogeneity [27] are caused by the integration of the customer into the service process. Additionally, services cannot be produced in advance and thus are non-storable and not easily patentable. In addition to the lines usually postulated by service marketing, *relationships and lines towards stakeholders* can determine the division of labor in PSS, as processes might be outsourced to external manufacturers or service providers.

Moreover, it is difficult for customers to assess the value of a service in advance of the service process, which makes marketing and pricing services especially challenging. Therefore, offering *value propositions* for customers can be supported by adequate modeling languages. This might be achieved by providing constructs to describe, and individually configure and price (*attributes of components*) customer solutions [6]. The combination of physical objects and services is a crucial factor to be considered here. For instance, characteristics of the physical good influence which services can be offered concurrently. This means that a physical good is often the platform of which services are offered to a customer. An example would be a smart phone (i.e., a physical good) on which various services, ranging from music downloads to location-based services, can be offered. On the other hand, the services offered to a customer can also determine the properties of a physical good.

As an example, a complex apparatus might be integrated with value-added services so tightly that the services contained in the bundle determine the structure of the physical good. For instance, the general outline of a machine might be determined by the underlying business model that is used to market the solution (e.g., function-oriented as opposed to offered as-a-service).

The *marketing lifecycle* of customer solutions and its components is important to consider, because customers often take services which are in the saturation phase (such as assembly or maintenance services) for granted and might be unwilling to pay for them. In contrast, rather innovative services (such as layout planning or resource optimization services) have only recently been introduced and are more likely to be paid for by customers (*cash flow*). *Service level agreements* (SLAs) might be offered to define the quality level of physical goods and services more consistently and convince customers that the value propositions offered to them will lead to the creation of high-quality solutions that will likely be beneficial for them. Dispatching qualified personnel (*personnel allocation, qualification of personnel*), promising low *failure rates* and short *process cycle times* can be some of the elements dealt with in SLAs.

Apart from these disciplines, customer solutions from an ENVIRONMENTAL RESEARCH standpoint are seen as a means to create customer solutions with less environmental impact [51, 62, 81]. Customer solutions, if offered in performance contracting business models by specialized providers, might allow for resources to be used more efficiently due to exploiting economies of scale. Therefore, value for customers can be created in an environmentally 'sustainable' way. Authors arguing from this point-of-view tend to explicitly take environmental aspects into their definitions of customer solutions and PSS (cf. the discussion in [5]).

Modeling languages for PSS should address some basic ideas that have significance from an ecological point of view. Most importantly, the *consumption of operand resources (ecological point of view)* during production and service processes should be taken into account, because it may entail some negative environmental impact, for example due to emissions. At the end of its lifecycle (*longevity*), a physical good might be refurbished or recycled. In both cases, information about the product structure and its components is necessary (*customer solutions subdivided into components*). If modules are to be refurbished, a *substitution of modules* takes place. If modules are recycled, their material might be reused to build other physical goods (*flow of materials, attributes of components*).

A division of customer solutions into sub-components and raw materials in connection with adequate attributes can help to quantify this impact, while a modular structure with reusable components can help to spare resources due to exploiting substitution effects and economies of scale. *Legal constraints* (such as WEEE: Waste Electrical and Electronic Equipment; European Union 2003) might be important to address from an ecological standpoint, since compliance with regulations may be a binding requirement.

Table 2.1 summarizes the modeling requirements for customer solutions derived from the four perspectives. The origin of each criterion is displayed, taking into account that several criteria stem from more than one discipline.

Table 2.1: Modeling requirements of customer solutions.

Modeling language	Engineering	SCM	Marketing	Environmental
Customer Solutions subdivided into components	X			X
Reuse of the components as modules	X			
Configuration rules to specify inclusion (may)	X			
Configuration rules to specify exclusion (must not)	X			
Substituation of modules or components	X			X
Attributes of components / products	X		X	X
Longevity	X		X	X
Activities	X	X		
Control flow	X	X		
Flow of materials		X		X
Flow of information		X		
Cash flow		X	X	
Process cycle time		X	X	
Failure rate		X	X	
Capacity	X	X		
Consumption of operand resources (economic point of view)		X		
Consumption of operand resources (ecologic point of view)				X
Legal constraints				X
Value proposition for customers			X	
Marketing lifecycle			X	
Line of visibility towards customers		X	X	
Line of interaction towards customers		X	X	
Relationships / lines towards other stakeholders		X	X	
Quality standards	X			
Service level agreements			X	
IT Systems and applications		X		
Data (e.g. master data)		X		
Personnel allocation		X	X	
Qualification of personnel		X	X	
Organisational units		X	X	
Resources to be introduced by customers		X	X	

2.3 A Review of the Status-quo of PSS Conceptual Modeling and Perspectives to Service Description

2.3.1 *Identifying Relevant Modeling Languages*

Several modeling languages might be applied for or are particularly catered to the description of customer solutions in the context of PSS. In this section we identify a choice from the multitude of approaches has that been proposed in previous research, which is listed in Table 2.2 From this list, we have selected eight modeling languages for a more thorough analysis with the help of the previously presented requirements catalogue (grey-shaded in Table 2.2). For a more detailed overview, cf. [7]). The selected approaches are discussed below.

Table 2.2: A compilation and selection (shaded) of modeling languages.

Source	Modeling Language
Belz [8]	Proplan
Bitner et al. [9]; Kingman-Brundage [39]; Fließ [28]); Shostack [72]	Service Blueprinting
Black et al. [10]	ITSM-Model
Bley et al. [11]	Integrated product and process model
Bossmann [12]	CAD
Botta [13]; Steinbach et al. [75]	PDD-Approach
Congram, Epelman [16]	Structured Analysis and Design Tech-nique (SADT)
Corsten, Gössinger [17]	Framework for integrative Modeling
Dadam et al. [18]	EPAT
Dietrich, Kirn [20]	EwoMacs
Emmrich [24]	Method for systematic development of product-oriented services
Gu [31]	General Product Modeling
Hartel [33]	Collaborative Blueprinting
Scheer et al. [65]; Klein [40]	ARIS / Model-based Service System Engineering
Klein, Schreiner, Seemann [40]	K3-Method
Kunau, Loser, Herrmann [43]	SeeMe
Manavazhi [47]	Hybrid Modeling Framework
Mason [48]; Pratt [56]; Koonce and Judd [41]; IAI [36]; International Standards Organization [37]	STEP/EXPRESS-G
Maussang, Zwolinski, Brissaud [49]	Sakao's Service Representation
Rainfurth, Tegtmeyer, Lay [61]	Industrial Service Blueprinting
Schmied [66]	ProMod
Schnieder [68]; Ahrens et al. [1]	GMA 7.21
Shostack [70, 71, 72])	Molecular Model
Winkelmann [87, 88]	Coloured Petri Nets for Service Simulation

Molecular Model The Molecular Model approach rests upon the four main "ingredients" of the "marketing mix." The approach assigns to any "market entity" a distribution strategy, a pricing strategy, advertising / promotion strategy and a hierarchical product structure. The "market entity" is the key concept of the approach. It is intended to provide a rough overview on service offerings, particularly including their composition and internal relationships.

Service Blueprinting Service Blueprinting is a family of related approaches, all of which focus on exposing the activities a service process comprises of in a chronological order. Its key analytical instrument is the "line of visibility" that separates activities. It distinguishes activities that are visible to the customer (onstage activities) from those that are not visible to the customer (backstage activities). Products, decisions and documents can be attached to the activities.

SeeMe SeeMe is also a process-centered modeling method. Apart from the representation of service processes it facilitates the assignment of organizational units, of roles, resources (e.g., documents, software programs, physical objects). Activities can be combined by connectors. The method particularly stands out by its support for assigning certain types of vagueness to roles, activities and resources.

Structured Analysis and Design Technique (SADT) SADT is grounded on the concept of Structured Analysis (SA) boxes that are intended to represent hierarchical relationships between components of a product and the idea of top-down system design. The ultimate goal of SADT is to provide a one single graphic language for blueprinting systems. In the 1980s the SADT approach was embedded into the design of the IDEF0 modeling standard. Today IDEF is part of the KBSI (Knowledge Based Systems, Inc.). Through the IDEF0 standard SADT benefits from wide adoption and extensive tool-support.

ARIS / Model-based Service System Engineering The Model-based Service System Engineering approach bases upon the ARIS modeling method. It distinguishes a product model, a process model, and a resource model to describe a service system. It extends the ARIS approach as it allows to model products from an internal perspective (product tree) as well as from a customer's perspective (product bundles). Further it introduces the concept of process module chains that shall facilitate a timely composition of service processes out of standardized components.

Colored Petri Nets for Service Simulation The approach is based on the Colored Petri Nets (CPN) technique. CPN is intended be used to model and simulate service processes. Activities are represented by transitions. Resources and events are represented by places. Properties, requirements, competencies, measures, etc. are represented by complex data types (color sets). By assigning complex expressions to the edges it is possible to estimate if an activity can be executed or which resources are required to execute it. Outgoing edges can be used to describe the output of activities in service processes.

Method for systematic development of product-oriented services The method distinguishes involved objects and activities of the service development process. It provides a modeling perspective for the (service) products that is based on the STEP approach. Further it provides reference processes that prescribe the product development process. Both are integrated in a procedure model that guides the integrated product and process development.

STEP/EXPRESS-G EXPRESS is a data modeling language that allows for the description of products structures. It is based on the standard for the exchange of product models (STEP). EXPRESS is standardized as ISO 10303-11. Behavior cannot be represented. Processes can be annotated as free text elements only. The specification comprises of 122 expressions and 318 syntax products. EXPRESS-G is a graphical notation for the EXPRESS method.

As can be inferred from this list, different modeling languages were taken into account. On the one hand, we identified general-purpose modeling languages that have been developed independently from any domain of application. Modeling languages belonging to this category are, for instance, the Event-Driven Process Chains (EPC) within the ARIS method and the Structured Analysis and Design Technique (SADT). On the other hand, we also included modeling languages that had been explicitly designed for modeling aspects in service systems, such as Service Blueprinting, the modeling techniques offered by the Model-Based Service System Engineering approach, or SeeMe. These domain-specific modeling languages provide modeling constructs that are custom-fit to the properties of services, whereas the general purpose modeling languages are generic and feature constructs that can be applied to services, however, on a more general level.

There exists a myriad of further related work, which cannot be reviewed in detail within the scope of this chapter. For example there are further approaches for standardizing the vocabulary and the processes of tendering service contracts. E.g., the Publicly Available Specification 1018 of the German Institute for Standardization [22] defines 14 stages of service provision from detecting the need of a service request to the actual fulfillment of a service contract. For each stage criteria are provided that can be used to specify a service. Each criterion is described by a definition / description of its content. The attributes are assigned either to the entire service (the header) or to a single position (position) of the service.

2.3.2 Review of PSS Modeling Languages

The features of each of the selected modeling languages were matched with the proposed modeling requirements of customer solutions. According to this exploration, the as-is capabilities of the analyzed modeling languages are depicted in Table 2.3. Results can be used to ascertain which features can already be displayed by current modeling languages, while other requirements can be shown to remain unaddressed. Gaps can be seen as potential areas for extensions with features incorporated from other modeling languages. We continue with a brief review.

Molecular Model / Service Blueprinting

Both approaches focus on a service marketing perspective. Shostack's Molecular Model allows only for roughly modeling the components of a value bundle as well

Table 2.3: As-is capabilities of modeling languages.

Modeling language	Molecular Model, Service Blueprinting	SeeMe	SADT	ARIS / Model-based Service System Engineering	Coloured Petri Nets for Service Simulation	Method for systematic devel. of product-oriented services	STEP / Express-G	Engineering	SCM	Marketing	Environmental
Customer Solutions subdivided into components	X	(X)	X	X	X	X	X	X			X
Reuse of the components as modules	X	-	X	X	X	X	X	X			
Configuration rules to specify inclusion (may)	-	-	-	-	-	-	X	X			
Configuration rules to specify exclusion (must not)	-	-	-	-	-	-	X	X			
Substituation of modules or components	-	-	-	-	-	-	X	X			X
Attributes of components / products	-	X	-	X	X	X	X	X		X	X
Longevity	-	-	-	-	(X)	X	X	X		X	X
Activities	X	X	X	X	X	X	(X)	X	X		
Control flow	X	X	X	X	X	X	(X)	X	X		
Flow of materials	-	-	-	X	(X)	(X)	-		X		X
Flow of information	-	X	X	X	-	-	-		X		
Cash flow	-	-	-	(X)	-	-	(X)		X	X	
Process cycle time	X	X	-	(X)	-	-	X		X	X	
Failure rate	(X)	-	-	(X)	X	-	X		X	X	
Capacity	-	-	-	(X)	X	X	X	X	X		
Consumption of operand resources (economic point of view)	X	X	X	(X)	X	X	X		X		
Consumption of operand resources (ecologic point of view)	-	-	-	(X)	-	X	-				X
Legal constraints	-	-	X	(X)	-	(X)	-				X
Value proposition for customers	-	-	-	-	-	-	-			X	
Marketing lifecycle	-	-	-	-	X	-	-			X	
Line of visibility towards customers	X	-	-	X	-	-	-		X	X	
Line of interaction towards customers	X	-	-	X	-	X	-		X	X	
Relationships / lines towards other stakeholders	(X)	-	(X)	X	-	(X)	-		X	X	
Quality standards	-	-	X	-	X	X	(X)	X			
Service level agreements	-	-	-	X	-	-	-			X	
IT Systems and applications	-	X	X	X	X	-	(X)		X		
Data (e.g. master data)	(X)	X	-	X	-	-	X		X		
Personnel allocation	(X)	X	X	X	X	X	-		X	X	
Qualifikation of personnel	-	(X)	X	-	-	X	(X)		X	X	
Organisational units	-	X	X	X	X	X	-		X	X	
Resources to be introduced by customers	X	(X)	(X)	X	X	X	-		X	X	

as related strategies for distribution, price and communication. Processes, resources and the interaction with stakeholders are not (explicitly) considered. The analyzed Service Blueprinting approaches allow for a more detailed modeling of the service processes' activities. These activities can be assigned to roles. Products and materials are mainly not in the focus — only Shostack's approach allows for representing them as specific objects. Resources and legal constraints are not considered as well as required qualification of personnel or quality attributes. Shostack's approach further provides attributes for benchmarking and exception handling of processes. Kleinaltenkamp highlights interaction aspects within the service company. Bitner et al. focus on the customer integration.

SeeMe

The SeeMe approach is also centered on the process view. It follows a modularization strategy that allows bundling activities that can be reused in different service modules. Physical products (and their structure) can also be represented. A specific representation of the service (product) is not supported.

SADT

SADT is a ubiquitous modeling approach. Accordingly, SADT models tend to suffer from complexity. Several models of the same objects would be required, e.g., in order to represent the resource flow and to facilitate a knowledge management or physical component perspective.

ARIS / Model-based Service System Engineering

Model-based Service System Engineering is an ARIS approach that is tailored to fit services. It provides constructs to meet a variety of our requirements. Particularly the service product description and the representation of the interfaces between products and services as well as the degree of integration could be addressed more explicitly.

Coloured Petri Nets for Service Simulation (CPN)

This method strives to facilitate the simulation of service processes. Obviously the method focusses on a process view on service systems thus it does not comprise of specific concepts for representing static structures. Service products could be integrated into the models as complex data types only. Also resource-based aspect of the service system can be described by such data types. Dedicated software support is available that eases reuse of model components. However, modifications and extensions of the modeling technique, versioning of models, and maintaining model variants are not supported.

Method for systematic development of product-oriented services

The approach has a manufacturing background. Accordingly it strives to suit engineering disciplines at first. While it emphasized a modular structure of service products, it disregards the relational processes of service value creation. Also, required IT systems and data objects are neglected.

STEP / EXPRESS-G

STEP and EXPRESS-G focus on representing static product structures (product definition, product representation, product presentation). However, modeling of processes and resources are not supported. Projects for further developing the method strive for filling that gap.

Although some modeling languages provide constructs for a variety of requirements, none of the modeling languages is capable of accounting for all the proposed modeling requirements. E.g., interfaces between physical goods and services as well as the configuration and offering of customer solutions are seldom addressed. Thus, in conceptual models derived by using these modeling languages it remains unclear, how intense the integration of physical goods and services is and how cooperation scenarios for offering and delivering customer solutions should be designed.

In addition to general deficiencies, the investigated modeling languages are unlikely to display features originating from other research areas than the one from which the modeling language emerged. We make the following observations in this regard:

- Modeling languages not originating from an engineering background usually lack a representation of a bill-of-materials and other product-related data, such as lifecycle information (referring to maintenance cycles) on a component level of detail. This information would nevertheless be helpful to guide service processes, for example by identifying components and parts that require service activities along their lifecycle.
- Modeling languages not originating from an SCM background tend not to display the IT systems as well as business units involved in service and manufacturing processes. As service processes tend to be labor intensive and require information to be delivered at the correct moment (i.e., the 'moment of truth'), providing these constructs seems to hold significant potential to assign resources and information on time.
- Modeling languages not stemming from a service marketing point-of-view are unlikely to address the type and intensity of customer integration, e.g., by displaying the line of interaction and line of visibility towards customers. Acknowledging customers as important members of PSS and as co-creators of value implies accounting for their information and resource input during the service processes. Hence, customers should be representable by conceptual models in PSS.
- Environmental aspects remain largely unaddressed by all the evaluated modeling languages. We could not identify any formal modeling language specially designed for this purpose.

2.4 Discussion, Limitation and Outlook

In this chapter we have introduced the creation and provision of customer solutions as a chance and challenge to manufacturing companies. PSS here serves as an example for the increasing importance of service from an economical as well as a managerial perspective. Designing and delivering customer solutions is a complex undertaking since it requires the cooperation of various business units, companies and customers in PSS. We exhibited that the description of customer-specific solutions is a crucial task in these efforts.

Conceptual modeling techniques can be used to better cope with this challenge. Conceptual modeling benefits from sound modeling languages, which provide constructs for formally representing business-related aspects of a service description. To analyze the status-quo of modeling languages for PSS, we derived an evaluation sheet drawing from some viewpoints emphasized by four relevant academic disciplines involved in service research. We applied the proposed evaluation sheet on a selection of eight modeling languages in the area of PSS, originating from these different research areas. We found that adequate support of conceptual modeling by modeling languages for PSS is lacking, as the modeling languages under consideration tended not to adopt an interdisciplinary point of view, often restricting themselves to the research discipline from which they originated. In the light of the need to comprehensively describe services from several points of view (which is a prerequisite to identify and invoke services on electronic service marketplaces) this underlines the need of developing more advanced conceptual modeling support for the description of services.

The approach taken in this chapter is subject to some limitations. First, the criteria developed for analyzing the expressive power of conceptual modeling languages cannot be exhaustive, but remain limited to the perspectives we have applied to identify them. In future work, a more thorough catalogue of evaluation criteria might be developed that comprises the foci of still other research disciplines involved with research in Service Science. Second, additional criteria originating from the four discussed research streams might be added to the list. Another limitation is the selection of the conceptual modeling languages themselves. Although we included all conceptual modeling languages that we came across into the analysis, still others likely exist and need to be analyzed with the proposed criteria to create a more comprehensive overview.

Therefore, the concepts presented in this chapter are meant to act as a starting point to develop a more advanced support of service description and with regard to customer PSS-offered customer solutions in particular. With the results we strive to propose directions for extensions to be made to existing modeling languages. This might help to encourage further interdisciplinary research activities carried out in the currently emerging disciplines of Service Science and Service Science Management and Engineering (SSME).

References

1. W. F. Ahrens, E. Schnieder, and M. Chouikha. Formale Prozessbeschreibungen - gestern, heute und morgen. *atp - Automatisierungstechnische Praxis*, 42(9):24–32, 2000.
2. R. Anderl and D. Trippner. *STEP: Standard for the exchange of product model data.* Gabler, Wiesbaden, Germany, 2000.
3. C. Y. Baldwin and K. B. Clark. Managing in an Age of Modularity. *Harvard Business Review*, September:84–93, 1997.
4. I. R. Bardhan, H. Demirkan, P. K. Kannan, R. J. Kauffman, and R. Sougstad. An Interdisciplinary Perspective on IT Services Management and Service Science. *Journal of Management Information Systems*, 26(4):13–64, 2010.
5. J. Becker, D. Beverungen, and R. Knackstedt. The challenge of conceptual modeling for product-service systems: status-quo and perspectives for reference models and modeling languages. *Information Systems and E-Business Management*, 8(1):33–66, 2010.
6. J. Becker, D. Beverungen, R. Knackstedt, and O. Müller. Model-Based Decision Support for the Customer-Specific Configuration of Value Bundles. *Enterprise Modelling and Information Systems Architectures*, 4(1):26–38, 2009.
7. J. Becker, R. Knackstedt, D. Beverungen, S. Bräuer, D. Bruning, D. Christoph, D. Jorch, F. Joß Bächer, H. Jostmeier, S. Wiethoff, and A. Yeboah. *Modellierung der hybriden Wertschöpfung: Eine Vergleichsstudie zu Modellierungstechniken.* University of Münster, Arbeitsberichte des Instituts für Wirtschaftinformatik, Nr. 125, Münster, 2009.
8. C. Belz. *Fit for Service: Industrie als Dienstleister.* Hanser Wirtschaft, St. Gallen, 1997.
9. M. J. Bitner, A. L. Ostrom, and F. N. Morgan. Service blueprinting: A practical technique for service innovation. *California Management Review*, 50(3):66–94, 2008.
10. J. Black, C. Draper, T. Lococo, F. Matar, and C. Ward. An integration model for organizing IT service management. *IBM Systems Journal*, 46(3):405–422, 2007.
11. H. Bley, M. Bernardi, B. Schmitt, and C. Zenner. Assembly planning of mini and micro products enhanced by an integrated product and process model. In *Proceedings of the International Precision Assembly Seminar (IPAS 2003)*, pages 31–38, Bad Hofgastein, Austria, 2003.
12. M. Bossmann. Feature-basierte Produkt- und Prozessmodelle in der integrierten Produktentstehung. 2007.
13. C. Botta. *Rahmenkonzept zur Entwicklung von Product-Service Systems.* Josef Eul Verlag, Lohmar, Germany, 2007.
14. H.-J. Bullinger. Service Engineering: Ein Rahmenkonzept für die systematische Entwicklung von Dienstleistungen. In H. .-J. Bullinger and A.-W. Scheer, editors, *Service Engineering - Entwicklung und Gestaltung innovativer Dienstleistung*, pages 51–82. Springer, Berlin et al, Germany, 2nd edition, 2003.
15. Bundeszentrale für politische Bildung (bpb). Datenreport 2008 - Ein Sozialbericht für die Bundesrepublik Deutschland. Technical report, Bonn, Germany, 2008.
16. C. Congram and M. Epelman. How to describe your service: An invitation to the structured analysis and design technique. *International Journal of Service Industry Management*, 6(2):6–23, 1995.
17. H. Corsten and R. Gössinger. Rahmenkonzept zur integrativen Modellierung von Dienstleistungen. 2003.
18. P. Dadam, K. Kuhn, M. Reichert, T. Beuter, and M. Nathe. ADEPT: Ein integrierender Ansatz zur Entwicklung flexibler, zuverlässiger kooperierender Assistenzsysteme in klinischen Anwendungsumgebungen. 1995.
19. N. P. Dalal, M. Kamath, W. J. Kolarik, and E. Sivaraman. Toward an integrated framework for modeling enterprise processes. *Communications of the ACM*, 47(3):83–87, Mar. 2004.
20. A. J. Dietrich and S. Kirn. Flexible Wertschöpfungsnetzwerke in der kundenindividuellen Massenfertigung — Ein serviceorientiertes Modell für die Schuhindustrie. In O. K. Ferstl, E. J. Sinz, S. Eckert, and T. Isselhorst, editors, *Wirtschaftsinformatik 2005*, pages 23–42. Heidelberg, Germany, 2005.

21. B. Dietrich. Resource Planning for Business Services. *Communications of the ACM*, 49(7):62–64, July 2006.
22. DIN. *Publicly Available Specification 1018: Essential structure for the description of services in the procurement stage*. German Standards Institute, Beuth Verlag, Berlin, 2002.
23. DIN. *Publicly Available Specification 1094: Product-Service Systems — Value Creation by Integrating Goods and Services*. German Standards Institute, Beuth Verlag, Berlin, 2009.
24. A. Emmrich. *Ein Beitrag zur systematischen Entwicklung produktorientierter Dienstleistungen*. Dissertation, University of Paderborn, Germany, 2005.
25. W. H. Engelhardt, M. Kleinaltenkamp, and M. Reckenfelderbäumer. Leistungsbündel als Absatzobjekte. Ein Ansatz zur Überwindung der Dichotomie von Sach- und Dienstleistungen. *Schmalenbachs Zeitschrift fuer betriebswirtschaftliche Forschung*, 45(5):395–426, 1993.
26. R. Ferrario, N. Guarino, C. Janiesch, T. Kiemes, D. Oberle, and F. Probst. Towards an Ontological Foundation of Services Science: The General Service Model. In *10th International Conference on Wirtschaftsinformatik*, Zürich, Switzerland, 2011.
27. J. A. Fitzsimmons and M. J. Fitzsimmons. *Service Management — Operations, Strategy, and Information Technology*. McGraw-Hill, Inc, Boston, MA, USA, 3rd edition, 2001.
28. S. Fließ. *Die Steuerung von Kundenintegrationsprozessen*. Deutscher Universitätsverlag, Wiesbaden, Germany, 2001.
29. W. Ganz. Germany: Service Engineering. *Communications of the ACM*, 49(7):79, July 2006.
30. R. Garud and A. Kumaraswamy. Technological and organizational designs for realizing economies of substitution. In R. Garud, A. Kumaraswamy, and N. Langlois, editors, *Managing in the Modular Age*, pages 45–77. Blackwell Publishing, Malden, MA, USA, 2003.
31. P. Gu and K. Chan. Product modelling using STEP. *Computer-Aided Design*, 27(3):163–179, Mar. 1995.
32. M. Hanna and W. Newman. *Integrated Operations Management: A Supply Chain Perspective*. Thomson/South-Western, Florence, KY, USA, 2006.
33. I. Hartel. Aufbau und Betrieb eines kooperativen Dienstleistungsmanagements in der Investitionsgüterindustrie. In R. Kreibich, editor, *Erfolg mit Dienstleistungen*, pages 47–54. Schäffer-Pöschel, Stuttgart, Germany, 2004.
34. J. Hausmann, R. Heckel, and S. Sauer. Dynamic Meta Modeling with Time: Specifying the Semantics of Multimedia Sequence Diagrams. *Software and Systems Modeling*, 3(3):181–193, Feb. 2004.
35. R. Holten. Entwicklung einer Modellierungstechnik für Data-Warehouse-Fachkonzepte. In H. Schmidt, editor, *Proceedings of the MobIS-Fachtagung*, pages 3–21, Siegen, Germany, 2000.
36. IAI. International Alliance for Interoperability: Data Modeling Using EXPRESS-G for IFC Development, 2008.
37. International Standards Organization (ISO). *Product Data Interchange using STEP (PDES) Part 11 - The EXPRESS Language Reference Manual*, 1995.
38. H. Katzan. *Service Science: Concepts, Technology, Management*. iUniverse, New York, NY, USA, 2008.
39. J. Kingman-Brundage, W. George, and D. Bowen. Service Logic: Achieving Service System Integration. *International Journal of Service Industry Management*, 6(4):20–39, 1995.
40. L. Klein, P. Schreiner, and C. Seemann. *Die Dienstleistungen im Griff — Erfolgreich Gründen mit System*. Fraunhofer IRB Verlag, Stuttgart, Germany, 2003.
41. D. Koonce and R. Judd. A visual modelling language for EXPRESS schema. *International Journal of Computer Integrated Manufacturing*, 14(5):457–472, 2001.
42. W. Kraemer and V. Zimmermann. Public Service Engineering — Planung und Realisierung innovativer Verwaltungsprodukte. In A.-W. Scheer, editor, *Rechnungswesen und EDV: Kundenorientierung in Industrie, Dienstleistung und Verwaltung, 17. Saarbrücker Arbeitstagung*, pages 555–580. Physica, Heidelberg, Germany, 1996.
43. G. Kunau, K.-U. Loser, and T. Herrmann. Im Spannungsfeld zwischen formalen und informellen Aspekten: Modellierung von Dienstleistungen mit SeeMe. In T. Herrmann, U. Kleinbeck, and H. Krcmar, editors, *Konzepte für das Service Engineering - Modularisierung, Prozessgestaltung und Produktivitätsmanagement*. Springer, Heidelberg, Berlin, Germany, 2005.

44. R. F. Lusch, S. L. Vargo, and M. Tanniru. Service, value networks and learning. *Journal of the Academy of Marketing Science*, 38(1):19–31, 2010.
45. S. Madhavaram and S. D. Hunt. The service-dominant logic and a hierarchy of operant resources: developing masterful operant resources and implications for marketing strategy. *Journal of the Academy of Marketing Science*, 36(1):67–82, Aug. 2007.
46. P. P. Maglio and J. Spohrer. Fundamentals of Service Science. *Journal of the Academy of Marketing Science*, 36(1):18–20, July 2008.
47. M. Manavazhi. Hybrid modelling framework for synthesizing virtual structures. *Construction Management and Economics*, 18(4):415–426, June 2000.
48. H. Mason. ISO 10303 - STEP: A key standard for the global market. *ISO Bulletin*, January:9–13, 2002.
49. N. Maussang, P. Zwolinski, and D. Brissaud. Design of Product-Service Systems. In *10th European Roundtable on Sustainable Consumption and Production (ERSCP)*, 2005.
50. Mercer Management Consulting. Mercer-Analyse — Ungenutzte Chancen im Servicegeschäft, 2003.
51. O. Mont. *Product-service systems: Panacea or myth?* Phd thesis, Lund University, 2004.
52. N. Morelli. Designing Product/Service Systems: A Methodological Exploration. *Design Issues*, 18(3):3–17, 2002.
53. OECD. Enhancing the Performance of the Services Sector. Technical report, June 2005.
54. A. Parasuraman, V. A. Zeithaml, and L. L. Berry. SERVQUAL: A multiple-item scale for measuring consumer perceptions of service quality. *Journal of Retailing*, 64(1):12–40, 1988.
55. A. F. Payne, K. Storbacka, and P. Frow. Managing the Co-creation of Value. *Journal of the Academy of Marketing Science*, 36(1):83–96, 2008.
56. M. J. Pratt. Introduction to ISO 10303 — the STEP Standard for Product Data Exchange. *Journal of Computer and Information Science and Engineering*, 1(1):102–103, 2001.
57. ProSTEP. iViP: Architektur und Aufbau, 2007.
58. J. B. Quinn, J. Baruch, and P. Paquette. Exploiting the manufacturing-services interface. *Sloan Management Review*, 29(4):45–56, 1988.
59. J. B. Quinn, T. L. Doorley, and P. C. Paquette. Beyond Products: Service-Based Strategy. *Harvard Business Review*, 68(2):58–67, 1990.
60. A. Rai and V. Sambamurthy. Editorial Notes — The Growth of Interest in Services Management: Opportunities for Information Systems Scholars. *Information Systems Research*, 17(4):327–331, 2006.
61. C. Rainfurth, S. Tegtmeyer, and G. Lay. Organisation produktbegleitender Dienstleistungen. In G. Lay and M. Nippa, editors, *Management produktbegleitender Dienstleistungen*, pages 99–119. Springer, Heidelberg, Germany, 2005.
62. T. Sakao and Y. Shimomura. Service Engineering: A novel engineering discipline for producers to increase value combining service and product. *Journal of Cleaner Production*, 15(2):590–604, 2007.
63. S. E. Sampson and C. M. Froehle. Foundations and implications of a proposed unified services theory. *Production and Operations Management*, 15(2):329–343, 2006.
64. A.-W. Scheer. *ARIS — Vom Geschäftsprozess zum Anwendungssystem*. Springer, Berlin et al., Germany, 4th edition, 2002.
65. A.-W. Scheer, O. Grieble, and R. Klein. Modellbasiertes Dienstleistungsmanagement. In H.-J. Bullinger and A.-W. Scheer, editors, *Service Engineering - Entwicklung und Gestaltung innovativer Dienstleistungen*, pages 19–52. Springer, Berlin et al., Germany, 2nd edition, 2006.
66. M. Schmied. *Themenheft Service Engineering*. Bonn, Germany, 2002.
67. K. Schneider, H.-J. Bullinger, and A.-W. Scheer. Service engineering — entwicklung und gestaltung innovativer dienstleistungen. In H.-J. Bullinger and A.-W. Scheer, editors, *Service Engineering*, pages 3–18. Springer Berlin Heidelberg, 2006.
68. E. Schnieder. Modellkonzepte in der Automatisierungstechnik. In G. Engels, A. Oberweis, and A. Zündorf, editors, *Modellierung 2001*, pages 7–17. Bonn, Germany, 2001.
69. D. R. Shaw and C. P. Holland. Strategy, networks and systems in the global translation services market. *The Journal of Strategic Information Systems*, 19(4):242–256, Dec. 2010.

70. G. L. Shostack. Breaking Free from Product Marketing. *Journal of Marketing*, 41(2):73–80, 1977.
71. G. L. Shostack. How to design a service. *European Journal of Marketing*, 1(16):49–63, 1982.
72. G. L. Shostack. Design services that deliver. *Harvard Business Review*, 62(1):133–139, 1984.
73. J. Spohrer, P. P. Maglio, J. Bailey, and D. Gruhl. Steps Toward a Science of Service Systems. *Computer*, 40(1):71–77, 2007.
74. J. Spohrer and D. Riecken. Service Science. *Communications of the ACM*, 49(7):31–34, 2006.
75. M. Steinbach, C. Botta, and C. Weber. Integrierte Entwicklung von Product-Service Systems. *Werkstatttechnik online*, 95(7/8):546–553, 2005.
76. F. Stille. Product-related Services — Still Growing in Importance. *DIW Economic Bulletin*, 40(6):195–200, 2003.
77. F. Sturm, A. Bading, and M. Schubert. *Investitionsgüterhersteller auf dem Weg zum Lösungsanbieter — Eine empirische Studie*. Fraunhofer IRB Verlag, Stuttgart, Germany, 2007.
78. C. S. Sturts and F. H. Griffis. Pricing Engineering Services. *Journal of Management in Engineering*, 21(2):56–62, 2005.
79. A. Tan, T. C. McAloone, and M. M. Andreasen. What Happens to Integrated Product Development Models with Product/Service-System Approaches. In *Proceedings of the 6th Integrated Product Development Workshop (IPD 2006)*, Magdeburg, Germany, 2006.
80. J. Teboul. *Service is Front Stage: Positioning Services for Value Advantage*. Palgrave Macmillan, Basingstoke, UK, 2006.
81. A. Tukker and U. Tischner. Product-services as a research field: past, present and future. Reflections from a decade of research. *Journal of Cleaner Production*, 14(17):1552–1556, 2006.
82. K. R. Tuli, A. K. Kohli, and S. G. Bharadwaj. Rethinking Customer Solutions: From Product Bundles to Relational Processes. *Journal of Marketing*, 71(3):1–17, 2007.
83. S. Vandermerwe and J. Rada. Servitization of business. *European Management Journal*, 6(4):314–324, 1988.
84. S. L. Vargo and R. F. Lusch. Evolving to a New Dominant Logic for Marketing. *The Journal of Marketing*, 68(1):1 – 17, 2004.
85. S. L. Vargo and R. F. Lusch. Service-dominant logic: continuing the evolution. *Journal of the Academy of Marketing Science*, 36(1):1–10, 2008.
86. S. L. Vargo, R. F. Lusch, and M. Archpru Akaka. Advancing service science with service-dominant logic — clarifications and conceptual development. In P. Maglio, C. Kieliszewski, and J. Spohrer, editors, *Handbook of Service Science*, Service Science: Research and Innovations in the Service Economy, pages 133–156. Springer, Berlin, Heidelberg, New York, 2010.
87. K. Winkelmann. *Prospektive Bewertung der kooperativen Erbringung industrieller Dienstleistungen im Maschinenbau durch Simulation mit Petri-Netzen*. Shaker Verlag, Aachen, Germany, 2007.
88. K. Winkelmann and H. Luczak. Prospective analysis of cooperative provision of industrial services using coloured petri nets. In S. Donatelli and P. S. Thiagarajan, editors, *Petri Nets and Other Models of Concurrency - ICATPN 2006, 27th International Conference on Applications and Theory of Petri Nets and Other Models of Concurrency, Turku, Finland, June 26-30, 2006, Proceedings*, volume 4024 of *Lecture Notes in Computer Science*, pages 362–380. Springer, 2006.
89. F. Wolff and U. Frank. A Multiperspective Framework for Evaluating Conceptual Models in Organisatinal Change. In *Proceedings of the 15th European Conference on Information Systems (ECIS 2005)*, St. Gallen, Switzerland, 2005.

Chapter 3
Service Network Approaches

Ivan S. Razo-Zapata, Pieter De Leenheer, Jaap Gordijn, and Hans Akkermans

Abstract This chapter discusses several approaches to design, analyze, describe and compose service networks. We analyze the technical and business-related aspects of these approaches, their evolution, and the trends they will be likely to follow. We further suggest how the two major trends driving these approaches (i.e., business and process orientation) can converge. The chapter concludes with a discussion of future lines of research in this area.

3.1 Introduction

Value Networks consist of relationships that generate tangible and intangible added value between individuals. Along the line of *service-dominant* logic in marketing (as opposed to traditional product-dominant logic), the notion of service is becoming a predominant instrument for actors to apply their specialized competencies (such as knowledge and skills), through deeds, processes, and performances for the benefit of another actor or the actor itself; hence create value [1]. A well-known example of an established value network is Microsoft and its elaborated partner ecosystem that provides various kinds of services around Microsoft products globally.

The Web also brought opportunities to establish short-term value networks without the actors even realizing they interact across boundaries such as time, scale, and geography. E.g., if you initiate the purchase of a second-hand book at Amazon, you are actually initiating the formation of a short-term value network that is medi-

Ivan S. Razo-Zapata, Pieter De Leenheer, Jaap Gordijn, and Hans Akkermans
VU University, Department of Computer Science, De Boelelaan 1081 A, 1081 HV Amsterdam, The Netherlands, e-mail: i.s.razozapata@vu.nl, e-mail: pieterdeleenheer@vu.nl, e-mail: j.gordijn@vu.nl, e-mail: elly.lammers@vu.nl

Pieter De Leenheer
Collibra nv, Brussels, Belgium

ated by Amazon between you (in the role of customer), the virtual bookshop, and a packaging and delivery partner that serves at your location.

In short, within a value network it is common to observe three types of relationships: 1) Customer to Customer (C2C), customers exchanging valuable outcomes with each other. 2) Business to Customer (B2C), when service suppliers exchanges valuable outcomes with the final customers, e.g., you and Microsoft. 3) Business to Business (B2B), service suppliers exchanging valuable outcomes with each other, e.g., Microsoft and the packaging and delivery partners [2].

As discussed in Chapter 1, Web services (in the technical sense) should not be confused with commercial services as found on the real-world market place. Web (a.k.a. software-based) services specify interfaces and communication protocols, and by doing so implement a specific computing paradigm. Web services are still much dominated by manual engineering, and predefined top-down composition.

Commercial services are simply not describable in terms of interface specifications, as their conception inherits from marketing and social studies. As explained above, in a value network, services create value. E.g., the composition of an educational commercial service bundle should seek maximum value creation for all stakeholders. Students get value from educational services by achieving certain learning objectives and get evidence (by means of a diploma or certificate) for the acquired competencies. Educational institutes, in return for their services get value usually in the form of money.

Functions of a commercial service might be partly supported by Web services such as, e.g., billing facilities. However, in order for these commercial services to find each other and automatically bundle into economically sustainable value networks, we must model and analyze their valuable outcomes as well. Analogously, customers should be able to express their needs — i.e., the value they expect from commercial services — in human-understandable terms. Examples of value aspects that should be taken into account include pricing, competitive service properties such as discounts or home delivery, agreements on the extent to which services are legally dependent or exclusive, and ratings based on user sentiment analysis.

In this chapter we present a survey of approaches addressing issues such as description, design, analysis, and composition of service networks (SNs). Although these approaches might have some similarities such as the notion of service or the use of standardized tools, each approach has different purpose, which influences and drives the way in which SNs are described. Therefore, in our survey we not only describe these frameworks but we also analyze how they can provide some insights on service value network (SVN) composition.

The chapter is organized as follows. Section 3.2 presents the foundations to understand the concept of SNs as well as the main elements to be considered when modeling such networks. In Section 3.3 we perform an analysis of several approaches to design, analyze and compose SNs. Finally, Section 3.4 provides a discussion.

3.2 Comparison Framework

3.2.1 Definitions

Value network methods and technologies are inspired by two main foundations: value chains and value networks that cooperate. In 1985, Porter introduced the notion of *value chain*, to conceive the combination of value-adding activities within a firm to provide value to customers [3]. These activities can be classified generally as either primary or support activities that all businesses must undertake in some form. The value chain is defined as follows:

Definition 3.1. A *Value Chain* is a directed sequence of activity relationships that generates tangible and intangible added value through bilateral static communication between organizations.

In 2000, Tapscott, introduced the notion of *b-web* [4]. He explains that the advent of the Web provides a new platform for business opportunities. He identifies typical roles stakeholders can play. Depending on how value is being created and the organizational dynamics of the b-web, he distinguishes five archetypes along all b-webs can be categorized. Later, in 2006, he introduced the notion of wikinomics, where he highlights the importance of collaboration/networking [5]. A *value network* is defined as [6]:

Definition 3.2. A *Value Network* is any web of activity relationships that generates tangible and intangible added value through complex dynamic communication between two or more individuals, groups, or organizations.

Within the notion of network marketing, Lovelock and Wirtz define good networkers as entities who are able to put individuals in touch with others who have a mutual interest, e.g., providing services to a market segment [1]. In this matter, a *service network* can be defined as follows:

Definition 3.3. A *Service Network* is a team of individuals who establish relationships among homogeneous peers to provide a specific service.

By homogeneity we mean that individuals have a common business objective, i.e., provide a service solution. Based on ideas from Hamilton [7], Verna Allee [6], Lovelock and Witz [1], and our point of view, a *service value network* is defined as follows:

Definition 3.4. A *Service Value Network* is a flexible and dynamic web of homogeneous enterprises and final customers who reciprocally establish relationships with other peers for delivering an added-value service to a final customer.

3.2.2 Service Network Criteria

In order to provide both expressiveness and usability, the description approaches must cover the following criteria, some of the aspects are proposed by ourselves whereas others are taken from a previous analysis [8].

1) B2C interaction: refers to whether the approach allows interaction between final customers and suppliers (or brokers) to co-create SNs. Since customers are not static participants, they must be provided with some mechanism for expressing not only service requests but also preferences and/or recommendations. In this sense, we determine three possible situations: The customer is involved, the customer is partially involved or the customer is not involved at all during the composition of the SN.

2) B2B relationships: modeling how suppliers relate to each other is a required step within SN composition. Such B2B relationships might be based on business rules, strategic alliances among other requests or constraints. The approaches being analyzed can model this aspect in different ways such as: inter and intra company relationships, inter-company relationships, intra-company relationships.

3) Network Definition: at this point we evaluate whether the approach provides a definition for an SN. It might be a formal or informal definition. A formal definition provides meaningful descriptions not only about the SN itself but also about the components and their relationships within the SN. In contrast, an informal definition only provides descriptions for an SN.

4) Visualization: since internal and external relationships among network participants might bring about hidden structures or patterns, providing a visual of the SN is also relevant [9]. Moreover, this visualization might be also exploited for analysis and synthesis tasks that lead to innovation based on discovered niches.

5) Orientation: the locus of attention differs from approach to approach. Some have a process-oriented view, others focus on business-oriented issues. By business orientation we mean that the approach takes into account economic relationships rather than work-flow properties. It has been discussed in previous work why business modeling is different from process modeling [10]. Briefly, an important goal in process modeling is to reach a common understanding about how activities should be carried out (e.g., in which order). On the contrary, business models are centered around the notion of *value*, therefore it is relevant to determine who is offering what of *value* to whom and what expects of *value* in return, i.e., economic reciprocity.

6) Tool support: this aspect describes whether the approach, design, analyze or compose SVNs using standard tools. When applicable, we also specify the tools being used.

3.2.3 ICT Support

Moreover, we have also identified six levels of ICT maturity on which all the approaches can be positioned:

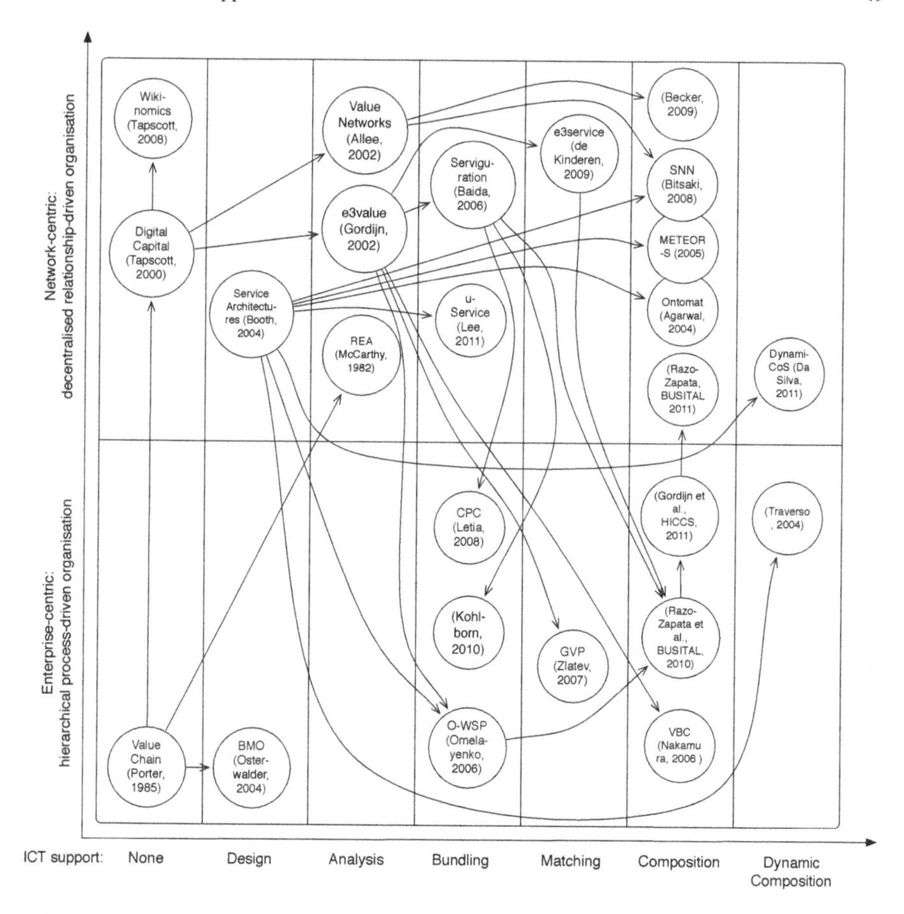

Fig. 3.1: Evolution of Service Network Approaches. Three dimensions: 1) Economic context. 2) ICT maturity. 3) Orientation: solid circles represent business-oriented approaches whereas dotted circles stand for process-oriented approaches. Arrows represent influences among approaches.

a) **Design.** In this level, building SNs is seen as a design task. The network is built by manually selecting enterprises as well as the services they offer or request.
b) **Analysis.** ICT tools allow to evaluate properties of a given SN such as profitability, misalignments with business models or potential risks.
c) **Bundling.** At this level ICT provides tools for combining services into bundles, which are the starting point for generating SNs. The generation of bundles might take into account the properties offered by services, business rules or pre-defined patterns. Indeed, this process mainly looks at the supplier perspective, i.e., what service bundles can be generated to cover more complex customer needs.

d) **Matching.** ICT tools provide support such that: 1) customers provide a formalization of their needs, and 2) the customer needs are matched with the offerings coming from the supplier side, then identifying possible solutions.
e) **Composition.** It combines the previous steps, bundling and matching. Once a customer need has been formalized and matched with a service bundle, the next stage is to solve the bundle's dependencies. For instance, a bundle providing an streaming music service to the customer might require a software-protection service to work safely. Actually, this is the case in the real-life bundle composed of Spotify[1] and Last.fm[2] which depends on the software-protection service provided by Morpher.[3] The customer only interacts with Spotify and Last.fm, nevertheless the complete SN also includes the Morpher service. Fixed templates can achieved this step by providing a clear description of the services to be found.
f) **Dynamic composition.** at this level ICT tools perform *on-the-fly* composition. Therefore, the SNs are composed from scratch based on a given customer need and a pool of service suppliers that can be dynamically combined. The challenge is to achieve *self-organization* among services so they can network themselves to cover a customer need. To address this issue approaches usually explore ideas coming from the Semantic Web area, so services can be described by means of ontologies linked to real-world service descriptions [11].

Fig. 3.1 shows how SN approaches fit within the economic context (vertical axes) and the ICT maturity (horizontal). Moreover, Fig. 3.1 also depicts how these approaches have been influenced by others. For instance, the definition of a Web service as defined by Booth *et al.* has influenced approaches such as OntoMat, METEOR-S among others [12, 13, 14].

1. (vertically) the economic context in which SNs thrive is evolving from an hierarchical process-driven organization to a decentralized and relationship-driven organization. This evolution is enabled by the increasing social and knowledge connectivity on the Web.
2. (horizontally) the support of ICT in the different activities towards finding the right SNs in these organizations.

3.2.4 Illustrative Example

The next section presents different approaches for either designing, analyzing or composing SNs. Furthermore, for each approach, an analysis is given, by means of a table, which explains whether the mentioned approach supports the aspects previously described in Section 3.2.2. Moreover, when applicable, we also provide an example of how the given approach depicts an SN.

---

[1] `http://www.spotify.com/`

[2] `http://www.last.fm/`

[3] `http://morpher.com/`

Our illustrative modeling example consists of four entities: a buyer, a seller, a tax office and an ad company. The real-world version of this example could be more complex, e.g., including a manufacturer, a company to deliver the good among other entities. However, due to space constrains we only consider these four entities. By making use of the facilities offered by each approach, we tried to model this example describing not only the entities but also the interactions among them.

3.3 Approaches for Modeling Service Networks

3.3.1 BMO

Osterwalder proposes the Business Modeling Ontology (BMO) to model inter-company relationships within business models [15]. Whereas Table 3.1 presents the aspects covered by BMO, Figure 3.2 depicts its main constructs. BMO provides four strategic areas: product, customer interfaces, infrastructure management and financial aspects, which allow to describe the business model of a firm. Briefly, product refers to the value propositions (products) offered to the market. The product area includes one building block: value proposition. The customer interface not only addresses issues about how the firm deliver the products or services to the customers but also how it builds relationships with them. Customer interface includes three building blocks: customer segments, distribution channel and customer relationships.

Table 3.1: BMO analysis.

Aspect	Analysis
B2C interaction	The customer is partially involved
B2B relationships	Inter-company relationships
Network Definition	None
Visualization	Graphical representation
Orientation	Business oriented
Tool support	Standardized tools: XML-based language. Non-standardized: Business Model Canvas

Infrastructure management refers to how the company performs infrastructural or logistical issues, with whom, and as what kind of network enterprise. Infrastructure management also includes three blocks: key resources, key activities and key partners. Finally, the financial aspects describe the revenue model, i.e., the cost structure and the business model's sustainability. Financial aspects includes two blocks: cost structure and revenue streams.

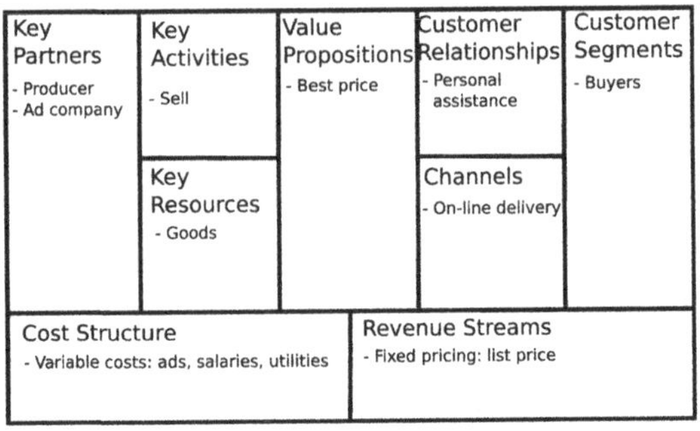

Key Partners
- Producer
- Ad company
Key Activities
- Sell
Key Resources
- Goods
Value Propositions
- Best price
Customer Relationships
- Personal assistance
Channels
- On-line delivery
Customer Segments
- Buyers
Cost Structure
- Variable costs: ads, salaries, utilities
Revenue Streams
- Fixed pricing: list price

Fig. 3.2: BMO Example.

3.3.2 REA

Initially proposed as a framework for accounting systems, the Resource-Event-Agent (REA) approach has evolved into an enterprise ontology that allows modeling enterprise-wide value chains [16]. Table 3.2 describes the aspects covered by REA, which relies on three economic concepts: resources, events and agents. Shortly, economic agents exchange economic resources by means of economic events. Examples of economic agents are: customers and suppliers. Examples of economic events are: exchanges and processes (Figure 3.3 shows four economic processes: Buy, Sell, Taxing and Advertising). Finally, examples of economic resources are: products and services, which in Figure 3.3 are depicted as Good, Payment, VAT, Legal Compliance and Audience.

Table 3.2: REA analysis.

Aspect	Analysis
B2C interaction	The customer is not involved
B2B relationships	Inter & intra company relationships
Network Definition	None
Visualization	Graphical representation
Orientation	Business oriented
Tool support	Standardized tools: ebXML and UMM meta-models

A central concept within the approach are REA activities which are either exchanges, trading resources between agents; or processes, consuming input resources and producing output resources [17, 18]. REA activities are connected by stock flows, which represent one resource moving from one activity to the next. In this sense, the REA information flow across multiple companies is as follows:

$$Process \rightarrow StockFlow \rightarrow Exchange \rightarrow StockFlow \rightarrow Process \qquad (3.1)$$

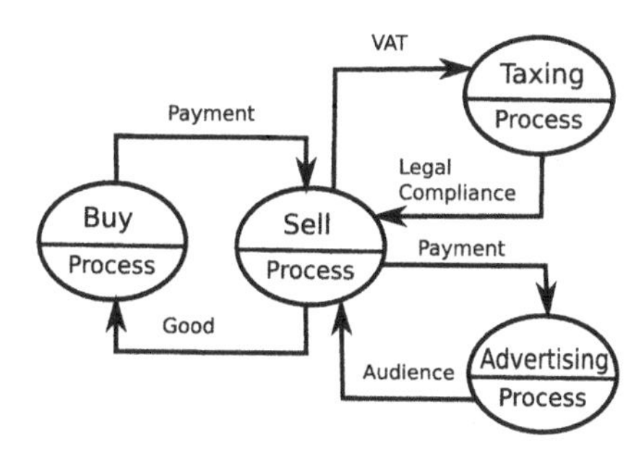

Fig. 3.3: REA Example.

Each event on an REA supply chain knows what other events it is connected to, via the stock flows, which gives also to agents an organizational structure since they know which events they control. The expression in (3.1) shows how REA focuses on supply chain's issues, i.e., how a directed sequence of activities exchange resources. Although Figure 3.3 depicts value exchanges, the expression in (3.1) also denotes a more process-oriented perspective, i.e., sequential steps. Finally, we can also conclude that REA models the supply chain taking into account only the supplier point of view which does not allow interaction with the customer.

3.3.3 Value Network Analysis (VNA)

Verna Allee argues that value networks are like living systems experiencing physical exchanges and interactions [6]. Table 3.3 presents the aspects covered by VNA. Allee proposes a graphical representation to describe these phenomena by means of tangible and intangible deliverables. The main argument for building such representation is that any value interaction is supported by some mechanism that enables it to happen, i.e., exchange of deliverables.

Table 3.3: VNA analysis.

Aspect	Analysis
B2C interaction	The customer is not involved
B2B relationships	Inter & intra company relationships
Network Definition	Formal definition
Visualization	Graphical representation
Orientation	Business oriented
Tool support	No information provided. Although VNA generates visuals in Microsoft PowerPoint and Visio and reports in PDF

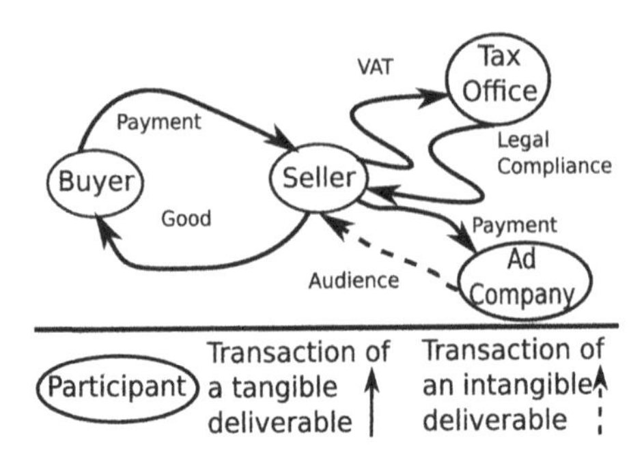

Fig. 3.4: VNA Example.

According to Verna Allee, three constructs are needed for representing value exchanges: participants, transactions and deliverables. Participants, graphically represented as ovals, are described as real people, they are the entities performing roles in the system. Software programs, databases or any other kind of technology are not considered as participants since they rely on people's decisions. In Figure 3.4 the participants are the buyer, the seller, the tax office and the advertiser.

Transactions[4] are depicted by arrows that represent the direction of something that is happening among participants. Deliverables are the real "things" that are exchanged from one participant to another, deliverables can be either tangible or intangible and are represented as labels on top of the arrows (transactions). In Figure 3.4 transactions dealing with tangible deliverables are depicted by solid arrows, whereas intangibles by dotted arrows. This basic representation allows to explore complex behavior inside the value networks such as value creation, cost/risk analysis, patterns of exchange among others [6].

3.3.4 O-WSP

One of the first attempts to achieve automatic service bundling was proposed by Omelayenko [19]. Table 3.4 describes the main aspects of the O-WSP approach. The author applies a Semantic Web approach called Open-World Skeletal Planning (O-WSP). In this way the bundling problem is solved through a planning-based reasoning which uses a skeleton to guide such process. Even though this approach solves a bundling problem, the skeletons are process-oriented templates overseeing value aspects. In this way, the skeletons only describe the services to be performed following a description very similar to a flow diagram. Among all the problems presented in his value models, the most important are the inconsistency with the skeleton and the unnecessary generation of failed plans.

Omelayenko's skeletons capture a basic representation of an SN [19]. Figure 3.5 shows a basic skeleton in which the activities to be performed are depicted. As can

[4] Also referred as activities by Verna Allee.

Table 3.4: O-WSP analysis.

Aspect	Analysis
B2C interaction	The customer is not involved
B2B relationships	Inter & intra company relationships
Network Definition	Informal definition
Visualization	Graphical representation
Orientation	Process oriented
Tool support	Standardized tools: RDF [20]. Non-standardized tools: UPML-S [21] and e^3*value* [22]

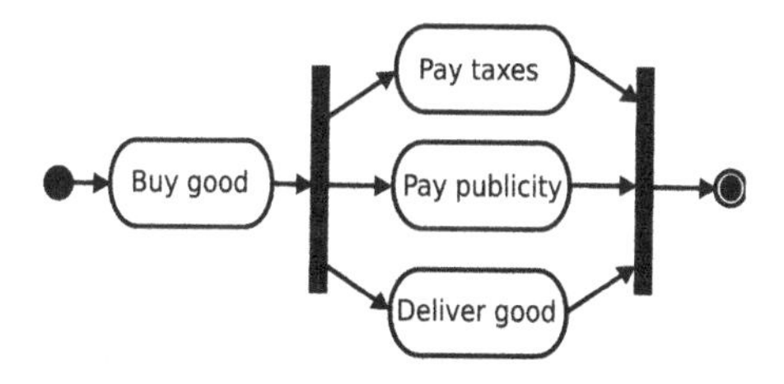

Fig. 3.5: O-WSP Example.

be observed, this kind of skeleton only describes what activities must be sequentially performed missing information about the resources exchanged among participants.

3.3.5 Serviguration

Baida proposes an ontological approach for service bundling called Serviguration [23]. Table 3.5 describes the main aspects of Serviguration. This approach uses three types of ontologies, the first two represent the demand and supply perspectives respectively, as a third ontology performs the composition process. Therefore, by matching the demand and supply perspectives, it is possible to generate a set of service bundles. An additional idea is that interactions among service suppliers can be modeled through a set of *dependencies*, which describe concepts such as enhancing, supporting, exclusion, optionality, bundling among others. In this way, services can be combined by following those dependencies. Since services can be combined in different ways, this approach generates alternative bundles for possibly matching customer needs. Serviguration is actually a guideline for combining services in a multi-supplier environment. One of the main disadvantages of Serviguration is the need for defining all those relationships among suppliers, especially when the number of suppliers grows. In addition, the approach lacks a selection mechanism to prioritize one bundle when more than one is generated.

Table 3.5: Serviguration analysis.

Aspect	Analysis
B2C interaction	The customer is partially involved
B2B relationships	Inter & intra company relationships
Network Definition	Informal definition
Visualization	Graphical representation
Orientation	Business oriented
Tool support	Standardized tools: RDF, e^3*value* ontology

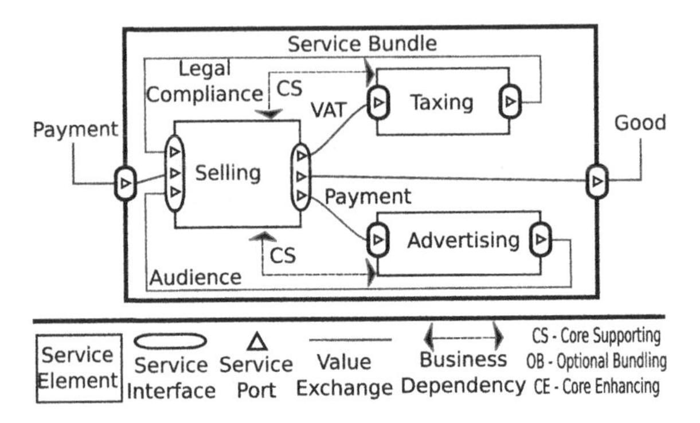

Fig. 3.6: Serviguration Example.

The Serviguration approach provides an idea about how SN participants can interact with each other by means of business dependencies. Besides, it also offers a visual representation of what can be a basic SN. Figure 3.6 depicts a Serviguration bundle where three services are described: selling, taxing and advertising. In this example the customer is supposed to provide a payment in exchange of a good.

3.3.6 The e^{3}family

In this section, we provide a short overview of the e^3*family* of ontologies for building SVNs. Table 3.6 presents the aspects covered by the e^3*family*.

3.3.6.1 e^3*value*

The central ontology of the e^3*family* ontologies is the e^3*value* ontology [22]. The e^3*value* ontology is a design time tool for exploring, analyzing and evaluating value networks. A value network is considered as a number of actors who exchange things of economic nature with each other. The ontology comes with a graphical editor. With the same tool, it is possible to assess economic profitability for all actors involved. The e^3*value* ontology is not aware of the notion of service, although a value object in e^3*value* can be considered as a service outcome, which is valued by a customer.

3.3.6.2 e^3*strategy*

The e^3*strategy* approach is used to analyze the strategic positioning of an actor in a network of enterprises, cf. [24]. In the network suppliers can be active as well as customers and competitors. These may exercise a force on the actor under consideration. Questionnaires are used to determine the exercised forces.

3.3.6.3 e^3*control*

For models constructed with e^3*value* we assume a perfectly honest world; that is no one is cheating and so is behaving honestly. In e^3*control* we relax this constraint of a perfectly honest world; we assume a sub-ideal world in which actors may misbehave. The e^3*control* approach offers constructs and methodologies to find misbehaving actors in a network of enterprises. Additionally, e^3*control* comes with a library of a patterns to address the misbehaving actors. The patterns include solution on the e^3*value* level, but also suggest solutions in terms of changed business processes.

3.3.6.4 e^3*alignment*

While exploring an e-business cases multiple perspectives are considered including the strategy perspective (e^3*strategy*), the value perspective (e^3*value*), the business process perspective and the IT perspective. The e^3*alignment* approach [25] ensures that all these perspectives are aligned, or detects misaligned perspectives.

Table 3.6: e^3*family* analysis.

Aspect	Analysis
B2C interaction	The customer is involved
B2B relationships	Inter & intra company relationships
Network Definition	Formal definition
Visualization	Graphical representation
Orientation	Business oriented
Tool support	Standardized tools: RDF-S [20]

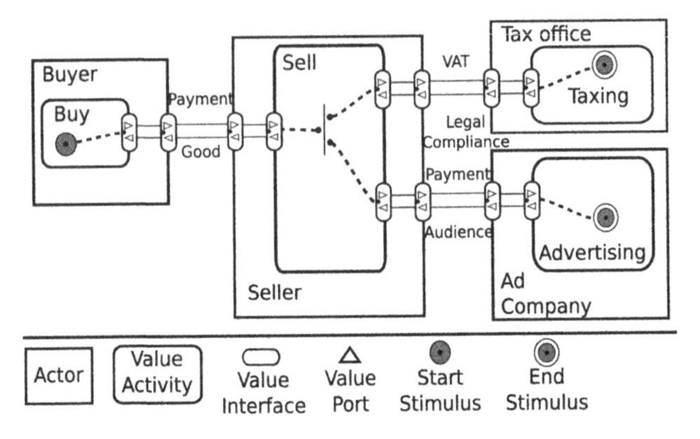

Fig. 3.7: e^3*family* Example.

3.3.6.5 e^3*service*

The e^3*value* ontology is the foundation of the e^3*family* of ontologies with on top of it e^3*service* as an ontology to be used for the composition of SNs, and for relating customer needs to services outcomes as available in the the market. The e^3*service* ontology comprises two separate ontologies for service modeling during design time, as well as configuration of services in *bundles*. Based on needs, a feature solution graph is constructed mapping needs or derived constructs onto services. Then a set of bundles that can potentially covered the given need is retrieved. The *serviguration* ontology, also known as the OBELIX ontology is used off-line for the bundling of services [23]. Although the e^3*service* ontology focuses on deriving a service bundle based on a given need, e^3*service* does not build the complete SVN to provision the service bundle. The second generation of e^3*service*, which is under development [26, 27, 28, 29], will integrate *serviguration* ideas with e^3*service*, such that the resulting ontology is not only compatible with the e^3*value* ontology but also dynamically composes a complete SVN. Figure 3.7 depicts an SVN in which four actors performing different activities can jointly work to provide what the customer needs.

3.3.7 VBC

Nakamura *et.al.* present the Value-Based Composition (VBC) approach [30]. Table 3.7 summarizes the aspects offered by VBC which is composed of three elements. 1) value models, 2) a value meta-model, and 3) a service broker. The idea is to achieve composition by allowing customers to interact with a service broker who has knowledge of service suppliers that can match customer requests. Besides, the broker is composed of nine components. The broker has two repositories, one for value models and one for process models. Both of them store templates, nevertheless the authors only give examples of the first one. The value templates are represented through value models and their Value-based Service Description Language (VSDL) representation. At this point it is very important to mention that the concept of value model is different from the one specified by the e^3*value* ontology. They conceive a value model as a hierarchical structure in which the value of a service is described at different levels. The broker can traverse this structure to evaluate whether the service matches the expected values requested by the customer.

Table 3.7: VBC analysis.

Aspect	Analysis
B2C interaction	The customer is involved
B2B relationships	None
Network Definition	None
Visualization	None
Orientation	Business oriented
Tool support	Standardized tools: Oracle BPEL [31]

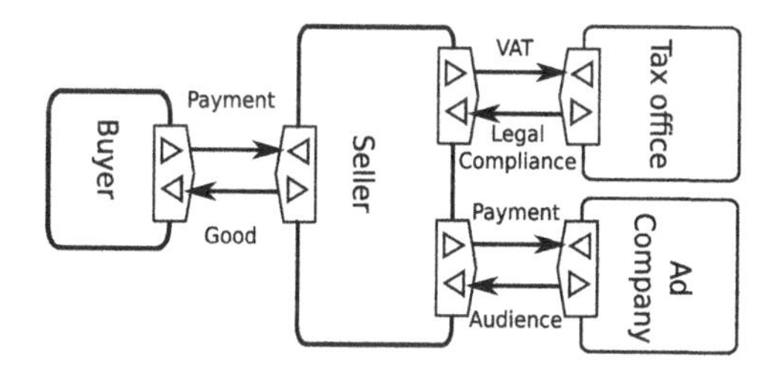

Fig. 3.8: VBC Example

They also present an interesting idea about composing services through a process of iteratively matching suppliers' interfaces. Regarding the rest of the elements within the broker, the authors do not present any example but just a description of what is supposed to occur at each step. Moreover, the output of their prototype shows an example where the performance of the whole process is only specified. Therefore, it seems to be a work in progress. In this sense, from our point of view, VBC shows two contributions: 1) the idea of a broker composed of a set of elements for matching customer requests with services, and 2) the iterative process for composing services. Although VBC does not provide insights on SN modeling, it provides an automatic approach for achieving composition [30]. Moreover, VBC also offers meta-models for describing B2C interaction. Finally, Figure 3.8 depicts how an SVN might look like under the VBC's approach.

3.3.8 GVP

Zlatev proposes the concept of patterns to reuse knowledge and perform a semi-automatic matching of needs, GVP [32]. Table 3.8 depicts the aspects covered by GVP which considers a pattern as a recurrent design fragment that solves a problem in a particular context. In this sense, Zlatev's work focuses on designing a library of patterns, these patterns are represented with value, process and goal models. In this way, before starting the matching, a goal model must be defined. This model represents what goals must be covered. Later on, parts of this goal model are matched with the goal-model representation of the patterns, which at the same time are linked to the value and process model representation.

Table 3.8: GVP analysis.

Aspect	Analysis
B2C interaction	The customer is partially involved
B2B relationships	Inter &. intra company relationships
Network Definition	None
Visualization	None
Orientation	Business oriented
Tool support	Non-standardized tools: e^3*value* ontology

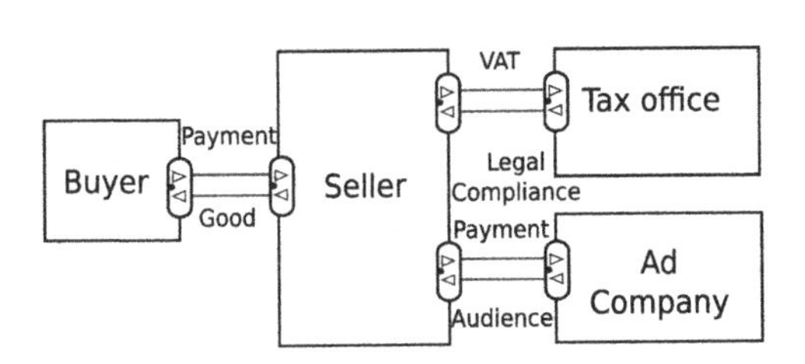

Fig. 3.9: GVP Example.

The whole process relies on four steps. Starting with the design of a goal model, following with the selection of patterns that match with the goal model, continuing with an evaluation where possible solutions are determined, and finalizing with a synthesis step. Even though this methodology looks like a good solution, the process requires a lot of human intervention, so there is neither automation nor implementation of this process. In the end, GVP presents a manual approach for service composition [32]. The main contributions of the approach are: 1) Business orientation, and 2) the matching of customer desires by means of goal models. Figure 3.9 depicts an GVP model, as can be observed, GVP makes use of e^3*value* models to represent an SVN. Nevertheless, GVP only describes the actors involved within the SVN omitting information about the internal activities performed by those actors.

3.3.9 Becker

Becker *et.al.* present a modeling language for bundling products and services [33]. Table 3.9 summarizes the aspects covered by Becker's approach that takes into account four main requirements (supplier's point of view, customer's point of view, bundle's functionality & structure and bundle's economic consequences) which are addressed by a meta-model that allows to specify three elements: the solution space, customer-specific instances from the solution space and the economic consequences for the customer (cf. also Chapter 2).

Table 3.9: Becker analysis.

Aspect	Analysis
B2C interaction	The customer is involved
B2B relationships	Inter & intra company relationships
Network Definition	Informal definition
Visualization	Textual representation
Orientation	Business oriented
Tool support	No information provided

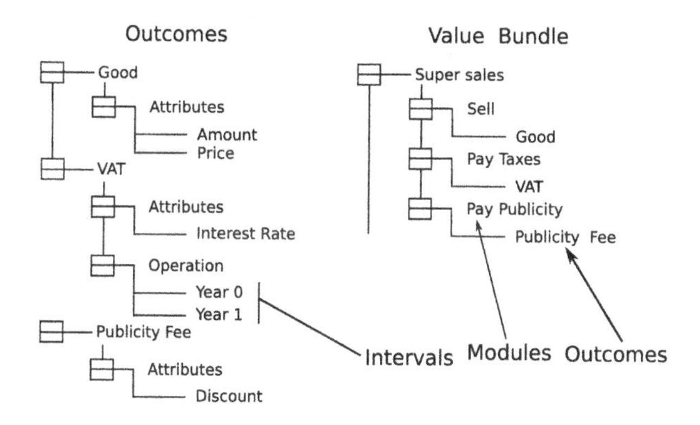

Fig. 3.10: Becker Example.

The generation of value bundles is performed based on the the three elements previously described. In this sense, three steps are required to generate a value bundle. First, modeling value bundle types. It involves modeling possible configurations of generic value bundles. Figure 3.10 illustrates how the bundles, their internal modules (services or products) and outcomes are defined. Second, customers configure individual value instances. The process requires to indicate preferences about parameters such as price, availability, delivery time among others. Once the preferences are indicated, the configurator can make recommendations about modules (services or products) to generate a bundle or automatically generate the bundle. Finally, the computation of the economic consequences for each bundle is performed. Based on this information, the most adequate value bundle can be selected.

3.3.10 Traverso

Traverso and Pistore propose an approach for automatically composing Semantic Web Services (cf. also Chapters 6 and 7), which can subsequently be transformed into executable processes [34]. Table 3.10 describes the aspects covered by this approach which relies on several components: OWL-S process models for each service, a composition goal and a Model Based Planner (MBP). The first step requires translating OWL-S process models into state transition systems (STSs) that describe the dynamic interactions within services.

Table 3.10: Traverso analysis.

Aspect	Analysis
B2C interaction	The customer is involved
B2B relationships	Inter-company relationships
Network Definition	Informal description
Visualization	None
Orientation	Process oriented
Tool support	Standardized tools: OWL-S

In the second step, a composition goal is defined. Afterwards, in the third step the BMP generates a plan that interacts with services in such a way that the composition goal is satisfied. Finally, the last step translates the plan to BPEL4WS executable code. Although, the approach performs an automating generation of plans that are supposed to deal with a specific goal, it does not say anything about the interactions among customers and service suppliers. Finally, the approach lacks a visual representation that allows analyzing structural properties of the composite service.

3.3.11 OntoMat-Service

Agarwal *et al.* present an approach to annotate, compose and execute Semantic Web Services [13]. The approach provides both a software piece and a four-step framework in which suppliers and users interact to generate execution plans where the active elements are Web services. Table 3.11 summarizes the aspects provided by OntoMat.

Table 3.11: OntoMat-Service analysis.

Aspect	Analysis
B2C interaction	The customer is involved
B2B relationships	Inter & intra company relationships
Network Definition	Informal definition
Visualization	Graphical representation
Orientation	Process oriented
Tool support	Standardized tools: WSDL

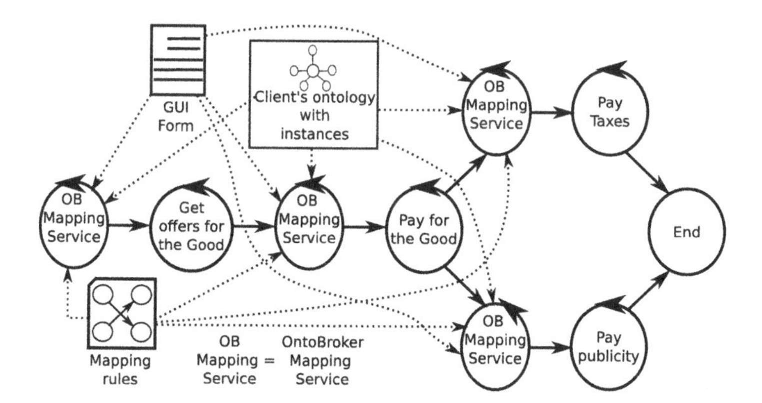

Fig. 3.11: Onto-Mat Example.

In the first step suppliers advertise their Web services by means of WSDL descriptions. The second step, which is called deep annotation, is performed by a user and it requires a mapping of rules between the WSDL-based advertised services and concepts in a client ontology. Later on, the user interconnects the desired Web services based on either data-flow or control-flow driven planning techniques. Finally, a software engine takes the plan and calls the Web services in the proper order. Figure 3.11 depicts an example of such service flow. As can be observed, the customer is all the time involved in the composition of this flow. The customer interact with the GUI form through which s/he has to do the mapping of rules.

Even though this framework allows basic interaction between suppliers and users, the composition process lies on the user side, which implies that the user must have a good understanding of what s/he is building up. According to Agarwal *et al.*, OntoMat-Service does not aim at completely automating the discovery, composition and invocation of Web services. Instead, the idea is to support users' intelligence and guides them to add semantic information such that only few valid paths remain to be chosen from by the user.

3.3.12 METEOR-S

Sivashanmugam et al. [14] provide the METEOR-S framework for Web service composition, which is composed of four components: the Process Builder, the Discovery Infrastructure, XML repositories and the Process Execution Engine. Table 3.12 describes the aspects supported by METEOR-S.

Table 3.12: METEOR-S analysis.

Aspect	Analysis
B2C interaction	The customer is not involved
B2B relationships	None
Network Definition	Informal definition
Visualization	None
Orientation	Process oriented
Tool support	Standardized tools: WSDL

The whole idea relies on designing Semantic Process Templates (SPTs). This is achieved by the Process Builder. Such templates are collections of activities that can be linked by using control flow constructs. The Discovery Infrastructure allows suppliers to advertise their services by means of a registries ontology which maintains a categorization of services according to their domains.[5] XML repositories are mainly used for managing (storing, searching, sharing) ontologies, SPTs and WSDL service interfaces. Afterwards, once a SPT is designed or selected from the repositories, the required services are discovered and added to the data flow according to the required activities. Later on, an executable process is generated based on the process template and the WSDL files of the participating services. Finally, the process is validated, deployed and it is ready for invocation on the Process Execution Engine. Since all the idea is around STPs, Sivashanmugam et al. actually aim at a framework for semantic web process composition in which a skilled designer or a domain expert creates templates for composite services that might be called later on.

3.3.13 Service Network Notation (SNN)

Bitsaki *et al.,* [35, 36] present an approach for modeling SNs focusing on relationships and exchanges of software services among the involved parties. The authors offer a formalism to model SNs by making use of graph theory and visual modeling constructs. The main concepts are: service offering, service description, service request, service providing, contract, service providing dependency and par-

[5] The authors claim that this categorization helps for finding the right services.

ticipant internal dependency. Table 3.13 summarizes the aspects that are covered by SNN. Figure 3.12 depicts an SVN based on SNN constructs. Although the approach models resource exchanges it focuses on temporal dependencies among services and does not provide the notion of economic reciprocity which is important for a business-oriented approach.

Table 3.13: SNN analysis.

Aspect	Analysis
B2C interaction	The customer is not involved
B2B relationships	Inter & intra company relationships
Network Definition	Formal definition
Visualization	Graphical representation
Orientation	Process oriented
Tool support	Standardized tools: BPMN [37]. Non-standardized tools: SN4BPM

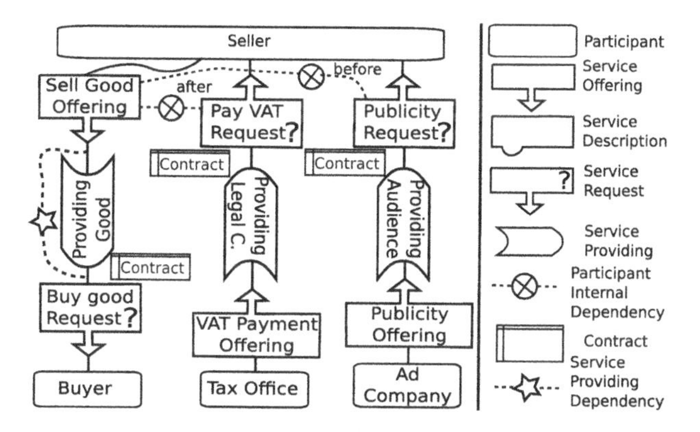

Fig. 3.12: SNN Example.

The authors also argue that the technology stack for enacting SNs consists of four elements: SN models, business processes, service compositions and services. Consequently, they propose a modification to the Business Process Management (BPM) life-cycle. In short, the traditional life-cycle involves six steps: analysis, modeling, IT refinement, deployment, execution and monitoring. In this sense, the authors propose to add an extra step called rationalization which deals with the modeling and analysis of SNNs models. As a matter of fact, the idea is to create a bond between the SNNs models and the abstract process models.

Once the SN is created, they map such representation to Business Process Modeling Notation (BPMN). The resultant BPMN process specifies the operational details to be carried on within the SN. Although the authors offer a novel approach for modeling SNs, they do not say anything about how such SN can be influenced by customers. Moreover, the transformation to BPMN is made by hand which constraint the possibility for developing an automatic process. Finally, since the design of SNs is also performed by hand, the approach is not suitable for automatically composing such networks.

3.3.14 DynamiCoS

Da Silva et al., [38] describe the DynamiCoS framework for runtime discovery, selection and composition of semantic services. The framework provides a service registry where services are semantically described by means of ontologies. Briefly a service is represented as a tuple $s = < ID, I, O, P, E, G, NF >$, where ID is the service identifier, I and O stand for input and output respectively, E are the service effects, G the goals that can be achieved by the service, and NF is the set of non-functional properties and constraint values. Table 3.14 summarizes the aspects covered by this approach.

Table 3.14: DynamiCoS analysis.

Aspect	Analysis
B2C interaction	The customer is involved
B2B relationships	None
Network Definition	None
Visualization	None
Orientation	Process oriented
Tool support	Standardized tools: OWL

Service developers publish services into the service registry following the s representation, an end-user also sends requests following the s representation.[6] Once an end-user defines a service request, DynamiCoS discovers the services that semantically match such request. Later on, a graph-based algorithm compose services to fulfill the service request. Overall DynamiCoS offers a runtime approach for composing semantic services matching user requests, nevertheless the authors does not say anything about how a user can select one composite service among the alternative options. Moreover, they also assume that a user is completely aware of the services he requires to satisfy his need, which is not always the case with users lacking of technical background. Finally, they also claim that whenever a user request cannot be matched with the services in the service registry, they can request the user to refine the details of his request, nevertheless along the paper there is no explanation about how they perform such action.

[6] The authors do not mention whether ID is also part of the final request, but we assume that is not the case.

3.3.15 u-service

Lee *et al.,* [39], describe a service bundling method which uses service complementary indexes. The idea is to bundle services according to a similarity measure. The approach relies on a Case Based Reasoning (CBR) algorithm to determine whether user satisfaction is met or not. Whenever user satisfaction is below a given threshold, several strategies can be followed. 1) If any individual single service can meet the requirements, according to the current context, select the individual service with the highest similarity. If not, 2) given the user context find a bundled service with the highest similarity and provide it to the user. 3) If there is not an existing bundled service, then the already provided service must be bundled with other service. The aspects covered by this approach are described in Table 3.15.

Table 3.15: u-service analysis.

Aspect	Analysis
B2C interaction	The customer is involved
B2B relationships	Inter company relationships
Network Definition	None
Visualization	None
Orientation	Process oriented
Tool support	Not mentioned

This bundling requires both services have high complementary indexes. In addition, once a new bundle is generated, it is added as a new case for the current user context. To sum up, Lee et al. propose an algorithm for bundling services based on user context and service complementary indexes.

3.3.16 CPC

In the field of service bundling, Letia *et.al.* propose the idea of allowing Client-Provider Collaboration (CPC) for achieving co-creation of service bundles [40]. Table 3.16 describes the aspects covered by CPS, which relies on a dialogue between the client and the supplier, such a dialogue takes the form of a persuasion whose dynamic object is the current best composition. In order to achieve this persuasion process, client and supplier use a common ontology for describing needs, where quality plays an important role. According to the authors, a need has some quality requirements that can be represented by using two sub-ontologies, one for properties and one for sacrifices. In this way in order to cover a need, a supplier should match some quality properties while a client should sacrifice something for getting the required quality.

Table 3.16: CPC analysis.

Aspect	Analysis
B2C interaction	The customer is involved
B2B relationships	None
Network Definition	None
Visualization	None
Orientation	Business oriented
Tool support	None

Therefore under CPC, the best bundle is the one providing a good trade-off between properties and sacrifices. The overall CPC process works as follows: once the client submits a request, the supplier generates a possible bundle for covering such request, giving at the same time a set of arguments supporting the proposed bundle. Later on, the client provides either pro or counter arguments for the given proposal. The supplier takes those arguments to either only counter argument or provide other proposal and then counter argument. In fact there are only two ways for finishing the dialogue, (1) the client cannot counter argument and then accepts the proposal, or (2) the supplier cannot counter attack or improve its proposal, therefore there is no acceptable bundle. Although CPC seems to be a good approach, the idea lacks a real test case. Therefore, as with VBC, the main contribution is only the idea of allowing client-supplier collaboration for bundling services. In addition, since all the interaction is between one client and one supplier, it is not clear if the approach will also work in an environment where more than one supplier compete for covering client needs.

3.3.17 Kohlborn

Kohlborn *et.al.* [41] propose a set of relationships for bundling services. The aspects covered by this approach are described in Table 3.17. In this approach, relationships are considered as connections among services. In addition, relationships can play two roles (or even both) as *enabler* and/or *constraint*. Furthermore, their approach distinguishes between two types of relationships, *generic* and *specific*. In this sense a relationship is influenced by two dimensions, the role it can play and its type. The main idea about defining roles is to allow strategic reasoning when bundling services.

In this way when a relationship plays the role of enabler, it can help to find services targeting the same type of needs. In contrast, as constraint a relationship can discard among services. On the other hand, the idea behind defining two types of relationships aims to constrain the solution space. Indeed, the authors claim that the process of building bundles should move through four stages. The first stage, called *Possible Bundles*, includes all the possible combinations among services regarding

Table 3.17: Kohlborn analysis.

Aspect	Analysis
B2C interaction	The customer is not involved
B2B relationships	Inter & intra company relationships
Network Definition	None
Visualization	None
Orientation	Business oriented
Tool support	None

validity or feasibility. The second stage, *Generic Bundles*, contains the set of bundles that can fulfill the requirements of generic relationships. In the same direction, the next stage, *Specific Bundles*, includes the bundles that can cover the requirements of specific relationships, such relationships are supposed to take into account a specific environment. Finally, the last stage, *Feasible Bundles*, applies domain specific knowledge to extract the set of bundles that can meet internal and external requirements. Internal requirements are features as quality, risk assessment, service level among others. External requirements deal with aspects such as customer demand, market saturation or legislation. Although Kohlborn *et.al.* present an interesting approach for bundling services, their work lacks from both case studies and applications, therefore the applicability of such approach remains more theoretical than practical.

3.4 Discussion

In this chapter we have analyzed approaches to design, analyze or compose SNs. Even though each approach models SNs for different purposes, they also share some aspects which are described in Section 3.2.2. Moreover, we have also provide an illustrative example for each approach to visualize how these approaches model SNs. However, there were approaches which we could not provide a visualization of the way SNs are modeled — either because of the lack of tools for modeling (CPC and Kohlborn) or the complexity to represent the SN (Becker, Traverso and u-service). Table 3.18 summarizes the aspects covered by each approach. As can be observed it is only VNA and SNN that provide a formal definition about networks. VNA provides a definition for value networks while SNN provides a definition for SNs, however VNA focuses on analyzing value networks and SNN on composition of SNs (cf. Figure 3.1).

We provide also tables to analyze business and implementation aspects for each approach (Table 3.19 and Table 3.20). Since half of the approaches consider the notion of value and economic reciprocity, in Section 3.2.1 we have provided a definition for SVNs that focuses on these aspects. In many approaches, value aspects of services are ignored simply because this information is usually implicit or not rep-

Table 3.18: Required Aspects. ✓ fully supported, ≈ partially supported.

	B2C interaction	B2B relationships	Network Definition	Visualization	Business Oriented	Tool Support
BMO [15]	✓	✓		✓	✓	✓
REA [16]		✓		✓	≈	✓
Allee [6]		✓	✓	✓	✓	≈
O-WSP [19]		✓	≈	✓		✓
Serviguration [23]		✓	≈	✓	✓	✓
e^3*family* [22]	✓	✓	≈	✓	✓	✓
VBC [30]	✓			≈	✓	✓
GVP [32]	≈	≈	≈	✓	✓	
Becker [33]	✓	✓		✓	✓	✓
Traverso [34]		≈	≈			✓
OntoMat-Service [13]	✓	✓	≈	✓		✓
METEOR-S [14]			≈			✓
SNN [35, 36]		✓	✓	✓	≈	✓
DynamiCoS [38]	✓	≈	≈	≈		✓
u-service [39]	✓	≈				
CPC [40]	✓				✓	
Kohlborn [41]		✓			✓	

resentative enough. Yet, both the Social Web and the Semantic Web have liberated critical amounts of linked data about people, needs, and services that provide an actionable foundation for a Service Web to emerge from dynamic communication within and between customer and supplier communities.

Table 3.19: Business aspects.

	B2C interaction		**B2B relationships**		**Business Oriented**	
	Customer Need	Dialogue	Inter-company	Intra-company	Value Perspective	Economic Reciprocity
BMO [15]	✓		✓		✓	✓
REA [16]			✓	✓	✓	✓
Allee [6]			✓	✓	✓	✓
O-WSP [19]			✓	✓		
Serviguration [23]	✓		✓	✓	✓	✓
e^3*family* [22]	✓	✓	✓	✓	✓	✓
VBC [30]		✓			✓	✓
GVP [32]	✓		✓	✓	✓	✓
Becker [33]	✓	✓	✓	✓	✓	✓
Traverso [34]			✓			
OntoMat-Service [13]	✓	✓	✓	✓		
METEOR-S [14]						
SNN [35, 36]			✓	✓	✓	
DynamiCoS [38]		✓				
u-service [39]		✓	✓			
CPC [40]	✓	✓			✓	✓
Kohlborn [41]			✓	✓	✓	✓

Indeed, currently services (such as iPhone apps) are too much pushed by suppliers; anticipating mainstream needs [42]. Yet as more and more customers can express and discuss their needs freely, niches will rather pull profitable service mar-

Table 3.20: Implementation aspects.

	Network Definition		**Visualization**		**Tool Support**		
	Informal	Formal	Textual	Graphical	Manual Composition	Meta-model Composition	Dynamic Composition
BMO [15]			✓		✓		
REA [16]				✓	✓		
Allee [6]		✓		✓	✓		
O-WSP [19]	✓			✓		✓	
Serviguration [23]	✓			✓		✓	
e^3*family* [22]	✓			✓		✓	
VBC [30]				✓	✓		
GVP [32]				✓	✓		
Becker [33]			✓			✓	
Traverso [34]	✓						✓
OntoMat-Service [13]	✓			✓		✓	
METEOR-S [14]						✓	
SNN [35, 36]		✓		✓	✓		
DynamiCoS [38]							✓
u-service [39]						✓	
CPC [40]							
Kohlborn [41]							

kets to thrive. As an example, consider a community of interest that can use collaborative tools to express, discuss, and detail their needs and value expectations, and finally publish a request for proposals to the self-organizing service market. Service suppliers can participate in proposing a reasonable offer to these needs, on which communities in turn can critique. Therefore, a dynamic interaction between customers and suppliers is a key aspect as depicted in Section 3.2.2. Nevertheless, as can be observed in Table 3.20, there are still several approaches that perform a manual or meta-model based composition for SNs, i.e., they follow a given template to perform the composition. Future efforts must focus also on providing a framework to allow dynamic composition for SNs.

Approaches such as customer interaction- or user sentiment-analysis can source now from big enough customer communities in order to give a representable image on how SNs can anticipate and evolve towards new expectations. Although the Wikinomics bubble in the top-left corner in Figure 3.1 is still isolated from the state of the art, Tapscott's work may inspire future SN composition approaches [4, 5]. Along this line, Pedrinaci *et al.* introduce the notion of Linked Services that tap from Linked Data initiatives to drive their composition [11]. Maamar *et al.* describe how service engineers can capitalize on Web services' interactions, namely, collaboration, substitution, and competition, to build social networks for service discovery [43].

The highly variable customer needs emerging from the social Web stretch the long tail of the market call; hence call for multi-supplier service bundles that cannot be offered by single parties. One example specifically for book reselling is Amazon, that also relies on many third-party book shops globally to have such a wide offering and quick delivery to its customer. This configuration problem can be partly automated, but will still involve a considerable amount of human-driven dialogue.

Key here is to model, analyze, and match divergent perspectives on the notion of value, including needs and services, by customers and suppliers, respectively [29]. In order to express this knowledge unambiguously, communities agree on common semantics of vocabularies and rules [44]. Though detailed enough to prevent misinterpretation, these standards should be as generative as possible to allow mass-scale adoption. Apart from FOAF and SIOC — which are standards to link social data — one recent successful example is GoodRelations,[7] a global initiative to leverage SMEs by publishing e-commerce data in a standardized fashion.

3.5 Acknowledgements

The authors would like to thank Dr. Carlos Pedrinaci for his valuable comments and suggestions. This paper has been partially funded by the NWO/Jacquard project VALUE-IT no 630.001.205 and the European FP7 project ACSI (no. 257593).

References

1. C. H. Lovelock and J. Wirtz, *Services Marketing: People, Technology, Strategy - 7th Edition.* Pearson Higher Education, 2010.
2. R. C. Basole and W. B. Rouse, "Complexity of service value networks: conceptualization and empirical investigation," *IBM Syst. J.*, vol. 47, pp. 53–70, January 2008.
3. M. E. Porter, *Competitive Advantage - Creating and Sustaining Superior Performance.* Free Press, New York, NY., 1985.
4. D. Tapscott, D. Ticoll, and A. Lowy, *Digital Capital - Harnessing the Power of Business Webs.* Nicholas Brealy Publishing, London, UK., 2000.
5. D. Tapscott and A. D. Williams, *Wikinomics: How Mass Collaboration Changes Everything.* Portfolio, 2006.
6. V. Allee, "A value network approach for modeling and measuring intangibles," in *Transparent Enterprise Conference*, 2002.
7. J. Hamilton, "Service value networks: Value, performance and strategy for the services industry," *Journal of Systems Science and Systems Engineering*, vol. 13(4), pp. 469–489, 2004.
8. J. Gordijn, A. Osterwalder, and Y. Pigneur, "Comparing two business model ontologies for designing e-business models and value constellations," in *BLED 2005 Proceedings*, AIS Electronic Library (AISeL), 2005. http://aisel.aisnet.org/bled2005/15.
9. R. C. Basole, "Visualization of interfirm relations in a converging mobile ecosystem," *Journal of Information Technology*, vol. 24(2), pp. 144–159, 2009.
10. J. Gordijn, H. Akkermans, and H. v. Vliet, "Business modelling is not process modelling," in *Proceedings of the Workshops on Conceptual Modeling Approaches for E-Business and The World Wide Web and Conceptual Modeling: Conceptual Modeling for E-Business and the Web*, ER '00, (London, UK), pp. 40–51, Springer-Verlag, 2000.
11. C. Pedrinaci and J. Domingue, "Toward the next wave of services: Linked services for the web of data," *Journal of Universal Computer Science*, vol. 16, pp. 1694–1719, jul 2010.
12. D. Booth, H. Haas, F. McCabe, E. Newcomer, M. Champion, C. Ferris, and D. Orchard, "Web services architecture," tech. rep., W3C Working Group Note 11, February. W3C Technical Reports and Publications, http://www.w3.org/TR/ws-arch/., 2004.

[7] http://purl.org/goodrelations/

13. S. Agarwal, S. Handschuh, and S. Staab, "Annotation, composition and invocation of semantic web services," *Journal of Web Semantics*, vol. 33, pp. 1–24, 2004.
14. K. Sivashanmugam, J. A. Miller, A. P. Sheth, and K. Verma, "Framework for semantic web process composition," *Int. J. Electron. Commerce*, vol. 9, pp. 71–106, January 2005.
15. A. Osterwalder, *The Business Model Ontology - a proposition in a design science approach.* PhD thesis, University of Lausanne, Ecole des Hautes Etudes Commerciales HEC, 2004.
16. W. E. McCarthy, "The REA Accounting Model: A Generalized Framework for Accounting Systems in a Shared Data Environment," *Accounting Review*, vol. 57, no. 3, p. 554, 1982.
17. R. Haugen and W. McCarthy, "Rea, a semantic model for internet supply chain collaboration.," in *Business Objects and Component Design and Implementation Workshop VI: Enterprise Application Integration*, 2000.
18. G. L. Geerts and W. E. McCarthy, "Using object templates from the rea accounting model to engineer business processes and tasks," *Review of Business Information Systems (RBIS)*, vol. 5, no. 4, 2001.
19. B. Omelayenko, *Web-Service Configuration on the Semantic Web: Exploring How Semantics Meets Pragmatics.* PhD thesis, Vrije Universiteit Amsterdam, 2005.
20. F. Manola and E. Miller, "RDF Primer." W3C Recommendation, Feb 2004. http://www.w3.org/TR/rdf-primer/.
21. B. Omelayenko, M. Crubézy, D. Fensel, R. V. Benjamins, B. J. Wielinga, E. Motta, M. A. Musen, and Y. Ding, "UPML: The Language and Tool Support for Making the Semantic Web Alive," in *Spinning the Semantic Web* (D. Fensel, J. A. Hendler, H. Lieberman, W. Wahlster, D. Fensel, J. A. Hendler, H. Lieberman, and W. Wahlster, eds.), pp. 141–170, MIT Press, 2003.
22. J. Gordijn and J. Akkermans, "Value-based requirements engineering: exploring innovative e-commerce ideas," *Requirements Engineering*, vol. 8, pp. 114–134, 2003. 10.1007/s00766-003-0169-x.
23. Z. Baida, *Software-aided service bundling.* PhD thesis, Vrije Universiteit Amsterdam, 2006.
24. M. Porter, *Competitive Advantage.* Free Press, New York, 1985.
25. V. Pijpers, *e3alignment: Exploring Inter-Organizational Business-ICT Alignment.* PhD thesis, Vrije Universiteit Amsterdam, 2010.
26. I. S. Razo-Zapata, A. Chmielowiec, J. Gordijn, M. V. Steen, and P. De Leenheer, "Generating value models using skeletal design techniques," in *5th international BUSITAL workshop*, 2010.
27. J. Gordijn, P. D. Leenheer, and I. S. Razo-Zapata, "Generating service value webs by hierarchical configuration: A case in intellectual property rights clearing," in *44th Hawaii International International Conference on Systems Science (HICSS-44 2011), Proceedings, 4-7 January 2011, Koloa, Kauai, HI, USA*, pp. 1–10, IEEE Computer Society, 2011.
28. I. S. Razo-Zapata, P. De Leenheer, J. Gordijn, and H. Akkermans, "Service value networks for competency-driven educational services: A case study," in *In Proceedings of the 6th international BUSITAL workshop*, 2011.
29. I. S. Razo-Zapata, J. Gordijn, P. De Leenheer, and H. Akkermans, "Dynamic cluster-based service bundling: A value-oriented framework," in *Commerce and Enterprise Computing (CEC), IEEE 13th Conference on*, 2011.
30. K. Nakamura and M. Aoyama, "Value-based dynamic composition of web services," *Asia-Pacific Software Engineering Conference*, vol. 0, pp. 139–146, 2006.
31. T. Andrews, F. Curbera, H. Dholakia, Y. Goland, J. K. F. Leymann, K. Liu, D. Roller, D. Smith, S. Thatte, I. Trickovic, and S. Weerawarana, "Business Process Execution Language for Web Services Version 1.1." Specification, Feb 2005. `http://www-128.ibm.com/developerworks/library/specification/ws-bpel/`.
32. Z. Zlatev, *Goal-oriented design of value and process models from patterns.* PhD thesis, University of Twente, 2007.
33. J. Becker, D. Beverungen, R. Knackstedt, and O. Müller, "Model-based decision support for the customer-specific configuration of value bundles," *Enterprise Modelling and Information Systems Architectures*, vol. 4, no. 1, pp. 26–38, 2009.
34. P. Traverso and M. Pistore, "Automated composition of semantic web services into executable processes," pp. 380–394, Springer-Verlag, 2004.

35. M. Bitsaki, O. Danylevych, W.-J. Heuvel, G. Koutras, F. Leymann, M. Mancioppi, C. Nikolaou, and M. Papazoglou, "An architecture for managing the lifecycle of business goals for partners in a service network," in *Proceedings of the 1st European Conference on Towards a Service-Based Internet*, ServiceWave '08, (Berlin, Heidelberg), pp. 196–207, Springer-Verlag, 2008.
36. O. Danylevych, D. Karastoyanova, and F. Leymann, "Service networks modelling: An soa & bpm standpoint.," *Journal of Universal Computer Science*, pp. 1668–1693, 2010.
37. D. Miers and S. A. White, *BPMN Modeling and Reference Guide — Understanding and Using BPMN.* Lighthouse Pt, FL: Future Strategies Inc., Aug 2008.
38. E. G. da Silva, L. F. Pires, and M. van Sinderen, "Towards runtime discovery, selection and composition of semantic services," *Comput. Commun.*, vol. 34, pp. 159–168, February 2011.
39. N. Y. Lee and O. Kwon, "A complementary ubiquitous service bundling method using service complementarity index," *Expert Syst. Appl.*, vol. 38, pp. 5727–5736, May 2011.
40. I. A. Letia and A. Marginean, "Client provider collaboration for service bundling," *Advances in Electrical and Computer Engineering*, vol. 8, pp. 36–43, 2008.
41. T. Kohlborn, C. Luebeck, A. Korthaus, E. Fielt, M. Rosemann, C. Riedl, and H. Krcmar, "Conceptualizing a bottom-up approach to service bundling," in *22nd International Conference on Advanced Information Systems Engineering (CAiSE'10)*, 2010.
42. H. Chesbrough and J. Spohrer, "A research manifesto for services science," *Commun. ACM*, vol. 49, pp. 35–40, July 2006.
43. Z. Maamar, P. Santos, L. Wives, Y. Badr, N. Faci, and J. de Oliveira, "Using social networks for web services discovery," *Internet Computing, IEEE*, vol. 15, pp. 48 –54, july-aug. 2011.
44. P. De Leenheer, S. Christiaens, and R. Meersman, "Business semantics management: A case study for competency-centric hrm," *Comput. Ind.*, vol. 61, pp. 760–775, October 2010.

Chapter 4
Service System Approaches
Conceptual Modeling Approaches for Services Science

Roberta Ferrario, Nicola Guarino, Romano Trampus, Ken Laskey, Alan Hartman, and G. R. Gangadharan

Abstract Over the last several years, *services science* has emerged as an effective means to understand services and the socio-technical systems in which they are deployed. This systemic view requires a genuinely interdisciplinary approach to the study of services. In this chapter, we review a number of significant approaches to analyze, understand and model service systems, with an emphasis on showing similarities and differences that highlight the many aspects of a rich service ecosystem. The goal of this chapter is to provide developers with an overall perspective on such rich service system models, as a basis for choosing those which mostly fit their own needs.

Roberta Ferrario, Nicola Guarino
ISTC-CNR, Laboratory for Applied Ontology, via alla Cascata 56C, 38123 Trento, Italy,
e-mail: roberta.ferrario@cnr.it, e-mail: nicola.guarino@cnr.it

Romano Trampus
University of Trieste, Piazzale Europa,1 34127 Trieste, Italy, e-mail: trampus@units.it

Ken Laskey
The MITRE Corporation, M/S H305, 7515 Colshire Drive, McLean, VA, 22102-7508, USA,
e-mail: klaskey@mitre.org
The author's affiliation with The MITRE Corporation is provided for identification purposes only, and is not intended to convey or imply MITRE's concurrence with, or support for, the positions, opinions or viewpoints expressed by the author.

Alan Hartman
IBM Research Lab, Haifa University Campus, Mount Carmel, Haifa, 31905, Israel,
e-mail: hartman@research.ibm.il

G. R. Gangadharan
IBM India Research, India, e-mail: gangadharan@in.ibm.com

4.1 Introduction

A relevant feature of services is that they are never given in isolation. They are typically conceived as being composable one with the other, and they always exhibit their effects in larger contexts. So, a new subject of inquiry has emerged in the recent years: service systems. As often happens, the expression "service system," while being relatively popular, is understood in different ways by the various communities. In some cases, it mainly refers to a set of interconnected services, while in other cases it is used to include other entities besides the service itself, i.e., people, artifacts, resources, the external environment. In these cases, a service system is a complex socio-technical system.

The latter view — service systems as complex socio-technical systems — is strongly advocated by the founders of a new science, services science [10], urging for the need of a radically interdisciplinary approach to model, understand and control in a systemic way all the economic and social aspects behind the notion of service. Of particular value note, the concept of co-creation is the sharing and distribution of labor, investments, expertise, risk, and — most of all — knowledge. In the last few years, the studies dedicated to this new field have multiplied ([20, 21, 26]), involving disciplines as diverse as economics, sociology, computer science, philosophy, psychology, and linguistics.

We shall review in this chapter the main service modeling approaches that borrow this wide services science perspective. After a recap of Steven Alter's view of service systems as work systems, we present Ferrario and Guarino's foundational ontology of services (General Service Model — GSM), which among other things proposes a unifying definition for the general notion of service, and clarifies the difference between services and service systems. We introduce then three different approaches which more or less build on the notion of service system, namely the TEXO Service Ontology, the OASIS SOA Reference Architecture Foundation (SOA RAF), and the IBM Service Design Model (SDM), discussing their differences and similarities among themselves and with respect to the GSM. While there are other examples in the literature (for example, [9]) that provide a representation of service systems, the ones discussed in this chapter emphasize an ontological approach, paying explicit attention to the nature and structure of service systems, the various entities involved with them, and their mutual relationships.

In order to facilitate understanding and comparisons, we will use a recurring example throughout the chapter: car washing. This is indeed a very popular example of service, thoroughly discussed in literature.

4.2 Alter's Framework: Service Systems as Work Systems

Before defining service systems, Steven Alter considers first possible independent definitions of services and systems, such as: "Services are acts performed for someone else, including providing resources that someone else will use," and "a system is a consciously designed combination of things or parts that perform useful work" [6]. Quickly, however, he realizes there are problems with such definitions, and proposes to go *beyond a definition of service*, suggesting to focus on the broader notion of service system: whatever services are,[1] they are produced through service systems. The core of Alter's position is that service systems are *work systems*, where human participants or machines perform work using information, technology, and other resources to realize products. So the emphasis is more on *how* services (and products) are produced and *who* is involved in the production and consumption process, and less on *what* services are. Customers and customers' issues are prominent throughout the analysis of systems.

So, we can conclude that, according to Alter, describing a service amounts to describing the work system where the service is produced. Indeed, for him every work system is a service system [5]. Under this assumption, he presents three interleaved frameworks to describe a service system. The *work system framework* [3] provides a system-oriented view of any system that performs work within or across an organization, described in terms of nine basic conceptual categories. The work system framework puts customers first in the service process, and aims to indicate a path to customer satisfaction. The *service value chain framework* [4] augments the work system framework by introducing further notions that are associated specifically with services. The *work system life cycle model* [20] looks at how work systems (and therefore service systems) change and evolve over time. The three frameworks are the basis for a comprehensive business-oriented analysis, intended to be also used by IT professionals [3].

The *work system framework* is based on four general categories: processes and activities, participants, information, and technologies. Five more specific categories help to fill out the picture: products and services, customers, strategies, environment, and infrastructure.

The *service value chain framework* outlines service-related activities and responsibilities of the main parties involved (service provider and service customer) in the form of *service responsibility tables*. This framework is based on a number of assumptions, among which we find:

1. understanding services requires recognition of activities and responsibilities;
2. services are often co-produced by service providers and customers;
3. the idea of a service is the same regardless of whether services are directed at internal or external customers;
4. customer satisfaction is affected by the complete set of activities, responsibilities, and experiences occurring within the service system, as acquiring, receiving, and benefiting from a particular service;
5. the service is delivered as based on negotiated commitments, under which the service may be requested and delivered repeatedly;

[1] The difference between products and services has not been analyzed by Alter, who adopts a general, business oriented definition for services provided by Vargo and Lusch [29]: a service is "the application of specialized competencies (knowledge and skills) through deeds, processes, and performances for the benefit of another entity or the entity itself."

6. the service value is captured by the leftmost and rightmost portions of the service value chain, and includes all parties experience in the service exploitation.

In order to model the service value chain framework, Alter introduces *service responsibility tables*, a conceptual tool to clarify the service's scope and context, focus attention on activities and responsibilities, identify job roles, bring customer responsibilities into the analysis, and identify service interactions between providers and customers.[2]

The *work system life cycle model (WSLC)* provides a dynamic view of how work systems change over time. It is an iterative model based on the assumption that a service system evolves through a combination of planned and unplanned changes. The framework distinguishes several phases, such as *operation and maintenance*, *initiation*, *development*, and *implementation*.

All the three frameworks above can be deployed to support the analysis, design, and improvement of service systems, helping their participants and stakeholders in different ways, according to the different roles they play. Alter distinguishes five of such roles:

Role 1. Executives can use the work system framework to check whether all the relevant business aspects of the service system are properly covered.

Role 2. Strategists can use the three frameworks to provide some kind of organized access to all the relevant design variables (e.g., for performance simulation or optimization purposes).

Role 3. Managers can use service responsibility tables to understand the essence of main steps of service workflow without requiring detailed modeling tools such as flowcharts or database schemes.

Role 4. Implementers of service system changes can exploit the work system life cycle model to understand the effects of changes.

Role 5. Consultants and IT professionals, who have to deal with a large number of technical details, can use the three frameworks to communicate effectively with the other roles, mapping technical choices to high level business aspects.

Important characteristics of a service system can be described using a number of service-related design dimensions. These are important in the analysis of customer-centric service systems, and dimensions such as "product-like features" vs. "service-like features" help understanding the different viewpoints of different stakeholders. The analysis should consider dimensions that take into account the customer's needs, the product vs. service balance, the personalization and the coproduction or self-service approach, the technology, the infrastructure and the environment, among others [6].

[2] Along similar lines, there are some works in business processing, as for instance the Linear Responsibility Charts (or RASCI matrixes) that associate to each task relevant members inside the organization that are either responsible/accountable, or must be informed and/or consulted with respect to it.

4.3 The General Service Model

At the ISTC-CNR Laboratory for Applied Ontology in Trento, two authors of the present chapter, Roberta Ferrario and Nicola Guarino, have explored the ontological assumptions behind the notion of service (cf. [12, 13, 14]), by developing an approach recently presented as the General Service Model. As this work is still ongoing, we briefly present here its most recent version.

The initial motivation behind this work was to develop an ontology of services suitable to be used in the e-government domain, where the problem of interoperability is particularly crucial, and multiple understandings of the word 'service' co-exist. By looking at the computer science literature, it was immediately evident that most of the available models adopt the *black box view* of services, describing them as transfer functions from an input to an output state, with a strong focus on the external service interface.[3] Under this view, the internal details concerning *how* the service is performed are kept hidden, despite their relevance from the business point of view. Business applications need not only specify what the service does, but also how the service is performed and when the various processes involved in a service occur. Moreover, contracts and service level agreements need to refer to internal and contextual details (i.e., how the service interacts with its environment). In other terms, one needs to be able to look both *inside* and *outside* of the box, i.e., we need to adopt a *glass box view*, where the box is in this case, as Alter suggests, the whole service system.

However, adopting a glass box view to model a service system forces us to face some fundamental questions: what is there inside the box? What's the difference between a service system and a service? And what is a service, after all? The main contribution of Ferrario and Guarino is that a service — as opposed to a good — always develops in time, i.e, it has an essential temporal nature: ontologically speaking, services are complex *events*, while goods are *objects*.

The complex internal structure of a service, as well as its relationship with the broader service system, is depicted in Figure 4.1, which is a revised version of a similar figure presented in [12]. The picture clarifies Alter's idea of the service system life-cycle, presenting it as a complex temporal entity involving three main components, that are necessarily always present: the Service Commitment, the Service Process, and the Service Value Exchange. In terms of the DOLCE [22] ontology of temporal entities, the Service Commitment is a *state*, holding as long as the provider is willing to offer the service; the other components are dynamic *processes*, involving a number of different activities. An ontological dependence relation holds between the service commitment and the service process, in the sense that the latter cannot exist without the former. The interplay between service commitment, service process and service as a whole is described by the following informal definitions, adapted from [12]:

A *service commitment* is an agent's explicit and enduring commitment to guarantee the execution of some type of *core actions*, on the occurrence of a certain

[3] For a detailed description of these approaches see Chapters 5, 6, and 7 of this book.

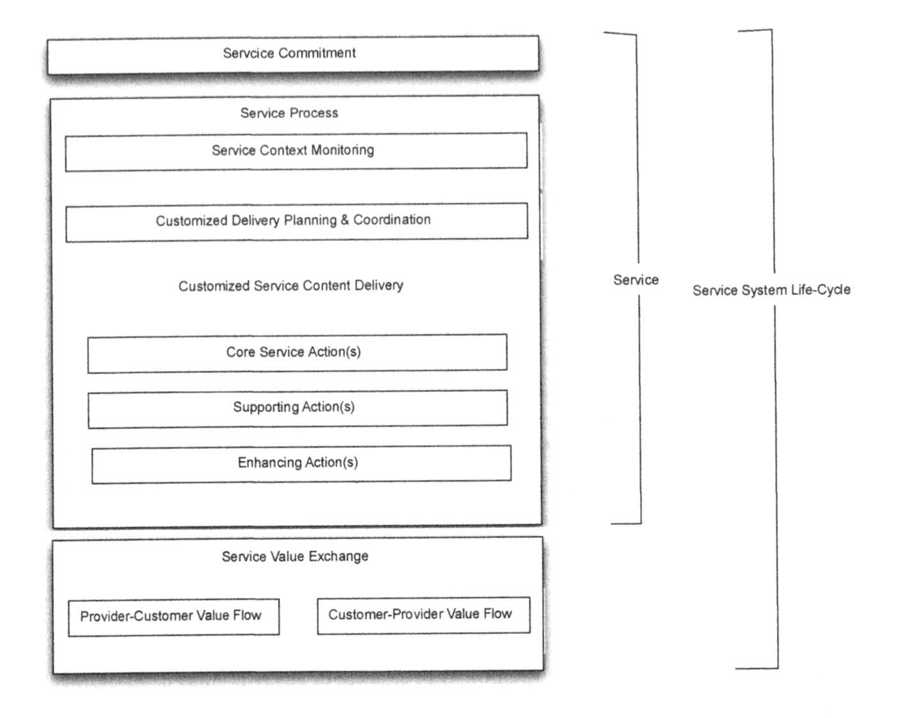

Fig. 4.1: Service and Service system.

triggering event, in the interest of another agent and upon prior agreement, according to a certain specification (*service description*) which constrains the way service actions will be performed. In most cases, two kinds of service commitment need to be distinguished: a *generic* commitment towards potential customers, whose service description is intended to facilitate service discovery, and a *specific* commitment towards a particular customer, where the service description takes the form of a binding *contract*, resulting from a negotiation process.

A *service process* is the actual implementation of a service commitment, consisting of a number of interdependent actions including those necessary to monitor the triggering events, the core actions mentioned in the commitment, and any further actions aimed at supporting or complementing the successful execution of such core actions. What actually happens in the service process is constrained by the *service description*, which defines and constrains the type of actions that must and/or can be executed in the service process.

A *service* is a complex temporal entity (a *complex event*)[4] consisting of a service commitment and the corresponding process.

[4] Generic temporal entities are called *perdurants* in DOLCE, and include *events, states,* and *processes*. However, sometimes "event" is also used as synonymous of perdurant, leaving the context to disambiguate between the generic use and the specific use of the term.

The *service system* is defined as the sum of all the objects anyhow involved in a service (through a *participation* relationship). In other words, while a service is a complex *event*, a service system is a complex *object*, consisting of all the objects somehow participating to any of the sub-events, processes or states constituting the service: typically, a service system includes the provider, the customers, the resources used to produce the service, and so on.[5]

The *service system life-cycle* is a complex temporal entity corresponding to the dynamics of a service system. So the difference between a service system and its life-cycle is like the one existing between a person and its life.

The *service value exchange* is a crucial part of the service system life-cycle. It is a complex process involving two symmetric value flows: the *provider-customer* value flow accounts for provider's costs in implementing the service process, and the corresponding benefits on the side of customers; the *customer-provider* value flow accounts for the costs customers incur in order to receive the service, and the corresponding benefits on the side of providers. Such value flows are also events, and, altogether, the service value exchange is also ontologically dependent on the commitment. Note that the service value exchange is not part of the service itself, since it involves activities occurring at the customer's side: it is rather part of the service system life-cycle.

Relating Figure 4.1 to the car washing example, the service commitment starts when the car wash owner goes to the chamber of commerce to attend all the bureaucratic practices that are necessary to start the commercial activity. Among these practices, there will be some signed official declaration in which the main features of the service are described. In this description, the car wash owner commits to certain business intentions (to be integrated with the content of the ads he or she publicly posts).

The service process is composed of various events, and sub-processes, including the events that trigger the service, e.g., a request by the customer who brings his or her car to the car wash. After the initiating event, we find the customized delivery planning and coordination; here we can imagine that the car wash offers a range of different possible implementations of the service, such as washing only the outside of the car, cleaning the inside, using particular products, such as specific shampoos or waxes etc. In the customized delivery planning phase, the customer and the car wash personnel agree to all these details.

With respect to the service delivery, the core action is washing the car; singling out supporting actions is a bit harder in the example, as there are many actions that are necessarily preparatory to the service but are not explicitly mentioned as constituting the service. Examining possible examples, we could say that the activity of removing loose items from the car in order to be able to clean the inside could be considered a supporting action, as well as buying the cleaning products. Enhancing actions are actions meant to augment the value of the service. Here we could think about an additional service that is connected but not strictly included in the service,

[5] To stress that the notion of service system really includes the context it is embedded in, the expression *service ecosystem* might be appropriate (see also Section 4.5). We shall stick however to service system in the following.

such as replacing air filters or, alternatively, we could think about a luxury service in which someone picks up the car at the customer's location, takes it to the car wash, washes it and then brings it back. The picking up and bringing back would be in this case enhancing actions.

Finally, all the activities connected to the flow of value, both from the customer to the provider (such as payment, loss of customer's time etc.) and from provider to customer (such as time, labor and resources implied in the service production) constitute the service value exchange. In this case, these are the transfer of money and the time spent to drive to the car wash, wait for the car to be washed and drive back for what concerns the customer-provider flow and the time, labor and materials used in washing the car for the provider-customer flow.

A UML diagram of the General Service Model is shown in Figure 4.2.[6] There are three main classes: *Service system, Service system life-cycle,* and *Service system description.* The elements of these classes have a different ontological nature (not shown in the figure): service systems and their parts are *objects*, service system life-cycles and their parts are *events* (generic temporal entities), service system descriptions and their parts are *informational objects.* We adopt specific relations to account for the way an object participates to an event, called "thematic relations" in linguistics [15, 13]. Typical thematic relations are:

Agent	pointing to the entity that plays an active role in the event
Theme/Patient	pointing to what undergoes the event; the patient changes its state, the theme does not
Recipient/Beneficiary	pointing to what receives the effects of the event
Instrument	pointing to what is used to perform the event

This choice allows the authors to propose a formal version of Alter's responsibility tables [4, 13] where rows represent specific service sub-events, and columns describe the specific structure of such events, in terms of thematic relations. Starting from the center of Figure 4.2, we see that a service system life-cycle has two mandatory parts, the service itself and the service value co-production process. In turn, a service has two essential parts: a commitment, and a process that realizes it. The commitment's theme is a *service description* that says what the service is supposed to do. In particular, such description constraints the *core actions* to be performed during the service process. The service description is part of a more general *service system description,* which accounts for the *service value co-production* between the customer and the provider, describing (among other things) the price policy and the legal constraints which limit or regulate the service's range of applicability. Participants to the service system life-cycle are all the parts of the service system, including the service system context (for instance the surrounding economic, legal, and social

[6] Note that Figure 4.2 provides a UML representation but additional work is ongoing to develop models that exhibit more semantic and logical expressiveness. An example would be a GSM model using a first order logic formalism and possibly translatable at least partially to OWL.

systems) and the various actors, such as the service provider, service customer, service producer, and service consumer.[7]

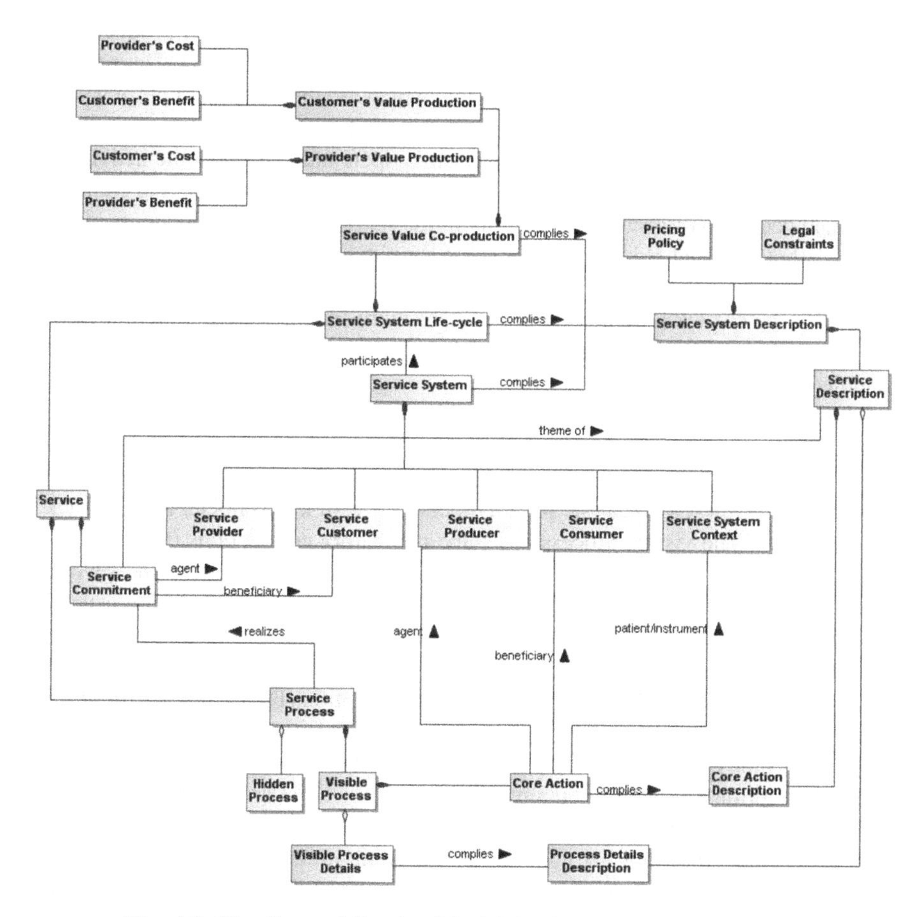

Fig. 4.2: The General Service Model (revised version from [14]).

The picture explicitly shows the thematic relations characterizing the structure of *service commitment*. The commitment's agent is the *service provider*, while the beneficiary is the *service customer*. In car wash example, the service provider is the car wash owner, and the beneficiary is a generic (possible) customer, while the chamber of commerce is, in a sense, acting on behalf of these possible customers. The service description is possibly contained in a document that is stored at the chamber of commerce and includes an explanation of the service. What is written there is what the owner of the car wash is promising to deliver and is what can eventually be handled by the customers in case what was promised is not then realized.

[7] We implicitly assume that participation is distributive with respect to parthood, so if the service system participates to the service system lifecycle all its parts do the same.

In very simple terms, if the description only says that the service merely consists in washing cars, the customer can protest just in case his or her car is dirty after the execution of the service; but if the description specifies, for instance, that only ecological products will be used and the customer finds out that other products are used, he or she can claim that the commitment has not been honored. The service commitment has also a duration and location, which are the period and place where the owner guarantees that the service will be available. For the duration, usually it starts the first moment in which the car wash is open and lasts until the activity is ceased, i.e., the car wash will finally be closed. According to the modeling choices, one could decide to restrict the availability of the service to the opening hours of the car wash, but, as usual, this depends on what is written in the service description. In this example the commitment location is not particularly meaningful as it is identified with the car wash location, but there are more interesting examples, such as fire extinguishing, where the area in which the service is active must necessarily be specified beforehand.

The *service process* realizes the commitment, i.e., it is the execution of the actions described in the service description, according to the constraints there stated and is composed of two parts: the visible process (mandatory) and the hidden process (optional); these two can be roughly identified with the front end and the back end processes. The visible process has some mandatory core actions (those that in a sense define the service for what it is, i.e., the core action is what the service fundamentally does) and some optional visible process details.[8] These are usually enhancing or supporting actions, that may equally be visible or invisible. Also, the core action has to comply with the core action description, while the visible process details have to comply with the process details description. The core action description and process details description are both part of the service description (though only the former is necessary). The hidden process does not have a correspondent in the description because it contains all those actions that are performed but not constrained by the description, i.e., the provider is free to perform such actions as he or she wishes since they are not ruled by the commitment.

Note that the core action's agent and beneficiary are the *service producer* and *service consumer*, respectively, who may or may not coincide with the provider and the customer, depending on the kind of service. In the car washing example, the core action is the washing itself, whose agent is the worker who actually washes the car; this may or may not coincide with the owner; the consumer is the guy who goes to the car wash for having the car washed (also this may or may not be the owner of the car: in the former case he or she is also the customer, in the latter case he or she is not, think about someone who goes washing the car that a friend has lent him or her for a period who, though being the customer, is not the final beneficiary, i.e., the consumer).

The duration of the core action coincides with the time that is taken to actually wash the car and the location is again the car wash itself. The instruments here are the water system, the sponges, the brushes, shampoo, wax etc.

---

[8] Here "visible" and "hidden" refer to the customer's perspective.

Finally, the upper part of Figure 4.2 describes the service value co-production process, which is constituted of two symmetric "flows" of value: from provider to customer, and from customer to provider. What happens is that there is no real *flow*, since increase or decrease of value are subjective events resulting from different evaluations (from the provider's or the customer's side) of the same objective phenomena. Consider again the car washing. While the physical action is performed, there is in parallel a cost event on the side of the provider, while there is a benefit event on the side of the customer, starting from the time the washing is completed, and lasting for a while. Symmetrically, there is a cost event (a sacrifice) on the side of the customer at the payment time, corresponding to a benefit on the side of the provider. Modeling sacrifices and benefits as temporal entities having a non instantaneous duration allows us to account for different kinds of service, depending on how value is produced at different times. So we can say that, for instance, paying for having your car washed is a bad deal if the roads are muddy, so that you can enjoy your car clean only for a short time.

4.4 The TEXO Service Ontology

The TEXO Service Ontology [24] has been developed in the framework of the THESEUS/TEXO project [28], and has taken inspiration in its latest phases from the ongoing work on the General Service Model, as well as from several adjacent ontologies for capturing information about service innovation, pricing, licenses, rating, etc. In order to understand the rationale of the TEXO Service Ontology, one has to have a look at the *service lifecycle* (Figure 4.3), which loops between the innovation, offering, matchmaking, usage, and feedback phases.[9] The *innovation phase* allows for new business models and new consumption and development paradigms. In the *offering phase*, services are supplied to the market. Once a service is designed and developed, it needs to be turned into a commercial offer. In order to create a commercial offer out of a service implementation, several parameters need to be described and published on a service marketplace. The *matchmaking phase* denotes the process of matching a service provider's service offer to a service consumer's service need, i.e., the central application of service description. The *usage phase* in the service lifecycle essentially comprises the delivery of services. Feedback for future iterations of the service lifecycle is channeled back to the service provider during the *feedback phase*. This includes the analysis of the feedback from the applications monitoring.

It is the goal of the TEXO Service Ontology to provide a general scheme for master and transactional data across all phases of the service lifecycle. *Master data* comprise data which seldomly change over time, e.g., a price plan or service licenses. In contrast, *transactional data* grow over time, e.g., data about the service value

[9] This is analogous to the notion of service system lifecycle discussed before, with an emphasis on the fact that the way a service is conceived (and hence the service commitment) can evolve in time, on the basis of customers' feedback.

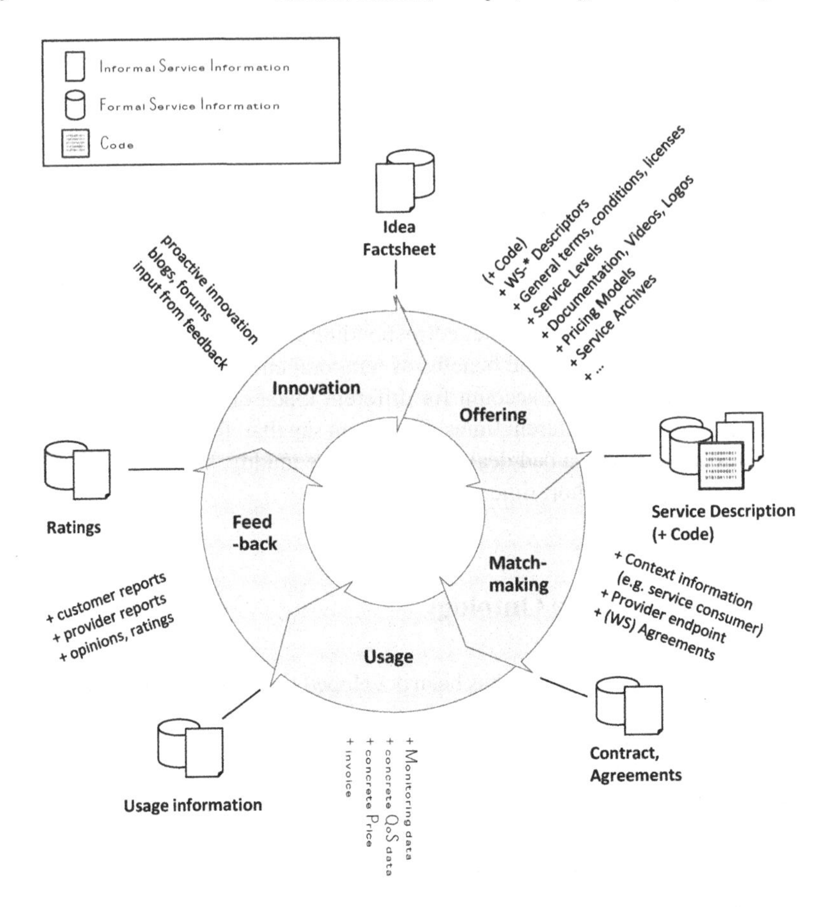

Fig. 4.3: Different phases in the lifecycle generate different kinds of service information.

exchange are generated whenever a service is consumed, and multiple contract data are generated (as transactional data) for one service master data description. Something similar happens in the usage and rating phases where specific monitoring data, concrete service levels, or ratings are generated.[10]

The resulting Service Ontology is depicted by a pyramid in Figure 4.4, which is a metaphor for the number of classes and relations that increases from top to bottom. The Service Ontology is specified in OWL-DL and consists of several modules. Each module basically coincides with an OWL file that imports other OWL files. The modules are depicted as parts of the pyramid. The ontology modules can be divided into four layers according to the requirements:

[10] Transactional data correspond to the *customized service content delivery* phase in the GSM, shown in Figure 4.1, which however, for the sake of simplicity, has not been considered in Figure 4.2.

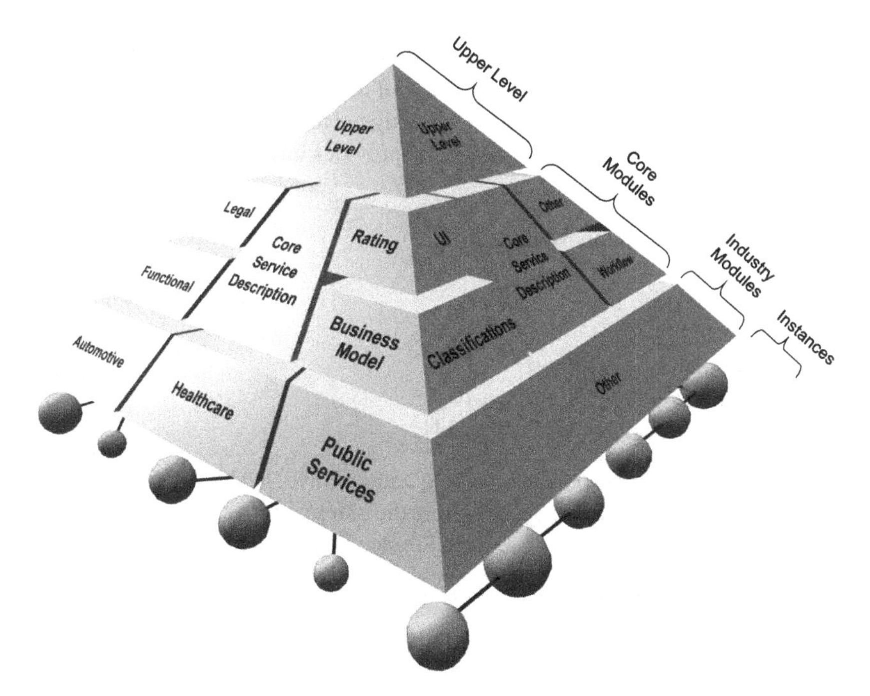

Fig. 4.4: The Service Ontology as a pyramid with increasing amount of classes and relations from top to bottom.

At the top layer, the *upper level* module consists of a concise foundational ontology providing us with a generic set of classes and relations as well as ontology design patterns. More specifically, the DOLCE foundational ontology [16] is applied, which serves the following purposes: (a) DOLCE can be used as a modeling starting point because it provides a basic set of generic classes and relations valid in any domain. Using a foundational ontology as a modeling basis means relating core classes and relations to some proposed invariant categories of human cognition. This prompts the ontology engineer to sharpen his/her notions with respect to the distinctions made in the foundational ontology. What is typically gained is an increased understanding of one's own ontology as well as a cleaner design. (b) DOLCE can also help defining general ontology design patterns as best practices for reoccurring modeling needs.

At the middle layer, a set of *core modules* is built around the Core Service Description module, which captures information common to every service (e.g., info on service provider, quality of service, etc.). Note that the Core Service Description Module does essentially coincide with the General Service Model described above.[11] In addition, different aspects of a service description (legal, busi-

[11] For this reason in this section we won't make use of the car wash example, as it would be essentially represented in the same way as in the GSM.

ness model, technical, rating, UI, etc.) are placed in separate modules and linked to the classes belonging to the Core Service Description module. All modules in this middle layer are aligned under the common roof of the DOLCE foundational ontology. So far, several core modules have been designed to a mature state and published in diverse literature. The enumeration below provides an overview and pointers to the corresponding resources.

- Core Service Description Module [16]
- Idea Module [25]
- Pricing Module [18, 17]
- Legal Module [8]
- Rating Module [23, Section 9]
- Classification Module [23, Section 10]
- Documentation Module [23, Section 11]

At the third layer, *industry modules* (e.g., automotive, healthcare, or public services modules) can be modeled by exploiting the aforementioned ontology modules. The core knowledge specified in the Core Service Description module and adjacent aspect-related core modules discussed above can be specialized for specific industries. Industries can define their own hierarchies of service categories. It is expected that the modules will be populated at run-time by industry consortia or the like.

Finally, *instances* of the classes and relations (depicted as a mesh below the pyramid) can potentially be distributed across the Web according to the principles of Linked Data.

4.5 The OASIS SOA Reference Architecture Foundation (SOA-RAF)

The OASIS SOA Reference Architecture Foundation (SOA-RAF) [2] is an abstract realization of the service-oriented architecture (SOA), focusing on the elements and their relationships needed to enable SOA-based systems to be used, realized and owned, while avoiding reliance on specific concrete technologies. By use, the SOA-RAF captures what it is meant to participate in a space in which stakeholders (human and non-human), processes, and machines act together to deliver the effects of business functionality through services. The space with its stakeholders and the environment (or context) within which they all operate taken together forms the *SOA ecosystem*. This is consistent with Alter's idea of the service system. The SOA ecosystem assumes service implementations utilize capabilities to produce specific (real world) effects that fulfill business needs.

In our car washing example, the capability is the collection of equipment and car washing knowledge to produce the real world effect of a washed car. The service is the access to this capability to wash the car of a specific customer.

From a software perspective, the emphasis is often on the implementation of such business functionality such that it is accessible through a well-defined interface; this is necessary but by no means sufficient. Our car washing example could proceed without an explicit mention of software (or, in general, automated) interactions. However, the customer may have made use of a browser-based interface to schedule a time for the car wash and the custom options desired. Also, payment may have been arranged through an electronic funds transfer (EFT). The capability of a bank to support EFT and the car wash's ability to access this capability fall in line with the previously discussed concept of supporting or complementary actions.

Both those using the services, and the capabilities themselves, may be distributed across ownership domains, with different policies and conditions of use in force. The role of a service in the SOA context is to enable effective business solutions in a distributed environment. SOA is thus a paradigm that guides the identification, design, implementation (i.e., organization), and utilization of such services.

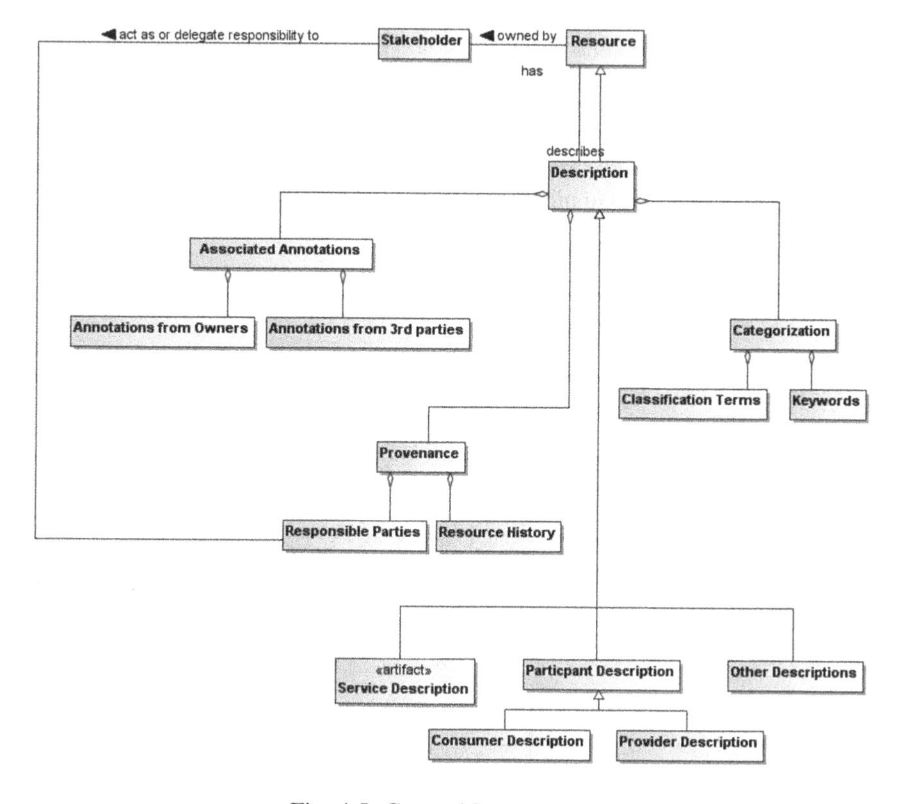

Fig. 4.5: General Description.

The SOA-RAF also discusses the realization and ownership issues involved in the SOA ecosystem. Realization relies heavily on service description, and this will be explored in detail below. Realization also requires sufficient visibility to establish willingness and communications among the participants, effective interactions, and support for policy and contract statement and enforcement. In the realm of effective ownership, elements of governance, security, management, and testing are explored. These aspects of realization and ownership are consistent with the ontological underpinnings described in the General Services Model.

Much of the service realization and aspects of ownership rely on an accurate and sufficiently complete service description. As discussed in the SOA-RAF, a service description is an artifact, usually document-based, that defines or references the information needed to use, deploy, manage and otherwise control a service. This includes not only the information and behavior models associated with a service to define the service interface but also includes information needed to decide whether the service is appropriate for the current needs of the service consumer. Thus, the service description will also include information such as service reachability, service functionality, and the policies associated with a service.

Interactions within a SOA ecosystem rely on many resources, and the SOA-RAF introduces a general *Description* class in Figure 4.5 to represent a number of description properties that are expected to be common among all specialized descriptions supporting a service-oriented architecture. A registry often contains a subset of the description instance, where the chosen subset is identified as that which facilitates mediated discovery. Additional information contained in a more complete description may be needed to initiate and continue interaction.

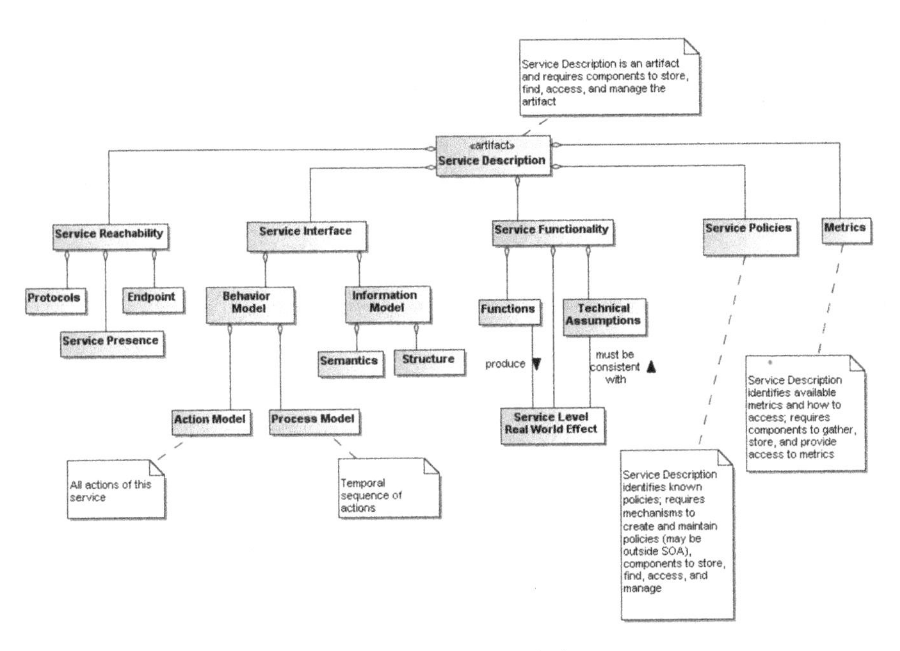

Fig. 4.6: Service Description.

The major description properties for the Service Description subclass follow directly from the areas discussed in the OASIS SOA Reference Model (SOA-RM)

[19] and are shown in Figure 4.6. In particular, the service description conveys the functionality of the service (including the real world effects that are realized through interaction with the service), the conditions of use defined through policies, the operational characteristics captured through metrics, the particulars of the service interface as defined by the service behavior and information models, and the endpoints and corresponding protocols through which message exchange with a present service is accomplished. In addition, provenance, characterization, and identity information and the ability to provide annotations are inherited from the general description.

If we assume we have awareness, i.e., access to relevant descriptions, the service participants must still establish willingness and presence to ensure full visibility as defined in [2] and to interact with the service. Service description provides necessary information for many aspects of preparing for and carrying through with interaction. Recall the fundamental definition of a SOA service in the SOA-RM is a mechanism to access an underlying capability; the service description describes this mechanism and its use. It lays the groundwork for what can occur, whereas service interaction defines the specifics through which occurrences are realized.

Figure 4.7 combines the detailed models for each descriptive element in Figure 4.6 to concisely relate *Action* and the relevant components of *Service Description.* The purpose of Figure 4.7 is to demonstrate that the components of service description go beyond arbitrary documentation and form the critical set of information needed to define the what and how of *Action.* In Figure 4.7, the leaf nodes from Figure 4.6 are shown in differently colored boxes.

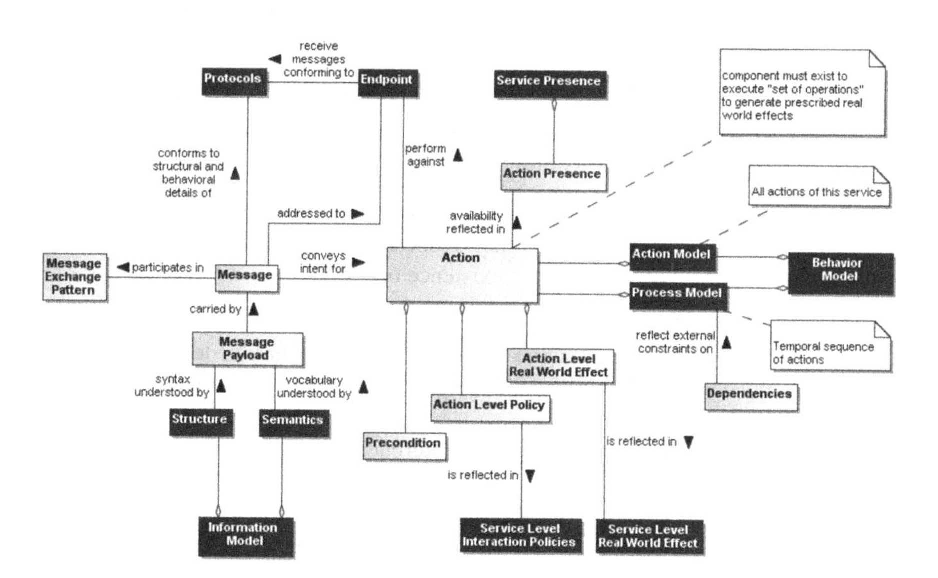

Fig. 4.7: Relationship between Action and Service Description Components.

An adequate service description *must* provide a consumer with information needed to determine if the service policies and the (business) functions and service-level real world effects are of interest and there is nothing in the technical assumptions that precludes use of the service.

Note at this level, the business functions are not concerned with the Action or Process Models. These models are detailed separately. The Actions in the Action Model and their temporal dependence as captured in the Process Model only apply to externally facing actions with which a service consumer would interact. Internal processes are only exposed to the extent they are reflected in policies and other conditions of use or metrics and other measures of operational characteristics. This is the counterpart to the GSM "glass box."

The service description is not intended to be isolated documentation but rather an integral part of service use. Changes in service description should immediately be made known to consumers and potential consumers.

The description of service description indicates numerous architectural implications on the SOA ecosystem:

1. Description will change over time and its contents will reflect changing needs and context. This requires the existence of:
   a. mechanisms to support the storage, referencing, and access to normative definitions of one or more versioning schemes that may be applied to identify different aggregations of descriptive information, where the different schemes may be versions of a versioning scheme itself;.
   b. configuration management mechanisms to capture the contents of each aggregation and apply a unique identifier in a manner consistent with an identified versioning scheme;
   c. one or more mechanisms to support the storage, referencing, and access to conversion relationships between versioning schemes, and the mechanisms to carry out such conversions.
2. Description makes use of defined semantics, where the semantics may be used for categorization or providing other property and value information for description classes. This requires the existence of:
   a. semantic models that provide normative descriptions of the utilized terms, where the models may range from a simple dictionary of terms to an ontology showing complex relationships and capable of supporting enhanced reasoning;
   b. mechanisms to support the storage, referencing, and access to these semantic models;
   c. configuration management mechanisms to capture the normative description of each semantic model and to apply a unique identifier in a manner consistent with an identified versioning scheme;
   d. one or more mechanisms to support the storage, referencing, and access to conversion relationships between semantic models, and the mechanisms to carry out such conversions.

3. Descriptions include reference to policies defining conditions of use. This requires the existence of:

    a. descriptions to enable the policy modules to be visible, where the description includes a unique identifier for the policy and a sufficient, and preferably a machine processable, representation of the meaning of terms used to describe the policy, its functions, and its effects;
    b. one or more discovery mechanisms that enable searching for policies that best meet the search criteria specified by the service participant; where the discovery mechanism will have access to the individual policy descriptions, possibly through some repository mechanism;
    c. accessible storage of policies and policy descriptions, so service participants can access, examine, and use the policies as defined.

4. Descriptions include references to metrics which describe the operational characteristics of the subjects being described. This requires the existence of (as partially enumerated under governance):

    a. the infrastructure monitoring and reporting information on SOA resources;
    b. possible interface requirements to make accessible metrics information generated or most easily accessed by the service itself;
    c. mechanisms to catalog and enable discovery of which metrics are available for a described resource and information on how these metrics can be accessed;
    d. mechanisms to catalog and enable discovery of compliance records associated with policies and contracts that are based on these metrics.

5. Descriptions of the interactions are important for enabling auditability and repeatability, thereby establishing a context for results and support for understanding observed change in performance or results. This requires the existence of:

    a. one or more mechanisms to capture, describe, store, discover, and retrieve interaction logs, execution contexts, and the combined interaction descriptions;
    b. one or more mechanisms for attaching to any results the means to identify and retrieve the interaction description under which the results were generated.

6. Descriptions may capture very focused information subsets or can be an aggregate of numerous component descriptions. Service description is an example of a likely aggregate for which manual maintenance of all aspects would not be feasible. This requires the existence of:

    a. tools to facilitate identifying description elements that are to be aggregated to assemble the composite description;
    b. tools to facilitate identifying the sources of information to associate with the description elements;

c. tools to collect the identified description elements and their associated sources into a standard, referenceable format that can support general access and understanding;
d. tools to automatically update the composite description as the component sources change, and to consistently apply versioning schemes to identify the new description contents and the type and significance of change that occurred.

7. Descriptions provide up-to-date information on what a resource is, the conditions for interacting with the resource, and the results of such interactions. As such, the description is the source of vital information in establishing willingness to interact with a resource, reachability to make interaction possible, and compliance with relevant conditions of use. This requires the existence of:

   a. one or more discovery mechanisms that enable searching for described resources that best meet the criteria specified by a service participant, where the discovery mechanism will have access to individual descriptions, possibly through some repository mechanism;
   b. tools to appropriately track users of the descriptions and notify them when a new version of the description is available.

4.6 The IBM Research Service Design Model

The Service Design model described in [11] was conceived as a meta-model for a service in the widest sense of the word. There was an explicit attempt to go beyond the world of Web services and include services with a large element of human involvement in their delivery and consumption. The examples that motivated this original meta-model came from the world of IT services — outsourced support of IT infrastructure, help desks, call centers, and the like. Subsequently, an attempt was made to incorporate the main elements of the formal model for service delivery described in [7], and to widen the context beyond IT Services to include other domains, including Public Services [27].

The initial model was intended primarily for recording the design parameters of a service without explicitly stating their values at the design phase. The design meta-model has been then extended to a solution meta-model where the parameter values were instantiated to create a specific service instance. This approach has been developed further with the incorporation of feature modeling techniques to describe the design model as a service product line using concepts from the world of software product lines [30]. The benefits of feature modeling are mainly in increasing the accessibility of modeling to non-technical users of the system. The emphasis in the research work has been on creating a well-defined formal representation using UML and BPMN notations which are amenable to analysis and automated transformation into service implementation and simulation artifacts. A further focus is on the facilitation of a participatory design methodology,[12] where all stakeholders provide input and feedback on the services being designed for them. A key challenge has always been to hide the complexity and formality from service domain experts, service consumers, and other non-technical stakeholders.

The tooling which is being developed to accompany the design model exposes simple form filling interfaces to gather the data necessary to populate the model, together with a logical organization of the forms which parallels the workflow. The artifacts produced by the tooling are presented to the stakeholders in the service through a deliberation platform to facilitate participatory design.

The Service Design Model consists of the following elements (cf. Figure 4.8):

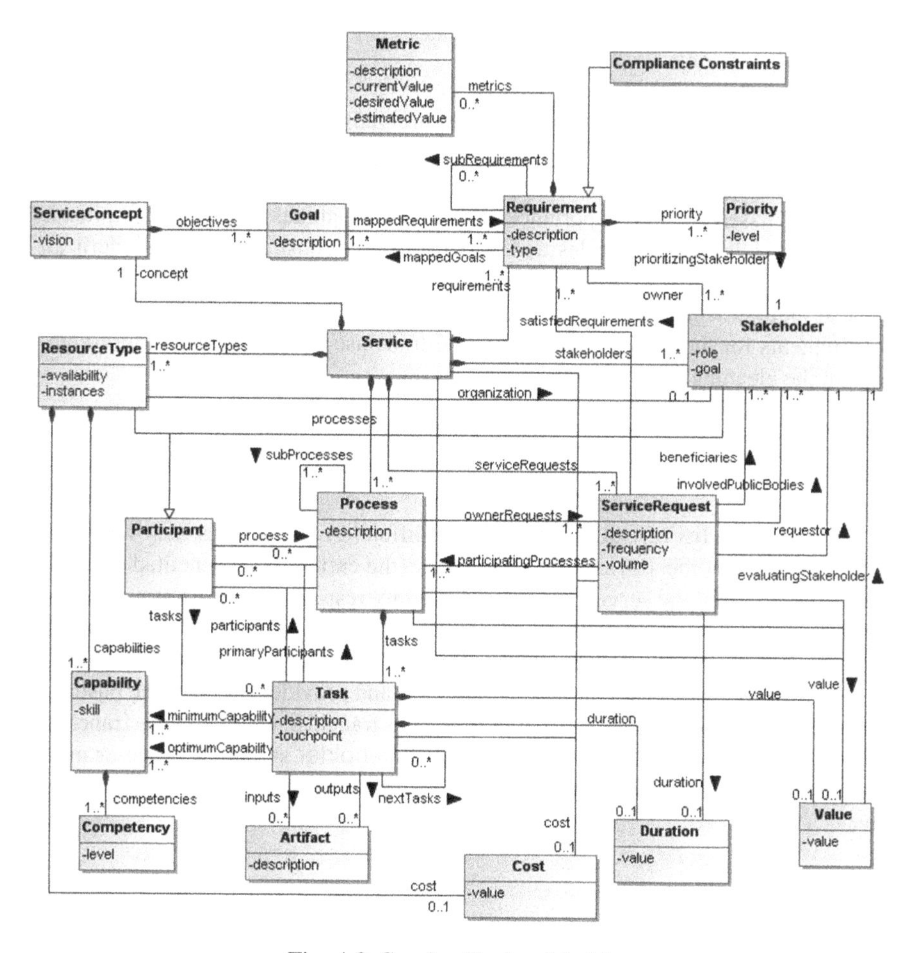

Fig. 4.8: Service Design Model.

[12] There is an extensive literature on participatory design: see [1] for a quick summary.

Service Instances and Service Concepts

A specific service is always an instance of a well-defined ServiceConcept which specifies the vision and the goals of the service. Each goal is mapped to one or more Requirements. In the case of the car wash service, the concept could be one of a franchised car washing company which provides machinery, raw materials (soap, brushes etc.) and premises to franchise holders who pay off the initial investment to the franchising company over a three year period. A service instance in this case is a particular franchise with employees, and local adaptation to the environment and customers at the franchise location.

Service Requirements

A Service has a set of Requirements that are gathered from all the Stakeholders. Each Requirement can be split into sub-requirements and may also specify Metrics to validate if the requirement is met. Stakeholders can prioritize these requirements based on their importance. A Service also abides by certain compliance constraints including legal rules and policies. These constraints which inherit all the properties of a Requirement are high priority requirements that must be met. Typical requirements for a car wash are safety constraints for the workers, income requirements for the franchise company and franchise owner, speed and thoroughness of the cleaning service for the customer.

Service Stakeholders

A Service has a list of Stakeholders with different roles such as the owner (the entity in charge of the service), the beneficiary (the entity that is benefited from the service result), and the service provider (the entity responsible for provisioning the service).[13] The distinction between owner and provider is mainly relevant to the domain of outsourced services where the provider delivers the service under a contract to the owner. In many cases, however, the owner and provider are the same business entity. In the car wash, key stakeholders are the franchise company, the franchise owner, the customer, and the employees. A Stakeholder should own one or more Requirements, and each Requirement must have at least one Stakeholder as its owner.

Service Request

A Service occurs as a response to ServiceRequests which are triggered by a Stakeholder (requestor) who may or may not be the beneficiary of the results.

[13] Owner and provider as defined here correspond respectively to provider and producer in the GSM.

A ServiceRequest invokes a Process — which may invoke other Processes during its execution. Each ServiceRequest has associated parameters such as cost and duration of execution and value derived by the beneficiary. These parameters are aggregated from the sub processes and tasks performed while the request is being handled by the service delivery system. The car wash service request is initiated by the customer who arrives at the franchise.

Service Process

A Service contains many Processes that capture the internal service workflow. A Process has a sequence of Tasks. Each Task is associated with two sets of Capabilities, at different Competency levels. The minimum Capability is the minimal set of skills needed to complete the Task. The optimum Capability of a Task is the set of Capabilities needed to complete the Task with best possible performance. The Participants of a Task can be Stakeholders (e.g., for a customer submitting an application) or Resources (e.g., for front desk employee validating the application, or a computer system needed to complete the Task). A Task provides different Values for different Stakeholders. As a result the containing Process, ServiceRequest and hence Service have Values associated with them by aggregating the Values provided by the constituent Tasks. The service designer uses this Value in order to provide different design alternatives and hence achieve a balance between the Values perceived by different Stakeholders. The car wash process consists of payment (providing value to the owner and operator), cleaning outside and inside the car, drying the car body, and cleaning the windows (each of which provides value to the customer). The capabilities of the employees may include a specialist cashier, cleaning staff, and a sales person who approaches the customer at the end of the process to get feedback and sell a subscription to a set of car washes. In some cases, the cashier may also have the capability to perform some subset of the cleaning work.

Resource Type

A Service has a set of ResourceTypes representing the resource units (human/IT) involved in the service delivery. These ResourceTypes have instances (the actual persons or machines) and availabilities associated with them. They also have a skill set in the form of Capabilities at different Competency level. Aside from the human resources involved in the car wash, there is also car washing machinery, a cash register, and a subscription database sitting on a computing resource.

4.7 Discussion

The presented approaches each reflect differences in viewpoint. Avoiding any exhaustive comparison, we shall focus here on the key features more or less present in all approaches, and on the way the different solutions proposed complement each other by focusing on different perspectives, giving altogether a reasonably complete picture of the most important aspects of a service system.

4.7.1 Service Definition

In Alter's works we find a definition of service system directly imported from Vargos and Lusch [5, 29] "the application of specialized competencies (knowledge and skills) through deeds, processes, and performances for the benefit of another entity or the entity itself."

The General Service Model (GSM) adds details to this informal definition by showing the relationship between service and service system, which is somewhat blurred in Alter's approach. The GSM defines a service as "a complex event composed of different sub-events," from the perspective of the whole service life-cycle. It also analyzes the notion of service value chain, and considers all the elements that may contribute to "value co-creation." The service *system* is instead viewed as a composite object that includes all the entities involved in the execution of interrelated services, such as the agents that participate in these service events, as well as the artifacts and resources that are used and possibly transformed by such services and that can be of different nature (physical, informational etc.). So, unlike Alter, the complex processes and actions at the core of the service notion are not considered as parts of the service *system*, but as parts of the *service* itself. According to the GSM, the difference between a service system and a service reflects the ontological difference between objects and events (endurants and perdurants in DOLCE): the service system is the complex object whose global behavior (i.e., the *service system lifecycle*) "produces" the service, so to speak: to have a service, you need a service system that produces it. To account for the dual nature of services, which always involves a value exchange interaction with the outer environment, Figure 4.1 shows how the *service system lifecycle* articulates into two main components: the service itself, and the service value exchange.

The TEXO Service Ontology imports the definitions of concepts constituting the core service description module directly from the General Service Model, but adopts a broader notion of service (system) lifecycle, which accounts for changes in the service commitment, as a consequence of interaction with the external context.

SOA-RAF discusses the inherent differences between services providing a business function and services considered as a software artifact in a SOA ecosystem. An important role in SOA-RAF is played by the notion of service description, defined as "an artifact, usually document-based, that defines or references the information needed to use, deploy, manage and otherwise control a service." This definition is close to the GSM definition, according to which service descriptions are informational artifacts whose contents are the constraints on the way the service is supposed to be delivered. However, SOA-RAF does not provide a crisp definition of what a service is, although the SOA-RM definition concerning the nature of services as "mechanisms to access an underlying capability" fits with the GSM idea of detaching the service commitment from the actual execution of the core service action. As a matter of fact, it seems that SOA-RAF finds a crisper definition more of a distraction than a clarification.

The IBM-SDM neither provides an explicit definition of service, nor of service system; however, the meaning of such notions is weakly constrained by the UML model reported in Figure 4.8. According to this diagram, a service is described as an aggregate of several heterogeneous entities, notably including resources, which are not explicitly mentioned in the other approaches. The inclusion of entities as disparate as processes, resources and concepts as *parts* of a service is however confusing from the ontological point of view, and hopefully the present analysis will help realizing the need of some cleanup, especially from the conceptual point of view. The main point of the IBM-SDM is to provide a clear path to practical simulation models of service delivery as described in [11].

4.7.2 Service Application Perspective

To better understand the various approaches presented so far, it is useful to discuss the different application perspectives they focus on, which determine the different service modeling goals and motivations.

Alter's framework models a service system as a work system, adopting explicitly a *business perspective*. His notion of *work system lifecycle* is pretty close to the *service system lifecycle* introduced in the General Service Model, whose viewpoint can however be better described as focusing on *service interactions*, i.e., on all interactions taking place in the service system (so among actors, artifacts, resources and the surrounding natural and institutional environment). Among other things, the GSM takes also into account the legal aspects of customer interaction, which are not considered in Alter's framework.

The TEXO Service Ontology's application perspective is on *service evolution*. The *service lifecycle* loops between the innovation, offering, matchmaking, usage and feedback phases. In each phase, both *master data* (that seldom change over time, such as licenses) and *transactional data* (that relate to a single service transaction, such as value exchange data) are exchanged.

SOA-RAF's focus is on the *integration between business needs and the available information technology*, as a key requirement to fully exploit a service-oriented architecture (SOA), intended mainly as "a paradigm for organizing and utilizing distributed capabilities (. . .)." While the goal of SOA-RM (the SOA Reference Model) is "to define the essence of service-oriented architecture, and emerge with a vocabulary and a common understanding of SOA"), the Reference Architecture Foundation (SOA-RAF) focuses "on the elements and their relationships needed to enable SOA-based systems to be used, realized and owned. So, the emphasis of SOA-RAF is on service systems as composed by distributed capabilities that belong to a larger "ecosystem." Both SOA-RAF and SOA-RM model the service system as a snapshot, without referring to a service lifecycle.

IBM-SDM's main application perspective is on *reuse of service artifacts, and on the use of simulation as a design tool.* Three key stakeholders are present in the service lifecycle: the owner, the beneficiary (sometimes also called customer), and the service provider. The distinction between owner and provider resembles very closely the GSM one between provider and producer, as the owner is the one in charge of the business management of the service, while the provider is like an operations manager, responsible for day-to-day delivery. The term customer is instead more, as it may sometimes refer to the requestor and sometimes to the actual beneficiary, depending on context. The service lifecycle loops between design, solution, transition, and delivery. But only design and solution have been modeled so far.

4.7.3 Service System Perspective

While the service application perspective concerns the service system as a whole, two different modeling perspectives can be isolated concerning how the service system is perceived and described from its two main players, namely the owner/provider and the customer/consumer. Here we shall label as inside-out an approach centered on the former perspective, and outside-in an approach centered on the latter perspective. The inside-out perspective focuses on how the service system satisfies the provider's needs; the outside-in perspective focuses on how the service system satisfies the customer's needs. Note that the choice to use the term "inside" for the provider perspective and "outside" for the customer perspective is arbitrary.

Obviously at some point in the service lifecycle the two visions converge, but they are initially different, since, for instance the customer awareness of the need for a service is a completely different starting point from the provider awareness that such a need exists. The customer often has a clear idea of what he or she wants, while possibly ignores what is really feasible or what is economically convenient, and vice versa for the provider. The IBM-SDM model attempts to converge between both outside-in and inside-out by promoting a participatory style of service design, taking into account not only the provider and customer, but also the needs and values of the delivery organization.

Note that this distinction turns out to be orthogonal to that between glass- and black-box models, as both the inside-out and the outside-in views can show internal activities vs. external behavior.

It is also different from the top-down vs. bottom-up distinction, as in the former case the two perspectives (outside-in and inside-out) are not related to how the modeling activity is performed: under both perspectives one can decide to analyze the service from details up or from the top into details.

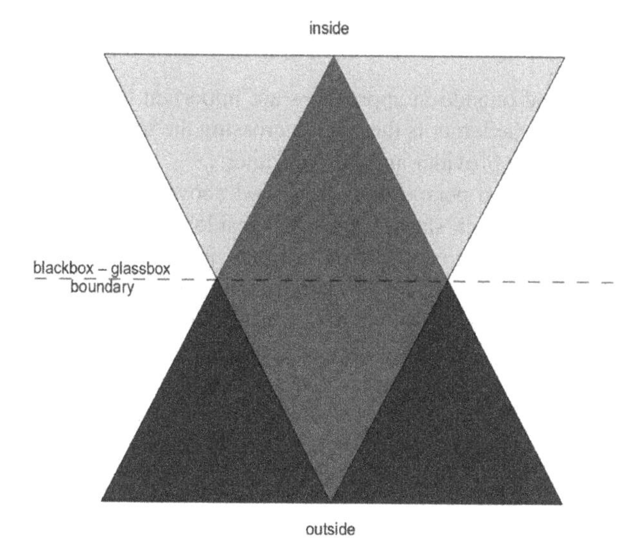

Fig. 4.9: In-out side approach vs black-glass box.

In Figure 4.9 the relation between the inside-out/outside-in choice and the black-box/glass-box choice is summarized. The upper (light) triangle represents the service system modeled according to the provider's point of view (inside-out perspective), while the bottom (dark) triangle represents the service system modeled according to the customer's point of view (outside-in perspective). The larger base of each triangle represents a larger amount of details about the particular point of view encoded in the model. The upper part of each triangle (crossing the "box boundary line") represents that part of the model that accounts for the other party's perspective.

Both approaches are compatible with the black- and the glass-box views. From the provider's view-point, the focus is on "internal" aspects of the service. The model has to be considered a black-box one when only the lowest part of the upper (light) triangle (of the service) is shown to the customer. It is instead a glass box when the whole (or almost the whole) upper triangle is visible to the customer. From the customer's viewpoint, a glass box model corresponds to the whole lower triangle, while a black box model corresponds to the upper part of the lower triangle, denoting a model which only picks up those customer's details which are of interest from the provider's point of view.

The intersection of the two triangles — the center diamond — is the common area of knowledge about the service; the four smaller triangles outside the intersection represent knowledge known to one side but not the other. The two small not overlapping upper triangles could refer to technical and strategic aspects that the provider wants to keep hidden to the customer. The two small not overlapping bottom triangles represent the customer's expectation on the service (what the service should do and why the customer wants to use the service in his or her own application context).

Both inside-out and outside-in approaches are important in modeling a service, as in a glass box view, the larger is the section crossing the service border, the better is the model for both the provider and the consumer.

With respect to the two perspectives described above, Alter's approach lies in the middle, because although starting from an "inside" point of view, in the work system framework (taking into account participants, information, technologies, and processes and activities in a context of infrastructure, strategies and environment), he puts the customer and his/her role in the "service value chain framework," investigating his/her responsibilities in the service process.

The GSM aims at being comprehensive, but although the customer's perspective is taken into account while modeling the value co-production process, it has probably a bias towards an inside-out approach, because the internal aspects of the service process are accounted for in some more detail.

The approach of the TEXO Service Ontology is an inside-out approach, although the distinction between master and transactional data is useful for the customer in the understanding of the service, too.

SOA-RAF has aspects of both: the inside-out approach tries to answer the question "what I have to build and how I promote the use of the service system with the customer/consumer?;" the outside-in approach looks for a consumer to evaluate whether the service adequately addresses a need in a manner consistent with acceptable conditions of use.

IBM-SDM is an inside-out approach, because its focus is on how the service is built and how it is presented (delivered) to the customer.

4.7.4 Service Science Readiness

An interesting way to compare service modeling approaches is to discuss, so to speak, their *service science readiness*, i.e., the extent to which they are suitable to be adopted within the large interdisciplinary perspective known with the term *service science*. To this purpose, we shall briefly discuss the presence of interdisciplinary aspects in the modeling proposals described so far.

Alter's framework does not explicitly adopt an interdisciplinary approach, as it mainly focuses on the work system under a business perspective, but certainly his work is in the spirit of a broad service science, and can be used by multiple categories of stakeholders, including non-business professionals in some of the proposed "roles."

The GSM approach is deliberately interdisciplinary, as it relies on theories belonging to philosophy, cognitive science, linguistics, and aims at including juridical and deontic notions as well as business process aspects and economic notions.

The TEXO Service Ontology adopts also an interdisciplinary approach, being based on different modules which account for business, legal, and industry aspects.

Industry modules are populated at run-time to match specialized service application contexts (automotive, healthcare, public services, ...).

SOA-RAF (SOA-RM) is partly interdisciplinary, in the sense that it is built from an IT perspective, but it is oriented towards ecosystems.

IBM-SDM initially lacked an interdisciplinary perspective, as it was mainly intended for IT services; nonetheless, the latest works focus on a more general notion of service, in which the organizational side is taken into account. The goal is to introduce engineering rigor in the design process, not to anchor concepts to a shared service ontology.

4.7.5 Value Modeling

Since a core aspect of services is their (co-)production of value, the ability to account for the value production (or value exchange) process is an important aspect to consider while comparing the different modeling approaches. According to Alter's and other approaches, value production is not an exclusive responsibility of the producer.[14] Of course, we should also observe that value (as well as cost) has not just a monetary nature, so the notions of benefit and sacrifice for the two parties involved (service provider and service beneficiary) should be somehow considered in a proper account of value.

Alter explicitly considers value in the *service value chain framework*, which augments the *work system framework*. Because the value of a service is co-produced by consumer and producer, a chain of responsibility for each activity involved in the service process can be identified.

One of the three key components of the service system according to the General Service Model is service value co-production, seen as a complex event in which symmetric events corresponding to provider's or customer's costs and benefits co-occur. The value that is exchanged during such event is the result of positive and negative import of value both from the side of the provider and of the customer (value co-production) throughout all the phases of the service life-cycle, which are inspired by Porter's value chain.

The TEXO Service Ontology explicitly considers value exchange as generated whenever a service is consumed.

SOA-RAF (SOA-RM) represents values as *real world effects*. It could be the response to a request or a change of state for some defined entities. SOA-RAF distinguishes between *social effects* and *physical effects* as goals for the reference architecture.

IBM-SDM associates value with every service object, as a numerical estimate of the value provided by that object to each stakeholder. The model estimates costs

[14] In the present chapter, value co-creation is seen as a process that involves the whole service system and is mainly analyzed as one among several components of the service system life-cycle. Chapter 3 of this book is expressly dedicated to approaches that focus on service value networks.

and values of services, focusing on the values perceived by all different stakeholders. More precisely, value is represented as a vector with one component for each stakeholder.

4.7.6 Service Contract

With the term "contract" we refer to an event in the service life cycle in which something is established that can be used by all involved parties to "judge" whether the value delivered matches all expectations. For such reason, "contract" and "value" are two different but related concepts, as the former could constitute a framework to evaluate the latter (although actual value may go beyond the contractual terms). An important part of a service contract is the Service Level Agreement (SLA). The contract may also be interpreted as a legal object, which can be used by any participant in the service process to enforce an action.

Alter's framework includes service awareness and negotiation in the *service value chain framework*. He assumes however that a service is always co-produced by customer and provider from early stages, independently of any notion of contract.

In the General Service Model there are two events that are connected with the idea of negotiation and contract: one is the service commitment, which in a sense constrains the provider to guarantee that the service is executed in a certain way. Note that such commitment is a generic one, as the service description is about types of action and does not refer to specific customers. But there is another event, customized planning and delivery, which is duplicated for each customer, and occurs after a negotiation, which results in a customization of the general constraints contained in the service description. So there is at least a clear room to include a notion of service contract.

The TEXO Service Ontology explicitly envisions service contracts between service customers and providers as part of the transactional data belonging to a particular service lifecycle. The notion of contract is however somewhat simplified, as it doesn't take into account the possibility that customer and consumer may be different entities.

SOA-RAF (SOA-RM) introduces *contract and policy* as one of the principal concepts involved. "A *policy* represents a constraint or condition on the use of the service." It is an assertion stated by one participant on the conditions of use of some resource. "A *contract* represents an agreement between two or more parties."

IBM-SDM is missing an explicit notion of service contract, apart from modeling it as a generic quality parameter that should be numerically represented in a service quality object. As the model allows representing service constraints, contract targets (as requirements on the quality of service) can be implemented as performance constraints.

4.7.7 Reusability

As it happens in any engineering process, a model is considered a solid one if it can be applied on a (relevant) number of application contexts. In the context of service science, with "reusability" what is meant is the capability to reuse the proposed modeling approach, rather than the capability of a model to reuse external components. Both viewpoints are important, of course, but since service systems involve knowledge and participants from different domains, it is even more straightforward that any proposed model has to reuse, at least, core concepts and components of each involved domain.

Alter's framework describes service systems from a business viewpoint, with no assumption whether IT is involved. The framework can be reused for any service system, automated or not.

Reusability is an implicit goal both of DOLCE and of the General Service Model as well. First of all, the level of generality is such that it is fairly straightforward to specialize the primitives to adapt them for more special tasks. On the other hand, the GSM can also be easily extended by adding elements built starting from DOLCE.

The TEXO Service Ontology is reusable in modeling any service lifecycle from the evolution viewpoint, as long as in the setup process of the service itself the innovation, the offering, the matchmaking, the usage, and the feedback phases are planned.

SOA-RAF (SOA-RM) is a reference architecture and is reusable by definition.

IBM-SDM reusability concerns configurability, variability, and extensibility of ServiceObjects used in the modeling process, at design level.

4.7.8 Service Time Frame

With the locution "Service Time Frame" we mean the overall period of time considered by a service model. For example, the time frame can span from the start of a contract to its end. Alternatively, the evolution of a service business model could start from initial design and proceed to final delivery. The different approaches described here differ in the time frame they assume. In particular, it is important to distinguish:

- the *service delivery time frame* concerning a service delivered to a specific customer, within the temporal validity of a specific contract;
- the *service commitment time frame* where a generic service commitment (involving multiple potential customers) continues to hold, satisfying the same generic description;
- the *service evolution time frame* concerning the evolution of a service system, where the service description changes in time to account for maintenance changes, new service policy choices, feedback from customers, and so on.

Alter's framework clearly adopts a service evolution time frame, as it focuses on service systems that "evolve through a combination of planned and unplanned changes." The temporal phases considered are "operation and maintenance," "development" and "implementation."

The GSM adopts instead a service commitment time frame, as the service system life cycle starts with the service commitment and ends with the service dismissal. The service delivery time frame is embedded in the service commitment time frame as a sub-event (indeed, a plurality of them, one for each customer contract), while the service evolution time frame is not modeled explicitly. There is a possibility to account for a service evolution time frame by allowing multiple services (at different temporal stages) within the same service system lifecycle, but what needs to be clarified, from the ontological point of view, is the notion of persistence of a service through time. Presently, in GSM a service loses its identity if the service commitment changes.

The TEXO Service Ontology focuses on the *service evolution time frame*, modeling a service whose generic description can change while it "loops between the innovation, offering, matchmaking, usage, and feedback phases," in a fashion very similar to Alter's model.

The SOA-RAF doesn't apparently adopt any time frame, as it models an evolving snapshot of a service system. For example, there are no time constraints concerning contracts and policies.

The IBM-SDM focuses also, such as the TEXO approach, on the service evolution time frame, whose phases are however slightly different from TEXO's ones: design, solution, transition, delivery and end of life.

4.8 Concluding Remarks

In this chapter, we have illustrated several approaches to service systems modeling. The key point is that while the various models emphasize different aspects of a larger system, there is consensus that the service conceptual models should account not only for the core service activities, but also for the business and social environment in which the service is used. While a complete comparison of the various approaches is not feasible, we have highlighted the differentiating points of focus of each of these approaches so as to enable designers to choose what is more suited to his/her own needs.

Abstracting away from the specificities of every single approach, in these concluding remarks we would like to make some considerations about the importance of service systems modeling.

First, in the services domain, it is important to have reference models (and several of the presented approaches can be considered as such) to facilitate sharing the fundamental primitives needed to define service ecosystems. The use of reference models enhances mutual understanding and improves the likelihood of interoperability.

Second, the shift from simply modeling services to modeling service systems helps to explicitly account for the social implications of the various aspects of service processes, making the whole service system transparent to designers and stakeholders.

Finally, it is noteworthy that most of these approaches make an explicit separation between design and realization level, distinguishing the service description and what is declared in contracts from the actual service execution. This allows one to verify the compliance of the executed actions with those specified in the service description, including any additional conditions agreed to in explicit contracts. Moreover, the interplay between these two levels is at the basis of complex chains of responsibilities, with duties and rights that can also be transferred through delegation. Models including the explicit representation of responsibility chains enable transparency and predictability, thus providing greater trustworthiness of services.

References

1. Wikipedia entry on participatory design. Accessed March 2011, `http://en.wikipedia.org/wiki/Participatory_design`.
2. Reference Architecture Foundation for Service Oriented Architecture 1.0. Committee Draft 2, OASIS, Oct 2009. `http://docs.oasis-open.org/soa-rm/soa-ra/v1.0/soa-ra-cd-02.pdf`.
3. S. Alter. *The Work System Method: Connecting People, Processes, and IT for Business Results*. Work System Press, Larkspur, CA, USA, Apr 2006.
4. S. Alter. Service responsibility tables: A new tool for analyzing and designing systems. In *Proceedings of the Thirteenth Americas Conference on Information Systems (AMCIS 2007) Keystone, Colorado, August 09 - 12 2007*, 2007.
5. S. Alter. Service system fundamentals: Work system, value chain, and life cycle. *IBM Systems Journal*, 47(1):71–85, 2008.
6. S. Alter. Viewing systems as services: A fresh approach in the IS field. *Communications of the Association for Information Systems*, 26(11), 2010.
7. G. Banavar, A. Hartman, L. Ramaswamy, and A. Zherebtsov. A formal model of service delivery. In P. P. Maglio, C. A. Kieliszewski, and J. C. Spohrer, editors, *Handbook of Service Science*, Service Science: Research and Innovations in the Service Economy, pages 481–507. Springer US, 2010.
8. C. Baumann and C. Loës. Formalizing copyright for the internet of services. In G. Kotsis, D. Taniar, E. Pardede, I. Saleh, and I. K. Ibrahim, editors, *iiWAS'2010 - The 12th International Conference on Information Integration and Web-based Applications and Services, 8-10 November 2010, Paris, France*, pages 714–721. ACM, 2010.
9. M. Böttcher and K.-P. Fähnrich. Service systems modeling: Concepts, formalized meta-model and technical concretion. In H. Demirkan, J. C. Spohrer, and V. Krishna, editors, *The Science of Service Systems*, Service Science: Research and Innovations in the Service Economy, pages 131–149. Springer US, 2011.
10. H. Chesbrough and J. Spohrer. A research manifesto for services science. *Commun. ACM*, 49(7):35–40, 2006.
11. K. A. Dhanesha, A. Hartman, and A. N. Jain. A model for designing generic services. In *2009 IEEE International Conference on Services Computing (SCC 2009), 21-25 September 2009, Bangalore, India*, pages 435–442. IEEE Computer Society, 2009.
12. R. Ferrario and N. Guarino. Towards an ontological foundation for services science. In J. Domingue, D. Fensel, and P. Traverso, editors, *Future Internet - FIS 2008, First Future*

Internet Symposium, FIS 2008, Vienna, Austria, September 29-30, 2008, Revised Selected Papers, volume 5468 of *Lecture Notes in Computer Science*, pages 152–169. Springer, 2008.

13. R. Ferrario, N. Guarino, and M. Fernandez-Barrera. Towards an ontological foundation for services science: The legal perspective. In G. Sartor, P. Casanovas, M. A. Biasiotti, M. Fernandez-Barrera, P. Casanovas, and G. Sartor, editors, *Approaches to Legal Ontologies*, volume 1 of *Law, Governance and Technology Series*, pages 235–258. Springer Netherlands, 2011.
14. R. Ferrario, N. Guarino, C. Janiesch, T. Kiemes, D. Oberle, and F. Probst. Towards an ontological foundation of services science: The general service model. In *10th International Conference on Wirtschaftsinformatik, 16th - 18th February 2011, Zurich, Switzerland*, pages 675–684, 2011.
15. C. Fillmore. Types of lexical information. In D. Steinberg and L. Jacobovitz, editors, *Semantics. An Interdisciplinary Reader in Philosophy, Linguistics and Psychology*. Cambridge University Press, London, UK, 1971.
16. A. Gangemi, N. Guarino, C. Masolo, A. Oltramari, and L. Schneider. Sweetening ontologies with dolce. In A. Gómez-Pérez and V. R. Benjamins, editors, *Knowledge Engineering and Knowledge Management. Ontologies and the Semantic Web, 13th International Conference, EKAW 2002, Siguenza, Spain, October 1-4, 2002, Proceedings*, volume 2473 of *Lecture Notes in Computer Science*, pages 166–181. Springer, 2002.
17. T. Kiemes and D. Oberle. Generic modeling and management of price plans in the internet of services. In K.-P. Fähnrich and B. Franczyk, editors, *Informatik 2010: Service Science - Neue Perspektiven für die Informatik, Beiträge der 40. Jahrestagung der Gesellschaft für Informatik e.V. (GI), Band 1, 27.09. - 1.10.2010, Leipzig*, volume 175 of *LNI*, pages 533–538. GI, 2010.
18. T. Kiemes, D. Oberle, and F. Novelli. Towards a reusable and executable pricing model in the internet of services. In G. Kotsis, D. Taniar, E. Pardede, I. Saleh, and I. K. Ibrahim, editors, *iiWAS'2010 - The 12th International Conference on Information Integration and Web-based Applications and Services, 8-10 November 2010, Paris, France*, pages 722–729. ACM, 2010.
19. C. M. MacKenzie, K. Laskey, F. McCabe, P. F. Brown, and R. Metz. Reference Model for Service Oriented Architecture 1.0. Oasis standard, OASIS, Oct 2006.
20. P. Maglio and J. Spohrer. Fundamentals of service science. *Journal of the Academy of Marketing Science*, 36:18–20, 2008.
21. P. P. Maglio, S. Srinivasan, J. T. Kreulen, and J. Spohrer. Service systems, service scientists, ssme, and innovation. *Commun. ACM*, 49(7):81–85, 2006.
22. C. Masolo, S. Borgo, A. Gangemi, N. Guarino, and A. Oltramari. Ontology Library (final). WonderWeb Deliverable D18, Dec 2003. `http://wonderweb.semanticweb.org`.
23. D. Oberle. Service ontology final report. Deliverable D.TEXO.9.3.2b, BMWi, Theseus Programme, Use Case Texo, NOV 2010.
24. D. Oberle, N. Bhatti, S. Brockmans, M. Niemann, and C. Janiesch. Countering service information challenges in the internet of services. *Journal of Business & Information System Engineering (BISE)*, 5, 2009.
25. C. Riedl, N. May, J. Finzen, S. Stathel, V. Kaufman, and H. Krcmar. An idea ontology for innovation management. *Int. J. Semantic Web Inf. Syst.*, 5(4):1–18, 2009.
26. J. C. Spohrer, P. P. Maglio, J. H. Bailey, and D. Gruhl. Steps toward a science of service systems. *IEEE Computer*, 40(1):71–77, 2007.
27. Y. Taher, W.-J. van der Heuvel, S. Koussouris, and C. Georgousopoulos. Empowering citizens in public service design and delivery: a reference model and methodology. In *Proceedings of the Service Modelling And Representation Techniques Workshop (2010)*, 2010.
28. O. Terzidis, A. Fasse, B. Flügge, M. Heller, K. Kadner, D. Oberle, and T. Sandfuchs. Texo: Wie THESEUS das Internet der Dienste gestaltet — Perspektiven der Verwertung. In L. Heuser and W. Wahlster, editors, *Internet der Dienste*, acatech diskutiert, pages 141–161. Springer, 2011.
29. S. L. Vargo and R. F. Lusch. Evolving to a New Dominant Logic for Marketing. *The Journal of Marketing*, 68(1):1 – 17, 2004.

30. E. Wittern and C. Zirpins. On the use of feature models for service design: the case of value representation. In *Proceedings of the Service Modelling And Representation Techniques Workshop (2010)*, 2010.

Chapter 5
SOA Approaches

Thomas Kohlborn and Marcello La Rosa

Abstract As the service-oriented architecture paradigm has become ever more popular, different standardization efforts have been proposed by various consortia to enable interaction among heterogeneous environments through this paradigm. This chapter will overview the most prevalent of these SOA approaches. It will first show how technical services can be described, how they can interact with each other and be discovered by users. Next, the chapter will present different standards to facilitate service composition and to design service-oriented environments in light of a universal understanding of service orientation. The chapter will conclude with a summary and a discussion on the limitations of the reviewed standards along their ability to describe service properties. This paves the way to the next chapters where the USDL standard will be presented, which aims to lift such limitations.

5.1 Introduction

Broadly speaking, a *Service-oriented Architecture (SOA)* is a paradigm for arranging and utilizing business capabilities and resources that may be under the control of different business domains [6], for the sake of producing business value. While SOA does not imply the use of technology, it can help structure how technology is deployed and organized within a particular organization, or across a consortium of organizations that need to interact with each other. That said, the underlying concepts of service-orientation are typically realized by developing technical (i.e., electronic) services that communicate with each other over the Web, as opposed to business services, which may also be carried out manually. In this regard, SOA can also be seen as a principle for designing software architectures which revolves around the notion of "*Web service*." A Web service is a self-contained, autonomous, reusable software

Thomas Kohlborn, Marcello La Rosa
Queensland University of Technology, GPO Box 2434, Brisbane, QLD 4001, Australia,
e-mail: `t.kohlborn@qut.edu.au`, e-mail: `m.larosa@qut.edu.au`

component encapsulating discrete functionality, which is distributed and accessible over the Internet.

Thus, the term Web service refers to a specific technology approach for implementing a SOA, when the channel of communication is the Web. The following definition of Web service has been proposed by the World Wide Web Consortium in 2004 [24]:

> *A Web service is a software system designed to support interoperable machine-to-machine interaction over a network. It has an interface described in a machine-processable format (specifically WSDL). Other systems interact with the Web service in a manner prescribed by its description using SOAP-messages, typically conveyed using HTTP with an XML serialization in conjunction with other Web-related standards.*

As SOA and Web services have been researched by academia and applied by practitioners for nearly a decade, multiple standards have emerged in this field. These standards propose how a service-oriented environment can be created, how basic service functionality can be described, and how services can be aggregated to provide high-level functionality. In this context, the major standardization bodies are the World Wide Web Consortium[1] (W3C), the Organization for the Advancement of Structured Information Standards[2] (OASIS), the Object Management Group[3] (OMG) and The Open Group.[4] This chapter will describe the most mature and widely-used SOA standards promoted by these four consortia, and discuss the shortcomings of these standards when it comes to a more universal understanding of service properties.

The chapter is organized according to Figure 5.1, which classifies these SOA standards based on their level of abstraction. Accordingly, Section 5.2 will start by presenting two different styles for describing basic service functionality, i.e., WS-* and REST. These styles manifest themselves into two main description languages, namely WSDL and WADL, which have reached different levels of maturity. Section 5.2 will also discuss SOAP as a standard message protocol to allow communication among services, and its extensions to model non-functional aspects of a service such as WS-Addressing and WS-Security. Section 5.3 will describe two different mechanisms that can be used to store and discover Web services described in WSDL, namely UDDI and WS-Discovery. Next, Section 5.4 will overview two different specifications (WS-BPEL and BPMN) for compositing services according to two complementary models: orchestrations and choreographies. Focusing on the top-level of abstraction of the SOA Stack, Section 5.5 will present a UML metamodel (SoaML) and two reference models (SOA-RM and SOA Ontology) that can be used to design service-oriented architectures and describe their constituent elements. The chapter will conclude with a summary and a discussion on the shortcomings of these standards along their ability to capture service properties.

---

[1] `http://w3.org`

[2] `http://www.oasis-open.org`

[3] `http://www.omg.org`

[4] `http://www.opengroup.org`

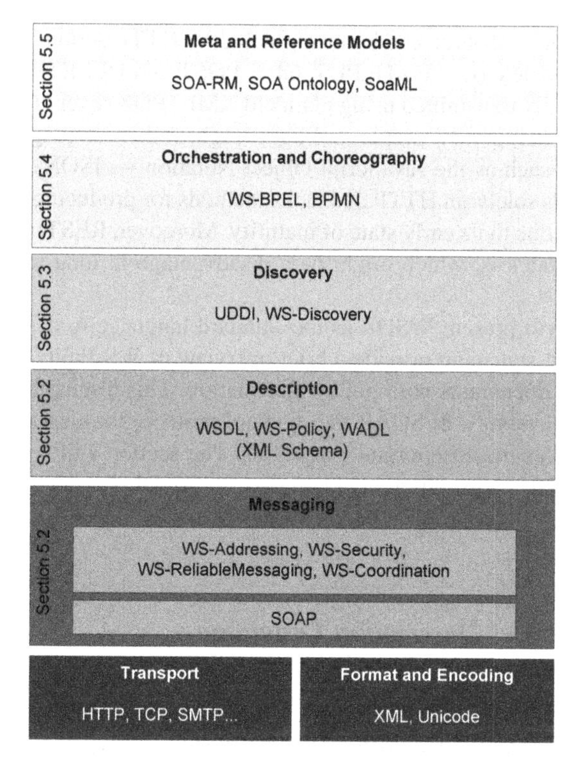

Fig. 5.1: The SOA Stack.

5.2 Service Description

There are two different architectural styles to describe Web services, namely *WS-** and *Representational State Transfer (REST)*. The WS-* style builds upon the idea that each service is accessible at one ore more locations or endpoints, each described by a Unique Resource Identifier (URI). Each service encompasses multiple operations, which in turn means that each operation is service-dependent. A message that is exchanged between operations or services consists of some metadata (header) and the message body (payload). This architectural style is typically realized by using WS-* standards over SOAP as the messaging protocol. However, other protocols such as HTTP or SMTP may be used. In fact, WS-* services can operate on top of different protocols, which enables the delivery of infrastructure services, such as security, transactions, routing and reliability. In scenarios where a service contract describing the service is already available, this architectural style can be well utilized. Nonetheless, on the downside it requires more specific infrastructure, which induces overhead and can easily be misused.

On the other hand, REST is a design idiom built around the idea of exposing services as resources identified by a URI, instead of exposing single service operations. The state of a resource can be manipulated by HTTP actions such as create, read, update and delete (i.e., HTTP PUT, GET, POST and DELETE). Data to/from a service is typically transmitted using plain old XML (POX) over HTTP, which has the advantage of being very simple and lightweight (other languages can be used in place of POX such as the JavaScript Object Notation — JSON). However, this paradigm depends solely on HTTP and lacks methods for producing well-described service contracts due to its early state of maturity. Moreover, RESTful Web services are completely stateless, which might be a disadvantage in long-running transactions.

This section will present WSDL as the standard language to describe Web services in the WS-* style, and provide a brief overview of WS-Polity, a specification to enrich WSDL documents with policy information. This discussion will be complemented by an overview of SOAP, the standard protocol for message exchange in this style. and of its most important extensions. The section will conclude with an overview of WADL — an alternative language to WSDL for describing RESTful services.

5.2.1 Web Services Description Language

The *Web Service Description Language (WSDL)* provides a means to describe a Web service contract by using XML [16]. The specification is driven by the W3C and is currently published as a recommendation in version 2.0 [27], while version 1.1 [23] is a group note, which does not have the same level of W3C endorsement. WSDL allows one to invoke the service's operations independently of the service's implementation, thereby sharing characteristics of classical interface definition languages that are commonly found in middleware technologies. Instead of utilizing naming and directory services, as commonly found in a middleware environment, Web services operate in rather decentralized environments. Therefore, in WSDL the specific location of the service needs be be explicitly defined, so that other services or applications can invoke the service's operations.

A WSDL 2.0 document is divided into two parts, namely an *Abstract part* and a *Concrete part* as shown in Figure 5.2.

The Abstract part contains one or more Interface elements and the Types element. An Interface element describes a set of operations offered by the service including the required input and output messages for each operation. For example, a Supplier service may offer service operations to create quotes, to process orders and to generate invoices. Each of these operations needs an input document and may produce an output document in reply. For example, the operation to create quotes accepts 'request for quote' documents as input and produces 'quote' documents as output (synchronous operation), while the operation to process orders only accepts 'order' documents as input but does not produce any output document (asynchronous operation).

The Types element describes the data types of each message used in an Interface using XML Schema [31]. One can either specify their own (complex) data types, such as a type to describe the 'quote' or the 'order' documents, or use built-in XML Schema simple data types, such as String or Date. However, WSDL is not bound to XML Schema, and different data type languages may also be used by using extension elements.

The Concrete part defines the implementation details necessary to access the service. This part contains one or more Binding elements, and one or more Service elements. A Binding describes the mapping between the various messages specified in an Interface and a transmission protocol such as SOAP (for WS-* services) or HTTP (for both WS-* and RESTful services). It also describes the binding style for each message, e.g., Remote Procedure Call (RPC) or Document-style. A Service element is used to link a Binding with a service endpoint, i.e., the URI where the service can be accessed. Since multiple Service elements can be defined, a service may be accessible via different endpoints, and using different transmission protocols.

WSDL 2.0 provides the capability of modeling eight different *message exchange patterns* between a service and its consumer, enriching those available in WSDL 1.1. Four patterns (In-Only, Out-Only, Robust-In-Only, Robust-Out-Only) consist of a single message being sent by the consumer to the service or vice versa, which may be replied by a fault message in case of an error. Another four patterns (In-Out, Out-In, In-Optional-Out, Out-Optional-In) consist of an initial message sent to either party which is (optionally) replied by a correlated message. These message exchange patterns abstract out binding-specific information like timing between messages, whether the pattern is synchronous or asynchronous (an aspect which was

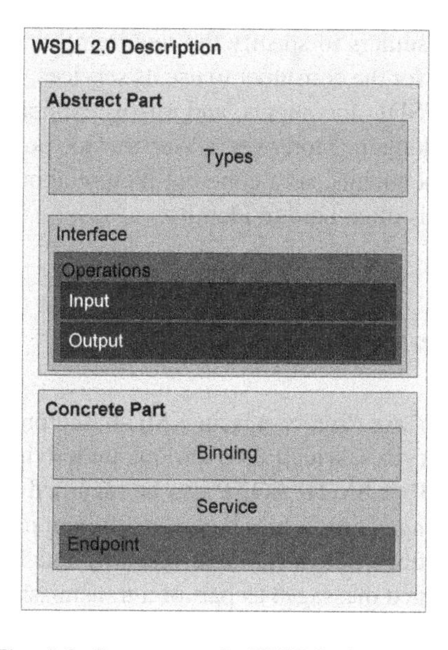

Fig. 5.2: Structure of a WSDL document.

instead predetermined in WSDL 1.1), and whether the message is sent over a single or multiple communication channels.

There are two generic approaches to develop Web services: *code-first* and *contract-first*. The former is a top-down approach requiring the development of classes in an object-oriented programming language, which can then be used to generate a WSDL document that describes the functionality offered by these classes as service operations. The latter is a bottom-up approach requiring the development of a WSDL document first, which is then used to create the skeleton of the required classes to be implemented. Existing web programming platforms offer ways to control the mapping between WSDL and object oriented languages. For JAVA, for example, open-source Web service platforms such as Apache AXIS 2 and Oracle Metro Stack can be used. An alternative in the .NET development environment is Windows Communication Foundation (WCF).

As an interface description language (IDL), WSDL offers a language-independent description of the structural aspects of a Web service, while at the same time being well supported by languages such as Java which need to provide or consume the operations described in a WSDL document. Nonetheless, WSDL is a very technical language which restricts its applicability to an IT audience only. Further, WSDL does not provide any capabilities to link service operations to business entities such as processes, objects and capabilities within an organization, nor does it provide any means to specify non-functional aspects of a Web service such as pricing, legal terms and conditions under which the service may be consumed, and service level agreements (response time, availability, etc.).

WS-Policy is a W3C recommendation [28] that aims to partly lift such limitations. It provides a mechanism for service providers to enrich WSDL artifacts with policies on various service aspects such as security or service level agreements. It can also be used by consumers to specify the requirements that must be met by a service provider in order for the consumer to use its services. However WS-Policy is too tightly-coupled to WSDL documents, and misses explicit connections to counterparts in the business domain. Moreover, a Web service policy is typically spread over different WSDL documents, so a coherent picture about what policy belongs to what kind of services is often hard to identify.

5.2.2 SOAP and Messaging Specifications

SOAP (Simple-Object Access Protocol) is an XML-based protocol for message exchange between Web services, which uses various underlying transport protocols, such as, e.g., HTTP, TCP or SMTP. SOAP can be regarded to as an asynchronous way to support the exchange of messages between different parties by implementing request-response interactions out of multiple one-way interactions. SOAP defines the format of the exchanged messages as part of a transmission between sender, receiver and potential intermediaries. However, due to the nature of the protocol, it is stateless and provides no semantic capabilities to interpret the meaning of the messages that are exchanged. W3C published SOAP in version 1.2 as a recommendation in 2007 [26].

A SOAP message consists of an envelope containing a Header and a Body as depicted in Figure 5.3:

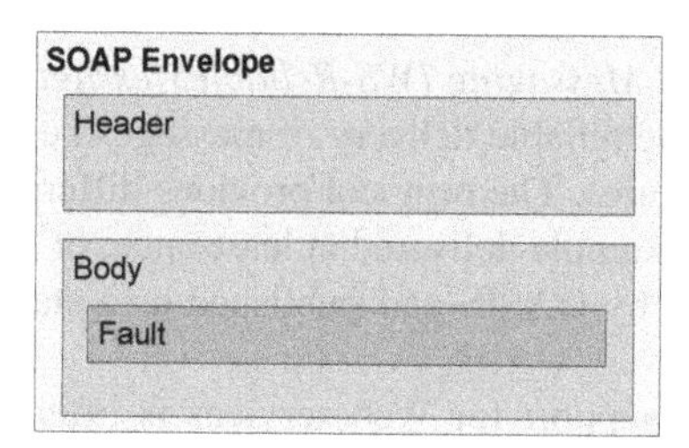

Fig. 5.3: The SOAP Envelope.

The Header carries the metadata for infrastructure services, such as security, transactions, routing and reliability. As such, it is extensible, but optional. The Body, which is mandatory, contains the payload of a message, i.e., its content, and can be interpreted by the targeted component, i.e., the consumer of a Web service or its provider. Additionally, it can contain optional fault elements which hold errors and status information for a SOAP message.

WSDL and SOAP by themselves are not enough: they just provide mechanisms to realize basic point-to-point communications. Major shortcomings are the inability to capture complex multi-party interactions, and to specify security, reliability, and transactional aspects of a service. To this purpose, a number of WS-* specifications have been defined to extend SOAP messages with various types of non-functional information. Notably, *Web Services Addressing (WS-Addressing)* provides a standard mechanism to specify message routing data within a SOAP Header. Using WS-Addressing, the network-level transport protocol (e.g., HTTP) becomes responsible only of delivering the message to a dispatcher indicated in the destination address, e.g., a Web service run-time server which can interpret the WS-Addressing metadata, and route the message to the right service instance. This information is contained in the Endpoint Reference, which is an XML structure defined in the SOAP Header. It includes the destination URI of the message and any parameter that is required to dispatch the message to the destination. It may also contain optional metadata (such as WSDL) about the service. Moreover, WS-Addressing allows request-response interactions to be decoupled from the lifetime of the underlying HTTP request/response protocol. This is achieved by specifying a special field (ReplyTo) in the SOAP Header which a service provider can use to reply to its requester. Thus, a service does not need to rely on the network-transport level to deliver a response message to the specified recipient. In this way, WS-Addressing enables long-running interactions that can span arbitrary periods of time. The specification has been standardized by the W3C Web Services Addressing Working Group in version 1.0 [25].

Another extension to the SOAP Header is *Web Services Security (WS-Security)*. This specification prescribes how to sign and encrypt SOAP messages to ensure integrity and confidentiality of a SOAP message. It also specifies how to attach security tokens such as Security Assertion Markup Language (SAML) [5] and Kerberos [19] to messages to ensure the sender's identity. The specification has been driven by OASIS and published as a standard in version 1.1 in 2006 [7].

Web Services Reliable Messaging (WS-ReliableMessaging) is also a SOAP extension that focuses on the reliable delivery of messages between distributed applications in the case of failures. The protocol provides different delivery assurances, so that a message is for example delivered at least once, or at most once. The specification has been driven by OASIS and published as a standard in version 1.1 in 2007 [9].

Finally, transactional support for Web services is offered by the *Web Service Coordination (WS-Coordination)* specification. In this context, a transaction defines two or more service operations that must all be performed with a specific workflow for the transaction to be committed successfully. WS-Coordination, published by OASIS in version 1.2 [12], provides an extensible framework for the support of coordination between distributed applications. Two alternative specifications, both building on top of WS-Coordination, can be used to define the boundaries of a transactional context. These are the *Web Service Atomic Transaction (WS-AtomicTransaction)* [10], more suitable for short-lived transactions, and the *Web Service Business Activity (WS-BusinessActivity)* [11], more suitable for long-running transactions, both published by OASIS.

Many other (minor) WS-* specifications exist, however their description falls outside the scope of this introductory chapter to SOA standards. For a consolidated overview of existing Web service standards until 2006, the interested reader is advised to visit `http://www.innoq.com/soa/ws-standards`.

5.2.3 Web Application Description Language

An alternative language to WSDL for the description of Web service contracts is the *Web Application Description Language (WADL)*. More generally, WADL aims to provide a machine-readable description of HTTP-based Web applications, which is platform- and language-independent. As such, this specification can be used to describe RESTful Web services. In this regard, WADL focuses on the resources that are needed and provided by a service and their interrelationships, contrarily to WSDL which focuses on the service operations.

A WADL document is described in XML, according to the structure illustrated in Figure 5.4.

The Grammar element provides a container for the data schemas used to describe the format of data exchanged by the service. These definitions can be included inline or referenced from an external document. While no specific data format is mandated,

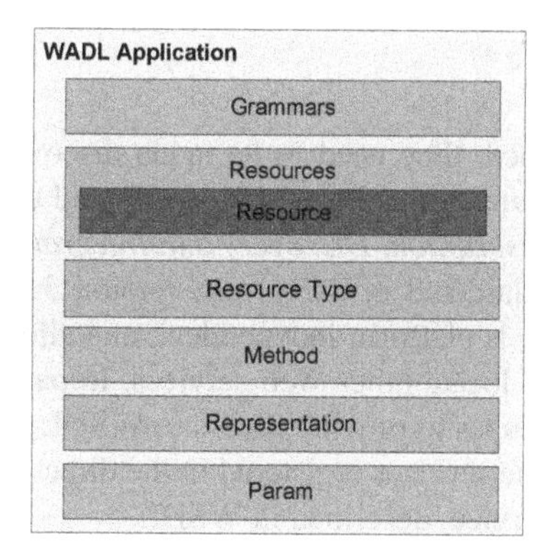

Fig. 5.4: The structure of a WADL document.

the WADL specification describes the use of XML Schema and RelaxNG [3] for this purpose.

The Resources element is where the various resources on offer or needed by a service can be specified. Each Resource is described by a URI where the resource is available, by an optional set of sub-resources, and by specifying the relationships with other (sub-)resources. Moreover, each resource has a resource type, described in the Resource Type element, which lists all the HTTP methods that can be applied to that resource (e.g., a GET or a POST), including the required inputs and outputs for each method. The available methods that can be associated with a resource are defined in the Method element, and pointed to from each resource type definition, while the formats of the input (i.e., HTTP request) and output (i.e., HTTP response) messages are those specified in the Grammar element. Each input message may have a set of HTTP parameters, which are defined in the Param element. Finally, the Representation element describes the representation of a resource's state, e.g., an XML document or a simple text document. The elements Resource Type, Method, Representation and Param are optional as their content can be directly defined within a Resource element. They are used to specify resource behavior that is expected to be supported by multiple resources.

WADL can be regarded to as being the REST counterpart to WSDL 1.1. However, WSDL 2.0 also supports the ability to describe RESTful Web services, thus the two specifications are competing with each other. For such reason, while WADL is currently a W3C member submission [30], the consortium itself has "no plans to take up work based on this submission" [1].

Similar to WSDL, there exist software packages that can generate client side software stubs from a WADL file, such as Glassfish WADL. However, WADL suffer from the same problems that affect WSDL when it comes to describing service properties. It just limits its scope to functional aspects of Web services, and addresses them at a technical level only. As such, WADL is also an IDL that can only be used by a technical audience.

5.3 Service Discovery

Once services are described, they need to be made discoverable in order for service consumers (i.e., humans or machines) to identify and use them. Moved by this purpose, the *Universal Description, Discovery and Integration (UDDI)* registry has been developed and standardized in its current version 3.0.2 by OASIS [4]. The registry is based on XML, is platform-independent, and allows service providers to register their services and locate other Web services. In particular, the registry can be queried by SOAP messages to provide access to WSDL documents, which point out how to interact with the services registered in the directory. Therefore, UDDI is tightly-related to WS-* service, described in WSDL.

The register consists of three components: i) *White Pages*, ii) *Yellow Pages* and iii) *Green Pages*. White Pages contain information about the provider of the service, such as their name and address. Thus, the registry can be queried for services for which the service provider is already known. Yellow pages categorize services based on underlying taxonomies, such as the Standard Industrial Classification,[5] the North American Industry Classification System,[6] or the United Nations Standard Products and Services Code.[7] Green Pages are comparatively technical in nature as they describe how to access the Web services being stored in the repository. In particular, they provide information about the service bindings. This includes the endpoint and parameters of the service as well as the references to specification of relevant Interfaces as described in a WSDL document. As services can have multiple bindings, a service can be related to multiple Green Pages. The OASIS Technical Committee that was responsible for developing UDDI completed their work in 2007. Thus this specification will not be further extended in the future [2].

UDDI is a technical discovery mechanism, which has no explicit links to elements in the business domain. And while it offers multiple extension mechanisms to address these lacunae, it provides limited insights into their concrete utilization. For example, pointers to other documents beyond WSDL that accompany a service can be specified in UDDI, but search capabilities that leverage those documents are limited due to a lack of standardization in regard to the required structure of these documents. In other words, UDDI does not prescribe how the information that is deposited in the White, Yellow, and Green Pages should be described.

An alternative discovery mechanism to UDDI is the *Web Service Dynamic Discovery (WS-Discovery)*, which has been standardized by OASIS in version 1.1 in 2009 [13]. Instead of utilizing a centralized registry, this specification proposes a multicast discovery protocol to locate services in a local network. In this way, services that match the requirements return a response directly to the requester. Unfortunately WS-Discovery remains a technical discovery mechanism which suffers from the same shortcomings of UDDI.

---

[5] `http://www.sec.gov/info/edgar/siccodes.htm`

[6] `http://www.census.gov/eos/www/naics`

[7] `http://www.unspsc.org`

5.4 Service Composition: Orchestration and Choreography

Often the consumption of a single service is not sufficient to fulfill a certain business goal. In this case, *service composition* becomes relevant. In service composition, services are no longer executed individually, but coordinated through business processes. A business process describes the logical and temporal order in which a number of activities have to be executed to achieve a given goal. For example, a business process for order fulfillment describes interactions among two parties: a Supplier and a Buyer, and examples of activities of this business process are "Make a request for quote," "Emit order," etc.

Business processes are typically described in a diagrammatic fashion by means of process models, where control-flow dependencies among activities can be enriched with information on the resources performing these activities and the data being exchanged by these activities. In service composition, the activities of a process model are performed by Web service operations. This means that the execution of an activity leads to the invocation of a specific Web service operation.

There are two types of process models, namely *choreography* and *orchestration*, which capture two different viewpoints in a service composition. A choreography model describes the global business process of the interactions that occur among all participants. In the order fulfillment example, this means capturing all communication activities (i.e., send and receive) that occur among the Carrier, the Suppler and the Buyer. An orchestration model is the projection of a choreography onto a single participant: it describes the order in which the various service operations need be invoked from the perspective of that participant. For example, if a communication activity in a choreography model is "Make a request for quote," this activity will correspond to the communication activity "Send request for quote" in the orchestration model of the Buyer, and to the communication activity "Receive request for quote" in the orchestration model of the Supplier. Furthermore, while choreography models focus on communication activities only, an orchestration model needs to also specify the internal activities that are required to create or consume the messages being exchanges by that particular participant. For example, activity "Send request for quote" in the Buyer's orchestration model needs be preceded by an activity "Prepare request for quote" to compile the 'request for quote' document to be sent, while activity "Receive request for quote" in the Supplier's orchestration model needs be followed by an activity "Emit quote" to create the quote that will be sent back to the Buyer.

The restriction of an orchestration model to communication activities only is called *behavioral interface*. This artifact, also known as *public process* or *business protocol*, describes the dynamic aspects of a Web service, i.e., the order in which its operations have to be provided or consumed, while hiding the internal activities which may contain business-sensitive information. Thus, a behavioral interface complements a WSDL or WADL document (called *structural interface*), which describes the static aspects of a Web service such as its operations and the content of its messages.

From a service design perspective, a choreography model could be used by business/IT analysts to frame a B2B collaboration among various parties, and then each involved party could create its own orchestration model by projecting the choreography to the activities that are related to that specific party, and then refining this model by adding internal activities. Vice versa, one could expose their own service's structural and behavioral interfaces, and once a prospective consumer for that service has been identified, a choreography involving both participants could be framed against these interfaces. However, as we will see in this section, languages such as WS-BPEL and BPMN that support orchestration and choreography modeling, do neglect important aspects such as information about delivery channels, payment details, obligations on interactions, service level agreements, and any rights that participants may have in regard to the global business process. As such, these languages alone are inadequate to be used in business contexts.

5.4.1 Web Services Business Process Execution Language

WS-BPEL (BPEL for short) is an XML-based language used to define behavioral interfaces and orchestration models, which has been standardized by OASIS in version 2.0 [8]. BPEL process models can be deployed to a BPEL engine and be automatically executed.

At its core, BPEL is an imperative programming language, since it allows the creation of process models which are essentially structured in blocks of instructions (barring a few exceptions). It supports the typical constructs of an imperative programming language such as Java. For example, one can create Scopes (like Java routines) and scoped variables (like Java local variables), Assigns (to capture internal activities that manipulate data), as well as to handle exceptions (like the Throw and Catch constructs in Java). Further, there are constructs to establish the order of activities, such as Sequence (to sequentialize activities, like the Java semicolon), Flow (to execute multiple activities at the same time, like Java Threads), and constructs to route control such as While, RepeatUntil and If-Then-Else.

In addition, BPEL has a few more features which are specifically designed to coordinate Web service operations through business process models, thus providing an advantage over traditional object-oriented languages in this respect. First, BPEL is built on top of WSDL 1.1. The language provides three communication activities (Invoke, Receive and Reply) which are directly mapped to Web service operations described in an underlying WSDL document. This is achieved by defining an extension to WSDL called *partnerLink type* where one can specify whether the BPEL process acts as a provider or consumer for the particular WSDL operation. Second, BPEL supports XML natively. BPEL variables are typed according to an XML Schema type which can be specified in an external XML Schema document or mapped to a WSDL message type (e.g., one can have a variable whose type is that of a 'request for quote' message). This eliminates the impedance between XML and object data structures, since there there is no need to unmarshal XML documents into objects (e.g., a Java object) to manipulate them. Third, BPEL offers a set of activities designed to model process-specific aspects, such as asynchronous interactions, multiple sequential or concurrent executions of a scope (ForEach), race conditions between incoming messages and timers (Pick), explicit modeling of parallelism and synchronization (Flow), transactional support (Compensate), and complex multi-party interactions (Correlations). Fourth, since BPEL relies on WSDL, it supports the full stack of WS-* specifications, such as WS-Addressing or WS-Security, which can be seamlessly integrated into BPEL processes. Finally, while the BPEL specification does not come with a visual notation to represent BPEL process models, many tools offer a diagrammatic view of a BPEL model which facilitates their understanding and editing.

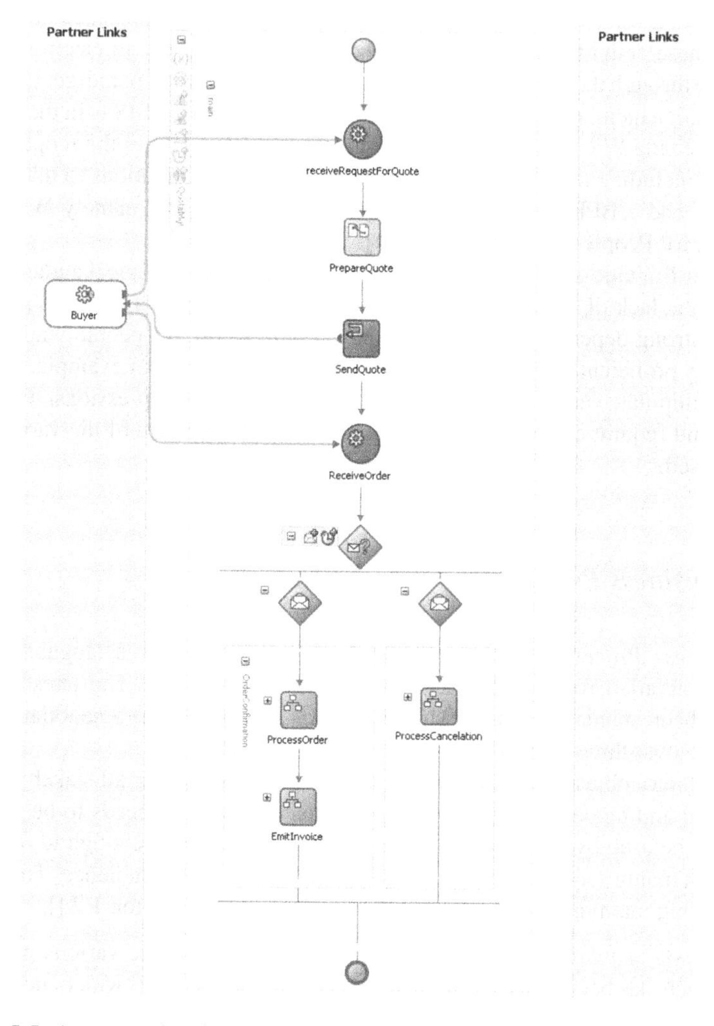

Fig. 5.5: An example of a BPEL executable process from Oracle JDeveloper.

BPEL differentiates between two levels of processes, namely *abstract process* and *executable process*. The former captures behavioral interfaces and as such it only contains Invoke, Receive and Reply activities, as well as routing constructs, and cannot be executed by a BPEL engine. The latter is used to model fully-fledged orchestrations, where communication and routing activities are interleaved with internal activities like Assigns to model the underlying process logic. As an example, Figure 5.5 shows an extract of the executable BPEL process for the Supplier participant in our order fulfillment example. For example, activity "prepareQuote" is an Assign (internal activity), while activity "SendQuote" is a Reply (communication activity). The notation used is that of the Oracle JDeveloper 11 BPEL editor.

BPEL lacks a mechanism to model user activities, i.e., those activities that require user input such as the preparation of a 'request for quote' document, or the approval of an 'invoice' document. During the execution of an instance of a BPEL process, these activities could be exposed to a process user via an electronic form accessible through the user's worklist application (an interaction paradigm typical of workflow management systems). To obviate this limitation, OASIS is in the process of standardizing WS-HumanTask [14], a specification that defines the scope of user activities, including their properties, behavior and a set of operations to manipulate their data, and a BPEL extension to support WS-HumanTask, namely WS-BPEL Extension for People (BPEL4People) [15].

Another limitation of BPEL is that it is only suitable to a technical audience. For example, the lack of a standard notation hampers its use among business analysts, while its strong dependence on XML makes it hard to use for people who are not sufficiently proficient with XML and its related specifications. For example, the only way to manipulate variables' data in BPEL is via XPath [22] expressions, which are verbose and require a deep understanding of the XML structure of the variables to be addressed.

5.4.2 Business Process Model And Notation

The *Business Process Model and Notation (BPMN)* provides a language and a graphical notation for the specification of business processes. The latest version (2.0) has been standardized by OMG in January 2011 [18] after a negotiation process lasted over three years.

The primary objective of BPMN 2.0 is to provide a means for all stakeholders to understand and model business processes. Thus, the standard needs to be intuitive enough to be used by business analysts while being expressive enough to represent complex semantics and implementation details for a technical audience. This led to the following main innovations in BPMN over its previous version 1.2 []:

- *executable semantics*, similar to BPEL, the semantics of the various modeling constructs has been formally defined, such that BPMN models which incorporate sufficient implementation details can be deployed to a native BPMN engine and be automatically executed;

- *interchange format*, the specification now provides a standard XML-based serialization format which is machine-readable, thus facilitating tool interchange while preserving semantic integrity, and automatic execution;
- support for both *orchestration and choreography modeling*, with the aim to facilitate proper integration between Business and IT people BPMN 2.0 supports new constructs specifically designed to capture various service composition viewpoints at different levels of abstraction.

Contrarily to BPEL, BPMN comes with its own visual notation, which resembles common flowcharting techniques which most users are already familiar with. However, its expressiveness goes well beyond flowcharts. The language itself is graph-oriented and not block-structured like BPEL, and provides over 100 modeling elements spanning from simple constructs (tasks, gateways, events, subprocesses) to more complex constructs (exception handling, compensation, transactional support, escalation, synchronization signals and more). BPMN also offers constructs to model organizational participants and their subdivisions (e.g., organizations, departments, roles, single persons) as well as business objects (including software systems, information and physical artifacts). Despite the richness of its meta-model, the specification defines four *conformance classes* with the purpose of facilitating the use of BPMN models and their exchange by different stakeholders and tools. These are:

1. *Process Modeling*, suitable at a business level for requirements analysis and communication purposes, this class does not contain implementation details;
2. *Process Execution*, suitable at a technical level for automation in a BPMN engine, this class must specify implementation details such as mapping to Web service operations, format of the messages being exchanged and communication protocols;
3. *BPEL Process Execution*, while the aim of BPMN is to be directly executable in its native XML format, the specification still provides a translation between BPMN modeling constructs and their BPEL counterparts. This class is a restriction of the previous class to remove those BPMN constructs that cannot be directly mapped onto BPEL constructs, due to BPEL's block-structure nature;
4. *Choreography Modeling*, suitable at a business level to frame B2B collaborations, this class contains the constructs for modeling choreographies.

In particular, BPMN offers different types of diagrams to model B2B collaborations at different levels of abstraction. *Conversation* diagrams sit at the top level of abstraction. They provide a simplified view on the order relation of the messages being exchanged among two or more business participants. Figure 5.6 shows an example of the conversation between the Supplier and the Buyer in our order fulfillment process. As we can observe, at this level of abstraction it is not possible to distinguish mandatory messages from optional ones, nor to group messages in interactions.

A Conversation diagram can be refined into a *Choreography* diagram, which specifies the logical and temporal dependencies among the various messages being exchanged, from the viewpoint of the single interactions. For example, at this level one can introduce data-driven decisions between messages, race conditions between

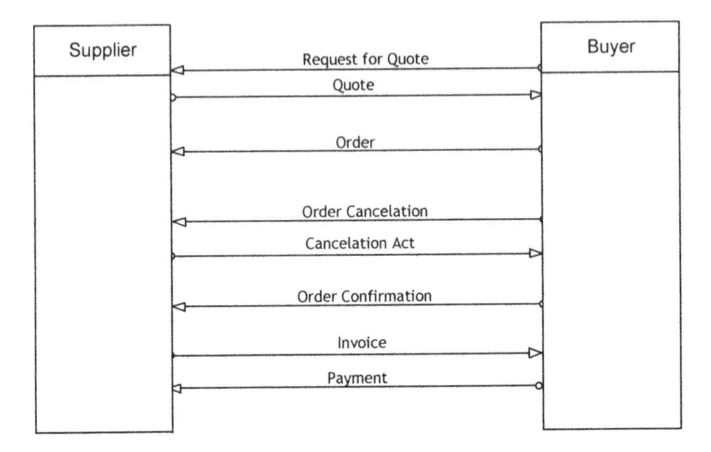

Fig. 5.6: An example of Conversation diagram in BPMN 2.0.

messages and a timer, and parallel messages. Figure 5.7 shows the refinement of the Conversation in Figure 5.6, where messages have been grouped into interactions (e.g., message 'Quote' is the reply to message 'Request for Quote'), and a race condition has been specified between messages 'Order confirmation' and 'Order cancelation.'

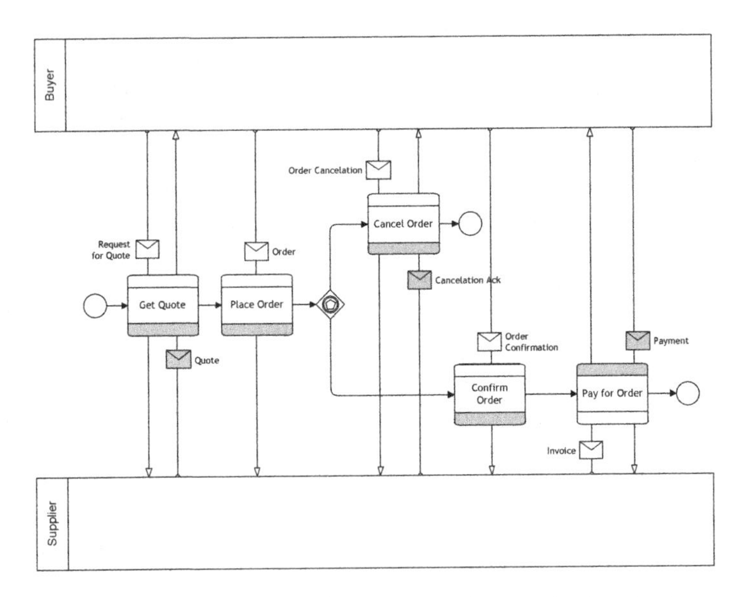

Fig. 5.7: An example of Choreography diagram in BPMN 2.0.

The next level of abstraction is represented by the *Collaboration* diagram, where the choreography of the messages being exchanged is matched by the activities that have to occur within each participant. The projection of a choreography to a specific participant can manifest itself into two types of diagrams within a Collaboration: *Public (or abstract) process* and *Private (or internal) process*. The former corresponds to a behavioral interface since it exposes communication tasks only, while the latter corresponds to an orchestration model where the internal tasks are also modeled. Figure 5.8 shows the Collaboration diagram . In particular, we can observe that the Buyer is represented as a private process (e.g., activity "Prepare request for quote" is internal) where the Supplier is represented as a public process, without showing any details of its internal realization.

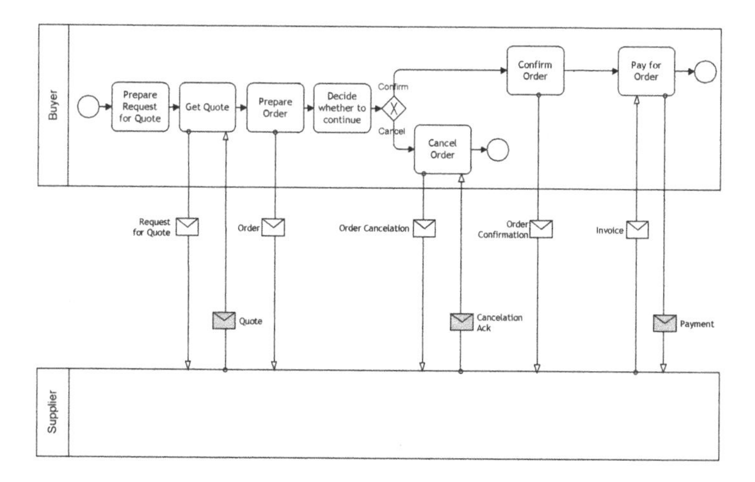

Fig. 5.8: An example of collaboration diagram in BPMN 2.0, with both private process (Buyer) and public process (Supplier).

BPMN addresses BPEL's main limitation of being skewed toward a technical audience, by providing different types of diagrams to model B2B collaborations at different abstraction levels. However, BPMN does not (yet) provide any concrete mechanism to automatically derive the abstract process model of a specific participant from a choreography model, or to automatically expose a set of choreography interactions from a private or abstract process model.

5.5 Meta and Reference Models for SOA

While by the mid 2000s the notions of service description, discovery and composition had reached a certain level of maturity, as evidenced by the proliferation of standards and research approaches in these fields, there was still little support for understanding and designing service-oriented architectures as a whole. Moved by this purpose, OASIS defined a *Reference Model for Service-oriented Architec-*

ture (SOA-RM) in 2006. Similarly, The Open Group drafted an alternative reference model in the form of an ontology for service-oriented architectures (*SOA Ontology*) in 2010. Finally, OMG developed the *Service-oriented architecture Modeling Language (SoaML)* in 2009 with the aim to facilitate the model-driven design of service-oriented architectures. These three initiatives will be the topic of this section.

A major shortcoming of these specifications, as it will be shown, is their limited support for describing complex (composite) services and processes that involve multiple interacting participants in a business network.

5.5.1 Reference Model for Service-oriented Architecture

SOA-RM is promoted by OASIS with the purpose to clarify the core notions in the SOA domain. The OASIS-RM Technical Committee was formed in 2003 with the objective to develop a model that could put clarity in the SOA domain, and meantime, foster the creation of specific service-oriented architectures. The outcome, the reference model, was approved for standardization in 2006 [6].

SOA-RM took the form of an abstract framework defining the significant relationships that exist among the entities involved in a service-oriented architecture. It consists of a minimal set of unifying concepts, axioms and relationships and is independent of specific technologies and concrete implementation details. According to this reference model, SOA is "*a paradigm for organizing and utilizing distributed capabilities that may be under the control of different ownership domains.*" Thus, SOA provides a mechanism for matching the needs of a service consumer to the capabilities offered by a service provider.

The central construct of the SOA-RM is the *service*, which is defined as "*a mechanism to enable access to one or more capabilities, where the access is provided using a prescribed interface and it is exercised consistent with constraints and policies as specified by the service description.*" Around this notion, three dynamic concepts are of relevance in SOA-RM:

1. Visibility: One has to ensure that the service provider and consumer are able to see each other. This is true for any type of relationship between service provider and consumer regardless of the specific instantiation of these entities (e.g., applications or humans). Visibility is influenced by three factors: awareness, willingness and reachability. The service consumer needs to be *aware* of the existence of the service provider; both need to be *willing* to interact with each other and the service provider needs to be *reachable* by the service consumer.
2. Interaction: The interaction is dependent on the visibility because service consumer and provider cannot interact if they do not see each other. In order to understand what is needed for interacting with a service, the description needs to cover an information model and a behaviour model. The former model details the information that will be exchanged when interacting with the service,

whereas the latter model depicts "*the actions that may be invoked against the service.*"
3. Real World Effect: The result of any interaction is a *real world effect*, which can instantiate itself as the information that is returned by the service or a change in the state of entities that are involved in the interaction.

The above concepts are considered to be dynamic aspects as they relate to the interactions with services. Besides, SOA-RM defines three additional aspects that relate to services themselves. These are:

1. Service Description: A service description contains all the information that is needed for a consumer to decide if the service is of relevance for the specific context. Additionally, the description contains information that is needed to actually use the service. Thus, a service description facilitates dynamic aspects such as visibility and interaction.
2. Contract and Policy: A contract is a formal agreement between two or more parties, whereas a policy constrains the use, deployment, or description of a specific entity as defined by its owner.
3. Execution Context: The execution context includes all technical and business elements that are somehow of relevance for the interaction between service providers and consumers, such as infrastructure elements and process entities, as well as any policies and contracts that may be in force. Thus, each interaction has a specific execution context.

SOA-RM also provides a notion of Process Model. Accordingly: "*the process model characterizes the temporal relationships and temporal properties of actions and events associated with interacting with the service.*" However this aspect is intentionally underspecified. For example, while orchestration and choreography may be part of a process model, the orchestration of multiple services is not addressed in the reference model. This is because the focus of this initiative is on modeling what services are and what key relationships are involved in modeling services.

5.5.2 Service-oriented Architecture Ontology

Moved by a similar purpose to SOA-RM, the SOA Ontology has been standardized by The Open Group in 2010 [21]. This standard relies on the following definition of *service*, which is purposefully agnostic to the context in which a service is applied, i.e., a business domain or an IT domain:

> *A service is a logical representation of a repeatable activity that has a specified outcome. It is self-contained and is a 'black box' to this customers.*

As such, the SOA Ontology aims to be used by different user categories: i) business people and system/software designers, to clarify on the SOA concepts and how they can be implemented within an organization; ii) solution architects, to provide metadata for architectural artifacts; and iii) architecture methodologists, to provide a component of SOA meta-models.

The ontology itself is specified in the Web Ontology Language (OWL) [29] and consists of a set of classes that capture the core concepts of SOA, their properties and interrelations. There are classes to define the *effects* of a service interaction; the various *elements* involved in an interaction (i.e., people, organizations, entire systems or single tasks), the *events* they generate, and the *policies* that may apply. Other classes describe specific aspects of a service, such as a *service interface* specifying what type of information a service can provide, and a *service contract*, specifying the effects of an interaction with that service, such that guaranteed service level agreements can be specified. Finally, a set of classes defines how services may be *composed* (i.e., via an orchestration, a choreography or a more abstract collaboration).

Although this ontology defines the relationships among the different classes, their specific application remains context-generic. Applied to a specific domain or context the ontology needs to be populated by "SOA OWL class instances of things in that domain." The ontology can be extended by importing other ontologies or classes.

5.5.3 Service-oriented Architecture Modeling Language

SoaML is an open-source UML profile and meta-model for designing service-oriented architectures, which is driven by OMG [17]. The objective of SoaML is to offer a comprehensive language to support service design within a model-driven development approach. This initiative was moved by the observation that existing enterprise architecture standards such as TOGAF [20], or general-purpose modeling languages such as UML, were insufficient for capturing all required concepts proper of a SOA in a standard and unambiguous way.

Instead of taking one specific perspective on service design, SoaML accommodates different viewpoints to offer a consistent and cohesive approach to describing service-oriented architectures. The meta-model includes constructs to identify services, their interdependencies and requirements, for specifying service functional capabilities, and the protocols and message exchange patterns, as well as non-functional aspects such as service consumers and providers, consumer expectations and the policies for using and providing services. The standard also provides an extensibility mechanism to integrate SoaML design artifacts with other OMG meta-models, such as BPMN 2.0, and a mechanism to define classification schemes for services.

In line with the model-driven development paradigm proper of UML, one can automatically generate code stubs from the various SoaML artifacts, e.g., deriving the code for a Web service provider or for its client. However, the specification neither limits SOA to be applicable to a purely technical level, nor a service to be purely realized by software components.

5.6 Summary and Discussion

This chapter provided an overview of the most mature standards in the SOA domain. First, it described two different architectural styles to design services, namely WS-* and REST, and their respective specifications, namely WSDL, SOAP and their extensions, and WADL. Subsequently, the chapter covered two discovery mechanisms for Web services, i.e., UDDI and WS-Discovery, before shifting the focus from single services to service composition, where multiple services can be organized together to provide business value. The chapter introduced different viewpoints in service composition, i.e., behavioral interfaces, orchestration and choreography models, and discussed two different standards that can be used for capturing such views, namely WS-BPEL and BPMN. The chapter completed the discussion on SOA standards by overviewing three specifications that provide guidance for the design of service-oriented architectures, namely SOA-RM, SOA Ontology and SoaML.

This chapter also highlighted the limitations of these standards with regard to their capability of describing service properties, and especially non-functional aspects of services and business processes, which are relevant in a business context. It is worth noting that there has been an effort of recent standardization initiatives such as SoaML and SOA Ontology, to move SOA away from a solely-technical domain, despite their standardization bodies traditionally have a technical focus. SoaML, for example, explicitly includes both technical and business aspects in its specification, and so it does the SOA Ontology, which even uses a car-wash example that is clearly non-technical. However, while providing valuable distinctions among the various concepts of a SOA, these specifications do not provide sufficient depth to allow a universal description of services. They do not consistently describe the functional aspects of a service, in light of the link with the various entities that are involved in the provision and consumption of such services, with the channels through which a service can be provisioned and consumed, and especially with non-functional aspects such as pricing, legal terms and service level agreements. These limitations have triggered the USDL initiative, which aims to provide a language for describing both technical and business services.

References

1. Lafon, Y.: Team Comment on the "Web Application Description Language" Submission (2009). URL http://www.w3.org/Submission/2009/03/Comment(RetrievedAugust2011)
2. McRae, M.: Closure of OASIS UDDI Specification TC. Web site (2008). URL http://lists.oasis-open.org/archives/uddi-spec/200807/msg00000.html
3. OASIS: RELAX NG Specification. OASIS Committee Specification (2001)
4. OASIS: UDDI Version 3.0.2. UDDI Specification Technical Committee Draft, OASIS (2004)
5. OASIS: Assertions and Protocols for the OASIS Security Assertion Markup Language (SAML) V2.0. OASIS Standard (2005). URL http://docs.oasis-open.org/security/saml/v2.0/saml-core-2.0-os.pdf

6. OASIS: Reference Model for Service Oriented Architecture 1.0. OASIS Standard, OASIS (2006). URL http://docs.oasis-open.org/soa-rm/v1.0/soa-rm.pdf
7. OASIS: Web Services Security: SOAP Message Security 1.1 (WS-Security 2004). OASIS Standard, OASIS (2006). URL http://www.oasis-open.org/committees/download.php/16790/wss-v1.1-spec-os-SOAPMessageSecurity.pdf
8. OASIS: Web Services Business Process Execution Language (WS-BPEL) Version 2.0. OASIS Standard, OASIS (2007). URL http://docs.oasis-open.org/wsbpel/2.0/OS/wsbpel-v2.0-OS.pdf
9. OASIS: Web Services Reliable Messaging (WS-ReliableMessaging) Version 1.1. OASIS Standard, OASIS (2007). URL http://docs.oasis-open.org/ws-rx/wsrm/200702/wsrm-1.1-spec-os-01.pdf
10. OASIS: Web Services Atomic Transaction (WS-AtomicTransaction) Version 1.2. OASIS Standard wstx-wsat-1.2-spec-os, OASIS (2009)
11. OASIS: Web Services Business Activity (WS-BusinessActivity) Version 1.2. OASIS Standard, OASIS (2009)
12. OASIS: Web Services Coordination (WS-Coordination) Version 1.2. OASIS Standard, OASIS (2009)
13. OASIS: Web Services Dynamic Discovery (WS-Discovery) Version 1.1. OASIS Standard, OASIS (2009)
14. OASIS: Web Services – Human Task (WS-HumanTask) Specification Version 1.1. OASIS Committee Specification (2010). URL http://docs.oasis-open.org/bpel4people/ws-humantask-1.1.html
15. OASIS: WS-BPEL Extension for People (BPEL4People) Specification Version 1.1. OASIS Committee Specification (2010). URL http://docs.oasis-open.org/bpel4people/bpel4people-1.1.html
16. OMG: Extensible Markup Language (XML) 1.1 (Second Edition). W3C Recommendation (2006). URL http://www.w3.org/TR/xml11
17. OMG: Service oriented architecture Modeling Languate (SoaML) - Specification for the UML Profile and Metamodel for Services (UPMS). OMG Finalisation Task Force Beta 2 document (2009)
18. OMG: Business Process Model and Notation (BPMN) - Version 2.0. OMG Standard (2011)
19. The Internet Society: The Kerberos Network Authentication Service (V5). RFC 4120 (2005)
20. The Open Group: The Open Group Architecture Framework (TOGAF) version 9 (2009). URL http://www.opengroup.org/togaf
21. The Open Group: Service-Oriented Architecture Ontology (2010)
22. W3C: XML Path Language (XPath) Version 1.0. W3C Recommendation (1999). URL http://www.w3.org/TR/xpath
23. W3C: Web Services Description Language (WSDL) 1.1. Web Site (2001). URL http://www.w3.org/TR/wsdl
24. W3C: Web Services Glossary – Web service. W3C Working Group Note 11 (2004). URL http://www.w3.org/TR/ws-gloss/
25. W3C: Web Services Addressing 1.0 – Core. W3C Recommendation (2006). URL http://www.w3.org/TR/2006/REC-ws-addr-core-20060509/
26. W3C: SOAP Version 1.2 Part 1: Messaging Framework (Second Edition). W3C Recommendation (2007). URL http://www.w3.org/TR/2007/REC-soap12-part1-20070427/
27. W3C: Web Services Description Language (WSDL) Version 2.0 Part 1: Core Language. W3C Recommendation (2007). URL http://www.w3.org/TR/2007/REC-wsdl20-20070626/
28. W3C: Web Services Policy 1.5 - Framework. W3C Recommendation (2007). URL http://www.w3.org/TR/ws-policy/
29. W3C: OWL 2 Web Ontology Language. W3C Recommendation (2009). URL http://www.w3.org/TR/owl-overview/
30. W3C: Web Application Description Language. W3C Member Submission (2009). URL http://www.w3.org/Submission/wadl/

31. W3C: XML Schema Definition Language (XSD) 1.1 Part 1: Structures. W3C Recommendation (2011). URL http://www.w3.org/TR/xmlschema11-1/

Chapter 6
Semantic Web Services Fundamentals

Stijn Heymans, Jörg Hoffmann, Annapaola Marconi, Joshua Phillips, and Ingo Weber

Abstract The research area of Semantic Web Services investigates the annotation of services, typically in a SOA, with a precise mathematical meaning in a formal ontology. These annotations allow a higher degree of automation. The last decade has seen a wide proliferation of such approaches, proposing different ontology languages, and paradigms for employing these in practice. The next chapter gives an overview of these approaches. In the present chapter, we provide an understanding of the fundamental techniques, from Artificial Intelligence and Databases, on which they are built. We give a concise, ontology-language independent, overview of the techniques most frequently used to automate service discovery and composition.

6.1 Introduction

Besides the SOA approach just discussed, a second thread of research on service descriptions and their exploitation is the field of *Semantic Web Services*. The basic idea is to describe services in the context of the Semantic Web, annotating them with a

Stijn Heymans, Joshua Phillips
SemanticBits, 13921 Park Center Road, Suite 420, Herndon, VA 20171, USA, e-mail: stijn.heymans@semanticbits.com, e-mail: joshua.philipps@semanticbits.com

Jörg Hoffmann
INRIA Nancy – Grand Est, Equipe MAIA, 615 rue du Jardin Botanique, 54600 Villers-les-Nancy, France, e-mail: joerg.hoffmann@inria.fr

Annapaola Marconi
Fondazione Bruno Kessler, via Sommarive 18, 38121, Povo, Trento, Italy,
e-mail: marconi@fbk.eu

Ingo Weber
The University of New South Wales, School of Computer Science & Engineering, K17, The University of New South Wales, Sydney, NSW 2052, Australia,
e-mail: ingo.weber@cse.unsw.edu.au

description using a formal ontology to express their precise mathematical meaning. This enables rich support for handling services, support that is not possible based on less information-rich descriptions such as WSDL. In other words, the ontological annotation serves to "explain" the services in more formal detail, and these details allow a higher degree of automation.

The Semantic Web Services research area started in the early 2000s, amongst others with the seminal (and emblematic) paper by McIlraith et al. [53]. As presented in that paper, the main goal of Semantic Web Services approaches is the automation of service discovery and service composition in a SOA. The literature of the last decade has seen a wide proliferation of such approaches. These differ in terms of the ontology languages proposed — there is a wide range of possible formalizations and implementations (language syntaxes) — and in terms of the paradigms proposed for employing these in practice. Chapter 7 is dedicated to providing an overview of these approaches, outlining their commonalities and differences. The present chapter provides an understanding of the fundamental techniques on which they are built. These techniques are drawn from a range of research areas, prominently including Artificial Intelligence and Databases.

We give a concise overview of the techniques most frequently used to automate service discovery and composition. To make this accessible, most of our discussion is informal. Where it is formal, we base it on simple mathematical notations common in the respective areas, thus disregarding the intricacies of practical ontology languages and their implementations.

We cover *Description Logics*, *Logic Programming*, *Planning for Service Chaining*, and *Planning for Service Interactions*. Table 6.1 provides an overview of these, along with their basic distinguishing properties.

Table 6.1: Survey of techniques underlying Semantic Web Services.

Approach	Purpose	Annotation: What is annotated?	Annotation: In which form?	Output of Approach
Description Logics	Discovery	WS input/output	1 logical formula	Set of WS matching query
Logic Programming	Discovery	WS input/output	1 logical formula	Set of WS matching query
Planning for Service Chaining	Composition	Input&prerequisites, output&effects	2 logical formulas	WS chain (composition template)
Planning for Service Interactions	Composition	WS Interface	State transition system	Executable WS composition

Discovery is similar to Web search: given a discovery query, the technology supports the detection of a (potentially ranked) subset of Web services matching the query. Description Logics and Logic Programming offer to improve the precision and recall of such discovery by formulating the query in a more precise way than with keywords. The discovery query will state, formalized in the respective logic, the kind of input the user can provide to the service, and the kind of output the user expects from the service. This query will be matched against Web services whose input/output is annotated in the same logic, thus allowing to find the subset of services relying on the available input, and delivering the desired output.

Composition is a more complex task, where we not only wish to find a suitable Web service, but where we wish to create, using a subset of already available Web services as atomic building blocks, a more complex Web service providing a more useful functionality. This is a form of programming, thus doing it automatically is quite a challenge. Planning for Service Chaining relaxes this challenge by viewing Web services as one-shot applications, taking into account their input/output typing and high-level properties (such as, available credit), but ignoring technical details such as their interaction patterns and any data transformations needed. Thus the approach provides only a composition template, pre-selecting and arranging a subset of relevant Web services. Planning for Service Interactions tackles the composition challenge in full, delivering an executable software artefact. Accordingly, the approach requires more detailed annotations, involving in particular a specification of how to interact with the Web service. This comes in the form of a transition system, i.e., a kind of abstract program similar to BPEL abstract processes [58].

In what follows, we focus in turn on Description Logics (Section 6.2), Logic Programming (Section 6.3), Planning for Service Chaining (Section 6.4), and Planning for Service Interactions (Section 6.5). For each, we provide a detailed summary at a non-technical level; some technical details are given in separate sub-sections that the reader not interested in such details may skip. We illustrate the approaches using examples, taken from applications where suitable.[1]

6.2 Description Logics

Description logics (DLs) is the most prominent formalization underlying the Semantic Web, and Semantic Web Services. In what follows, we first provide an approach synopsis giving the main facts in an informal way, then we include a detailed treatment of the basic DL formalities. The reader not interested in technical details may skip the latter sub-section.

6.2.1 Approach Synopsis

Description Logics (DLs) are a family of logical formalisms widely used for knowledge representation, e.g., for the representation of terminologies in application domains such as healthcare — witnessed by terminologies such as OpenGALEN[2] and SNOMED CT®.[3] Its basic language features include the notions of *concepts* and

[1] Since the approaches are quite different in their underlying intention and scope, there is no one unifying example suitable to illustrate them all.

[2] `http://www.opengalen.org/`

[3] `http://www.ihtsdo.org/snomed-ct/`

roles which are used to define the relevant concepts and relations in some (application) domain. Different DLs can then be identified, among others, by the set of constructors that are allowed to form complex concepts or roles.

The combination of a formal well-understood semantics and the availability of practical reasoners,[4] has led to the adoption of DLs as the formal underpinning of ontology languages such as OWL [59] on the Semantic Web or in the use of Semantic Web Services [53]. In the context of Semantic Web Services they are used with different purposes, e.g., to express the background domain ontologies, as the language for pre- and post-condition, input and output descriptions, ...

We give a small example showing the benefits of using (simple) Description Logics in the context of Web services. Further note that we make the simplifying assumption that Web services have one operation and that when given an input, they just give an output (no choreography).

Consider two services *S1-cure* and *S2-cause*. *S1-cure* takes as input a particular *Allergy* (one can see this as a SNOMED CT® concept) and returns as output a *Substance* (again a SNOMED CT® concept) that can alleviate the symptoms of the allergy. We write this as follows (using notation from [42]):

S1-cure:
INPUT *x Allergy*
OUTPUT *y Substance*

In words, given an allergy *x*, return the substance *y* that could resolve its symptoms. Note that *Allergy* and *Substance* are simple DL concept names. The service *S2-cause* takes exactly the same input and output but returns a substance that could be a cause of the allergy symptoms.

S2-cause:
INPUT *x Allergy*
OUTPUT *y Substance*

Two immediate issues arise:

1. Assume a user has a request *Q* for a service that inputs a certain allergy and would like to know a possible cause. Both Web services (described by WSDL if you want, where *Allergy* and *Substance* would be specific message types), are described in identical ways but do 2 entirely different things. Only *S2-cause* would be a suitable Web service satisfying the user's request, but both *S1-cure* and *S2-cause* would be returned as suitable Web services for the user based on the input and output types. Note that according to [42] this occurs commonly in the biomedical domain where services often have as input and output just strings.
2. Assume a user has a request *Q* that involves finding a service that takes a penicillin allergy as input and gives the resolving substances. Even though every penicillin allergy is an allergy and thus *S1-cure* would be able to resolve the

[4] Several reasoners for expressive DLs have emerged since the 80s, e.g., Racer [34], FaCT [41], Pellet [70], and HermiT [69].

issue, it would not match as the input type of *S1-cure* does not correspond to the input of the request.

Using a background domain ontology that indicates that *PCNAllergy* is a subclass of *Allergy* (as does SNOMED CT®), in addition to the above descriptions of *S1-cure* and *S2-cause*, would rightfully propose *S1-cure* as a solution to the request Q where Q is as follows:

Q:
INPUT x *PCNAllergy*
OUTPUT y *Substance*

Indeed, every *PCNAllergy* would be an *Allergy* according to the ontology, such that *S1-cure* which takes allergies as input could propose substances. Formally, a statement "every *PCNAllergy* is a *Allergy*" is called a *DL (subclass) axiom*, similar to a simple *is-a* specification in conceptual modeling.

Note that also *S2-cause* is applicable: the knowledge that the user needs a resolution and not a causing substance is not made explicit yet (neither is it explicit that *S1-cure* provides a curing substance nor that *S2-cause* provides a causing substance). Making this explicit can be done by pre- and post-conditions.

6.2.2 AI Formalism

We introduce a canonical version of a Description Logic; the reader can easily skip this technical section if need be.

The semantics of DLs is given by first-order style interpretations $\mathcal{I} = (\Delta^{\mathcal{I}}, \cdot^{\mathcal{I}})$ where $\Delta^{\mathcal{I}}$ is a non-empty domain and $\cdot^{\mathcal{I}}$ is an interpretation function. The basic building blocks are concept names, role names (abstract or concrete, possibly inverted in case of the former), data types, and nominals. For more details, we refer the reader to [8].

Based on those building blocks, we define *concept expressions* as in Table 6.2, where A is a concept name, R, S are abstract roles, T is a concrete role name, $d \in \mathbf{D}$ is a data type, and C, D are concept expressions.

A DL *knowledge base* is a set of *axioms*, where an axiom is of one of the following three types, respectively indicating subset relations between concept expressions, subset relations between roles, and transitivity of roles.

- *terminological axioms* $C \sqsubseteq D$ with C and D concept expressions,
- *role axioms* $R \sqsubseteq S$ where R, S may be inverse roles with the underlying roles both abstract or both concrete, and
- *transitivity axioms* $\mathsf{Trans}(R)$ for an (inverse) abstract role.

Table 6.2: Syntax and Semantics of DL Constructs

Construct name	Syntax	Semantics
concept conj.	$C \sqcap D$	$(C \sqcap D)^{\mathcal{I}} = C^{\mathcal{I}} \cap D^{\mathcal{I}}$
concept disj.	$C \sqcup D$	$(C \sqcup D)^{\mathcal{I}} = C^{\mathcal{I}} \cup D^{\mathcal{I}}$
negation	$\neg C$	$(\neg C)^{\mathcal{I}} = \Delta^{\mathcal{I}} \setminus C^{\mathcal{I}}$
exists restriction	$\exists R.C$	$(\exists R.C)^{\mathcal{I}} = \{x \mid \exists y : (x,y) \in R^{\mathcal{I}} \text{ and } y \in C^{\mathcal{I}}\}$
value restriction	$\forall R.C$	$(\forall R.C)^{\mathcal{I}} = \{x \mid \forall y : (x,y) \in R^{\mathcal{I}} \Rightarrow y \in C^{\mathcal{I}}\}$
atleast restriction	$\geq nS.C$	$(\geq nS.C)^{\mathcal{I}} = \{x \mid \#\{y \mid (x,y) \in S^{\mathcal{I}} \text{ and } y \in C^{\mathcal{I}}\} \geq n\}$
atmost restriction	$\leq nS.C$	$(\leq nS.C)^{\mathcal{I}} = \{x \mid \#\{y \mid (x,y) \in S^{\mathcal{I}} \text{ and } y \in C^{\mathcal{I}}\} \leq n\}$
data type exists	$\exists T.d$	$(\exists T.d)^{\mathcal{I}} = \{x \mid \exists y : (x,y) \in T^{\mathcal{I}} \text{ and } y \in d^{\mathbf{D}}\}$
data type value	$\forall T.d$	$(\forall T.d)^{\mathcal{I}} = \{x \mid \forall y : (x,y) \in T^{\mathcal{I}} \Rightarrow y \in d^{\mathbf{D}}\}$

Traditionally, a knowledge base contains also assertional statements[5] such as $C(a)$ (or $R(a,b)$) which intuitively means that the individual a is an instance of C (a is related to b by means of the role R).

Terminological and role axioms express a subset relation: an interpretation $\mathcal{I}$ *satisfies* an axiom $C_1 \sqsubseteq C_2$ ($R_1 \sqsubseteq R_2$) if $C_1^{\mathcal{I}} \subseteq C_2^{\mathcal{I}}$ ($R_1^{\mathcal{I}} \subseteq R_2^{\mathcal{I}}$). An interpretation satisfies a transitivity axiom $\mathsf{Trans}(R)$ if $R^{\mathcal{I}}$ is a transitive relation. An interpretation is a *model* of a knowledge base Σ if it satisfies every axiom in Σ. A concept C is *satisfiable* w.r.t. Σ if there is a model $\mathcal{I}$ of Σ such that $C^{\mathcal{I}} \neq \emptyset$.

As indicated above, DLs are useful for expressing knowledge in the healthcare domain. For example, the *Systematized Nomenclature of Medicine–Clinical Terms* (SNOMED CT®) [1] is a reference terminology for clinical data that can be seen as a particular DL knowledge base.

Example 6.1. Consider the example knowledge base Σ in Table 6.3, loosely inspired by SNOMED CT®.

Table 6.3: SNOMED CT® Fragment Amoxicillin

(1)	$\mathit{AmoxicillinTablet} \sqsubseteq \exists \mathtt{hasActiveIngredient}.\mathit{Amoxicillin}$
(2)	$\mathit{Amoxicillin} \sqsubseteq \mathit{Penicillin}$
(3)	$\mathit{SafeForPCNAllergies} \sqsubseteq \forall \mathtt{hasActiveIngredient}.\neg \mathit{Penicillin}$

The example indicates in axiom (1) that amoxicillin tablets have an active ingredient that is an amoxicillin, which in turn is a penicillin by (2). According to (3), we collect in the concept *SafeForPCNAllergies* all elements that have only active ingredients that are not penicillins. A possible interpretation is $\mathcal{I}$ with $\Delta^{\mathcal{I}} = x, y$ and

[5] The assertional statements in a knowledge base are also referred to as the *ABox*, while the non-assertional statements, the *terminological* statements, are referred to as the *TBox*.

with $AmoxicillinTablet^{\mathscr{I}} = x$, $\texttt{hasActiveIngredient}^{\mathscr{I}} = (x,y)$, $Amoxicillin^{\mathscr{I}} = Penicillin^{\mathscr{I}} = y$, $SafeForPCNAllergies = \emptyset$. This interpretation is clearly a model of Σ. If we are interested in knowing whether amoxicillin tablets are safe to take when you have a penicillin allergy, one could check whether $\Sigma \models AmoxicillinTablet \sqsubseteq SafeForPCNAllergies$, i.e., does *SafeForPCNAllergies* subsumes *AmoxicillinTablet* w.r.t. Σ. As we have a model $\mathscr{I}$ where $AmoxicillinTablet^{\mathscr{I}} \not\subseteq SafeForPCNAllergies^{\mathscr{I}}$, we can answer this negatively.

We could use such reasoning over a domain ontology (SNOMED CT®) in the context of discovery of Web services. For example, a user goal might be to find all services that output medication that is safe for patients with penicillin allergies. A service that has its output modeled using the concept *SafeForPCNAllergies* would satisfy this exactly. One can for example deduce using standard DL reasoning that services that output Amoxicillin tablets are not matching this goal (as *AmoxicillinTablet* is not subsumed by *SafeForPCNAllergies*) above.

6.3 Logic Programming

Logic Programming (LP) is the main alternative to DL, for handling Semantic Web Services. As before, we first provide an approach synopsis, then include a detailed treatment of the basic LP formalities (which the reader not interested in technical details may skip).

6.3.1 Approach Synopsis

Whereas expressing knowledge in Description Logics revolves around atomic concepts and roles to construct more expressive concept expressions, Logic Programming traditionally has as its basic building blocks first-order atoms. In particular, writing a DL-concept *4Door* as an atom, results in $4Door(X)$ where X is a variable and *4Door* is in that context a predicate for example identifying an item X as a 4-door car. Similarly, we have that DL roles such as *car_available* correspond to atoms $car\_available(X,Y)$ indicating that rental company X has a car Y available.

Whereas traditional DLs have only concepts and roles as basic building blocks, LP allows usually as well for n-ary atoms such as $travels(From, To, Name)$ indicating that *Name* travels from *From* to *To*.

Combining these building blocks in DLs is done by defining syntactical structures such as exist restrictions $\exists car\_available.4Door$ (the members of which all have a 4-door car available); in LP on the other hand we combine atoms by simple conjoining or disjoining them. For example, $car\_available(avis,X), 4Door(X)$ indicates that *avis* has some (X) car available that is a 4-door car.

Actually expressing knowledge is then done very similarly in LP as in DLs by means of an IF-THEN structure (recall the example in the previous section that an

PCNAllergy is-a particular *Allergy*):

$$rentAt(X) \leftarrow car\_available(X,Y), 4Door(X)$$

which indicates that if X has a 4-door car Y available, then one wants to rent at X. Note that such rules would be typical rules in a Virtual Travel Agency discovery scenario to express the preferences of a prospective renter.

In the presence of negation, different semantics have been proposed historically for such sets of rules (viz., *logic programs*), two of the prominent ones being *well-founded* semantics [28] and the *answer set semantics* [29].

Logic Programming, in particular the answer set methodology has been successfully applied in problem areas such as planning [47], configuration and verification [72], diagnosis [22], and database repairs [6]. Moreover, several *answer set solvers*, i.e., systems that return the answer sets of the program, have reached a mature stage of development. E.g., SMODELS [57] and DLV [46].

Moreover, as the envisioned basis of future information systems, the Semantic Web is a fertile ground for deploying AI techniques, and in turn raises new research problems in AI. As a prominent example, the combination of rules with Description Logics (DLs), which is central to the Semantic Web architecture, has received high attention over the past years, with approaches such as *Description Logic Programs* [33], *DL-safe rules* [56], *r-hybrid KBs* [67], $\mathcal{DL}$+*log* [68], *MKNF KBs* [55], *Description Logic Rules and ELP* [43], and *dl-programs* [24].

In the area of Semantic Web Services, Logic Programming plays an important role in for example the Web Service Modeling Language (WSML), [21]. As the backbone of WSML it plays a similar role as Description Logics, the backbone of OWL-DL. Indeed, it is used in the expression of background ontologies, the expression of goals, pre-conditions, post-condition, capabilities of services etc.

We summarize some differences between the paradigms of Description Logics and Logic Programming in Table 6.4. Note that there a lot of variations of the both, so Table 6.4 should be seen more as the general case than as covering all cases.

- Logic Programming in general has a minimal model semantics. In other words, one is interested in the minimal model (w.r.t. the subset relation) of a set of rules.
- A consequence of the minimal model semantics for LP, is that reasoning in LP is *nonmonotonic* vs *monotonic* in DLs. In Logic Programming, entailments of a logic program might not hold any longer after rules or facts where added to that logic program.
- Logic programming has a *closed domain assumption*: only the constants appearing in a logic program are relevant for the construction of models. This in contrast with Description Logics, which are generally speaking a fragment of first-order logic and have an *open domain assumption* where any non-empty domain can potentially serve as the universe/domain of the knowledge base.

Table 6.4: Differences Description Logics and Logic Programming.

DL	LP
model semantics	minimal model semantics
monotonic	nonmonotonic
open domain	closed domain

6.3.2 AI Formalism

In order to make the exposition as simple as possible, we define the language of answer set programming as a canonical example of a logic programming paradigm; the reader can easily skip this technical section if need be.

Note that in the case of logic programs without negation, the answer set semantics coincides with the canonical minimal model semantics, and in the case of stratified logic programs, the well-founded semantics coincides with the answer set semantics.

Terms, atoms, literals are defined as usual (see [9] for details); an *extended literal* is a literal l or a *naf-literal* not l, i.e., a literal preceded with the *negation as failure* symbol *not*.

A *(logic) program (LP)* is a countable set of *rules* $\alpha \leftarrow \beta$, where α and β are finite sets of extended literals, respectively called the *head* and *body* of the rule. The body of a rule is considered to be a conjunction of extended literals (denoted as a comma-separated list) and the head as a disjunction of extended literals (denoted as a $\vee$-separated list). The *positive part* of a set of extended literals γ is $\{\gamma^+ \equiv l \mid l \in \gamma, l \text{ literal}\}$ and the *negative part* is $\{\gamma^- \equiv l \mid not\ l \in \gamma\}$.

A *ground* atom, (extended) literal, rule, or program does not contain variables.

All following definitions in this section assume ground programs and ground (extended) literals. answer set of $gr(P)$. The *Herbrand Base* $\mathscr{B}_P$ of a program P is the set of all ground atoms that can be formed using the language of P. An *interpretation* I of P is any consistent subset of $\mathscr{L}_P$, where $\mathscr{L}_P$ is the set of all ground literals that can be formed using the language of P, i.e., $\mathscr{L}_P = \mathscr{B}_P \cup \neg\mathscr{B}_P$. For a literal l, we write $I \models l$, if $l \in I$, which extends for extended literals *not* l to $I \models not\ l$ if $I \not\models l$. In general, for a set of extended literals X, $I \models X$ if $I \models x$ for every extended literal $x \in X$. A rule $r : \alpha \leftarrow \beta$ is *satisfied* w.r.t. I, denoted $I \models r$, if $\exists l \in \alpha I \models l$, for some extended literal l, whenever $I \models \beta$, i.e., r is *applied* ($\exists l \in \alpha I \models l$ and $I \models \beta$) whenever it is *applicable* ($I \models \beta$). The set of satisfied rules in P w.r.t. I is the *reduct* P_I.

For a *simple* program P (i.e., a program without *not*), an interpretation I is a *model* of P if I satisfies every rule in P, i.e., $P_I = P$; it is an *answer set* of P if it is a minimal model of P, i.e., there is no model J of P such that $J \subset I$. We define answer sets for programs with *not* in terms of a reduction to simple programs. The *GL-reduct*[6] w.r.t. an interpretation I is the simple P^I, where P^I contains $\alpha^+ \leftarrow \beta^+$

[6] Named after its inventors M. Gelfond and V. Lifschitz.

for $\alpha \leftarrow \beta$ in P, $I \models \alpha^-$, and $I \models not\ \beta^-$. Thus, given an interpretation I of literals — the items that one supposes true — the GL-reduct contains those rules for which the negative part is consistent with the beliefs in I. If there is a naf-literal in the body that is not true in I, then the rule is not in the GL-reduct since its whole body is then false and cannot be used to deduce literals. If all naf-literals in the body are true, the rule stays in the GL-reduct (depending on the naf-literals in the head), but with the naf-literals removed (they are known to be true). A similar reasoning holds for the head of a rule: if there is a naf-literal in the head that is true w.r.t. I, we have that the rule is automatically true and can be removed; if all naf-literals in the head are false, then we remove them and leave the rule in the GL-reduct.

I is an *answer set* of P if I is an answer set of P^I. Thus, given an interpretation I, one calculates the GL-reduct, and checks that the minimal model of the GL-reduct is I; an answer set is thus *self-motivating* or *stable*.

For more details, we refer to [9, 20].

6.4 Planning for Service Chaining

The AI formalism we review now allows to describe services in terms of "preconditions" and "effects," serving to automatically compose service chains respecting (amongst possibly other things) the input and output behavior of the services.

The following sub-sections provide: a brief formalism summary; a more detailed description of the formalism (the reader not interested in technical details may skip this sub-section); a summary of the application to service composition; an example application; and a brief discussion of variants and their merits.

6.4.1 Formalism in a Nutshell

Planning is one of the long-standing sub-areas of AI, originating in the 1960s. In a nutshell, the approach allows the user to describe, in a high-level language, a problem involving an *initial state*, a *goal*, and a set of *actions*. The AI tool — the *planning system* — then automatically finds a schedule of actions — the *plan* — transforming the initial state into a goal state.

Actions are described in terms of two logical formulas, the *precondition* and the *effect*. The former states the condition that must hold for the action to be applicable, in a given state. The latter states the condition that holds after the action has been applied, i.e., it specifies how the action changes the state.

6.4.2 Formalism Details

Over the past four decades, the planning literature has come up with a plethora of frameworks for planning. For a comprehensive introduction into the field, we recommend the recent book by Ghallab et al. [31]. We describe in what follows one of the simplest planning formalisms, called *planning with finite-domain variables* [35]. A brief discussion of the wider literature follows below.

A finite-domain variable *planning task* is a 4-tuple (X, s_0, s_*, O). X is a finite set of *variables*, where each $x \in X$ is associated with a finite domain D_x. A *partial state* over X is a function s on a subset X_s of X, so that $s(x) \in D_x$ for all $x \in X_s$; s is a *state* if $X_s = X$. The *initial state* s_0 is a state. The *goal* s_* is a partial state. O is a finite set of *operators*. Each $o \in O$ is a pair $o = (\text{pre}_o, \text{eff}_o)$ of partial states, called its *precondition* and *effect*. Partial states are identified with sets of variable-value pairs, referred to as *facts*. The *state space* of the task is the directed graph whose vertices are all states over X, with an arc (s, s') iff there exists $o \in O$ such that $\text{pre}_o \subseteq s$, $\text{eff}_o \subseteq s'$, and $s(x) = s'(x)$ for all $x \in X \setminus X_{\text{eff}_o}$. A *plan* is a path in the state space, leading from s_0 to a state s with $s_* \subseteq s$.

Illustrating this with a simple example, say that we have a variable expressing the status of a flight booking, *Pending* vs. *Confirmed*. The booking is currently pending, we wish it to be confirmed, and the only operator we have is a service confirming the booking. Expressing this in the above formalism, we get: $X = \{\mathit{flightStatus}\}$ with $D_{\mathit{flightStatus}} = \{\mathit{Pending}, \mathit{Confirmed}\}$; $s_0 = \{(\mathit{flightStatus}, \mathit{Pending})\}$; $s_* = \{(\mathit{flightStatus}, \mathit{Confirmed})\}$; O contains a single operator taking the form $(\{(\mathit{flightStatus}, \mathit{Pending})\}, \{(\mathit{flightStatus}, \mathit{Confirmed})\})$ where $\{(\mathit{flightStatus}, \mathit{Pending})\}$ is the precondition and $\{(\mathit{flightStatus}, \mathit{Confirmed})\}$ is the effect. The state space contains the two vertices $\{(\mathit{flightStatus}, \mathit{Pending})\}$ and $\{(\mathit{flightStatus}, \mathit{Confirmed})\}$, the only arc going from the former to the latter. The plan traverses that arc, thus confirming the flight.

Let us match this formalism against the summary given above in Section 6.4.1. "States" are formalized in terms of (state) variables, where an assignment to all variables defines the state. Note that the number of states is exponential in the number of variables. Thus this language allows to describe compactly a large number of possibilities, and deciding whether or not there exists a plan is computationally hard (**PSPACE**-complete). The precondition/effect "formulas" are restricted here to the simplest possible notion of listing a subset of required/effected variable-value pairs. "Plans" are simple sequences of actions mapping the initial state into a state that complies with the list of variable-value pairs given in the goal.

Traditionally, planning formalisms have been rooted in propositional logics, restricting the state variables to be Boolean, having exactly two possible values: *True* and *False*. The simplest and most wide-spread formalism of this kind is called "STRIPS" [26], which is exactly like the finite-domain variable planning tasks above except that all variables are Boolean. Other formalisms, such as *ADL* [60], allow more complex specifications of pre, eff, and s_*, involving conditional effects and arbitrary 1st-order logic formulas (quantifiers ranging over a finite universe).

In the context of the International Planning Competition[7] that has been held biennially since 1998, a common input syntax for planning systems has been defined. This input language is called *PDDL* — Planning Domain Definition Language — and has a range of variants encompassing STRIPS and ADL [52], numeric and temporal planning [27], and a number of other extensions [38, 30].

Finally, it is relevant in our context to mention the field of *planning under uncertainty* (e.g., [17, 73]). The above formalisms all assume perfect knowledge about the initial state, and deterministic behavior of actions. These assumptions often do not hold in real-world applications, including many applications of service composition. Planning under uncertainty relaxes these assumptions in a variety of ways, of course at the cost of increased (theoretical and practical) computational costs.

6.4.3 Application to Services

The application of planning to service composition has first appeared as an idea in the late 1990s (e.g., [13]) and has been intensively researched since the early 2000s (e.g., [54, 64, 19, 44, 63, 48, 37, 40]). The different approaches differ widely in intention and scope, as well as underlying formalisms (see some details in Section 6.4.5 below). The lowest common denominator is that preconditions/effects allow to specify service behavior at an abstract level where they are understood as atomic one-shot operations. In the OWL-S service description framework (e.g., [2, 18]), this abstraction level is called the "service profile;" in the WSMO framework (e.g., [25]), it is referred to as the "service capability;" see also Chapter 7.

The service profile/capability encompasses of course its input and output behavior, i.e., the typing of the respective service parameters. As a simple example, the service input may be a customer-data object, and the output may be a reservation-object. One can further express additional prerequisites or consequences that the service may have, in a logics-based representation of the relevant surroundings. A precondition "$credit \geq 500$" may require the sufficient availability of money to cover at least the cost of 500 money units, and an effect "$credit := credit - 500$" may reduce the available money by that amount.

Planning is computationally hard in general, but AI research has come up with a range of rather effective approaches (e.g., [39, 65]). Provided that the plans to be created are not too large (up to 20 or so actions), practical runtime/memory performance when using these techniques typically is not a big issue, plans being found within a matter of seconds (e.g., [37, 40]).

A plan in this setting — some scheduling of the atomic services — is not necessarily an executable software artifact. This is because we do not take into account many technical aspects of the services involved. Most importantly, we disregard the order of interactions required for communicating with them (making a reservation typically involves several steps, rather than being a one-shot interaction), and we

[7] See `http://ipc.icaps-conference.org/`

disregard the actual underlying data structures or WSDL schemata (what exactly is a "customer-data object"?).

Since plans are not guaranteed to be executable, the planning facility provided typically accomplishes only a part of the service composition design. A typical perspective is that a human user is responsible for the overall design, and uses the planning facility for easing the task of selecting and arranging a useful subset of services from a large services database or the internet (e.g., [3, 40]). Another possibility is to use this abstract planning as a pre-process to more accurate — but more computationally expensive — service composition techniques (as will be described in Section 6.5). This reduces the size of the input to choose from, and thereby the computational resources required to come up with a composition [10].

In order to run the planning facility, services need be described as actions in the first place, and the user needs to enter the initial state and goal of the desired composed service. These are non-trivial tasks which need to be addressed in a clever way in order to keep the modeling overhead at bay. Several frameworks have appeared that allow users to conveniently design the models via user-friendly interfaces (e.g., [66, 32]). An alternative approach [40] suggests to instead exploit pre-existing models of software behavior, thus getting the planning model "for free," respectively sharing the modeling effort with other software design activities. We will now look at this approach in a little more detail, as an application example.

6.4.4 Example

SAP widely employs model-driven software engineering. The more modern developments are designed as service-oriented architectures (SOAs). Some of the models created in their development aim at describing the behavior of service operations. To achieve this, the *Status and Action Management (SAM)* models exist for each of over 400 business objects (BOs). BOs are software objects which have an actual correspondence in common business scenarios, such as, e.g., a customer quote, a sales order, an invoice, etc.

Each BO can have (vast) data structures and offers functionality on this data as service operations in the SOA. SAM models capture the relation between the status of a BO and the actions (operations) it offers: when can these actions be enacted, and how do they affect the BO? Concretely, SAM models consist of a set of finite-domain status variables and a set of actions that describe which status values for which variables are a precondition to an action, and how the status values may change as a result of a service execution. The original purpose of SAM is code generation: code skeletons check if preconditions are fulfilled at runtime, and update the status variables accordingly.

SAM is based on common business terms. For instance, the status variable "approval" may include values such as "approved" and "rejected." Hence, its expressions are understandable to business users. This is key to practical planning-based service composition, where business users describe in terms of SAM what the com-

Table 6.5: A SAM-like example, modeling the behavior of "customer quotes" CQ.

Action name	precondition	effect
Create CQ		(CQ.approval:necessary OR CQ.approval:notNecessary) AND CQ.acceptance:notAccepted AND CQ.archiving:notArchived AND CQ.submission:notSubmitted
CQ Approval	CQ.approval:necessary	CQ.approval:approved OR CQ.approval:rejected
Submit CQ	CQ.archiving:notArchived AND (CQ.approval:notNecessary OR CQ.approval:granted)	CQ.submission:submitted
Mark CQ as Accepted	CQ.archiving:notArchived AND CQ.submission:submitted	CQ.acceptance:accepted
Archive CQ	CQ.archiving:notArchived	CQ.archiving:archived

position should accomplish (what's its initial state and goal). The planning functionality is implemented as a prototypical research extension to the SAP NetWeaver BPM Process Composer.

The only important difference between SAM and the basic formalism introduced in Section 6.4.2 is that effects can be non-deterministic, i.e., the action has one out of a set of possible outcomes. For the business context, it is quite obvious that this is necessary. Any sensible order, booking, check, or approval action will have at least two possible outcomes — positive vs. negative.

For illustration, Table 6.5 gives a SAM-like model for a particular BO called "customer quote (CQ)." For confidentiality reasons, the shown object and model are artificial, i.e., they are *not* contained in SAM as used at SAP. In the figure, by "CQ.X:Y" we denote the atomic proposition stating that variable X of the customer query has value Y. We are using a standard planner [39] — modified to appropriately handle SAM's non-deterministic effects — on this input.

For presentation to the user, a simple post-process transforms plans into BPMN diagrams. Figure 6.1 shows a plan in that representation, for the above example. Note in particular the side effect of the create operation (top): approval can be either necessary or not. Depending on the actual outcome, an additional approval step is executed. In turn, this approval step may have a positive or negative outcome. Only in the positive outcome, the process continues by submitting the quote. No procedure is specified for the negative outcome — exception handling is needed in this case, and SAM does not contain any information about what that handling should be, so the planner cannot compose it automatically. Clearly, the plan is also incomplete in that not every customer quote should be approved, submitted, etc. straight after being created. Thus the plan serves merely as a template and needs to be completed by hand.[8]

[8] Note also that the standard process is often implemented in the standard system. The value of service composition here lies in convenient creation of variants of the standard.

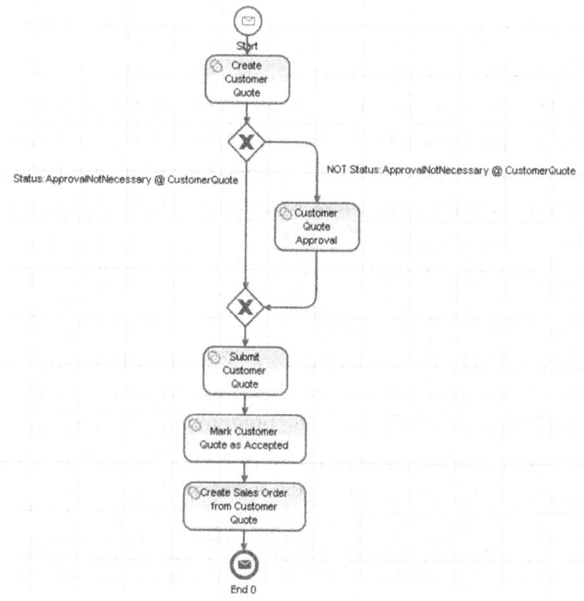

Fig. 6.1: Screenshot of the SAP NetWeaver BPM Process Composer with an automatically composed process of five non-deterministic services).

6.4.5 Discussion

As indicated, the different approaches to planning-based service composition differ widely in intention, scope, and formalization. Let us mention a few of the better-known works. Some authors compile service composition into more or less standard planning formalisms, e.g., [64]. Two frameworks allow the user to provide a plan skeleton whose control structures will be filled in by the planner [54, 71].

A large strand of work addresses the handling of "input/output typing" in a rich Semantic Web framework where the surroundings of the services are described in a formal logics, most often in some variant of Description Logics (DL) as discussed in Section 6.2. A main issue here is that DL contains non-trivial axiomatizations of the domain. These axioms make it difficult to define and compute the "outcome state" resulting from applying an action.[9] One body of work circumvents this problem by applying the axiom inferences only to the service outputs, i.e., to individuals that didn't exist prior to action application ([19, 48, 37]). A few works address the

[9] For example, say the ontology contains the axiom $A \sqsubseteq \neg(B \sqcap C)$ stating (intuitively) that an element of type A cannot be in the intersection of B and C. Say that we are in a state where we have an individual o so that $o \in A$ and $o \in B$. Say we apply an action whose effect is $o \in C$. What is the outcome state? Simply adding $o \in C$ results in a conflict with the axiom. Thus we need to "repair" this outcome state, by removing one of the previous facts, $o \in A$ or $o \in B$. Which one should we remove? This is difficult to answer in a principled manner. Most known answers have severe consequences on the computational complexity of computing outcome states [36].

problem in full, living with the computational costs incurred (e.g., [23]). Yet others address more feasible sub-classes of the problem (e.g., [14]).

A fundamentally different approach is the one described in Section 6.5, whose output is actual executable code [63]. The more abstract approach described in the section at hand can be used as a filtering method in front of process-level composition, reducing the choice of services to compose from, and thus the empirical performance of the more fine-grained composition [10].

In summary, planning for service chaining allows to automatically compose services at an abstract level taking into account their input and output typing, as well as any other prerequisites/effects they have relative to a high-level formalization of their surroundings. The method can be computationally quite feasible, depending on the specifics of the surroundings and on the complexity of any (DL-) axiomatizations that should be taken into account. The outcome of planning is a composition-template which most often is imperfect, but delivers useful input to either humans or other more detailed composition techniques. A key problem in practice is the modeling overhead. Recent work [40] suggests a connection to model-driven software engineering that could be exploited to reduce this overhead significantly.

6.5 Planning for Service Interactions

The AI planning approach presented in this section deals with the problem of automatically composing an executable Web service that, interacting with a set of component services, satisfies a given composition requirement.

Similarly to Section 6.4, the following sub-sections provide: a brief formalism summary; a more detailed description of the formalism (the reader not interested in technical details may skip this sub-section); a summary of the application to service composition; an example application; and a brief discussion of variants and their merits.

6.5.1 Formalism in a Nutshell

The approach is based on planning as model checking [15, 16]. It adopts "state transition systems" for the representation of the individual entities to be composed into a plan. This allows to represent stateful processes implementing a complex interaction protocol, exchanging asynchronous messages, and exhibiting a partially observable and non-deterministic behavior. The input to the planning problem is a set of such processes, as well as a "composition requirement" (a combined functionality we wish to achieve). The solution to the planning problem then is a "controller," another state transition system, such that executing the controller results in an orchestration of the processes achieving the composition requirement.

6.5.2 Formalism Details

Each service is encoded as a *state transition system* (STS from now on) $\Sigma = \langle \mathscr{S}, \mathscr{S}^0, \mathscr{I}, \mathscr{O}, \mathscr{R}, \mathscr{L} \rangle$ that can be in one of its possible states $\mathscr{S}$, a subset of which are *initial* $\mathscr{S}^0$, and can evolve to new states as a result of performing some actions. Actions are distinguished in *input actions* $\mathscr{I}$, which represent the reception of messages, *output actions* $\mathscr{O}$, which represent messages sent to external services, and a special action τ, called *internal action*. The action τ is used to represent internal evolutions that are not visible to external services, i.e., the fact that the state of the system can evolve without producing any output, and independently from the reception of inputs. The *transition relation* $\mathscr{R}$ describes how the state can evolve on the basis of inputs, outputs, or of the internal action τ. Finally, a *labeling function* associates to each state the set of properties in $\mathscr{P}rop$ that hold in the state. These properties will be used to define the composition requirements. For a complete description of the translation from Web services (described in terms of their BPEL and WSDL specification) to STS please refer to [49].

The *behavior* of an STS is represented by its set of possible *runs*, i.e., of sequences $s_0, a_0, s_1, a_1, \ldots$ such that $s_0 \in \mathscr{S}^0$ and $(s_i, a_i, s_{i+1}) \in \mathscr{R}$. In general, such runs may be finite or infinite. A run σ is said to be *completed* if it is finite, and if its last state is final. A state $s \in \mathscr{S}$ will be said *reachable* if there exists a run $\sigma = s_0, a_0, \ldots, a_{n-1}, s_n, \ldots$ such that $s_n = s$. We will denote with $Reachable(\Sigma) \subseteq \mathscr{S}$ the set of reachable states of Σ.

The composition problem has two inputs: the formal composition requirement ρ and the set of component services $\Sigma_{W_1}, \ldots, \Sigma_{W_n}$. The component services $\Sigma_{W_1}, \ldots, \Sigma_{W_n}$ evolve independently, and in fact represent, under a planning perspective, the domain to be controlled. Such domain $\Sigma_{\|}$ is obtained as the first step of the composition, by combining $\Sigma_{W_1}, \ldots, \Sigma_{W_n}$ by means of a *parallel product* operation. The system representing (the parallel evolutions of) the component services $W_1, \ldots, W_n$ is formally defined as $\Sigma_{\|} = \Sigma_{W_1} \| \ldots \| \Sigma_{W_n}$.

In a composition problem, the composite service is defined as a "controller" Σ_c (also described as a STS), which interacts with the domain Σ, orchestrating the component services. The STS $\Sigma_c \triangleright \Sigma$, describing the behaviors of system $\Sigma = \langle \mathscr{S}, \mathscr{S}^0, \mathscr{I}, \mathscr{O}, \mathscr{R}, \mathscr{L} \rangle$ when controlled by $\Sigma_c = \langle \mathscr{S}_c, \mathscr{S}_c^0, \mathscr{O}, \mathscr{I}, \mathscr{R}_c, \mathscr{L}_\emptyset \rangle$, is defined as $\Sigma_c \triangleright \Sigma = \langle \mathscr{S}_c \times \mathscr{S}, \mathscr{S}_c^0 \times \mathscr{S}^0, \mathscr{I}, \mathscr{O}, \mathscr{R}_c \triangleright \mathscr{R}, \mathscr{L} \rangle$. The transition relation $\mathscr{R}_c \triangleright \mathscr{R}$ is such that Σ_c and Σ either evolve independently by executing internal τ transitions, or evolve concurrently by executing input and output actions.

Due to the asynchronous nature of Web service interactions, and in order to guarantee a correct behavior of the composite service, there is the need to rule out explicitly the cases where the sender is ready to send a message that the receiver is not able to accept. According to [62], a state s is able to accept a message a if there exists some successor s' of s, reachable from s through a (possibly empty) sequence of τ transitions, such that an input transition labeled with a can be performed in s'. This intuition is captured by the notion of τ-closure(s), defining the set of states reachable from s through a chain of τ transitions.

In [62], the composition problem for domain Σ and composition goal ρ consists in generating a STS Σ_c that controls Σ so that its behavior satisfy the requirement ρ (according to a formal notion of requirement satisfaction).

Intuitively, a controller is a *solution* for a requirement ρ if it guarantees that ρ is achieved. That is, if every run σ of the controlled system $\Sigma_c \triangleright \Sigma_{\|}$ ends up in a state where ρ holds.

6.5.3 Application to Services

The approach has been implemented in the ASTRO framework. Similarly to Section 6.4, the basic idea is that existing services can be used to construct the planning domain, composition requirements can be formalized as planning goals, and planning algorithms can be used to generate plans that compose the published services. As outlined above, in ASTRO each service is represented as a state transition system. These can be obtained from abstract WS-BPEL protocols, and thus are easy to come by in practice.

The ASTRO framework has been widely adopted to deal with the different aspects of the Web service composition problem. In particular, ASTRO has been enabled to specify complex data-flow [50] and control-flow composition requirements [11] and an abstraction-based approach for composing services that manipulate complex, infinite-range data domains [61]. The framework has been implemented as a prototype automated composition tool, namely WS-Compose, and integrated in the ASTRO Toolset [7], a toolkit providing an integrated environment for the composition of Web services.

6.5.4 Example

The example we present in this section is taken from the *e-Bookstore* composition scenario [51] where the aim is to automatically synthesize an application that allows to order books through the Amazon E-Commerce Services and buy them via a secure credit card payment transaction offered by Banks of Monte dei Paschi di Siena Group (MPS). This composition scenario is particularly challenging since all component services are real Web services exporting complex interaction protocols and handling structured data in messages.

In particular, we will consider the Amazon Virtual-Cart (AVC) Web service and show its encoding as a STS. For a complete description of the e-Bookstore scenario please refer to [49].

Figure 2(a) represents a compact representation of the abstract WS-BPEL protocol of the Amazon Virtual-Cart (AVC) service. According to this process, once the AVC receives a request to create a new cart and the operation is successful, the client can start to add items and eventually checkout its shopping cart. If the

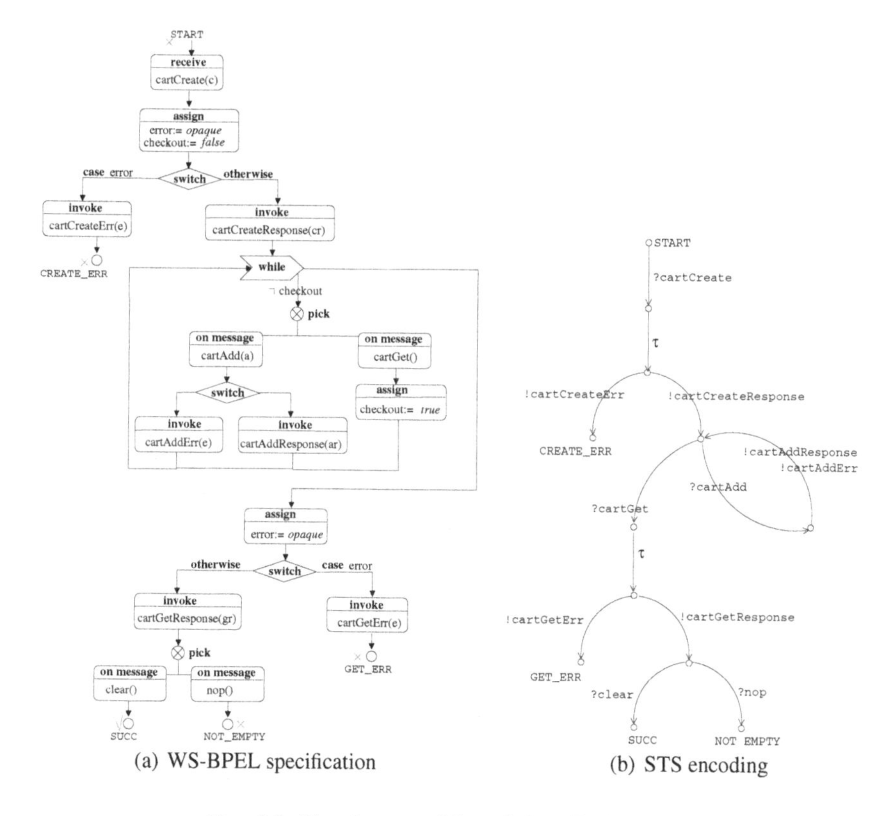

(a) WS-BPEL specification (b) STS encoding

Fig. 6.2: The Amazon Virtual-Cart Web service.

checkout is successful, the client can either clear the cart or keep its content for future use. In all these interactions if something goes wrong the AVC sends an error message describing the reason of the fault. The final states of the protocol are marked either as successful (symbol $\surd$) or as failing (symbol $\times$) states. These minimal semantic annotations are necessary to distinguish those executions that lead to a successful completion of the interaction from those that are failed. As explained in [51] this information are exploited in the definition of the control-flow composition requirements.

Figure 2(b) presents the STS encoding of the AVC service. Please refer to [12] for the complete description of a translation procedure that allows to encode WS-BPEL processes as state transition systems.

6.5.5 Discussion

The formal framework presented in this Section is primarily motivated by the necessity to provide an automated composition approach that is able to tackle real world Web service composition problems. Among the most important characteristics are (i) the possibility to specify both control and data flow composition requirements, (ii) the ability to consider complex stateful processes as component services, and (iii) the capability to produce an executable, ready to be deployed, composite service. We briefly summarize some of the details involved in tackling these challenges.

In [50] the authors propose to specify requirements on the data flow through a set of constraints that explicitly define the valid routings and manipulations of messages that the new composite service can perform. In particular, data flow composition requirements are defined through an intuitive graphical notation, the *data net*, i.e., a graph where the input/output ports of the existing services are modeled as nodes, the paths in the graph define the possible routes of the messages, and the arcs define basic manipulations of these messages performed by the composed service. Finally, the authors show how to encode these data constraints within the composition domain in an efficient compositional way.

Even though the data net approach allows to specify complex data flow composition requirements, it does not encode data within the composition domain (the states of the domain simply model the evolution of the processes). As a consequence, it is not possible to reason on data when searching for a solution and, in particular, on the conditions ruling the flow of operations and interactions of the component services. To face the data challenge, in [61] the authors propose an abstraction-based approach for handling data, which possibly ranges over an infinite domain, in a finite, symbolic way.

A limitation of the presented approach is the fact that the specification of control-flow and data-flow composition requirements may be time consuming and reduces the applicability of the approach in more dynamic applications (e.g., requiring to substitute/adapt on-the-fly the service components to be used). A significant step in this direction is done by the work in [11] through the usage of business objects that allow to detach the specification of composition requirements from the component service implementations. With respect to the specification of data flow requirements, an idea can be to add semantic annotations to the data used in the component services (similarly to what is done in [5, 4, 45] and then apply semantic matching and reasoning techniques to automatically derive the data links between message parts in order to obtain a first version of the data net diagram that can then be refined by hand.

Another limitation of this approach is that it scales-up well if we assume that we already selected the subset of relevant component services participating to the composition. As shown in [10], the approach described in Section 6.4 can be efficiently used to filter the component services before applying the automated synthesis proposed in here.

6.6 Conclusion

Automation in the handling of Web services requires their annotation with suitable information about their content. Research into this has come up with a broad range of approaches, drawing on several research areas, prominently on Artificial Intelligence and Databases. We have herein introduced four of the most wide-spread formalizations. Description Logics and Logic Programming serve to annotate service input/outputs for better service discovery; Planning for Service Chaining and Planning for Service Interactions require more detailed information about the service behavior, and provide a form of automatic programming composing atomic services to more complex units. All of the approaches involve a form of reasoning and are thus computationally costly in the worst-case; but practical methods can be designed.

Together, the paradigms just described form the basis of service discovery and composition in the Semantic Web Services area. The next chapter gives an overview of that area.

References

1. International health terminology standards development organisation - SNOMED CT. `http://www.ihtsdo.org/snomed-ct/`.
2. A. Ankolekar et al. DAML-S: Web service description for the semantic web. In *ISWC*, 2002.
3. V. Agarwal, G. Chafle, K. Dasgupta, N. Karnik, A. Kumar, S. Mittal, and B. Srivastava. Synthy: A system for end to end composition of web services. *J. Web Semantics*, 3(4), 2005.
4. R. Akkiraju, B. Srivastava, A. Ivan, R. Goodwin, and T. Syeda-Mahmood. Semaplan: Combining planning with semantic matching to achieve web service composition. In *Proc. of IEEE International Conference on Web Services (ICWS'06)*, 2006.
5. J. L. Ambite and D. Kapoor. Argos: a framework for automatically generating data processing workflows. In *Proc. of the 8th annual international conference on Digital government research (dg.o'07)*, 2007.
6. M. Arenas, L. Bertossi, and J. Chomicki. Specifying and Querying Database Repairs using Logic Programs with Exceptions. In *Proc. of the 4th International Conference on Flexible Query Answering Systems*, pages 27–41. Springer, 2000.
7. ASTRO. Project ASTRO: Supporting the Composition of Distributed Business Processes - http://astroproject.org.
8. F. Baader, D. Calvanese, D. McGuinness, D. Nardi, and P. Patel-Schneider. *The Description Logic Handbook*. Cambridge University Press, 2003.
9. C. Baral. *Knowledge Representation, Reasoning and Declarative Problem Solving*. Cambridge University Press, 2003.
10. P. Bertoli, J. Hoffmann, F. Lecue, and M. Pistore. Integrating discovery and automated composition: from semantic requirements to executable code. In *Proceedings of the IEEE 2007 International Conference on Web Services (ICWS'07)*, 2007.
11. P. Bertoli, R. Kazhamiakin, M. Paolucci, M. Pistore, H. Raik, and M. Wagner. Control Flow Requirements for Automated Service Composition. In *Proc. of the IEEE International Conference on Web Services (ICWS09)*, 2009.
12. P. Bertoli, M. Pistore, and P. Traverso. Automated composition of web services via planning in asynchronous domains. *Journal of Artif. Intell.*, 174:316–361, March 2010.

13. S. Biundo, R. Aylett, M. Beetz, D. Borrajo, A. Cesta, T. Grant, L. McCluskey, A. Milani, and G. Verfaillie. PLANET Technological Roadmap on AI Planning and Scheduling. http://planet.dfki.de/service/Resources/Roadmap/Roadmap2.pdf, Dec. 2003.
14. D. Calvanese, G. D. Giacomo, D. Lembo, M. Lenzerini, and R. Rosati. Tractable reasoning and efficient query answering in description logics: The *l-lite* family. *Journal of Automed Reasoning*, 39(3):385–429, 2007.
15. A. Cimatti, F. Giunchiglia, E. Giunchiglia, and P. Traverso. Planning via model checking: A decision procedure for ar. In *Proc. of the 4th European Conference on Planning*, pages 130–142, 1997.
16. A. Cimatti, F. Giunchiglia, E. Giunchiglia, and P. Traverso. Planning as model checking. In *Proc. of ECP*, pages 1–20, 1999.
17. A. Cimatti, M. Roveri, and P. Bertoli. Conformant planning via symbolic model checking and heuristic search. *Artificial Intelligence*, 159(1–2):127–206, 2004.
18. T. O. S. Coalition. OWL-S: Semantic Markup for Web Services, 2003.
19. I. Constantinescu, B. Faltings, and W. Binder. Large scale, type-compatible service composition. In *2nd International Conference on Web Services (ICWS-04)*, pages 506–513, 2004.
20. E. Dantsin, T. Eiter, G. Gottlob, and A. Voronkov. Complexity and Expressive Power of Logic Programming. *ACM Computing Surveys*, 33(3):374–425, 2001.
21. J. de Bruijn and S. Heymans. A semantic framework for language layering in wsml. In M. Marchiori, J. Z. Pan, and C. de Sainte Marie, editors, *Proceedings of the First International Conference on Web Reasoning and Rule Systems (RR 2007)*, volume 4524 of *Lecture Notes in Computer Science*, pages 103–117. Springer, 2007.
22. T. Eiter, W. Faber, N. Leone, and G. Pfeifer. The Diagnosis Frontend of the dlv System. *AI Communications*, 12(1-2):99–111, 1999.
23. T. Eiter, W. Faber, N. Leone, G. Pfeifer, and A. Polleres. A logic programming approach to knowledge-state planning, II: The DLVK system. *Artificial Intelligence*, 144(1-2):157–211, 2003.
24. T. Eiter, G. Ianni, T. Lukasiewicz, R. Schindlauer, and H. Tompits. Combining answer set programming with description logics for the Semantic Web. *Artificial Intelligence*, 172(12-13):1495–1539, 2008.
25. D. Fensel, H. Lausen, A. Polleres, J. de Bruijn, M. Stollberg, D. Roman, and J. Domingue. *Enabling Semantic Web Services– The Web Service Modeling Ontology*. Springer-Verlag, 2006.
26. R. E. Fikes and N. Nilsson. STRIPS: A new approach to the application of theorem proving to problem solving. *Artificial Intelligence*, 2:189–208, 1971.
27. M. Fox and D. Long. PDDL2.1: An extension to PDDL for expressing temporal planning domains. *J. Artificial Intelligence Research*, 20:61–124, 2003.
28. A. V. Gelder, K. Ross, and J. S. Schlipf. The well-founded semantics for general logic programs. *JACM*, 38(3):620–650, 1991.
29. M. Gelfond and V. Lifschitz. The Stable Model Semantics for Logic Programming. In *Proc. of International Conference on Logic Programming (ICLP 1988)*, pages 1070–1080. MIT Press, 1988.
30. A. Gerevini, P. Haslum, D. Long, A. Saetti, and Y. Dimopoulos. Deterministic planning in the fifth international planning competition: Pddl3 and experimental evaluation of the planners. *Artificial Intelligence*, 173(5-6):619–668, 2009.
31. M. Ghallab, D. Nau, and P. Traverso. *Automated Planning: Theory and Practice*. Morgan Kaufmann/Elsevier, 2004.
32. A. Gonzalez-Ferrer, J. Fernandez-Olivares, and L. Castillo. JABBAH: a java application framework for the translation between business process models and htn. In *Proceedings of the 3rd International Competition on Knowledge Engineering for Planning and Scheduling (ICKEPS'09)*, 2009.
33. B. N. Grosof, I. Horrocks, R. Volz, and S. Decker. Description logic programs: Combining logic programs with description logic. In *Proc. WWW 2003*, pages 48–57. ACM, 2003.
34. V. Haarslev and R. Moller. Description of the RACER system and its applications. In *Proc. of Description Logics 2001*, 2001.

35. M. Helmert. Concise finite-domain representations for pddl planning tasks. *Artificial Intelligence*, 173(5-6):503–535, 2009.
36. A. Herzig and O. Rifi. Propositional belief base update and minimal change. *Artificial Intelligence*, 115(1):107–138, 1999.
37. J. Hoffmann, P. Bertoli, M. Helmert, and M. Pistore. Message-based web service composition, integrity constraints, and planning under uncertainty: A new connection. *J. Artificial Intelligence Research*, 35:49–117, 2009.
38. J. Hoffmann and S. Edelkamp. The deterministic part of IPC-4: An overview. *J. Artificial Intelligence Research*, 24:519–579, 2005.
39. J. Hoffmann and B. Nebel. The FF planning system: Fast plan generation through heuristic search. *J. Artificial Intelligence Research*, 14:253–302, 2001.
40. J. Hoffmann, I. Weber, and F. M. Kraft. SAP speaks PDDL. In *Proceedings of the 24th AAAI Conference on Artificial Intelligence (AAAI'10)*, 2010.
41. I. Horrocks. The FaCT system. In *Automated Reasoning with Analytic Tableaux and Related Methods: International Conference Tableaux'98*, number 1397 in LNAI, pages 307–312. Springer, 1998.
42. D. Hull, E. Zolin, A. Bovykin, I. Horrocks, U. Sattler, and R. Stevens. Deciding Semantic Matching of Stateless Services. In *Proceedings, The Twenty-First National Conference on Artificial Intelligence (AAAI) and the Eighteenth Innovative Applications of Artificial Intelligence Conference, July 16-20, 2006, Boston, Massachusetts, USA*, pages 1319–1324, 2006.
43. M. Krötzsch, S. Rudolph, and P. Hitzler. Description logic rules. In *Proc. ECAI*, pages 80–84. IOS Press, 2008.
44. U. Kuter, E. Sirin, D. Nau, B. Parsia, and J. Hendler. Information gathering during planning for web service composition. *J. Web Semantics*, 3(2-3):183–205, 2005.
45. F. Lecue, A. Delteil, and A. Leger. Applying abduction in semantic web service composition. In *Proc. of IEEE International Conference on Web Services (ICWS'07)*, 2007.
46. N. Leone, P. Rullo, and F. Scarcello. Disjunctive Stable Models: Unfounded sets, Fixpoint Semantics, and Computation. *Information and Computation*, 135(2):69–112, 1997.
47. V. Lifschitz. Answer Set Programming and Plan Generation. *Artificial Intelligence*, 138(1-2):39–54, 2002.
48. Z. Liu, A. Ranganathan, and A. Riabov. A planning approach for message-oriented semantic web service composition. In *22nd National Conference of the American Association for Artificial Intelligence (AAAI'07)*, 2007.
49. A. Marconi and M. Pistore. Synthesis and composition of web services. In *Formal Methods for Web Services*, pages 89–157. Springer Berlin / Heidelberg, 2009.
50. A. Marconi, M. Pistore, and P. Traverso. Specifying Data-Flow Requirements for the Automated Composition of Web Services. In *Proc. of Fourth IEEE International Conference on Software Engineering and Formal Methods (SEFM06)*, 2006.
51. A. Marconi, M. Pistore, and P. Traverso. Automated Web Service Composition at Work: the Amazon/MPS Case Study. In *Proc. of IEEE International Conference on Web Services (ICWS'07)*, 2007.
52. D. McDermott et al. *The PDDL Planning Domain Definition Language*. The AIPS-98 Planning Competition Committee, 1998.
53. S. McIlraith, T. Son, and H. Zeng. Semantic web services. *Intelligent Systems*, 16(2):46–53, April 2001.
54. S. McIlraith and T. C. Son. Adapting Golog for composition of semantic Web services. In *Proc. of the 8th Int. Conf. on Principles and Knowledge Representation and Reasoning (KR-02), Toulouse, France*, 2002.
55. B. Motik and R. Rosati. A faithful integration of description logics with logic programming. In *Proc. IJCAI*, pages 477–482, 2007.
56. B. Motik, U. Sattler, and R. Studer. Query answering for OWL-DL with rules. *Journal of Web Semantics*, 3(1):41–60, July 2005.
57. I. Niemelä and P. Simons. SMODELS - An Implementation of the Stable Model and Well-founded Semantics for Normal Logic Programs. In *Proc. of the 4th International Conference on Logic Programming and Nonmonotonic Reasoning (LPNMR 1997)*, volume 1265 of *LNAI*, pages 420–429, 1997.

58. OASIS. *Web Services Business Process Execution Language Version 2.0*, Apr. 2007.
59. W. OWL Working Group. *OWL 2 Web Ontology Language: Document Overview*. W3C Recommendation, 27 October 2009. Available at `http://www.w3.org/TR/owl2-overview/`.
60. E. P. Pednault. ADL: Exploring the middle ground between STRIPS and the situation calculus. In *Principles of Knowledge Representation and Reasoning: Proceedings of the 6th International Conference (KR'98)*, pages 324–331, 1998.
61. M. Pistore, A. Marconi, P. Bertoli, and P. Traverso. Automated Composition of Web Services by Planning at the Knowledge Level. In *Proc. IJCAI'05*, 2005.
62. M. Pistore, P. Traverso, and P. Bertoli. Automated Composition of Web Services by Planning in Asynchronous Domains. In *Proc. ICAPS'05*, 2005.
63. M. Pistore, P. Traverso, P. Bertoli, and A. Marconi. Automated synthesis of composite BPEL4WS web services. In *3rd IEEE International Conference on Web Services (ICWS-05)*, 2005.
64. S. Ponnekanti and A. Fox. SWORD: A developer toolkit for web services composition. In *11th International World Wide Web Conference (WWW-02)*, 2002.
65. S. Richter and M. Westphal. The LAMA planner: Guiding cost-based anytime planning with landmarks. *J. Artificial Intelligence Research*, 39:127–177, 2010.
66. M. D. Rodriguez-Moreno, D. Borrajo, A. Cesta, and A. Oddi. Integrating planning and scheduling in workflow domains. *Expert Systems Applications*, 33(2):389–406, 2007.
67. R. Rosati. On the decidability and complexity of integrating ontologies and rules. *Journal of Web Semantics*, 3(1):41–60, 2005.
68. R. Rosati. DL+log: Tight integration of description logics and disjunctive datalog. In *Proc. KR*, pages 68–78, 2006.
69. R. Shearer, B. Motik, and I. Horrocks. HermiT: A Highly-Efficient OWL Reasoner. In *Proceedings of the 5th International Workshop on OWL: Experiences and Directions (OWLED 2008)*, 2008.
70. E. Sirin, B. Parsia, B. C. Grau, A. Kalyanpur, and Y. Katz. Pellet: A practical owl-dl reasoner. *Web Semantics: Science, Services and Agents on the World Wide Web*, 5(2):51 – 53, 2007. Software Engineering and the Semantic Web.
71. E. Sirin, B. Parsia, D. Wu, J. Hendler, and D. Nau. HTN planning for web service composition using SHOP2. *J. Web Semantics*, 1(4), 2004.
72. T. Soininen and I. Niemelä. Developing a Declarative Rule Language for Applications in Product Configuration. In *Proceedings of the First International Workshop on Practical Aspects of Declarative Languages (PADL 1999)*, number 1551 in LNCS, pages 305–319. Springer, 1999.
73. H. Younes, M. Littman, D. Weissman, and J. Asmuth. The first probabilistic track of the international planning competition. *J. Artificial Intelligence Research*, 24:851–887, 2005.

Chapter 7
Semantic Web Services Approaches

Carlos Pedrinaci, Maria Maleshkova, Maciej Zaremba, and Maryam Panahiazar

Abstract Semantic Web Services aim to better support the life-cycle of Web services and service-based applications by exploiting semantic descriptions of services. Research in this field has been considerably active and has produced a large number of ontologies, representation languages, and integrated frameworks supporting the discovery, composition and invocation of services among other tasks. In this chapter we provide a thorough, albeit necessarily brief, overview of the conceptual models devised so far, giving the reader a perspective on the relationships, coverage and applicability of each of them together with pointers for gathering further insights and details about these solutions and related software.

7.1 Introduction

Semantic Web Services (SWS), previously introduced in Chapter 6, were proposed in order to pursue the vision of the semantic Web presented in [9] whereby intelligent agents would be able to exploit semantic descriptions in order to carry out complex tasks on behalf of humans [34]. This early work on SWS was the meeting point between Semantic Web, Agents, and Web services technologies. Gradually, however, research focussed more prominently on combining Web services with semantic Web technologies in order to better support the discovery, composition, and

Carlos Pedrinaci, Maria Maleshkova
Knowledge Media Institute (KMi), The Open University, Walton Hall, Milton Keynes, Buckinghamshire, MK7 6AA, UK, e-mail: c.pedrinaci@open.ac.uk, e-mail: m.maleshkova@open.ac.uk

Maciej Zaremba
Digital Enterprise Research Institute (DERI), National University of Ireland, Galway, Ireland, e-mail: maciej.zaremba@deri.org

Maryam Panahiazar
Kno.e.sis Center, Wright State University, Dayton, OH, USA, e-mail: mary@knoesis.org

execution of Web services, leaving aspects such as systems autonomy more typical of agent-based systems somewhat aside, see Chapter 6 and [40] for a broader survey.

Research on SWS has been active and fruitful over the years leading to a number of ontologies, representation languages, as well as to a plethora of software components and even integrated execution frameworks that cover diverse tasks within the life-cycle of Web services and service-oriented applications. Some of these frameworks and techniques are detailed in Chapter 6. This chapter aims at providing a thorough, albeit necessarily brief, overview of the conceptual models devised so far, giving the reader a perspective on the relationships, coverage and applicability of each of the models together with pointers for gathering further insights and details about these solutions and related software. This chapter does also provide some basis that could help the reader carry out a deeper and more detailed analysis of the different approaches depending on the objectives sought. We do not provide a detailed comparison between all the approaches for the very reason that this would require addressing them from the perspective of the many tasks that can be supported with Semantic Web Services descriptions (i.e., discovery, invocation, composition, etc) and would also require taking into account the engines and frameworks developed.

The remainder of this chapter is organized as follows. In Section 7.2 we introduce some preliminaries for Semantic Web Service approaches and present briefly the vast majority of those that have been proposed over the years split into top-down and bottom-up approaches (Sections 7.3 and 7.4). Finally, in Section 7.5 we provide both general information upon which one could drive a more thorough comparison and targeted evaluation. This section also provides a chronological perspective over the state of the art in Semantic Web Service approaches that can better help understand the evolution of the field.

7.2 Preliminaries

Early on, Sheth et al. [48, 46] introduced the four main types of semantics, that corresponding semantic descriptions can capture:

1. **Data semantics**: The semantics pertaining to the data used and exposed by the service.
2. **Functional semantics**: Semantics pertaining to the functionality of the service.
3. **Non-Functional semantics**: Semantics related to the non-functional aspects of the service, e.g., Quality of Service (QoS), security or reliability.
4. **Execution semantics**: Semantics related to exceptional behaviors such as run time errors.

The essential characteristic of SWS is therefore the use of languages with well-defined semantics covering the subset of the mentioned categories that are amenable to automated reasoning. On the basis of these semantic descriptions, SWS technologies seek to automate the tasks involved in the life-cycle of service-oriented applications which include the discovery and selection of services, their composition, their execution and their monitoring among others [40]. The different conceptual frameworks proposed make particular emphasis on certain aspects such as better supporting the decoupling and scalability of the solutions, on supporting the interactions between agents, or on reaching a solution that appropriately fulfils humans expectations.

In this chapter we do not aim at comparing all the different approaches with respect to their capabilities for covering each of the tasks mentioned. Instead we limit ourselves to introducing the approaches, indicate their core characteristics and objectives sought and provide pointers for the interested reader to follow up. Although, we cannot claim this list to be exhaustive, this represents to the best of our knowledge, the largest collection of approaches which we have identified through some publication or public document online. Based on the two main trends in the area, the approaches covered in this chapter are divided into *top-down* approaches, which are driven by high-level views over Web services, and *bottom-up* approaches, which provide instead an incremental enrichment of existing Web services technologies. Within each subgroup the approaches are listed in chronological order of appearance.

7.3 Top-Down Approaches

Semantic Web Services are a technology anchored on the use of formal definitions of services and as a consequence a large body of work in the area is essentially driven by high-level ontologies defining the main aspects that are considered relevant for services. Most of the early work in this area thus focused on devising high-level ontologies and developing machinery that could exploit these descriptions. Despite the good results obtained, there has been an evident difficulty in obtaining rich descriptions and on providing scalable solutions due to the complexity of this kind of Semantic Web Services descriptions from a knowledge acquisition and computational perspectives. In recent times, although ontologies continue to be produced and refined, one can notice a trend towards simpler models typically based on simpler knowledge representation languages.

UPML

The Unified Problem-solving Method Development Language (UPML) is both an approach and its corresponding language for describing and developing knowledge-based systems based on libraries of generic problem-solving components [17]. UPML largely originated from the body of research on Knowledge Engineering systems and Problem-Solving Methods (PSMs) research, see [45, 51]. Although UPML was not directly aimed at modeling services and developing service-based systems it was the basis upon which the IBROW project developed an intelligent broker which is the precursor of considerable research on Semantic Web Services, and has a significant overlap with service-oriented computing.

UPML provides a meta-ontology for describing knowledge-based systems. The ontology introduces four main components, namely, *Tasks*, which define the problems that should be solved; *PSMs*, which define the reasoning process of a Knowledge Based System in domain-independent terms; *Domain models*, which describe the domain knowledge of the KBS; and *Ontologies*, which provide the terminology used in tasks, PSMs and domain definitions. Each of these components is supposed to be developed independently and therefore UPML identifies the notion of *adapters* to adjust the (reusable) parts to each other and to the specific application problem. UPML provides in particular two types of *adapters*: *bridges* and *refiners*. The former explicitly model the relationships between two distinct parts of an architecture, e.g., between domain and task or task and PSM. The latter can be used to express the stepwise specialization of a class of elements of a specification, e.g., a task is refined or a PSM is refined.

DAML-S / OWL-S

OWL-S (previously known as DAML-S [4]) was the first Semantic Web Service framework which identified key conceptual elements required to describe Web services. OWL-S is an ontology specified using the Web Ontology Language (OWL) [7]. Depending on the required expressivity and decidability the following languages can be used: *OWL-Lite*, *OWL-DL* and *OWL-Full*. The high-level objective of OWL-S is to [33]: (1) provide a general purpose Semantic Web Service framework, (2) support automated service usage and management by software agents, (3) build on existing Web and Semantic Web standards, (4) support a complete service lifecycle.

The main OWL-S elements include:

- **Service Profile** contains name of the Web service, information about the Web service provider, non-functional properties, textual description and most importantly the formal specification of the service functionality in terms of *inputs*, *outputs*, *precondition* and *results*. It contains functional description of what can be achieved by executing the Web service. There is no explicit notion of a service requester in OWL-S therefore a Service Profile can be used both as a specification of a requested service functionality as well as a description of a provided service.
- **Service Grounding** gives the details on how to access the service linking the *OWL-S* service abstract definitions to an underlying Web Services Description Language (WSDL) document. It specifies protocol and message format details on how the underlying service should be accessed.
- **Service Process** describes in terms of the process how to interact with the service. Three types of processes are available in OWL-S, namely: Atomic, Simple and Composite. An Atomic Process is used to describe a service that expects a single message and provides a single message in response. On the other hand, a Composite process maintains a state of conversation allowing for many correlated messages to be exchanged between service requester and service provider. Activities within a Composite Process are chained using workflow constructs such as Sequence, Join, Split-Join, Any-Order and Choice. A Simple Process expands to a Composite Process by providing an abstraction over the functionality offered within a Composite Process. Each process is described in terms of *inputs*, *outputs*, *precondition* and *results*.

Capabilities Model

The authors [36] present an approach for explicitly describing what a service or an agent can do. They introduce a model for capturing the capabilities in order to be able to advertise and discover Web services. The description form is based on a set of requirements specifying the service's capabilities including the behavioral aspects of a service, all the words and terms that it refers to, and providing sufficient information to enable the identification of alternative services without human intervention.

The capability description is presented as an Object Role Model (ORM), which is very suitable for visualizing conceptual data by employing modeling technology in combination with conceptual querying. Services are described in terms of entity types (capabilities) and the roles played by the entities. Web services that are captured is such a way can be discovered and queried by using a conventional query language such as ConQuer. In addition, existing natural language processing tools can be used to ease the generation of the descriptions.

myGrid **Ontology**

myGrid [55] is a project that applied Semantic Web Services to the bioinformatics domain in order to better support scientists in carrying out their analyses and experiments by facilitating the discovery of services they could use to transform or process the data they have. To this end the project produced the *myGrid* ontology designed to support Web service discovery and composition in the bioinformatics domain. The ontology contains 710 classes and 52 properties as two distinct components, the service ontology and the domain ontology. The domain ontology acts as an annotation vocabulary including descriptions of core bioinformatics data types and their relationships to one another, whereas the service ontology describes the information model and functional characteristics of Web services, such as, inputs and outputs. This ontology has also been adopted by other projects and approaches such as SADI, which additionally provides a set of principles and infrastructure enabling the enrichment and discovery of data by means of services [53].

The service ontology provided by *myGrid* solely focusses on the high-level aspect of services which could be relevant to scientists, leaving technical issues aside. The core entity of the service ontology in the model is the *operation*, which represents a unit of functionality. Operations can be grouped into units of publication represented by a *service*. An operation has one or more *input* and *output parameters*. In turn, each input and output parameters have a *name*, a *description* and belongs to a certain namespace denoting its *semantic domain type* and *domain-specific format*. Additionally, the domain ontology covers key concepts for data and data structures in the domains of informatics, bioinformatics and molecular biology as well as it provides a hierarchy describing generic tasks an operation can perform.

Active SWS

Active Semantic Web Services (ASWS) [43] is a platform based on a programming model and language called View Design Language (VLD) and an architecture that takes into account interaction with both human users and other services, based on an execution model for agents. The main idea of this approach is provide proactive software components whose actions, interaction capabilities and data manipulated are explicitly captured. ASWS have access to the semantics of their actions, are compatible with Web services standards, can interact with humans through HTML web pages, and they can answer questions about their actions in natural language.

In order to cater for this ASWS are based on an active component programming language called VDL and an execution framework based on XML tree rewriting through dynamic analysis of the document. VDL allows one to describe the operational part of an agent or service allowing humans to interact with the ASWS and figure out their abilities and state. In order to simplify this interaction, the framework also provides a query language which allow humans to easily create new questions about the activity and behavior of the VDL agents.

DIANE

DIANE is a framework for automating the discovery, composition, binding and invocation of services [23, 27]. The framework is based on DIANE Elements which is a specialized ontology language for describing service elements, and DIANE Service Description (DSD). DIANE Elements is based on object-orientation and exploits the notions of attributes and fillers from frame languages, builds upon the the primitive data types from XML Schema, and reuses the clean separation between schema and instances promoted by description logics. DIANE elements includes special constructs to describe service such as declarative and fuzzy set as well as variables.

DSD uses the specialized constructs provided by DE to describe services. DSD distinguishes three layers: the upper service ontology, category ontologies, and domain ontologies. The upper ontology used in DIANE is largely based on OWL-S and thus includes the notion of Service Grounding and Service Profile. The latter is defined in terms of inputs, outputs, preconditions, effect and non-functional properties. The category layer divides the space of services into clusters with similar state transitions. Finally, the third layer are domain ontologies. They conceptualize a certain application domain such as books, locations, files, etc. They are used to concretize the attributes of the state change, especially to fill the entity which the state refers to.

Semantic MOBY

Semantic MOBY first introduced in [44] as an extension of the BioMoby project [54] focuses on providing semantic descriptions to the already captured body of services for the biomedical domain part of BioMoby. Semantic Moby is centered on three main tasks, namely the provision of a common syntax and semantics for services, as well as the development of generic infrastructure that could exploit these descriptions to provide better service discovery. The vision pursued by Semantic MOBY is one whereby software agents interchange RDF graphs containing the data to be manipulated by the services. Subsequently this work has continued as part of SSWAP covered later in this chapter.

SWSO

The Semantic Web Services Ontology (SWSO) is part of the Semantic Web Services Language (SWSL) [6], which consists of formal conceptual definitions as well as individual Web services. SWSO provides a description of the services ontology, and a description of a first-order logic (FOL) axiomatization (FLOWS - the First-order

Logic Ontology for Web Services), which defines the model-theoretic semantics of the ontology. The goal of FLOWS is to enable reasoning about the semantics underlying Web services, and how they interact with each other and with the "real world". Its goal is not to achieve a complete representation of Web services, but rather an abstract model that is faithful to the semantic aspects of service behavior. As a result, the so created service descriptions enable automated discovery, composition, and verification, as well as the creation of declarative descriptions of a Web service, that can be mapped to executable specifications.

The structure of FLOWS is very similar to the one proposed in OWL-S and also consists of three main components: Service Descriptors, Process Model and Grounding. Service Descriptors provide basic information about a Web service (such as name, author, contact information, URL, identifier, version, etc.) to be used during service discovery and matching with client preferences. The Process Model adapts PSL processes generic ontology to provide a useful framework for describing Web services. This modular nature of the Process Model in SWSO enables Web services developers to define and use their own extensions to FLOWS-Core. Indeed, a main design objective in SWSO is to ease the integration with other Web service workflow models, for instance BPEL. Finally, the Grounding allows to declare the low-level details of Web service definitions: message formats, transport protocols, network addressing, etc. This is the mission of WSDL language, so SWSO implements mechanisms to map high-level ontology descriptions to WSDL.

USDL

Universal Service-Semantic Description Language (*USDL*)[1] "is a language for formally describing the semantics of Web services" [5]. *USDL* is based on OWL and it uses WordNet as a common basis for understanding the meaning of services. It is perhaps the first attempt to capture the semantics of web-services in a universal, yet decidable manner [5]. *USDL* is a formal service documentation, which helps to model conceptually and search for Web services. The design of *USDL* is based on two languages: WSDL and OWL. *USDL* defines a generic class called Concept, which is used to define the semantics of parts of messages. The *USDL* Concept class denotes the conceptual objects constructed from the OWL WordNet ontology.

For most purposes, message parts and other WSDL constructs are mapped to a subclass of *USDL* Concept so that useful concepts can be modeled as set theoretic formulas of union, intersection, and negation of basic concepts. These subclasses of Concept are Basic Concept, Qualified Concept, Inverted Concept, Conjunctive Concept, and Disjunctive Concept. Inclusion of the *USDL* description, makes service directly semantically searchable. The component services can be discovered from existing services using their *USDL* descriptions. Once we have the component services, the OWL-S description can be used to generate the new composed service [5].

WSMO

WSMO [42, 13] is a member submission to W3C of an ontology that aims at de-

[1] *USDL* in this case is not to be confused with the Unified Service Description Language which is at the core of this book. To distinguish between both we shall use italics in this chapter to refer to the one briefly introduced in this section as part of the related work.

scribing all relevant aspects for the partial or complete automation of discovery, selection, composition, mediation, execution and monitoring of Web services. WSMO has its roots in the Web Services Modeling Framework [16] and in Problem-Solving Methods [45, 51], notably the Unified Problem Solving Method Development Language (UPML) [17].

WSMO identifies four top-level elements as the main concepts, namely *Ontologies*, *Web Services*, *Goals* and *Mediators*. *Ontologies* provide the formal semantics for the terminology used within all other WSMO components. Essentially WSMO establishes that all resource descriptions and all data interchanged during service usage should be semantically described. *Web Services* are computational entities that provide some value in a given domain. *Goals*, represent clients' perspective by supporting the representation of users' desires for certain functionality. Finally, *Mediators* represent elements that handle interoperability problems between any two WSMO elements. In fact, one core principle behind WSMO, is the centrality of mediation as a means to reduce the coupling and deal with the heterogeneity that characterizes the Web.

The two core WSMO notions for semantically describing Web services are a *capability* and service *interfaces*. A capability defines the functionality offered by a service by means of *pre-conditions*, *assumptions*, *post-conditions*, and *effects*. A service *interface* defines how the functionality of a service can be achieved by means of a *choreography* and an *orchestration*. A *choreography* describes the behavior of a service from the client's point of view. An *orchestration* captures the control and data-flow within a complex *Web Service* by interacting with other *Web Services*.

Finally, *Goals* are derived from the notion of task prevalent in previous work including Knowledge Acquisition and Documentation Structuring (KADS) [45], UPML [17], whereas *Mediators* in WSMO handle heterogeneities which can occur when two software components are put together.

The Semantic SOA Reference Ontology being devised under OASIS [37] essentially revisits WSMO in the light of the OASIS SOA Reference Model [31] and provides an RDF(S) serialization instead. The work is still ongoing and the differences are minimal for it to be included as a separate model. The technical committee also envisaged producing reference conceptualizations for the main components conforming Semantic Web Services frameworks but at the time of this writing there is no material in this respect.

Core Ontology of Web Services (COWS)

The Core Ontology of Web Services [38] is based on the Core Ontology of Software Components, which is in turn based on the Core Software Ontology [38]. The fundamental concepts are separated in a core ontology to facilitate reuse and enable the extensibility. The Core Ontology of Software Components consists of fundamental concepts and associations such as software, data, users, policies and so on. On top of that, COWS supports modeling requirements of service profiles and service taxonomies. In particular, it includes a formalization of the term "Web service," while *Analyzing Message Contexts*, *Selecting Service Functionality*, *Relating Communication Parameters*, *Aggregating Service Information* and *Quality of Service* are defined as part of the modeling of service profiles and taxonomies. The developed ontology supports the semantic management of middleware and is realized as part of an inference engine that enables querying and reasoning with the semantic descriptions of components and services.

Web Service Ontology
The aim of the Web Service Ontology [22] is to support ubiquitous environments and enable composition of Web services based on a description capturing their functional and semantic aspects. The ontology is based on extracting semantic information about actions and objects of Web services published in open repositories such as UDDI. In particular, it enables capturing the service domain information, preconditions on input parameters and post-conditions on output parameters. The descriptions are divided into two parts, including functional descriptions and capability descriptions. Services described in such a way are used for performing composition in ubiquitous environments, where data transformation and mapping between connected services are realized with the help of mediators.

Fusion Ontology
The FUSION Ontology [26] defines different types of concepts used for modeling a service including: exchanged messages, functional categorization and service behavior in stateful process execution. The FUSION Ontology is used in the FUSION Service registry where the main aim is to support the discovery of services using their functional and non-functional annotations.

FUSION Services are described using Functional Profile which contains the following elements: *hasCategory* (service categorization), *hasInput* (represents parameters that the service expects to receive), and *hasOutput* (represents parameters that the service produces upon invocation). There are two kinds of Functional Profiles: Advertisement Functional Profiles (AFPs) that are used for describing services and Request Functional Profiles (RFPs) that are used for describing service requests.

Minimal Service Model
The Minimal Service Model introduced together with hRESTS [24] aims to cover for the fact that Web API descriptions do not typically have any structure in terms of the resources handled nor the operations exposed. MSM is a simple RDF vocabulary covering what can essentially be considered the core of WSDL. In particular, it defines basically *Services* as having a number of *Operations* which have an *Input*, an *Output*, and *Faults*. Although it was originally introduced to provide structure to hRESTS, it has subsequently been used as a means to integrate heterogeneous services (i.e., WSDLs and Web APIs), and together with WSMO-Lite it has been used as a means to provide a common framework covering the largest common denominator of the most used SWS formalisms on the Web. On the basis of MSM and WSMO-Lite generic publication and discovery machinery has been developed that supports SAWSDL, WSMO-Lite, hRESTS/MicroWSMO, and OWL-S services [41]. The services exposed by this infrastructure, are referred to as Linked Services [39], a new breed of Semantic Web Services where the emphasis lies on reducing the complexity of conceptual models and integrating services with existing Linked Data [11]. This integration serves both as a means to simplify the creation and management of Semantic Web Services through reuse, as well as it provides a

new view over Semantic Web Services understood as a means to support generating and processing Linked Data on the Web.

Subsequent work around the Minimal Service Model has focused on supporting the invocation of Web APIs. Notably, an ontology has been defined for capturing Web APIs authentication mechanisms, which is often a necessary requirement for supporting the invocation of Web APIs [32]. This ontology thus covers the main types of authentication mechanisms, the way to create credentials and how to send them at authentication time. Additionally, there has been work on capturing the grounding information, i.e., technical details concerning the actual invocation of a Web API, including the HTTP method used, the distribution of parameters across the HTTP packet, etc. This model has been validated with a generic Web API invocation engine [29].

ServONT

ServONT is an ontology-based hybrid approach where different kinds of matchmaking strategies are combined together to provide an adaptive, flexible and efficient service discovery environment [10]. Semantic-enriched frameworks are considered a key issue to enforce timely discovery and dynamic composition of services. It is based on keyword-based approach of UDDI Registry to obtain a semantic matchmaking approach based on the use of: a domain ontology, that provides the general knowledge about concepts of the business domain in which services are used; a service ontology, where services are organized by means of semantic relationships at multiple levels of abstraction [10].

Different matchmaking models are: (1) a deductive model to determine the kind of match; (2) a similarity-based approach, exploiting retrieval metrics to measure the degree of match between services and (3) a hybrid model combining the previous ones to mix deductive precision with similarity-based flexibility. The use of different matchmaking models aims at improving searching results and can be used in conjunction with optimization and ranking strategies [10]. Services are described, according to a description logic formalism, in terms of their functional interface. A domain ontology is used to express semantics of service description elements. Moreover, a three-layer service ontology is defined to organize services at different levels of abstraction. Matchmaker is able to recognize not only exact matches but also the degree of similarity between a service request and a service advertisement that do not match exactly. In summary, Ontologies are used to extend the functional description of services with semantic knowledge.

ServFace

The ServFace project aims at creating a model-driven service engineering methodology for an integrated development process for service-based applications.[2] In a nutshell the approach pursued it to add UI-related annotations to service descriptions, notably WSDLs, in order to better support building interactive service-based applications. Driven by the user interface annotations the project is devising new algorithms for the composition of annotated services to build interactive service-based applications. Additionally, ServFace is designing a methodology enabling the

[2] `http://www.servface.eu/`

development of interactive applications involving the annotation of Web services, the presentation-oriented composition of services to form complex applications, and the eventual generation of executable applications from these composites [21].

SSWAP

Simple Semantic Web Architecture and Protocol (SSWAP) is the driving technology for the iPlant Semantic Web Program.[3] It combines Web service functionality with an extensible semantic framework to satisfy the conditions for high throughput integration [18]. SSWAP originates from the Semantic MOBY project, which is a branch of BioMOBY project [54]. Under the umbrella of BioMOBY, Semantic MOBY defines the fundamental model for a semantic Web approach, while MOBY Services provides the Web services approach commonly referred to as BioMOBY. Semantic MOBY project was followed by The Virtual Plant Information Network (VPIN) that eventually turned into SSWAP. It is a lightweight, document-centric protocol and an architecture for the Semantic Web. Using SSWAP, users can create scientific workflows based on the discovery and execution of Web services and RESTful services. SSWAP utilizes OWL ontologies to describe the features and capabilities of Web services and standard HTTP methods to invoke the services.

The SSWAP architecture is based on five basic concepts Provider, Resource, Graph, Subject, and Object. Provider is an organization that owns and publishes resources. Resources can be Web pages, ontologies and databases, which are used to describe services offered on the Web. Graph defines a mapping from a SSWAP Subject (input) to a SSWAP Object (output) [18]. The match between a service and a query is based on the semantic relations between the resource, subject, and object types. In particular, a discovery server is expected to return all resources that are of the same class or a subclass of the query graph's Resource, and superclass of the query graph's Subject, and subclass of the query graph's Object.

ReLL

Resource Linking Language (ReLL) [3] is a language describing interlinked REST resources, and thus the service that can be accessed by interacting with those resources. It is based on a service description meta-model consisting of a service that provides one or more resources that have optionally a URI pattern. The URI pattern does not represent a fixed structure for a URI but instead describes the constraints for resource unique identifiers. Each resource may have representations, which are the serialization of the resource in some syntax. In turn, each representation can contain links relating one resource to another target resource. The explicit description of links between resources is a unique characteristic of ReLL. The links are typed and can be retrieved from representations through selectors that can be specified for example through XML Path Language (XPath). In addition, links follow the rules specified by a protocol, including the method to be used for the request, plus additional information.

ReLL is implemented as part of a crawler (RESTler) [2] that uses ReLL-based descriptions as a format for traversing a RESTful service and produces a typed graph of the crawled resources and the links connecting them. Services are described by

[3] `http://sswap.info`

ReLL generating XML documents that direct the steps of a Web crawler. While crawling, a translator component is invoked for generating RDF triples. The result is a graph based on the links between the crawled resources. The authors envision that ReLL may serve for guiding automated clients in the process of composing new services (e.g., mashups) and providing a framework for programming in a strictly resource-oriented scenario.

ROSM

The Resource-Oriented Service Model (ROSM) [14] ontology is a lightweight approach to the structural description of resource-oriented (RESTful) services, compatible with WSMO-Lite annotations. It enables the annotation of resources belonging to a service. In turn the resources can be described as being part of collections and having addresses (URIs) serving as endpoints for access and manipulation. The organization of resource in collections, which again belongs to a service, allows capturing an arbitrary number of resources and attaching service semantics to them following the SAWSDL approach.

Furthermore, resources can have certain (HTTP) methods associated with them, which define how it is possible to interact with a resource, which are connected through an operation. These operations are modeled in a much-more fine-grained way, since they basically only have to support the uniform interface of HTTP. In addition, ROSM[4] enables the explicit modeling of requests and responses with their associated aspects (e.g., parameters, response codes, etc.). In summary, ROSM represents a simple ontology for describing resource-centered services, in terms of resources, collection of resources, addresses and HTTP methods. The specific semantics of the ROSM annotations can be described by linking to the WSMO-Lite ontology.

7.4 Bottom-Up Approaches

After a few years of research on Semantic Web Services essentially driven by a top-down view over services, researchers started providing approaches that take instead as a starting point the underlying Web services. These approaches are predicated on the extension of existing standards and technologies with semantic annotations rather than on modeling the services entirely in terms of some ontology. Indeed, the annotations refer to ontologies of a more or less rich nature which thus provide a conceptual, top-down, view over services. In some cases, e.g., WSMO-Lite, it is unclear how one should categorize it. Nonetheless, since the path followed and the technical means differ, we herein contemplate all the approaches that start from an underlying traditional service or process definition document as being bottom-up.

WSDL-S

WSDL-S is a lightweight approach to associating semantic annotations with Web services. WSDL-S was proposed as a member submission to the W3C in November

[4] `http://www.wsmo.org/ns/rosm/0.1/`

2005 [1]. "The key innovation of WSDL-S lies in the use of extensibility in elements and attributes supported by WSDL specification" [40].

WSDL-S is independent from the language used for defining the semantic models and explicitly contemplates the possibility of using WSML, OWL and UML as potential candidates [1].WSDL-S provides a set of extension attributes and elements to associate the semantic annotations. The first one is the model reference, which allows one to specify associations between a WSDL entity and a concept in a semantic model. Model reference is used for annotating XML Schema complex types and elements, WSDL operations and the extension elements precondition and effect. Precondition and effect are two new children elements for the WSDL operation element, which facilitate the definition of the condition that hold before executing an operation and the effect of that [40]. WSDL-S also includes the category extension attribute on the interface element in order to define categorization information for publishing Web services in registries as defined by the Universal Description, Discovery and Integration (UDDI) specification for example [20].

GPO/PSAM

The General Process Ontology (GPO) and the Process Semantic Annotation Model (PSAM) [30] define business process annotations. The main aim of this work is to enable business process discovery and sharing within and across enterprises considering process heterogeneities. GPO/PSAM approach proposes four annotation sets for describing process models, namely: *profile annotations*, *meta-model annotations*, *model annotations*, and *goal annotations*. Profile annotations are basic process description and include the following groups: administrative (e.g., creator, publisher), descriptive (e.g., title, category), technical (modeling language), preservation (documentation) and use (e.g., used in). Meta-model annotations include typical business process constructs such as: *Activity*, *Actorrole*, *Input*, *Output*, *Merge*, *Join*, and others. Model annotations use domain specific ontologies to annotate various parts of business processes. Goal annotations are used to specify aims of business process activities with distinction on *local* and *global* goals.

QuASAR / ISPIDER

The goal of Quality Assurance of Semantic Annotations for Services (QuASAR)[5] is to support the full life-cycle of Web service annotations and to ensure trustworthiness and accuracy of annotations. A beta version of the QuASAR verification tool is available. QuASAR uses ISPIDER ontologies [8] to annotate services.

The following distinct ontologies which cover different aspects of service annotations are distinguished: *Domain ontology*, *Representation ontology* and *Extend ontology*. Domain ontology describes common concepts relevant within a given domain. Domain ontologies are referred to in service annotations from similar domains (e.g., biomedical services, geolocation services and others). Representation ontology describes the representation format of service parameters. Parameters specified in WSDL are augmented with information specified in the Representation ontology. Extend ontology describes scopes of values of service parameters. Information

[5] `http://img.cs.man.ac.uk/quasar`

about scopes of values helps to detect incompatibilities between seemingly well-formed services.

BPEL4SWS

BPEL4SWS [35] is a language for Semantic Web Service orchestration based on Business Process Execution Language (BPEL). BPEL is the de facto standard for Web service orchestration. BPEL4SWS extends BPEL with support for Semantic Web Services. BPEL4SWS can be used for orchestration of both Web services and Semantic Web Services. Semantic annotations can be attached to any part of BPEL4SWS descriptions. However, semantic annotations of conversations (message exchange) are advocated.

BPEL4SWS uses SAWSDL for handling data lifting and lowering. SAWSDL introduces *modelReference* (identifies the semantic concept that the XML message can be lifted to), *liftingSchema* (defines how XML message can be lifted to the semantic representation), and *loweringSchema* (defines how the semantic representation can be lowered to the XML message).

Support for the execution of BPEL4SWS is implemented in BPEL4SWS Execution Engine which is based on the open source Apache ODE BPEL engine.[6] It supports both invocations of Web services and WSMO Semantic Web Services.

SAWSDL

The Semantic Annotations for WSDL and XML Schema is nowadays the only Semantic Web Services standard. SAWSDL, which is essentially derived from WSDL-S, supports adding semantic annotation to Web services using a set of extension attributes for WSDL and XML Schema [15, 40]. Adding semantic annotations to WSDL documents paves the way for providing additional details concerning the service or for pointing to ontological definitions that capture the semantics of that service [40]. There are two kinds of extension attributes defined by SAWSDL. The first one, *modelReference*, is used to specify the association between a WSDL or XML Schema component (e.g., operation, and input) and a concept in some semantic model such as an ontology. The other kind of extension attributes are *schemaMappings*. In particular, *liftingSchemaMappings* capture how to lift data represented in XML to its semantic counterpart. Conversely, *loweringSchemaMappings* capture how to carry out the inverse process, that is to transform data represented semantically into the XML representation the service expects. SAWSDL does not impose nor restrict in any manner the languages that can be used to specify the schema mappings nor the semantic models. SAWSDL constitutes a lightweight and incremental approach (compared to OWL-S and WSMO) to annotating WSDL services.

In summary SAWSDL does not specify a language for representing the semantic models. Instead it defines some minimal extension mechanisms for WSDL and XML Schema that allow one to define the semantics of a service or its parts by pointing to concepts from externally defined semantic models. SAWSDL is a restricted and homogenized version of WSDL-S including a few changes trying to give a greater level of generality to the annotations and disregarding those issues,

[6] `http://code.apache.org`

notably the definition of conditions and effects, for which there existed no agreement among the community at the time the specification was created [40].

YASA

Yet Another Semantic Annotation for WSDL (YASA) [12] proposes an extension of SAWSDL. It uses two types of ontologies: a *Technical Ontology* that contains concepts defining service semantics and a *Domain Ontology* that covers a business domain. YASA introduces a new attribute called *serviceConcept* that is used together with SAWSDL *modelReference* attribute. *serviceConcept* attribute references concepts modeled in Technical Ontologies. YASA claims that introducing *serviceConcept* attribute makes SAWSDL descriptions more expressive and allows to explicitly capture information on service pre-, post-conditions and effects.

MicroWSMO/hRESTS

In comparison to WSDL-based services, which have a somewhat longer history of research on semantic descriptions and annotation approaches, research in the area of semantic RESTful services is newer and therefore relatively limited. MicroWSMO [25] is a formalism for the semantic description of RESTful services, which is based on adapting the SAWSDL approach. MicroWSMO uses microformats for adding semantic information on top of HTML service documentation, by relying on hRESTS (HTML for RESTful Services) [24] for marking service properties and making the descriptions machine-processable. hRESTS enables the annotation of service operations, inputs and outputs, HTTP methods and labels, by inserting HTML tags within the HTML. In this way, the user does not see any changes in the Web page, however, based on the tags the service can be automatically recognized by crawlers and the service properties can directly be extracted by applying a simple XSL transformation.

hRESTS provides the syntactical structuring of the service, while MicroWSMO [25] enables adding semantic annotations on top. It uses three main types of link relations: 1) *model*, which can be used on any service property to point to appropriate semantic concepts identified by URIs; 2) *lifting* and 3) *lowering*, which associate messages with appropriate transformations (also identified by URIs) between the underlying technical format such as XML and a semantic knowledge representation format such as RDF. Therefore, MicroWSMO, based on hRESTS, enables the semantic annotation of RESTful services in the same way in which SAWSDL, based on WSDL, supports the annotation of Web services.

In addition, MicroWSMO can be complemented by the WSMO-Lite service ontology specifying the content of the semantic annotations. Since both RESTful and WSDL-based services can have WSMO-Lite annotations, this provides a basis for integrating the two types of services. Therefore, WSMO-Lite enables unified search over both WSDL-based and RESTful services and tasks such as discovery, composition and mediation can be performed based on WSMO-Lite, completely independently from the underlying Web service technology (WSDL/SOAP or REST/HTTP).

WSMO-Lite

As described in the previous section SAWSDL provides simple hooks for pointing to semantic descriptions from WSDL and XML elements. SAWSDL does not advocate a particular representation language for these documents nor does it provide any specific vocabulary that users should adopt. WSMO-Lite continues this incremental construction of a stack of technologies for Semantic Web Services by precisely addressing this lack [52]. WSMO-Lite identifies four main types of semantic annotations for services which are a variant of those in [46], that is, functional semantics, non-functional semantics, behavioral semantics, and the information model.

WSMO-Lite provides a minimal RDFS ontology and provides a simple methodology for expressing these four types of semantic annotations for WSDL services using SAWSDL hooks. In particular, to specify the annotations over a concrete WSDL service, WSMO-Lite relies on the SAWSDL modelReference attribute for all four semantic annotations, except for information models where liftingSchemaMapping and loweringSchemaMapping might also be necessary.

WSMO-Lite offers two mechanisms for representing functional semantics, namely simple taxonomies and more expressive preconditions and effects. In order to distinguish functional classifications from other types of *modelReference* annotations, WSMO-Lite offers the RDFS class *wsl:FunctionalClassificationRoot*, where wsl identifies the namespace for WSMO-Lite. Whenever more expressivity is necessary, WSMO-Lite offers the possibility to enrich functional classification with logical expressions defining conditions that need to hold prior to service execution, and capturing changes that the service will carry out on the world by means of the classes *wsl:Condition* and *wsl:Effect* respectively.

Non-functional semantics in WSMO-Lite are represented using external ontologies capturing non-functional properties such as the quality of service and price. To do so WSMO-Lite includes the class *wsl:NonfunctionalParameter* so that ontologies defining concrete non-functional properties for a service, can refer to this class and let the machine know what kind of information it contains.

Behavioral semantics, i.e., how the client should communicate with a service, although they are identified are not explicitly addressed by WSMO-Lite. Instead, clients are expected to use the existing knowledge concerning operations functionality in order to determine how a service's operations should be combined.

Finally, the information model is directly covered by SAWSDL modelReference mechanism.

SA-REST

Semantically Enhancing RESTful Services and Resources is an open, flexible, and standards-based approach to adding semantic annotations to RESTful services and Web APIs [28, 40, 47]. SA-REST was accepted by the W3C as a member submission on April 2010 [19]. SA-REST is a poshformat for adding additional meta-data to REST API descriptions in HTML or XHTML. In a nutshell it allows one to embed meta-data coming from different models such an ontologies or taxonomies into Web documents describing Web APIs. The main goal pursued is to improve search, facilitates data mediation and even helps to integrate different Web APIs [19] by exploiting these annotations.

SA-REST defines three basic properties for the non-intrusive annotation of HTML/ XHTML documents [19]. The *domain-rel* property allows one to provide domain information descriptions for a resource. This annotation is essentially thought for providing coarse grained categorizations of the HTML elements by pointing to general domains such as "Business." The *sem-rel* property, which evolves from the popular rel tag, is used to capture the semantics of a link within an HTML document. This kind of annotation is thus supposed to be used solely within an anchor element (<a>). Finally, the *sem-class* property can be used for annotating a single entity within a resource [19].

Semantic annotation based on ER model

Usually, Semantic Web Services are based on the definition of an ontology that captures key domain information. Since the construction and maintenance of such an ontology requires considerable effort and the involvement of domain expects, the authors in [56] propose to use ER (entity-relationship) models instead. ER models are widely adopted in database management and provide a good starting point for constructing domain ontologies, with which the Semantic Web Services are annotated, reasoned and composed. In particular, the approach is based on taking the semantic information that can be retrieved from the ER model and mapping it to OWL or OWL DL, specifically, which has maximum expressiveness while retaining computational completeness and decidability. The authors construct SWS with OWL-S-like descriptions using the ER model, by creating a query service for single entity, similar to a single-table query in SQL, which is then used as part of composite processes based on multi-table queries or nested queries. The introduced Semantic Web Service annotation based on ER models is particularly targeted at supporting service composition.

Linked Data Services

Linked Data Services (LIDS) [49], are part of the new trend on Linked Services [39] that aim to bridge the Linked Data and services. LIDS, in particular, focuses on integrating existing data services exposed through Web APIs, with Linked Data principles [11]. LIDS are based on well-established Web standards such as HTTP, RDF and SPARQL. LIDS consume and produce RDF triples and are available over HTTP protocol. LIDS is a lightweight service description model where service inputs and outputs are specified using SPARQL graph patterns. A simple LIDS vocabulary[7] is defined. LIDS service descriptions are mapped to and executed as SPARQL CONSTRUCT queries. LIDS descriptions based on SPARQL are familiar to Linked Data community and RDF triples produced by LIDS services can directly contribute to Linked Open Data landscape.

Approach taken by LIDS follows a recent service trend of creating lightweight semantic service descriptions using well-established Web standards. Existing non-semantic services can be wrapped and described as LIDS by a typical Web developer. LIDS service modeling does not require background in logics in contrast to approaches such as WSMO and OWL-S. LIDS can be directly used by Linked Data consumers and there is no need for data lifting and lowering as it is assumed that services consume and produce RDF.

[7] `http://openlids.org/vocab`

7.5 Atlas and Evolution of Semantic Web Services Approaches

In the previous section we have briefly introduced the different models proposed in the state of the art, providing a basic description and pointers for the interested reader. Although, we can't claim this list to be exhaustive, to the best of our knowledge it represents the largest collection gathered of published models in the area. We have purposely not made any qualifications and comparisons of these models for both reasons of space but most importantly for such an analysis would require addressing each of the models from the perspective of the tasks they enable or could enable together with an educated guess or evaluation of existing support by corresponding engines. Given the breadth and depth of this kind of analysis it certainly cannot be covered within this chapter. In this section we provide instead a categorization of the models with respect to the kinds of semantics they cover and the tasks they aim to support. On the basis of this categorization one could carry out a more thorough and deeper analysis on a per task basis. Additionally, we provide a general view on these models in terms of the language they are represented in, their conceptual influences and the year they were proposed in. This information is then used to provide a map on existing models that can help better understand the evolution of the field over the years (cf. Figure 7.1 on page 180).

Table 7.1: General characteristics of Semantic Web Service approaches.

Approach	Year	Conceptual Input	Representation Lang.	Main Purpose
UPML	1999	PSMs	UPML/Lisp	Knowledge-Based Systems development
DAML-S / OWL-S	2001	Agents, Planning	DAML+OIL/OWL	Semantic annotations of WS
Capabilities Model	2003	DAML-S	ORM, ontologies and rules	Discovery, Composition
myGrid & SADI	2003	DAML-S	DAML+OIL / OWL	Discovery of services in bioinformatics
ASWS	2004	Agents, Rewriting Logics	VLD/XML	Agents, Human-Agent interaction
DIANE	2004	OWL-S, WSMO	DIANE Elements	Discovery, Invocation and Composition
Semantic MOBY	2004	Moby, RDF	RDF	Discovery and Invocation
SWSO	2005	OWL-S	SWSL	Discovery, Invocation, Composition and Orchestration
USDL	2005	OWL-S	OWL	Language for formally describing the semantics of Web services
WSDL-S	2005	WSMO, OWL-S, WSDL	XML Schema	Linking semantic annotations to Web services
WSMO	2005	WSMF, UPML	WSML, RDF (Semantic SOA Reference Ontology)	Discovery, Invocation, Mediation, Composition and Orchestration
COWS	2006	DOLCE	OWL	Semantic management of middleware
GPO	2006	Unified Enterprise Modeling Ontology	OWL	Process modeling
QuASAR / ISPIDER	2006	*myGrid*	OWL	Integrated platform enabled as Grid and Web services for the storage, dissemination and management of proteomic data
Web Service Ontology	2006	OWL-S, WSMO, WSBPEL	OWL	Service composition
BPEL4SWS	2007	BPEL4WS, WSMO, SAWSDL	XML Schema	Orchestration
SAWSDL	2007	WSDL-S	XML Schema	Semantic Annotations for WS WSDL and XML Schema
FUSION Ontology	2008	SAWSDL, UDDI	OWL-DL	Service registry
YASA	2008	SAWSDL	XML Schema	Extension of SAWSDL, service discovery

MicroWSMO/ hRESTS	2008	hRESTS/WSMO-Lite	HTML with micro-format tags	Semantic annotations of RESTful services and Web APIs
MSM	2008	WSDL, WSMO-Lite, hRESTS, Linked Data	RDF(S)	Discovery, Invocation
ServONT	2008	OWL	OWL-DL	Service discovery
WSMO-Lite	2008	SAWSDL, OWL-S, WSMO	RDF(S)	Discovery, Invocation, Composition
SA-REST	2009	SAWSDL, hRESTS	RDFa	Semantic annotations of RESTful services and Web APIs
ServFace	2009	WSDL	XML Schema	For adding of UI-related Annotations to Web service Descriptions (WSDL)
SSWAP	2009	HTML, Semantic MOBY	OWL	Data and service integration in Biology
ER Model	2010	ER, BPEL	ER / OWL DL	Discovery, Composition
Linked Data Services	2010	HTTP, Linked Data	SPARQL	Bridging the gap between data services and Linked Data principles. Lightweight composition
RELL	2010	REST	RDF / OWL	Description of resource-centered Web APIs in terms of resources
ROSM	2010	WSMO-Lite, REST	RDFS, SPARQL	Description of resource-centered Web APIs (RESTful services)

Table 7.1 provides some general aspects of the approaches described in the previous section. This includes the year of publication, conceptual input, representation language and the main goals pursued. The entries are ordered with respect to the year of publication and for those within the same year in alphabetical order. Table 7.2 on the other hand presents some more detailed aspects such as the intended coverage in terms of kinds of services (Web services and/or Web APIs) and the different kinds of semantics captured. On the basis of both tables we expect the interested reader to be able to perform more detailed and targeted comparisons driven, say, by the tasks, kinds of services to cover, or general expressivity.

Table 7.2: SWS Approach Characteristics.

Approach	For WS	For Web APIs	Functional	Non-Functional	Informational	Behavioral	Processes
UPML	no	no	yes	no	yes	yes	yes
DAML-S / OWL-S	yes	no	yes	yes	yes	yes	yes
Capabilities Model	yes	yes	yes	no	yes	yes	no
myGrid & SADI	yes	no	yes	yes	yes	no	no
ASWS	yes	no	yes	no	yes	yes	no
DIANE	yes	no	yes	yes	yes	yes	no
Semantic MOBY	yes	no	no	yes	yes	no	no
SWSO	yes	no	yes	yes	yes	yes	yes
USDL	no	no	no	no	yes	no	service composition
WSDL-S	yes	yes	yes	no	yes	yes	no
WSMO	yes	no	yes	yes	yes	yes	yes
COWS	yes	no	yes	yes	yes	yes	yes
GPO	no	no	yes	yes	yes	yes	yes
QuASAR / ISPIDER	yes	no	yes	no	yes	no	yes
Web Service Ontology	yes	no	yes	no	yes	yes	yes
BPEL4SWS	yes	no	yes	yes	yes	no	yes
SAWSDL	yes	no	not explicitly	not explicitly	yes	no	no
FUSION Ontology	yes	no	yes	yes	yes	no	no

YASA	yes	no	Via SAWSDL	Via SAWSDL	Via SAWSDL	Via SAWSDL	no
MicroWSMO/hRESTS	no	yes	Via WSMO-Lite	Via WSMO-Lite	Via WSMO-Lite	Via WSMO-Lite	no
MSM	yes	yes	no	no	yes	no	no
ServONT	no	yes	yes	no	yes	yes	not explicitly
WSMO-Lite	yes	no	yes	yes	yes	no	no
SA-REST	no	yes	not explicitly	not explicitly	yes	no	no
ServFace	no	yes	yes	no	yes	yes	yes
SSWAP	no	yes	yes	no	no	yes	not explicitly
ER Model	yes	no	yes	yes	yes	yes	yes
Linked Data Services	no	yes	yes	no	yes	no	no
RELL	no	yes	no	no	no	no	no
ROSM	no	yes	Via WSMO-Lite	Via WSMO-Lite	Via WSMO-Lite	Via WSMO-Lite	no

7.6 Conclusions

After almost a decade, research on SWS has produced a wealth of conceptual models, languages, architectures, algorithms and engines that highlight the potential of these technologies both for enterprise settings and for the Web. In this chapter we have provided an initial characterization of these efforts along a number of dimensions. Given the breadth of the field, notably in terms of the tasks that could be supported by means of SWS descriptions, this chapter is to be understood as an initial step allowing to better target state of the art analyses (e.g., according to tasks covered) and comparison in the field.

Despite the wealth of models proposed, however, the vision initially proposed in [34] when SWS were first proposed is still to be achieved. Research on SWS so far has focussed mostly on extending existing Web services technologies with semantics. Yet, recent analyses estimate that the number of publicly available Web services on the Web is around 28,600[8] which contrasts with the number of pages available on the Web and with the number of services that big companies have internally (e.g., approximately 1,500 for Verizon) [50]. Hence, despite their name, Web services seem to be essentially enclosed within enterprises. On the Web the use of SWS is even scarcer and it seems that the appearance of intelligent Web agents that act of behalf of users remains an elusive target. Still, the demand for services on the Web exists as indicated by the proliferation and popularity of publicly available Web APIs and RESTful services.

SWS are, in many cases, particularly demanding from a knowledge acquisition perspective. Creating a rich semantic description of a Web service requires the use of domain ontologies, the use of services taxonomies, the definition of lifting and lowering mechanisms, and in some cases the inclusion of complicated logical expressions. This tedious and complex annotation process has arguably hampered the

[8] According to `http://webservices.seekda.com/` as of August 2011.

adoption of SWS technologies especially in the early days of the semantic Web when publicly available ontologies and semantic information were scarce. As a consequence, the application of SWS technologies within real applications is usually limited to simple annotations that simplify the retrieval of services. This leaves aside more advanced features such as the automated selection of services and therefore reduces the potential benefit that can be obtained.

Both the results obtained but also the objectives yet unattained by SWS research are highly valuable for directing future efforts in automating the management of services and thus should drive the efforts around the technology described in this book. One fundamental limitation of past efforts has clearly been on the acquisition of service descriptions, may this be due to the lack of publicly offered services, or to the complexity of this endeavor. Recent efforts are aiming at addressing this by reducing the complexity of the models and the acquisition task, notably by using simple RDF(S) vocabularies and Linked Data. These new approaches present preliminary promising results that could certainly be beneficial for the Unified Service Description Language presented in this book.

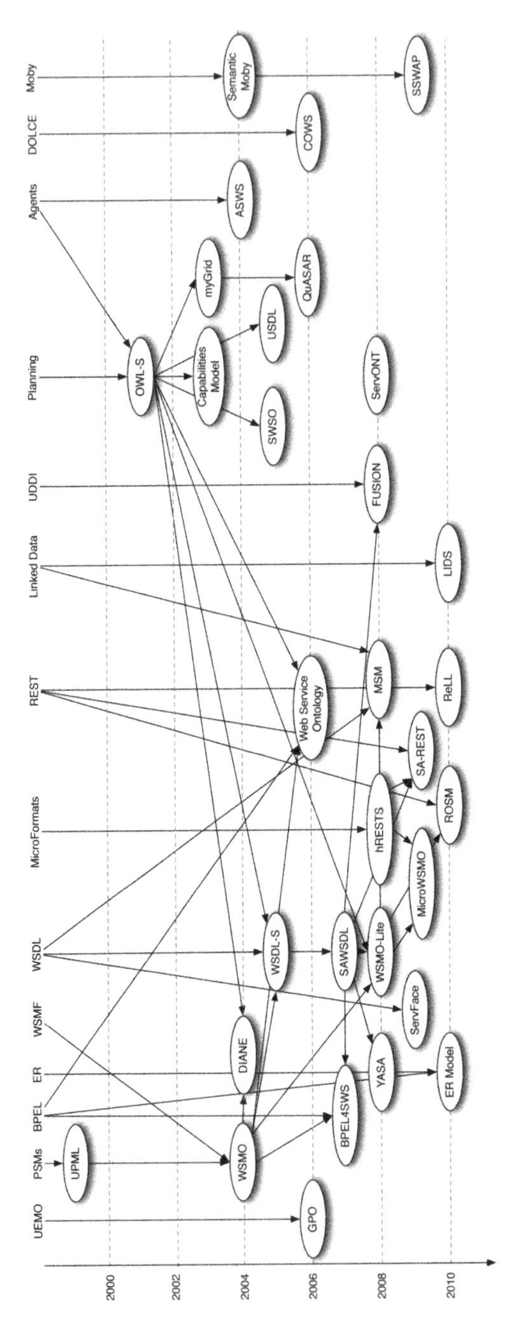

Fig. 7.1: Semantic Web Services Atlas.

References

1. R. Akkiraju, J. Farrell, J. Miller, M. Nagarajan, M.-T. Schmidt, A. Sheth, and K. Verma. Web Service Semantics - WSDL-S. http://www.w3.org/Submission/WSDL-S/, November 2005. W3C Member Submission.
2. R. Alarcón and E. Wilde. From RESTful Services to RDF: Connecting the Web and the Semantic Web. *CoRR*, abs/1006.2718, 2010.
3. R. Alarcón and E. Wilde. RESTler: crawling RESTful services. In *Proceedings of the 19th international conference on World wide web*, WWW '10, pages 1051–1052, New York, NY, USA, 2010. ACM.
4. A. Ankolekar, M. H. Burstein, J. R. Hobbs, O. Lassila, et al. DAML-S: Semantic Markup for Web Services. In *Semantic Web Working Symposium (SWWS)*, 2001.
5. A. Bansal, S. Kona, L. Simon, and T. D. Hite. A universal service-semantics description language. In *Proceedings of the Third European Conference on Web Services*, ECOWS '05, pages 214–, Washington, DC, USA, 2005. IEEE Computer Society.
6. S. Battle, A. Bernstein, H. Boley, B. Grosof, M. Gruninger, R. Hull, M. Kifer, D. Martin, S. McIlraith, D. McGuinness, J. Su, and S. Tabet. Semantic Web Services Language (SWSL). Member submission, W3C, 2005.
7. S. Bechhofer, F. van Harmelen, J. Hendler, I. Horrocks, et al. OWL Web Ontology Language Reference, W3C Recommendation, 2004.
8. K. Belhajjame, S. M. Embury, N. W. Paton, R. Stevens, and C. A. Goble. Automatic annotation of Web services based on workflow definitions. *Transactions on the Web (TWEB)*, 2(2), 2008.
9. T. Berners-Lee, J. Hendler, and O. Lassila. The Semantic Web. *Scientific American*, (5):34–43, May 2001.
10. D. Bianchini, V. Antonellis, and M. Melchiori. Flexible semantic-based service matchmaking and discovery. *World Wide Web*, 11:227–251, June 2008.
11. C. Bizer, T. Heath, and T. Berners-Lee. Linked Data – The Story So Far. *International Journal on Semantic Web and Information Systems*, 5(3):1–22, 2009.
12. Y. Chabeb and S. Tata. Yet Another Semantic Annotation for WSDL (YASA4WSDL). In *IADIS WWW/Internet 2008 Conference*, 2008.
13. J. de Bruijn, C. Bussler, J. Domingue, D. Fensel, M. Hepp, U. Keller, M. Kifer, B. Koenig-Ries, J. Kopecky, R. Lara, H. Lausen, E. Oren, A. Polleres, D. Roman, J. Scicluna, and M. Stollberg. Web Service Modeling Ontology (WSMO). Member submission, W3C, June 2005.
14. F. F. and N. B. D3.4.6 MicroWSMO v2 – Defining the second version of MicroWSMO as a systematic approach for rich tagging. Soa4all project deliverable.
15. J. Farrell and H. Lausen. Semantic Annotations for WSDL and XML Schema (SAWSDL). Recommendation, W3C, August 2007.
16. D. Fensel and C. Bussler. The Web Service Modeling Framework WSMF. *Electronic Commerce Research and Applications*, 1(2):113–137, 2002.
17. D. Fensel, V. Richard, B. Enrico, and M. Bobwielinga. UPML: A Framework For Knowledge System Reuse. In *In Proceedings of the International Joint Conference on AI (IJCAI-99*, pages 16–23. Morgan Kaufmann, 1999.
18. D. Gessler, G. Schiltz, G. May, S. Avraham, C. Town, D. Grant, and R. Nelson. SSWAP: A Simple Semantic Web Architecture and Protocol for semantic web services. *BMC Bioinformatics*, 10(1):309, 2009.
19. K. Gomadam, A. Ranabahu, and A. Sheth. SA-REST: Semantic Annotation of Web Resources. Member submission, W3C, April 2010.
20. L. C. A. Hately, C. von Riegen, and T. Rogers. UDDI Specification Version 3.0.2, 2004.
21. P. Izquierdo, J. Janeiro, G. Hubsch, T. Springer, and A. Schill. An annotation tool for enhancing the user interface generation process for services. In *Microwave Telecommunication Technology, 2009. CriMiCo 2009. 19th International Crimean Conference*, pages 372 –374, sept. 2009.

22. Y. S. Jeon, E.-H. Song, M. Guo, L. T. Yang, Y.-S. Jeong, J.-T. Choi, and S.-K. Han. Ontology-based composition of web services for ubiquitous computing. In G. Min, B. D. Martino, L. T. Yang, M. Guo, and G. Rünger, editors, *ISPA Workshops*, volume 4331 of *Lecture Notes in Computer Science*, pages 559–568. Springer, 2006.
23. M. Klein, B. König-Ries, and M. Mussig. What is needed for semantic service descriptions? a proposal for suitable language constructs. *Int. J. Web Grid Serv.*, 1:328–364, December 2005.
24. J. Kopecký, K. Gomadam, and T. Vitvar. hRESTS: an HTML Microformat for Describing RESTful Web Services. In *The 2008 IEEE/WIC/ACM International Conference on Web Intelligence (WI2008)*, Sydney, Australia, November 2008. IEEE CS Press.
25. J. Kopecky, T. Vitvar, C. Pedrinaci, and M. Maleshkova. *REST: From Research to Practice*, chapter RESTful Services with Lightweight Machine-readable Descriptions and Semantic Annotations. Springer, 2011.
26. D. Kourtesis and I. Paraskakis. Combining SAWSDL, OWL-DL and UDDI for Semantically Enhanced Web Service Discovery. In *European Semantic Web Conference (ESWC)*, 2008.
27. U. Küster, B. König-Ries, M. Klein, and M. Stern. Diane: A matchmaking-centered framework for automated service discovery, composition, binding, and invocation on the web. *Int. J. Electron. Commerce*, 12:41–68, December 2007.
28. J. Lathem, K. Gomadam, and A. P. Sheth. SA-REST and (S)mashups: Adding Semantics to RESTful Services. In *ICSC '07: Proceedings of the International Conference on Semantic Computing*, pages 469–476, Washington, DC, USA, 2007. IEEE Computer Society.
29. N. Li, C. Pedrinaci, M. Maleshkova, J. Kopecky, and J. Domingue. Omnivoke: A framework for automating the invocation of web apis. In *Fifth IEEE International Conference on Semantic Computing*, Palo Alto, CA, USA, 2011.
30. Y. Lin, D. Strasunskas, S. Hakkarainen, J. Krogstie, and A. Sølvberg. Semantic annotation framework to manage semantic heterogeneity of process models. In E. Dubois and K. Pohl, editors, *CAiSE*, volume 4001 of *Lecture Notes in Computer Science*, pages 433–446. Springer, 2006.
31. C. M. MacKenzie, K. Laskey, F. McCabe, P. F. Brown, and R. Metz. OASIS Reference Model for Service Oriented Architecture V 1.0. Technical report, OASIS, July 2006.
32. M. Maleshkova, C. Pedrinaci, J. Domingue, G. Alvaro, and I. Martinez. Using semantics for automating the authentication of web apis. In *International Semantic Web Conference (ISWC)*, Shanghai, China, 2010.
33. D. L. Martin, M. H. Burstein, D. V. McDermott, S. A. McIlraith, et al. Bringing Semantics to Web Services with OWL-S. pages 243–277, 2007.
34. S. McIlraith, T. Son, and H. Zeng. Semantic web services. *Intelligent Systems, IEEE*, 16(2):46 – 53, Jan 2001.
35. J. Nitzsche, T. van Lessen, D. Karastoyanova, and F. Leymann. BPEL for Semantic Web Services (BPEL4SWS). In *On the Move to Meaningful Internet Systems (OTM Workshops)*, 2007.
36. P. Oaks, A. H. M. T. Hofstede, and D. Edmond. Capabilities: describing what services can do. In *Proceedings of the 1st International Conference on Service Oriented Computing, 15–18*, 2003.
37. OASIS Semantic Execution Environment (SEE) TC. OASIS Semantic Execution Environment Specification. http://www.oasis-open.org/committees/tc_home.php?wg_abbrev=semantic-ex, 2007.
38. D. Oberle, S. Lamparter, S. Grimm, D. Vrandecic, S. Staab, and A. Gangemi. Towards ontologies for formalizing modularization and communication in large software systems. *Applied Ontology*, 1(2):163–202, 2006.
39. C. Pedrinaci and J. Domingue. Toward the Next Wave of Services: Linked Services for the Web of Data. *Journal of Universal Computer Science*, 16(13):1694–1719, 2010.
40. C. Pedrinaci, J. Domingue, and A. Sheth. *Handbook on Semantic Web Technologies*, volume Semantic Web Applications, chapter Semantic Web Services. Springer, 2010.
41. C. Pedrinaci, D. Liu, M. Maleshkova, D. Lambert, J. Kopecký, and J. Domingue. iServe: a Linked Services Publishing Platform. In *Proceedings of Ontology Repositories and Editors for the Semantic Web at 7th ESWC*, 2010.

42. D. Roman, U. Keller, H. Lausen, J. de Bruijn, R. Lara, M. Stollberg, A. Polleres, C. Feier, C. Bussler, and D. Fensel. Web service modeling ontology. *Appl. Ontol.*, 1:77–106, January 2005.
43. N. Sabouret. Active semantic web services: a programming model for agents in the semantic web. In *First European Workshop on Multi-Agent Systems (EUMAS)*, December 2003.
44. G. Schiltz, D. Gessler, and L. Stein. Semantic MOBY. In *W3C Workshop on Semantic Web for Life Sciences*, 2004.
45. G. Schreiber, H. Akkermans, A. Anjewierden, R. de Hoog, N. Shadbolt, W. V. de Velde, and B. Wielinga. *Knowledge Engineering and Management: The CommonKADS Methodology*. MIT Press, 1999.
46. A. Sheth. Semantic Web Process Lifecycle: Role of Semantics in Annotation, Discovery, Composition and Orchestration. Invited Talk at WWW 2003 Workshop on E-Services and the Semantic Web, May 2003.
47. A. Sheth, K. Gomadam, and J. Lathem. SA-REST: Semantically Interoperable and Easier-to-Use Services and Mashups. *Internet Computing, IEEE*, 11(6):91 – 94, Nov 2007.
48. K. Sivashanmugam, K. Verma, A. Sheth, and J. Miller. Adding semantics to web services standards. In *Proceedings of the 2003 International Conference on Web Services (ICWS'03)*, pages 395–401, Las Vegas, NV, June 2003.
49. S. Speiser and A. Harth. Taking the LIDS off data silos. In *I-SEMANTICS*, 2010.
50. M. Stollberg. *Scalable Semantic Web Service Discovery for Goal-driven Service-Oriented Architectures*. PhD thesis, Faculty of Mathematics, Computer Science and Physics Leopold-Franzens University Innsbruck, Austria, March 2008.
51. R. Studer, R. Benjamins, and D. Fensel. Knowledge Engineering: Principles and Methods. *Data Knowledge Engineering*, 25(1-2):161–197, 1998.
52. T. Vitvar, J. Kopecky, J. Viskova, and D. Fensel. WSMO-Lite Annotations for Web Services. In M. Hauswirth, M. Koubarakis, and S. Bechhofer, editors, *Proceedings of the 5th European Semantic Web Conference*, LNCS, Berlin, Heidelberg, June 2008. Springer Verlag.
53. M. Wilkinson, B. Vandervalk, and L. McCarthy. SADI Semantic Web Services - cause you can't always GET what you want! pages 13–18, Dec. 2009.
54. M. D. Wilkinson and M. Links. Biomoby: An open source biological web services proposal. *Briefings in Bioinformatics*, 3:331–341, 2002.
55. K. Wolstencroft, P. Alper, D. Hull, C. Wroe, P. W. Lord, R. D. Stevens, and C. A. Goble. The mygrid ontology: bioinformatics service discovery. *Int. J. Bioinformatics Res. Appl.*, 3:303–325, September 2007.
56. C. Xu, P. Liang, T. G. Wang, Q. Wang, and P. C.-Y. Sheu. Semantic web services annotation and composition based on er model. *Sensor Networks, Ubiquitous, and Trustworthy Computing, International Conference on*, 0:413–420, 2010.

Part II
USDL — Meta-Model

Chapter 1 made clear that an Internet of Services requires a way of describing a service to wrap and expose it in a novel way. A service should not only be considered as the invocation of a technical interface, but rather as an economic or social transaction with a broader context. Therefore, it is essential to describe the price scheme, the service level agreement, or the terms and conditions when consuming the service and paying for it. With the many roles involved in service networks, the need for a common standard in the description of operational and commercial meta-data becomes important.

The previous part surveyed a plethora of different service description approaches, with a multitude of languages, techniques, methods, architectures and frameworks, demonstrating that there is no shortage of existing proposals. Therefore, attempts to a service description language that cover services from purportedly all walks of life is fraught with the "yet another silver bullet" syndrome. Is a uniform conception of services across political, economic, business, entertainment, technological, individual and other spheres of the largest sector in the world, possible, or even desirable? Our answer is a qualified *yes*, and the following part is therefore devoted to the proposed *Unified Service Description Language (USDL)*.

Chapter 8 starts off by providing the positioning and design rationale of USDL leading into subsequent chapters of this part that provide details of the USDL meta-model. Chapter 9 discusses the scientific background of the USDL *Pricing Module* as a comprehensive, applicable, executable, and non-proprietary endeavor. As such, the chapter elicits what makes modeling and engineering the price of a service transaction less straightforward than in the case of a product sale. Chapter 10 covers the modeling of licensing aspects (i.e., the USDL *Legal Module*) according to two different jurisdictions, viz., the copyright acts of Germany and the USA. Thus, the chapter addresses the need for legal certainty and legal compliance when trading services in marketplaces and service networks. Chapter 11 draws attention to describing: (1) *what* the service does, i.e., which functionality it provides, (2) *where* the service resides, i.e., where it can be accessed and via which means it can be consumed, and (3) *how* the service behaves, i.e., how to interact with the service in order to properly consume it. In order to capture these aspects in an all-embracing manner, USDL defines three modules (*Functional*, *Technical*, and *Interaction Module*). The modules are commonly designed to provide a unifying description structure that abstracts from details and allows for the re-use and integration of existing as well as upcoming standards, thereby maintaining flexibility and extensibility of USDL. Chapter 12 covers the scientific background of the *Service Level Module* of USDL. In addition to general service level concepts, the chapter expands on two specific service level fields: security and trust. Finally, Chapter 13 provides the modeling foundation of USDL. In particular, the central *Service Module* that ties together the remaining modules, the *Participants Module* for capturing information about the different roles in service networks, and the *Foundation Module*.

Chapter 8
Design Overview of USDL

Alistair Barros, Daniel Oberle, Uwe Kylau, and Steffen Heinzl

Abstract Enabling Web-based service networks and ecosystems requires a way of describing services by a "commercial envelope" as discussed in Chapter 1. A uniform conception of services across all walks of life (including technical services) is required capturing business, operational and technical aspects. Therefore, our proposed *Unified Service Description Language (USDL)* particularly draws from and generalizes the best-of-breed approaches presented in Part I. The following chapter presents the design rationale of USDL where the different aspects are put in a framework of descriptions requirements. This is followed by the subsequent chapters of this part that provide details on specific aspects such as pricing or legal issues.

8.1 Introduction

Any attempt to design a service description language that covers services from purportedly all walks of life is fraught with the "yet another silver bullet" syndrome. Is a uniform conception of services across political, economic, business, entertainment, technological, individual and other spheres of the largest sector in the world,

Alistair Barros
Queensland University of Technology, GPO Box 2434, Brisbane, QLD 4001, Australia,
e-mail: alistair.barros@qut.edu.au

Daniel Oberle
SAP Research Karlsruhe, Vincenz-Priessnitz-Str. 1, 76131 Karlsruhe, Germany,
e-mail: d.oberle@sap.com

Uwe Kylau
SAP Research Brisbane, Building A4 Level 7, 52 Merivale Street, South Brisbane QLD 4101, Australia, e-mail: uwe.kylau@sap.com

Steffen Heinzl
SAP Research Darmstadt, Bleichstrasse 8, 64283 Darmstadt, Germany,
e-mail: steffen.heinzl@sap.com

possible, or even desirable? Moreover, as described through the previous part surveying the myriad of different approaches, with a plethora of languages, techniques, methods, architectures and frameworks therein, no shortage of proposals exists. Our answer, as stated at the outset of this book, is a qualified *yes*, and the onus of this chapter is on providing the positioning and design rationale of USDL, leading into subsequent chapters of this part that provide its details.

We have observed that the need for more comprehensively describing services is becoming pressing, with services increasingly being exposed and delivered outside company firewalls, on the Internet and through mobile communications, as seen in the rise of on-demand applications, cloud computing, and business networks. While the present day services being pushed into open market through the Web and mobile channels are mostly in form of simple and instantly consumable "Apps," more mainstream industries are following suit. Consider, for instance, just some of the services that add to the unfolding "on-demand" phenomenon: carrier bookings and track-and-trace of shipments, world-wide tariff look-ups, news events, health insurance comparisons, loan organizations and servicing, business formation, enterprise software services, water/energy utility monitoring, platform services such as business process management and enterprise services bus, and virtualized IT infrastructure services.

The prospect of opening up these sorts of services, on a mass scale, to a variety of channels, deployments and provisioning resources (e.g., cloud hosting, integration through B2B gateways) requires, as we have contended, a *unified* way of describing services. A rich, unified service description language would extend the capabilities of present day service discovery, composition, provisioning, delivery and access technologies such that they leverage more intricate details of service functionality: ownership and provisioning, availability, pricing, conditions of use, conditions of delivery, time of completion or fulfillment etc. Only then could the power of third-parties be unleashed for realizing the much anticipated "network effect" of services and the ambitious vision of the Internet of Services.

Service descriptions have been the subject of significant research in service sciences and computing. SOA languages (Chapter 5) have focused on describing services and their interactions in a uniform way, for leveraging heterogeneous technologies. Semantic Web Services (Chapter 7) have anchored these essentially programmatic descriptions of services with conceptual meaning, through ontologies. The prospect of improving automatic discovery, interaction and composition of services has additionally spurred efforts to conceptualize the wider context in which services are accessed, seen through SOA reference models. Ultimately, pragmatic methodologies focused on business insights, as with the product-service system approaches (Chapter 2), promise to broaden the conception of services from technology abstractions, paving the way for developing them as semantically rich business artifacts. The earlier example was the work of O'Sullivan [23] which conceptualized the "fine-print" of everyday services, from sources such as terms and conditions and newspaper advertisements, in detailing the comprehensive insight into non-functional, business properties of services for its time. Additionally, service system approaches (Chapter 4) have conceived services and their behavior in wider socio-technical systems. Service network approaches (Chapter 3) can be seen as a particular specialization of service system approaches focusing on customer-to-customer, business-to-customer, and business-to-business interactions.

The *Unified Service Description Language (USDL)* has been developed across several research institutes and publicly funded projects across Europe and Australia, and this now extends to the Americas as part of a standardization push (through a W3C incubation group). The scope of service has been on services as understood for business and supportive IT provisioning, i.e., the socio-technical sense of services. In this respect, purely human, purely automated and mixed human/automated services were considered, that have a boundary of cognizance that is available through the tasks of service provisioning, discovery, access and delivery. The overarching philosophy of development has been inspired from the design science approach of [18]. In addition, USDL is built in a collaborative and interdisciplinary way and thoroughly evaluated.

In light of the efforts and insights of multiple institutes and disciplines involved — comprising business management, information systems, IT and computer science (including SOA, security and cloud), economics and law — a *design rationale* has been established for USDL. It consists of a "universe of discourse," providing a consensual background of services in the socio-technical systems that they operate in. This established the fact that despite the diversity of perspectives and varying insights from previous and existing efforts and use cases, a consensual design of USDL is possible. It is recognized that USDL should contain:

- concepts (either well-established or new and agreed upon) that are essential to service descriptions;
- concepts to align with other artifacts, e.g., USDL should relate to, not overlap with, languages dedicated to other organizational phenomena (e.g., business processes, organizational resources, WSDL and other SOA aspects);
- concepts that can support domain-specific (e.g., industry-specific) specializations, since a "silver bullet" language for all service domains is infeasible.

To guide the design of USDL concepts, language requirements encountered in other conceptual modeling languages were identified. These are: *conceptualization*, *expressive power*, *comprehensibility*, *formal foundation*, *extensibility* and *modularity*. For the "grey area" of the suitability of concepts for services, particular requirements are proposed and framed against the service discourse. These address:

- the *organizational embedding* of services into wider systems so that service description languages are not "techno-centric" and aligned with other techniques such as enterprise architectures and business process modeling;
- *cognitive sufficiency* so that services can be understood coherently, across all concepts and modules as well as those outside USDL;
- *service information hiding* requiring that services are preserved by the encapsulation principle when being composed (e.g., service "wrapping" of business processes, data mash-ups and service bundles);
- *deployment symmetry* so that the contextual constraints of services are preserved when they are deployed in new settings;

- *execution (or delivery) resilience* requiring that in addition to functional aspects, all non-functional and exceptional parts be covered to ensure that a service is deliverable.

Collectively, USDL aims at wider systemic alignment and is open to specialization and extension. This reflects the constructivist approach that has been used in the philosophical treatment of information systems [17] and the development of conceptual modeling techniques of information systems [13].

The chapter is structured as follows. We first provide a discourse on services (Section 8.2). Next, the language requirements framed against the discourse are discussed (Section 8.3). Following this, we provide an overview of the structure of USDL (Section 8.4) and how the requirements have been addressed through its current state (Section 8.5). Then, we provide a discussion on how USDL was developed across several projects and use cases (Section 8.6). Section 8.7 introduces a running example in the logistics domain for the remaining chapters of this part. Finally, we summarize the chapter and future development for USDL by way of a contrasting the strengths and limitations of USDL (Section 8.8).

8.2 Universe of Discourse

The design rationale of USDL starts with a "universe of discourse" that ties together ideas and phenomenal features of services that we have elicited and generalized from existing languages, techniques, reference and standardization efforts. With the discourse as the context, we identify concrete requirements for service description languages (in the subsequent Section 8.3). The key ideas of the services discourse have been used in a Delphi study (cf. Chapter 20) with 20 organizations to determine the priorities of service conceptualization and utilization in current practice. They were also used to consolidate developments on USDL, across many services science research projects internationally.

8.2.1 Services

The foremost challenge of describing services lies in the fact that a wide range of artifacts, across economic, political, commercial, community and individual settings, are referred to as services. The surveyed approaches of service description efforts (from the Part I of the book) address IT services with progressively greater sensitivity to business aspects and the organizational settings (seen in the product-service system and service network efforts). Indeed, most services, readily observable in different walks of life, lie on a human-to-automation continuum. They range from purely human (e.g., consultancy services), to purely automated (e.g., message store/-forward system), with a mixture of human and automated resources used for others in between (e.g., purchase order requisitions).

Across their diverse forms, services share a general characteristic. They provide functionality aimed at delivering value to consumers in terms of expected outcomes, subject to delivery constraints, e.g., availability, pricing, copyright and disclaimers. In doing so, they alleviate consumers with ownership of resources, costs or risks. This is consistent with other definitions of a service.[1] Unlike goods, services are intangible and therefore cannot be stored, inventoried or distributed by either providers or consumers.

Seen across practically all our surveyed approaches, services of different sorts evince a common demonitor: they fundamentally involve operations or actions, exposed to consumers. In academic and standards-based literature for services, these are often referred to as *capabilities*. Service capabilities aim at generating value, measured in different ways, for expectations manifested as requests. The value is reached when the outcomes or responses meet the expectations of requests.

8.2.2 Service Agents and Networks

Services are provided and executed variously through organizational contexts, regardless of their degree of formality, complexity or visibility of the organization (e.g., a commercial organization, a conglomerate, an individual). A challenge is what aspects of organizational context are relevant for services and what to distill from the myriad of aspects and artifacts inherent in services, e.g., those related to business processes, applications, objects and resources constituting the enterprise phenomena surrounding services. In short, where do we draw the line when describing services as a distinct phenomenon, albeit related to other artifacts?

Unlike applications or business processes, services are encapsulated and exposed functionality, drawing from these core artifacts, as pure interfaces over implementations. Hence, services share similarities with these, such that they are sometimes regarded synonymously. This is especially the case with business processes. Business processes are composed of actions, with assigned resources (human or automated) for each, operating on business operations, applications and business resources. While being similar to services, their main gravitas is in the internal details of organizations and their systems, i.e., how requests, actions and responses are processed to fulfill consumer goals. In contrast, services are mostly targeted for external consumers and focus on interactions, between consumers and providers, for exposed capabilities; the resultant actions are internally executed. Where business process activities are orchestrated across collaborating resources, service capabilities are *delivered* to *consumers* by *providers*.

With service delivery increasingly taking place in wider settings such as business networks, a distinction has emerged between service agents, e.g., owners of services and the providers who owners outsource core delivery responsibility to.

[1] As an example, according to ITIL (version 3): "A service is a means of delivering value to customers by facilitating outcomes customers want to achieve, but without the ownership of specific costs and risks."

For example, a bank owns account management services. However, its subsidiaries in different parts of the world act as providers of the services. In relation to this, the notion of stakeholder is important since certain partners in a business network have an interest or involvement in a service without having any full responsibility for its provisioning and delivery. An example is a government regulation authority with a stake in compliance, such as customs and carbon footprint for commercial services.

Furthermore, the provisioning (creation and extension) of a service is often thought as being the responsibility of a service provider. However, the rise of service marketplaces, cloud computing, and business processing hubs are extending the distribution of provisioning to third parties. In service marketplaces, for example, providers from diverse agencies expose services, prices and conditions of delivery. The marketplace can then act as a broker between consumers and providers, specifically needs and service capabilities. Other *intermediaries* are emerging such as cloud providers who can offer third-party hosting, B2B gateways that can offer message translation as a service, third-party aggregators that extend the value and capabilities of services through innovative repurposing, bundling etc., and channel partners that allow services to be interacted with through new user interfaces and experience.

Since there is a variety of organizational settings for service delivery, the notions of provider, consumer, stakeholders and intermediary should be generally understood within customer-to-customer, business-to-customer or business-to-business situations. A particular implication is that consumers may be businesses that could be running local business processes, interacting with business processes of outside companies. Indeed, there might be shared value derived through different services of businesses, and reciprocal interactions, so that the provider-consumer relationship is hard to see. As seen from service network approaches in particular, provider and consumer is ultimately relative to an intention of a set of interactions: a consumer requests a service capability and a provider responds, even if there are several exchanges, and even if the roles shift for subsequent interactions.

The surveyed SOA and Semantic Web Services approaches have anchored their conception of services around service capabilities between service consumer, broker, and provider roles. Some of the service network and service system approaches consider more roles. However, they do not explicitly address these new trends of third-party service provisioning and intermediation that could be better supported in Web services architecture and SOA platforms (e.g., see extended roles for service delivery in [10]). Proposals of the product-service system approaches have by far the strongest conceptions of services however they were developed before the onset of intermediaries. This is a similar problem in the wider conception of services within socio-technical environment of the service systems approaches — i.e., the distribution of service provisioning also needs clarity as to what cloud, B2B gateway, channel partner etc., extended services entail.

8.2.3 Service Dependencies and Composition

To provision a service means it has to be prepared in a way that its capabilities can be delivered to consumers. This means that a service has all the necessary parts required for generating its capabilities and all the supportive resources are available so that it can be executed and delivered through its operational environment. Especially for loosely-coupled environments often encountered in practice, the need for different parts needs to be explicitly understood and the risks of failure for these need to be mitigated.

The necessary components yielding service capabilities can be understood for distinct purposes. Some parts constitute a service's functionality and directly relate to the essence of its capabilities. For example, a purchase order service makes using of stock checking and order management services drawn from an enterprise solution. In turn, the order management service may be a composite service consisting of elementary services in order creation, order modification, order cancelation and shipment request. Further services may be added to improve decision making during these services, such as tariff rate look-ups, carrier/container availability look-ups, tax calculations, carbon footprint checks, as well as environment and health policy look-ups concerning transportation of sensitive goods. We can see from this example that services can be put together from different parts, on different levels, through different techniques. Each service, within its own right, therefore carries a structural *granularity*, classified in complex enterprise software solutions such as SAP's as core (elementary and typically not externally exposed capabilities), compound (capabilities relating to a single business object) and composite services (capabilities relating to multiple business objects).

Composition may entail putting together different applications, business objects or business processes as encapsulated artifacts in services or it may involve composition of prior services. Classical service composition techniques are generally applicable for either purpose e.g., BPM techniques for allowing activities to use different functional parts to be orchestrated through ordered steps, or Service Component Architecture (SCA) for wiring components. In this regard, it is interesting to note that despite the advances in service composition, the subject of considerable development under SOA, Semantic Web Services, product-service system, service system, and service network approaches, the distinction between services vis-*à*-vis other enterprise artifacts is sometimes blurry. For example, services and business processes are sometimes regarded synonymously so that it is unclear whether functional dependencies are on the level of services or the workflows. In any case, proposals in the product-service system and services network approaches situate services explicitly for aggregations that are unambiguously characteristic of services, i.e., service bundling (addressing marketing and revenue generation requirements).

Other parts involved in a service may be supportive for achieving the desired functionality, i.e., use of platform, infrastructure, device and other ancillary resources. Indeed applications themselves could be regarded as supportive, e.g., metering and billing engines, in the sense of bringing particular services into view that can then be composed into the essential functionality of the service. While the distinction between essential and supportive functionality may not always be clear, the mechanisms of bringing these into the regime of the service is. As just discussed, essential functionality can be provisioned through classical service composition, while supportive functionality requires other means of specifying dependencies. Services exposed by corporate environments, for example, have complex dependencies in their business and IT landscapes where there could be thousands of resources and services contributing to capabilities of exposed services. Indeed, there are different forms of service dependencies for different purposes (e.g., dependencies must be in place or are incompatible). The treatment of service dependencies across application, platform and infrastructure levels has become explicit through the area of cloud provisioning [21].

8.2.4 Service Delivery

The delivery of services is subject to constraints so that expectations of requests and responses are firmly in place for consumers, providers and other agencies involved, such as intermediaries and stakeholders. Constraints include pricing, licensing, temporal availability, geospatial availability, dependent resources with necessary capabilities, and so on. Given the range and complexity of services across varying durations, the delivery of services entails rights, obligations, and penalties on the part of providers, consumers, and others. These are described in formal documents such as service contracts and service level agreements. The risks of contention, negligence or malpractice are considered significant. Hence, non-functional properties need to be made explicit for service delivery. This is in contrast to business process orchestration where pricing and similar information tends to be regarded as tacit.

Pricing concerns charging for a service prior to consumption — so to speak, the price tag of a service on the catalogue. It is distinct from the payments that arise for a specific instance of a service, after it has been discovered, minimally after a commitment has been made by a consumer to use a service.

Conditions arise from use of services at different times, with different degrees of coverage, the needed capabilities, the channels of consumption, the relationship of the consumer with companies delivering services, metered usage, and so on. The different conditions entail different costs associated with supply balanced against the dynamics of demand (i.e., the availability of a service varies as a scarce resource, to draw from economics, depending on the time of use and the "stock" of resources it uses). Different service offers have different price plans. The different plans have different components and conditions associated to different aspects of service usage, down to individual capabilities of services. The notion of pricing segmentation addresses multiple levels of pricing conditions.

To date, the work of O'Sullivan [23] has been one of the most comprehensive proposals on the conceptual level for business-focussed, non-functional properties of services. However, O'Sullivan's conceptual model has not been validated for service networks under the emergence of intermediaries (such as cloud providers, B2B gateways, channel partners etc). Indeed, delivery constraints under distributed service provisioning — where a channel partner exposes a service, brokered through a service marketplace, whose hosting parts are federated across a backend service provider and a cloud provider — are incrementally built up over the delivery chain of contributing partners. By implication, pricing and other delivery constraints are relative to other parties in the delivery chain. For example the price of channeling is a portion of the overall price, although the delivery needs the broker and hosting services in place. The (true) price for the consumer of the service involves prices of the different partner services. Such constraints in complex delivery chains are not apparent in any of the surveyed efforts.

8.2.5 Service Consumption

When services are consumed, they need to be delivered across all their dimensions — technical, human, business and material. As elaborated above, services are interactional and they have effects on consumers and other entities that they interact with. They need to be understood in terms of all interactions that can occur at different time intervals, in different locations, with different entities be they consumers, stakeholders, intermediaries, or providers.

Services are accessed through designated points of consumption known as channels. Business channels (e.g., mobile banking, whole-of-government citizen services point) are forms of business resources which have consuming applications that allow access to services or specific operations of services. They may be pre-configured to access particular services or they may allow run-time discovery, selection and ordering. Consumers may directly interact with these applications or the applications may entail automated interactions with services. In the case of the latter, it is crucial for interactions and their data contexts to be described in a way that permits the interoperation of consuming applications and services.

For end consumer interactions, different operations of a service may be required in different channels. For example, a passport application requires full authentication of a consumer through a counter channel. Progress in processing the passport may involve a self-serve channel while final handover of the passport can only occur through the counter. Channels may have time and geographic constraints in which services are accessed. They can support different technical channels for different types of devices that are permitted when interacting with services.

Existing SOA and Semantic Web Services efforts have the notion of the external behavior of services although they do not explicitly support channel constraints for particular capabilities, e.g., which capabilities require human interaction for full authentication and which may be accessed through mobile devices. Related efforts address service consumption requirements as part of non-functional properties of service delivery (e.g., the geospatial and temporal constraints of where services are consumed). However, they do not make reference to an explicit behavioral view of services for the service consumer — upon which the functionality, resources and constraints of a service are suitably comprehended by consumers (cf. external service model of the enterprise architecture framework, Archimate [19, 20]).

8.3 Language Requirements for USDL

Based on the general discourse of services described above, we identify the following key requirements to guide the formation of service description languages. These requirements are split into generic language requirements and service-specific concept formation requirements.

8.3.1 Generic Language Requirements

With respect to the generic language requirements discussed below, approaches of the SOA strand are largely technical. Their capabilities in all the generic requirements should be understood from a programming paradigm, and not on the level of service descriptions suitable for business practitioners. UDDI does address service descriptions more explicitly though as a way of supporting service description schemes concerning Web services, leaving open what the details of business-focussed attributes of services should be. Semantic Web Services approaches (e.g., WSMO or OWL-S) appeal to the conceptual level (through ontologies). Unless a domain is thoroughly conceptualized, a semantic Web service tends to be understood through a complementary description, based on ontologies and programming interfaces. In any case, Semantic Web Services approaches require designers to provide most of the service description schema (such as a detailed pricing scheme). Efforts of the product-service system and service network approaches address most of the generic language requirements. Where they are of questionable value is in the suitability of particular concepts for describing services. This is the subject of Section 8.3.2.

8.3.1.1 Conceptualization

First and foremost, a comprehensive description of services, across all the facets and perspectives of the wide variety of stakeholders, is best supported on the conceptual level. In general, the conceptual level, as described by the well-known Conceptualization Principle [27], focuses specifications on essence, allowing users to understand these free of implementation details.

8.3.1.2 Expressive Power

Correspondingly, a service description language should have a sufficient expressive power so that full conceptualization is possible. The expressive power of a language is the computational measure of language's capability to capture its target. When considering the different aspects of services such as service pricing and contract documents encountered in commercial trade, it is clear that the concepts, relationships, constraints and behavior needed for full service description are large and complex.

8.3.1.3 Modularity

Given the size and complexity of service related information, a service description language should explicitly support modularity in order to improve maintainability and extensibility of service descriptions. When describing services as diverse as purely professional/human services such as project management and cloud-based infrastructure services, not all concepts/attributes are applicable. It should be possible to use only those sets of service descriptions or modules that are relevant.

8.3.1.4 Extensibility

Given the diverse industries and domains involving services, a "one-size-fits-all" service description scheme is infeasible. Even on a generic level for aspects such as service ownership, pricing and availability, new and unforeseen requirements for descriptions should be expected. In other words, service description languages should be extensible so that they can be used in specific business contexts involving new requirements that have not been factored into the supported set of service descriptions.

8.3.1.5 Comprehensibility

In light of the size and complexity of service related information, and the fact that service domain experts are non-technicians, service description languages should be comprehensible. The conventional ways for making large and cumbersome models comprehensible include the use of graphical representation, model views, and stepwise decomposition from abstract to detailed levels. Since services are conveyed through textual form, different and multi-modal forms of service description should be supported. For example, it should be possible to view service contracts in structured, semi-structured or unstructured forms (textual narratives).

8.3.1.6 Formal Foundation

As with languages used for other applications, it is crucial that a service description language has a formal foundation. In particular, it should be assigned a formal semantics, in order to support for the full repertoire of concepts and features, alluded to in the aforementioned requirements, in a clear and unambiguous way.

8.3.2 Service Concept Formation Requirements

In information systems' conceptual modeling literature, the requirement of the *suitability* of a language or technique for its intended domain is prominent. Different domains require different language concepts and features suitable for them. The issue of suitability gets to the heart of a constructivist synthesis of concept formation for a service description language. In [11], principles were proposed to guide the identification of suitable modeling concepts for business processes. Since business processes are closely related to services, we adapt these requirements for the suitability of service description languages.

8.3.2.1 Organizational Embedding

In reality, services are organizational systems or subsystems and so all concepts of services relate directly or indirectly to organizational concepts. In that sense, the concepts, as with those of other types of organizational artifacts such as business processes, objects and resources, are said to be organizationally embedded [17]. More specifically, services are grounded in organizational functional structures. The requirement entails alignment of service descriptions with organizational concepts, whether they reference organizational concepts elsewhere maintained (e.g., LDAP directories) or whether they provide corresponding concepts. A significant aspect of organizational alignment for information systems and thus services is resourcing. In business processes, resources are allocated from within organizations to undertake ordered activities whereas at the level of services, the emphasis is on the delivery partners involved: owners, providers, stakeholders, intermediaries, and consumers. Each of these may be an individual or a role within an organization. The extent to which partners are described in a service language aligns with their organizational roles. Further aspects of service concepts that should be organizationally embedded include delivery aspects such as pricing, availability as well as functional composition and bundling, whose constraints are ultimately determined by organizational policies.

Figure 8.1 provides an abstract illustration of organizational embedding. In the middle part of the figure is an abstract concept dependency structure of an organization. At the business level, this could consist of the functional structure, broken down into organizational capabilities and, in turn, artifacts (resources, products, ser-

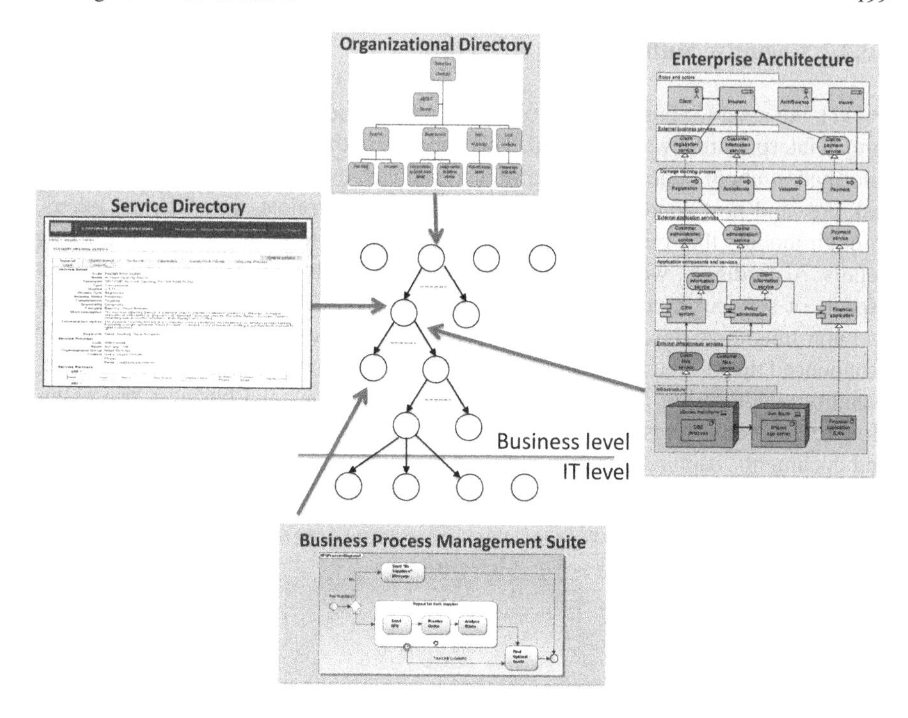

Fig. 8.1: Requirement for Organizational Embedding.

vices processes). At the IT level, IT automated services, applications, platforms and infrastructure would be incorporated. A service could be at the business or IT level, and its dependent organizational artifacts across business and IT levels would be traceable. The figure shows a service, whose details are captured and stored in a service directory and which is associated with the abstract organizational structure. The figure also shows other resources supporting techniques or methods capturing models related to the organizational embedding, at different levels.

As apparent from Figure 8.1, there are several key implications for the design of service description languages arising out of organizational embedding. Firstly, concepts can be justified in terms of wider systemic phenomena (e.g., policies or goals of the enterprise or its units). Secondly, the languages can be aligned with other languages, techniques or methods also concerned with organizational systems, i.e., service description languages do not override the concerns of these other languages, e.g., service description languages relate to, and should not be confused with, business process modeling and organizational resource languages. Thirdly, service description languages can be used in larger organizational methods such as enterprise architecture frameworks where they can be coopted with wider operational, tactical and strategic concerns for organizations, e.g., strategic planning, marketing. This is in line with the proposal of the product-service system approaches (cf. Chapter 2).

8.3.2.2 Cognitive Sufficiency

In [11], the need for business process models to provide a sufficient cognizance for human interpretation was identified. This need led to the requirement for integrating information about control and data flow, resources, events and message exchange, task pre- and post-conditions including temporal constraints, task UIs and others, through models that reduce the need for assumptions to be made due to missing details. Accordingly, we require that practitioners reliably understand the different aspects of services including models of their different artifacts, documents, code, and non-functional descriptions. For services, the challenge of cognizance is more pronounced than for business processes since services are the exposed units of functionality yielded from prior artifacts such as applications and business processes. How to coherently present a logical service entity across its different sub-models and documents, captured through different techniques and languages, is unclear.

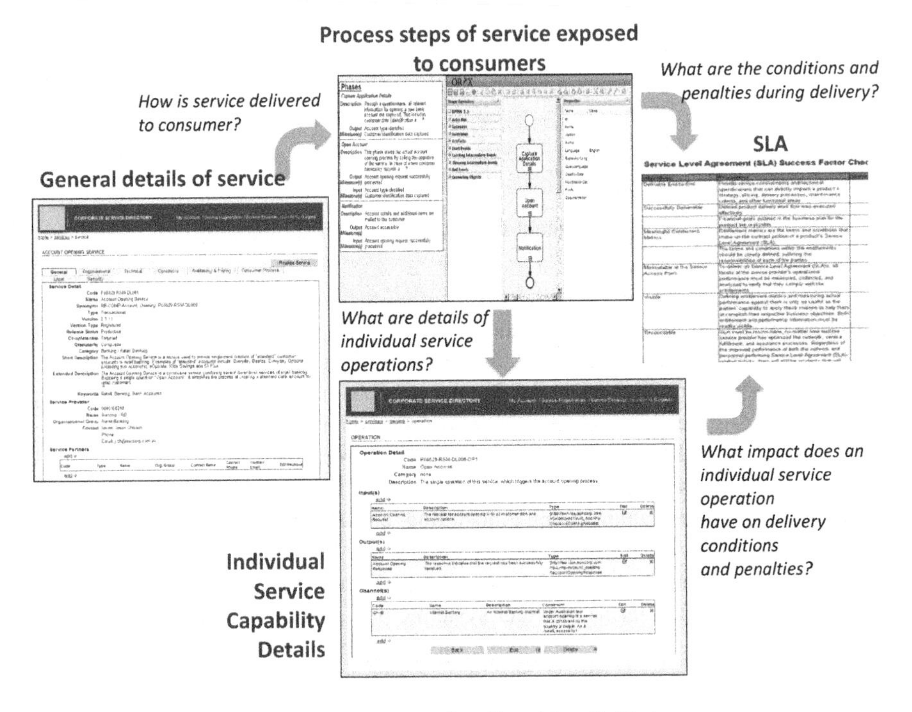

Fig. 8.2: Requirement for Cognitive Sufficiency.

In [16] (cf. also Section 4.6), an integrated modeling environment is described where services models and descriptions can be captured, viewed and extended together with the models of their supportive business processes and business objects. The environment uses an integrated meta-model supporting the techniques for capturing the different artifacts of services, in particular business processes and business objects. In [1], the range of techniques is opened up for a potentially arbitrary set of modeling techniques for different artifacts related to services (e.g., SCA, BPEL and BPMN support). The underlying framework does not aim at integration of the different techniques but on how core aspects of techniques are mapped into a structure that supports a common expression of a variety of service extension requirements that are also expressible through dedicated techniques. For example if a service's interface, business process or UIs need to be integrated with other applications, then their core elements are mapped to the core structure that also permits applications to be integrated, e.g., an application operation is extended by integrating it with a service's operation, or a business process is extended by integrating it with a activity from a service's business process. The framework, as such, allows different sub-models of services to be viewed together and the core elements to be extended irrespective of the technique used to capture the service artifact.

However, this only addresses the *models* of services, not the logical service entity per se. A service, as we described above, logically consists of capabilities implemented by operations — those operations are the ones related to different artifacts (such as business process activities or update transitions of business objects). Capabilities are subject of delivery and are subject to constraints of ownership, provisioning, pricing, availability etc. If such constraints are separately captured, e.g., in stove-piped SLA languages, then the association is lost as to which parts of services relate to specific constraints and dependencies etc.

Figure 8.2 illustrates the cognitive sufficiency requirement for a service across different representations of its details, attributes, models and documents. The figure shows the collective viewing of the details of a service for two different levels of concern. The general details of a service (name, functionality, version, owner etc.), on the left hand side, lead to the details of how the service is delivered from the perspective of a consumer, on the top. Interestingly, this part combines attribute descriptions of the different phases of delivery and a model diagram. The details of service level agreement (SLA) on the right can be viewed, as a document, for the service as a whole. However, individual operations (their inputs, outputs and channels where they are accessed), seen below, can be used to highlight different parts of the SLA document related to that operation.

It can be seen that the requirement of cognitive sufficiency can be fully supported by a language that abstracts the core functionality of a service (capabilities and operations) through its descriptions. It is then possible to provide more fine-grained correlation with different details of service, allowing delivery constraints of SLAs, pricing, etc., to be referred to. The intuitive meaning of a service is then conveyed coherently without requiring practitioners to make mental correlations across different service documents.

8.3.2.3 Service Information Hiding

The well-known principle of *Information Hiding* is intrinsic to services since they are encapsulated, reusable and deployed units of functionality. Current service languages support well-defined service interfaces while current service composition languages allow services composed on business process (e.g. BPMN), service operation (e.g., SCA) and UI levels (e.g., widget composition) to be hidden behind interfaces. However, services have complex structures and can recursively consist of other services of different sorts and with different types of artifacts.

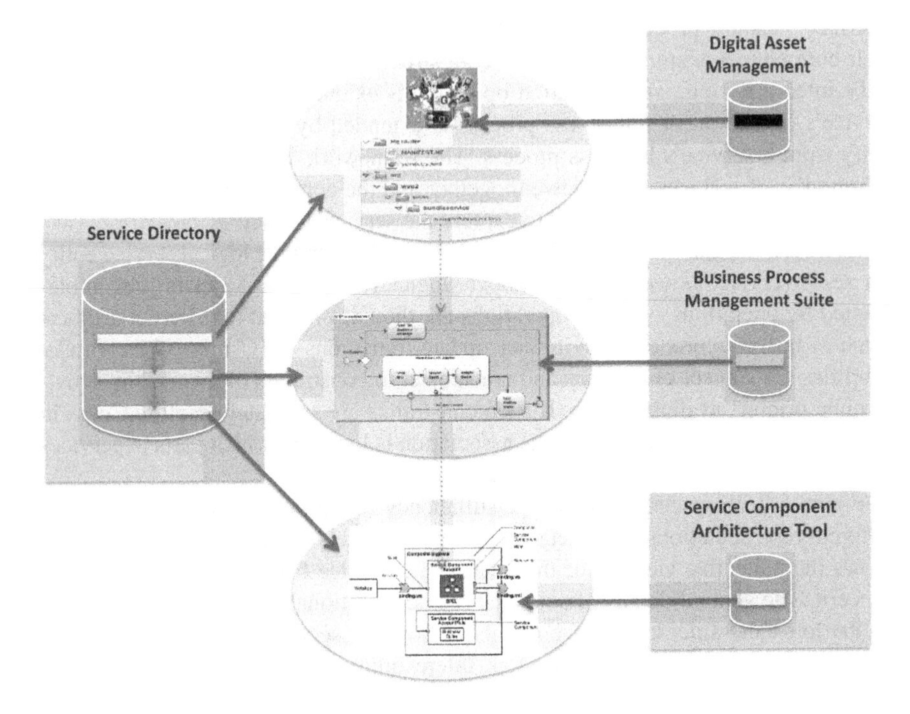

Fig. 8.3: Requirement for Service Information Hiding.

Consider for example, a service for business formation that is implemented through a long-running business process orchestrating many other processes for fulfilling license provisioning, with value-added user interface components available at different stages for data sensing (e.g., market monitoring of different business lines, visualization of local supply chains). Furthermore, consider that for certain types of business licenses, other services are bundled as part of a fulfillment stage, such as monthly tax reporting and office/HR support, through software-as-a-service applications with different business processes operating through them. The question is how such a complex service package, with its different parts and dependencies, can be made transparent through service description languages, without revealing internal details of artifacts and replicating the techniques and languages used for describing these (e.g., a business process modeling language).

To illustrate the point, if a provider deploys a service implemented through a business process, then such a service could be described as single service with different capabilities and operations. Its activities would not be represented as services. If, on the other hand, the business process makes use of other services that have been deployed (and catalogued as different services), the fact of those services being composed together should be exposed through a service description language (as service dependencies). Indeed, composed services should be transparent through service description languages while their related artifacts (e.g., BPMN models) are pointed to. Similarly, the dependencies of services through other forms of service composition, such as data correlation (widgets and service operations), implementation inheritance (package imports) and bundling (unordered and uncorrelated), should be exposed through service description languages without replicating the details available through composition languages.

Figure 8.3 illustrates the point. It shows services descriptions captured for service records maintained in a service directory, on the left. The service records contain references to external files containing models of different artifacts related to services — correspondingly, those artifacts are encapsulated as services. There are three artifacts. A service bundle, on the middle top, combines an iPhone and a business process. The iPhone is catalogued through a digital asset management repository. The business process used in the bundle, in the center, is also exposed as a business process. This is indicated by the service directory. The business process has a model managed through the business process management suite. An individual activity in the business process makes use of a composing a composed service, on the bottom. It is maintained through a service component architecture tool. The composed service is also maintained in the service directory, note, however, its constituent services are not explicitly catalogued. Taken together, the service directory makes the services across different artifacts and business operations, and keeps a track of their dependencies, thus promoting flexible reuse.

8.3.2.4 Deployment Symmetry

As discussed above, the provisioning and delivery of services can be extended to third-parties such as service brokers and cloud hosts. For example, service brokers allow services to be advertised and delivered through their (marketplace) platforms. They "on-board" services so that their technologies for discovery, orchestration, ordering/CRM, payment etc., can be exploited. In doing so, new service *offers* can be created with new monetization models, e.g. advertising and rewards can be used to offset prices set by providers. Different service offers could be created for the same service with different pricing, reliability and risk mitigation constraints. Since delivery and execution of new service offers shifts to intermediaries, we require that languages allow service descriptions to be consistent and coherent, i.e., *symmetric*, regardless of how they have been extended and where they have been deployed.

In other words, as services are extended, there should be no loss of information in terms of their functional and non-functional aspects. As extensions of existing services, core descriptions and constraints set in place by providers concerning future extensions should not be overridden, e.g., a cloud or channeled version of a service

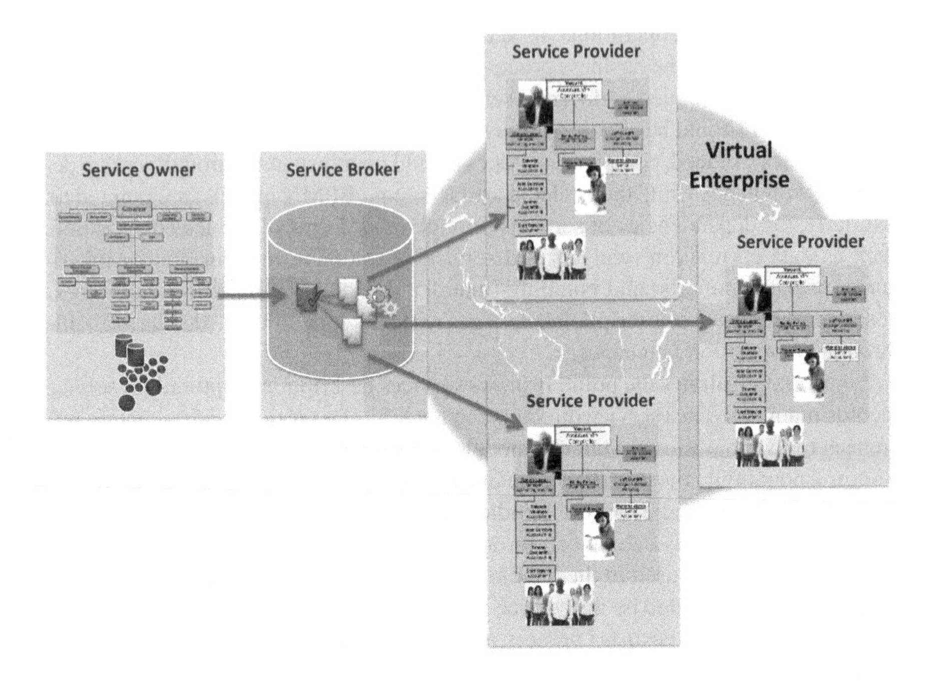

Fig. 8.4: Requirement for Deployment Symmetry.

does not change the core functionality of a service, although its price, response time and point of consumption does. As extensions are created, services should be traced back to their previous versions and extensions should occur with reference to prior constraints and core service details.

Consider two implications of this requirement for service description languages. The first concerns the organizational context of services. Take for example, a business formation service, traditionally operated through the government, opened up to allow the private sector to repurpose and target the service to different market segments, e.g., supporting start-ups, retail outlets or building trades. A variety of service specializations may be created where private sector agencies take up the task of overall orchestration of the service. However, the government retains the core responsibility of evaluating and granting individual business licenses. Moreover, the government requires that certain sensitive tasks such as settlement may be delegated to outside agencies, they need to be accredited/certified, however. Such constraints of delegation and retention of control when services are provisioned to third-parties can be made transparent through well-defined organizational contexts supported in service description languages. In general, the context and structure of services should be defined *abstractly* so that concrete bindings can be configured in particular specializations of services. A symmetric description of services, regardless of their lifecycle of provisioning and where they are deployed and operated, supports crucial compliance needs.

A second implication of the deployment symmetry requirement is pricing. As services are extended, e.g., to run in a cloud environment, new pricing and availability constraints should be incremental and not displace those put in place by the originating providers. For example, if providers require a certain fee for a service, new offers of the service created by brokers or other intermediaries (e.g., broker fees) result in additional pricing constraints. Some parties provide credits, e.g., advertising revenue. In general, a symmetric description of service pricing for multiply extended services sees debit or credit pricing constraints, or price apportions, in the overall pricing schedule. Deployment symmetry has hardly been addressed by the surveyed efforts.

Figure 8.4 provides an illustration of one aspect of deployment symmetry. It shows that a service owner advertises a service together with the core part of its "org. structure" through a service broker. The "org. structure" consists of designated roles required for undertaking the service as policies, for the purposes of regulatory compliance (e.g., with a government). The service can then be ordered through the broker with a target organization needing to comply with the constraints set in place by the service owner. The new setting distributes the required delivery through different services providers (i.e., in a virtual enterprise). Irrespective of whether the service is run through its originating operational environment or new ones, the service descriptions set in place contain constraints so that service delivery conforms to required constraints.

8.3.2.5 Execution Resilience

Functional descriptions of services generally describe what is entailed in the delivery, however further issues also impact delivery. One is non-functional constraints such as pricing. Another is deployment settings including business and technology resources needed for ensuring delivery. A critical concern for ensuring that a service is deliverable is exception handling.

The natural place for exception handling is in detailed specifications of service tasks, for example in pre- and post-conditions and action specifications. Therefore, a service description language should support the handling of exceptions, so that the execution resulting from a specification can be validated as being resilient. This includes rollbacks, cancelation policies, contingencies etc. These issues, of course, relate back to other non-functional issues. For example, "cooling off" periods can be built into pricing should there be uncertainty about customer satisfaction concerning the fulfillment of a service. In addition, service level agreements need to correlate with exceptions so that compliance of delivery can be assessed when things go wrong.

The requirement of execution resilience has been addressed in service/process orchestration and composition languages, such as WS-BPEL, BPMN and SCA, through technical exception handling considerations. However, exceptional issues at the business level, such as an explicit support for policies for cancelations and contingencies are only dealt with implicitly through SLA languages. A comprehensive treatment is not available through surveyed efforts.

8.4 Structure of USDL

Given the complexity of the service domain, seen from our discourse, USDL has been designed with the requirements of *Conceptualization* and *Modularity* in mind. With respect to the former, UML class models are used for capturing USDL. More specifically, the Eclipse Modeling Framework (EMF) [24] was chosen (Chapter 14 discusses representational issues in detail).

With respect to modularity, the distinction in business, operational, and technical information put forward in Section 1.4 carries the main idea of USDL, namely the unification and interconnection of service information from all areas of the spectrum. Yet, the distinction in business, operational, and technical aspects proved to be too coarse-grained for an adequate structuring. Instead, USDL is split into several packages (according to UML terminology). Each package represents one USDL "module" and contains one class model.

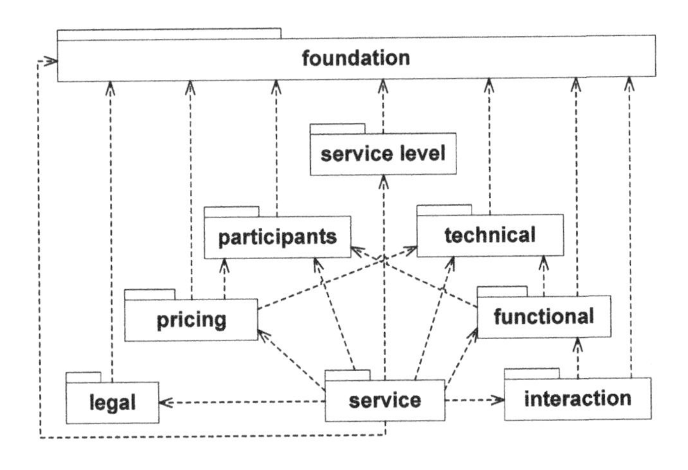

Fig. 8.5: UML package diagram of USDL (dependencies between packages are represented as arrows).

The resulting split in modules follows from prominent non-functional aspects such as pricing and legal constraints, how services are interfaced with for delivery and service level agreements, which partners have responsibility for the service and details about service functionality. Consequently, Figure 8.5 introduces the Functional, Interaction, Participants, Pricing, Service, Legal, Technical, and Service

Level Module and their dependencies. The Foundation factorizes common modeling elements as a consistent application of modularization. The modules are briefly explained in the following.[2] Note that the remaining chapters of Part II discuss the modules in more detail. Besides, detailed specifications of each module are available (cf. [3, 4, 8, 7, 5, 6, 9, 2, 12]).

Service Module (Chapter 13): The Service Module focuses on the essential structure of a service, i.e., the building blocks of a service, without detailing its functionality, responsible participants, and the other concerns. These are the subjects of other USDL modules. The dedicated focus on the structural make-up of a service is because of the complexity entailed. This arises from the conceptual diversity of service types, across the human-to-automation continuum, which brings services of different "shapes and sizes" into view. This makes the task of designing a unifying service description language challenging and warrants as a foremost challenge, the clarity of service structure. As we discussed in Section 8.2.3, there are different degrees of granularity for services, from simple information look-ups and atomic transactions, to orchestrations of different service capabilities in application packages or executable, BPEL-style business processes, coordinating the efforts of companies on an international scale, such as the transportation of goods. Ultimately, services can be bundled through market-competitive packaging seen in telephony bundles, infrastructure provisioning or professional consultancies. Also related to the structure of services, we have discussed (in Section 8.2.3), the orthogonal and adjacent issues of service dependency, related to the resources/services supporting a service's delivery, and composition, concerning the different parts put together for a service's functionality. Both these involve multiple, recursive use of services — of diverse types in and across different levels of abstraction. In the Service Module, services encapsulate functionality, from prior instrumental artifacts on a business or technical level. A Service can be concrete, namely manual, semi-automated and fully automated, or abstract, providing the template for concrete realization in a particular setting. One or more ServiceVariants allow access to predefined subsets of services. The packaging concept of services, ServiceBundle, allows a number of services (including abstract services and service bundles) to be grouped without any execution relationship. A CompositeService, as a specialization of Service, is similar, except that the different parts of functionality, from potentially different services (including abstract services, composite services and service bundles) have an execution relationship. This can be an ordering of steps (such as business processes), unordered steps (such as program import) or data dependency (such as widget mash-ups). In line with the *Service Information Hiding* requirement, composition should not duplicate language-specific composition artifacts (such as BPEL processes), but should apply when services, at the USDL level, have been composed (as discussed in Section 8.3.2.3). A major structural application of services is provisioning. The NetworkProvisionedEntity generalizes both Service and ServiceBundle to this end, and associated descriptive classifications

[2] Classes and relations of the USDL modules are written in sans serif font in the following.

(e.g., terms and keywords), instrumental artifacts and shared variables for execution. Importantly, a NetworkProvisionedEntity has dependencies for concrete services, abstract services, service bundles and resources, through the Dependency element. These relate to service delivery and reflect different dependency forms: Requires, Includes, Enhances, Mirrors, canSubstitute and canConflict.

Participants Module (Chapter 13): The Participants Module captures the organizational actors that are important for the provisioning, delivery and consumption of a service, thus factoring in a major aspect of the *Organizational Embedding* requirement. Since services captured in USDL should not expose or repeat the details of their instrumental artifacts, as required by *Service Information Hiding*, the details of organizational resources as stored in, for example, organizational directories, are not central to USDL's concerns. However, sufficient details should be exposed in USDL so that a correspondence can be made to organizational actors where they directly concern service provisioning, delivery and consumption. The Participant Module connects NetworkProvisionedEntity (Services or ServiceBundles), with the participants in provisioning, delivery and consumption Roles, resonating the wider setting of a service ecosystem, out to the Internet of Services vision (see Chapters 1 and [10]). As per the service discourse, the participant Role covers: service *owners* (cost center owners typically having governance responsibility of services), service *providers* (having service delivery responsibility), *stakeholders* (having regulatory, commercial or other designated interests in the service), *intermediaries* (having specialist provisioning, such as a broker or cloud provider, beyond the original provisioning), and end *consumer*. Basic organizational details are captured for these roles.

Functional Module (Chapter 11): The Functional Module allows the capture of service functionality at an abstract level, regardless of the proximity of the service on the human-to-automation continuum, and free from technical implementation details. As a unified language, it allows service functionality for a wide range of services to be expressed, so that users, designers and engineers, alike, can understand comprehensively how a service is provisioned, delivered and consumed. Conventionally, service functionality is captured as a set of operations. In contrast, USDL supports the capture of service functionality in different layers, for different levels of concern. For instance, service providers can see the detailed functionality of services aligned to organizational resources and objects accessed (a "white-box" view). Intermediaries can be limited to a less intimate view of functionality but sufficiently detailed so that they can configure third-party delivery functionality (a "grey-box" view). Consumers would only see a view of the service focused on their interactions (a "black-box" view). Accordingly, the Functional Module combines well-established concepts of capability modeling and functional decomposition. The element that abstracts service capabilities is Function, which is understood as representing a course of action that can be decomposed into sub-functions. A Function may feature one or more input and output Parameters, as well as one or more Faults (related to exceptions). A Function has preconditions and produces postconditions (effects), as well as references to context variables, i.e., held in the context of a service and are affected when the function is performed. Two types of resources are defined for a Function, namely those used for performing (utilizedResources), e.g., tools or organizational roles, and those manipulated (affectedResources), e.g., business objects. The decomposition of Function supports different degrees of detail for different concerns of providers, intermediaries and consumers.

Interaction Module (Chapter 11): The Interaction Module captures the behavioral aspect of services, complementing the Functional Module and Technical Module with their structural focus. The behavioral aspects of services principally concern the way services are interacted by the different participants involved. A large number of description/modeling languages have been proposed for modeling the external behavior of services as interaction protocols, and standards are now in wide used. As with the Technical Module, our design choice was not to replace existing efforts but to leverage them through the richer semantic setting of USDL. The central concept in the module is InteractionProtocol, which groups the set of mandatory and optional interactions taking place between the participants. The SimpleInteractionProtocol is used to define a single sequence of interactions. An Interaction models an act of communication between the consumer (user) of the service and one or more other participants that have a responsibility in delivery. These parties must be roles defined in the context of the service (feature involvedRoles). For more complex and long-running services, interactions can be grouped into phases of service delivery. A ComplexInteractionProtocol has a set of Phases. A Phase holds a sequence of Interactions and requires as preconditions, and yields as postconditions, a set of Milestones. A Milestone, in turn, is defined as the (formal or informal) description of a state of objects that are affected by the service. It thus describes achievements, and Phases may require certain Milestones to be reached before they can start (feature preconditions). The interaction protocol of USDL maps straightforwardly to those in languages and techniques dedicated to service interaction behavior. Language-specific models as the actual implementation specification can be linked with InteractionProtocol through Artifact.

Technical Module (Chapter 11): The Technical Module supports a common way of describing the technical interfaces (access mechanism) of services. It is based on two significant styles used in practice, with extensibility for the possibility of further support of styles that emerge. Following Semantic Web Services (cf. Chapters 6 and 7), the Technical Module serves to semantically associate technical interface descriptions with elements of USDL via the Functional Module. The two concrete interface classes of the Technical Module mostly perform the function of a container for "interface element" objects (e.g., objects of classes OperationBasedInterfaceElement and ResourceBasedInterfaceElement). These objects reference a particular element in an external interface description, e.g., an operation or parameter definition. A link to the actual interface description artifact (e.g., a WSDL) that should contain such definitions is provided through implementation-

Specifications. A link is also provided to the USDL object implementing the interface FunctionalElementRef.

Pricing Module (Chapter 9): The Pricing Module concerns the charging of services as mutually understood by those who own or deliver services and those who consume them. The key challenge for service pricing is the range of services covered by USDL and the variety of conditions that apply to a single one of these. Different offerings of the service, for different times, targeted at different customer profiles, in different industries — all contribute to pricing variations. In pricing theory, this is referred to as the segmentation of pricing, i.e., the rules governing when and how different consumers are charged different prices. For USDL, all segmented-pricing practices from the literature and those that were considered relevant for USDL were assessed. It turned out that price plans offered by mobile network providers offered among the most complex of pricing schemes. For instance, fixed-fee and pay-as-you-go policies are offered through different plans having different features, allowances and fees. In addition, different conditions apply, such cheaper rates outside business hours, cheaper rates when the provider's own network is called, or more expensive for calls made overseas. Accordingly, USDL has a hierarchical structure for service pricing, allowing different levels of a service that characterize the diversity of pricing schemes in both professional and ICT services. A PricePlan relates to different offerings of services or bundles, allowing a consumer to choose different plans for the same service. A PricePlan contains one or more PriceComponents, related to different service capabilities (Function), concerning different aspects of pricing, e.g., handset charge, domestic calls under standard coverage, calls in excess of standard coverage, overseas calls. Each PriceComponent has a monetary value specified through a PriceLevel which may be fixed per measurement (AbsolutePriceLevel) or proportional to a certain base (RelativePriceLevel). A PriceLevel has a fundamental PriceMetric as a measure upon which pricing is based. Discounts, surcharges, taxes and so on are supported through PriceAdjustments. Dynamic variations on pricing such as rewards status of customer, bundled deals and other accepted negotiations with the customer, are supported through PriceFences. Collectively, the USDL Pricing Module supports the representation of virtually any price structure encountered in the service industry, from simple single-tariff ones to complex price schedules involving multiple price plans. Segmenting conditions, which dictate when certain price elements apply, are also part of the module capabilities. Moreover, alternative possibilities are given to model the same set of charges.

Service Level Module (Chapter 12): The Service Level Module is intentionally kept completely generic, as it does not specify how concrete service levels on concrete aspects should be specified. Instead, its main purpose is twofold. First, it provides a proper glue between abstractly specified service level issues of other USDL concepts. For example it specifies to which elements of a function a certain service level shall apply and who is the related participant. Second, it should allow for incorporation of arbitrary attribute and expression languages. The Service Level Module concerns service level agreements (SLAs), and, thus, concepts capturing guarantees regarding quality of service operation, as claimed/requested by different participants involved in the delivery and consumption of a service. They are established as part of service provisioning and could be updated during delivery as new conditions arise for ensuring *Execution Resilience*. SLAs in practice are specified at the top-level interface between a service provider to monitor whether the actual service delivery complies with the agreed SLA terms. The USDL Service Level Module features a more fine-grained support of requirements for SLA management out to the business level. A ServiceLevel captures a single service level objective related to an offered, negotiated or agreed service. It is either related to state (GuaranteedState) or action (GuaranteedAction) and has a ServiceLevelExpression for capturing assertions. A single participant, viz., ObligatedParty (cf. Participant Module), is obligated to enforce the service level. Service levels for given service are realized via ServiceLevelProfile. A set of service level specifications is combined into one profile and offered, negotiated, or agreed upon as a whole. Different profiles can be used to specify different options of how service levels may be specified and grouped (e.g., as gold, silver, bronze profile). A ServiceLevelProfile resembles the concept of a Service Level Agreement Template as for example specified in WS-Agreement.

Legal Module (Chapter 10): The Legal Module addresses the need for legal certainty and compliance in service networks and in trading services on marketplaces. Participants need to know about the terms of usage of a particular service, for example, liability, privacy, or copyright. However, this information is rarely provided in a machine-processable manner but rather as informal text. Such informal representations of legal clauses are not accessible, for example, to search engines — one might think of a specific search for services which can be re-sold — or an semi-automated analysis for legal consequences such as if a transfer of copyright is legally allowed (cf. our research in [22]). Therefore, the Legal Module covers the modeling of licensing aspects according to two different jurisdictions. The class Work is central to the Legal Module since it represents the subject matter which can be licensed, i.e., a Service, CompositeService, or the service output. UsageRights can be granted for a Work according to different UsageTypes. The latter defines a specific, well-defined, economic manner of how to use a Work, e.g., the right to distribute. In the Legal Module pricing is considered as a Reward for the rights holder for allowing other entities the usage of his work. Therefore, the class PricePlan of the Pricing Module represents *one* possibility to describe a Reward. The class Function in the Functional Module describes details about the usage of an artifact. This determines the actual UsageRight and its UsageTypes. The class Role in the Participants Module is used to describe a licensor. This also means that the licensor has to be specified in the Participants Module: he is either the Provider of the service or has to be listed as a Stakeholder.

Foundation (Chapter 13): The Foundation factorizes common parts of the remaining modules as a consistent continuation of modularization. Because of its basic character, all other modules depend on the Foundation meaning they reference one or more of its elements.

8.5 How USDL Supports Requirements

With the overview of USDL modules, we discuss, in this section, how USDL addresses the requirements for service description languages aimed at the different types of services, on the human-to-automation continuum (cf. Section 8.3.2). A summary is given in Tables 8.1 and 8.2. Note that the individual modules' chapters further address the requirements in detail.

Table 8.1: Summary of how USDL supports the generic requirements.

Generic Language Requirement	Support
Conceptualization	USDL is positioned at the conceptual level by using UML class models and pointers to external artifacts.
Expressive Power	UML class models without the use of OCL.
Modularity	USDL is split in different packages (modules) along well-established lines of conceptual cohesion, e.g., pricing, legal, service level, functionality, etc.
Extensibility	Variant management (cf. Chapter 17).
Comprehensibility	No specific graphical representation of USDL is prescribed. USDL implementations make use of form-based representations of USDL attributes.
Formal Foundation	Currently, USDL lacks a formal foundation.

Concerning the generic language requirements (cf. Section 8.3.1), USDL has clearly been positioned at the conceptual level, free of implementation details, given its UML-based class models and abstraction of business to technical aspects. In addition, it does not subsume related artifacts of services, such as technical WSDL interfaces, but rather points to these so that it serves as a unifying, but not an all encompassing, language. This is a further indication of its adherence to *Conceptualization*. USDL has the *Expressive Power* of UML, although the majority of the language definition is largely structural in nature — and thus has not made use of UML's OCL, for example, for dynamic constraints. The *Comprehensibility* of USDL is at the same level as UML class models used to describe many languages and modeling techniques, and the wide range of concepts and different foci of services led to a dedicated treatment of *Modularity*, seen through the various modules. Indeed, the grouping of concepts has resonance with recognized sub-areas of services in existing service description languages, e.g., service, functional, interaction, pricing, service and legal, are treated as distinct conceptual areas in different languages as well as other unified service description proposals (e.g. [23]). With respect to *Extensibility* of USDL, different variants of USDL are required for different contexts. This is already shown by the Legal Module (cf. Chapter 10) which requires different contents depending on the jurisdiction of a country. The issue aggravates if more and more parameters are relevant to determine the correct variant. Chapter 17 presents one possible solution for managing variants of USDL consisting of a canonical data model, a context driver mechanism, governance processes, and appropriate tooling. Under current development, USDL lacks a *Formal Foundation*.

Table 8.2: Summary of how USDL supports the service concept formation requirements.

Service Concept Formation Requirement	**Support**
Organizational Embedding	USDL concepts correspond to well-known organizational concepts which align with behavioral and structural concepts such as business processes and organizational actors/roles. Thus, USDL could be integrated with enterprise architecture frameworks, BPM and other methods or techniques used in organizations. USDL also supports network-based organizations through network-style participants and services being extended by these.
Cognitive Sufficiency	USDL provides the interconnection of different service aspects across different modules. The interconnection allows for correlation of different service-related information at different levels, e.g., pricing components to service capabilities, or pricing levels to service offerings.
Service Information Hiding	The explicit concept of service allows encapsulation of functionality including that contained in artifacts related to services but captured through other languages (e.g., business process descriptions). Composition on the level of services is possible using different forms of composition and dependency relationships, e.g., recursive bundles.
Deployment Symmetry	Abstract services (cf. AbstractService in Service Module) allow service needs in different organizations to be supported. Different concrete services conforming to these could be used. However, the organizational context is not abstracted, meaning that the providers cannot pre-define the operational context constraints needed when their services are elsewhere deployed.
Execution Resilience	USDL supports the handling of exceptions by providing specific non-functional properties (pricing, legal, functional and interaction) so that the execution resulting from a specification can be validated as being resilient.

Concerning the service concept formation requirements (cf. Section 8.3.2), *Organizational Embedding* is inherent in USDL, seen through the various elements directly corresponding to organizational concepts, e.g., (human or business level) services, resources and participants. Indeed, the organizational setting of services extends beyond an intra-organizational scope to business networks, where roles and activities are outsourced to third-party organizations such as intermediaries. In general, all USDL concepts relate, directly or indirectly, to concepts readily identified as organizational concepts. Furthermore, a number of concepts have dependencies such that an explicit vertical alignment can be supported, from the business to technical levels, e.g., as we have seen with the Service and Resource elements. This means that services across business and IT landscapes of an organization, and in business networks, can be captured through USDL and supported through services directories, repositories or registries. As we discussed in Section 8.3.2.1, a comprehensive resource for managing service descriptions could then support the many different resources concerning organizational and technical artifacts. For example, an organization's enterprise architecture tool that captures information and organizational and technical models could conveniently make reference to a USDL service repository. It could show how services are used in different parts of an organization and on different levels with other artifacts such as business policies and capabilities, organizational actors, roles and resources, products, business processes and business objects. At the other end of the spectrum, a technical software registry could make reference to the USDL service repository for linking IT applications, platforms and infrastructure resources with both the technical and, ultimately, the business services that they support. The reader is referred to Figure 8.1 on page 199 for an elaboration of this point.

With a wide range of service concepts addressed by USDL and with different degrees of detail involved, a major design challenge of USDL has been on an effective alignment and correlation of concepts, such that the collective understanding of services, is imparted. Different users, be they involved in service provisioning, delivery or consumption, can have complex lines of enquiry on services. As we discussed in the requirement of *Cognitive Sufficiency* in Section 8.3.2.2, these cannot be expected to be isolated to any one perspective or module of service descriptions. In this regard, a number of different instances show the sorts of design challenges that confronted USDL, and how they were tackled. As we have discussed, service functionality broadly involves structural and behavioral aspects, with the complexity of details addressed through abstraction and decomposition. Service structure (the *what*) has been addressed through the Functional Module while service behavior has been addressed by the Technical Module (*where*) and Interaction Module (*how*). Dedicated artifacts described in well established languages are referenced, not duplicated in USDL. Thus, a service provisioning tool could place descriptive attributes and dedicated models (e.g., based on BPMN, BPEL or SCA) side by side. Correspondences could be established at the tooling level such that highlighted attributes (e.g., functions) could lead to highlighted parts of models. Of course, structural and behavioral aspects could be viewed side by side so that the service functions, interfaces and interactions in view, leads to a cognitively sufficient what-where-how understanding of service functionality. Indeed, tooling could be developed so that as further details of a function are viewed, corresponding details of interfaces and interaction protocols are brought into view. Stretching it further, we have seen that non-functional properties of services, such as pricing, legal, and SLA aspects, are also inter-linked. For instance, a particular service offer's price plan could be viewed together with functional, interface and interaction details. Moreover, the details of price components and fences as well as legal issues such as copyright clauses could be detailed down to more detailed functions of a service. Taken together, we can see that by carefully considering cognitive sufficiency in USDL, next-generation editing tools for service provisioning and delivery could be considerably enhanced by bringing together service functionality with non-functional aspects. Moreover, consumer comprehension would be greatly enhanced by combining these aspects that are relevant to ordering, accessing and tracking non-trivial services. Figure 8.6

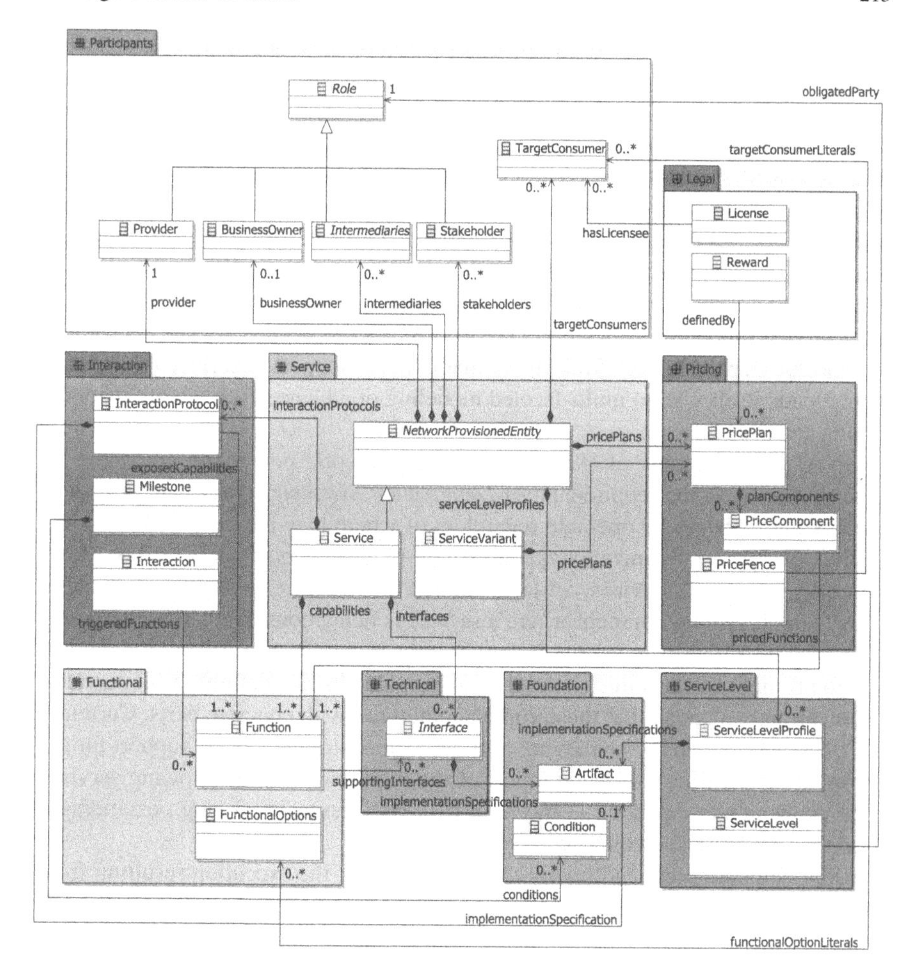

Fig. 8.6: Meeting the requirement of *Cognitive Sufficiency* leads to interconnected modules. Note that only interconnected elements of the individual modules are shown to give an impression.

provides a summary of conceptual correlations across different USDL modules for further insights into USDL's cognitive sufficiency.[3]

Although it concerns services and therefore the encapsulation of the internal functionality and reuse, there are wider considerations that USDL addresses concerning the *Service Information Hiding* requirement. Service functionality, by way of functions, interfaces and interaction protocols, is abstractly described in USDL

[3] In addition to direct relations between modules, there are "connecting interfaces." They are used as a non-intrusive method of referencing when the referenced target can be one of many USDL elements. The intent is to reduce module dependencies which would otherwise be introduced, because targeted elements are likely to be scattered across multiple modules (cf. Section 13.5.1.4).

but concretely bound to external artifacts such as WSDL and BPEL files. Such a two-tiered approach promotes a form of encapsulation and reuse. For instance, new functions can be added to services by reusing prior artifacts. Conversely, services within artifacts can reuse services described through USDL. For example, BPEL processes or SCA compositions can make use of services described through USDL. To promote reuse in large environments, artifacts themselves could be wrapped as USDL services so that they can be discovered and assessed fully for functional and non-functional compatibility with consuming services. Reuse then occurs on the level of USDL where the required functions of a service could be composed with other service functions or they used as part of service dependencies (i.e., Requires or Includes dependencies). We can see that USDL's support of service information hiding lends itself well to multi-faceted modeling of advanced service provisioning tooling envisaged above.

Given the positioning of USDL in a service network, potentially out to the Internet of Services, the requirement of *Deployment Symmetry* is relevant. Recall, it requires that services be operable regardless of where they are deployed including abiding by *a priori* organizational contexts set by providers of services. For example, roles, resources, services, artifacts defined for a service should be available in a newly deployed environment where a service is being consumed. The question is what happens if these are similar but differently named in the new environment. A way to promote reuse in this situation is to provide abstract definitions of services, resources and so on, so that these can be configured with concrete parts. Currently in USDL, services and resources are abstract, demonstrating a basic support for deployment symmetry. A further support of the requirement is by allowing services to be conveniently extended at the level of a business network, e.g., extending a provider service to be channeled.

USDL supports the handling of exceptions so that the execution resulting from a specification can be validated as being resilient. As discussed in Section 8.3.2.5, the requirement for *Execution Resilience* relates to non-functional properties readily addressed in several modules. For example, the Pricing Module allows to represent "cooling off" periods by price fences should there be uncertainty about customer satisfaction concerning the fulfillment of a service. The Legal Module can capture non disclaimers as put in place by service providers that will not be overridden when third parties extend and deliver services elsewhere. The Functional Module supports the definition of faults that can be mapped to technical fault messages (Technical Module) and exception handling procedures (Interaction Module).

8.6 Construction of USDL

Work on USDL started as early as 2007 when a team of researchers at SAP Research manifested their ideas for a *Service Description Framework*. This coincided with the start of the German publicly funded THESEUS/TEXO research project [25] lead by

SAP Research. These purely internal ideas were documented by informal diagrams and by a Wiki only.

The next significant progress happened in September 2008 when the first version of a *Universal* Service Description Language was completed. This first version was directly modeled in a simple XML Schema which was structured according to the segments in the circular USDL logo (cf. Fig. 1.4 on page 12). Accordingly, the XML Schema was composed of business, operational, and technical elements. This version lead to a project internal position paper and was published in [15]. A version 2.0 followed in December 2008 which however only added some additional elements to the XML Schema required for internal use cases. The version 2.0 was published in [14].

A significant change happened throughout 2009 in the lead up to the Future Internet Assembly in Stockholm (November 23-24, 2009) [26] when USDL was prominently disclosed to the public. It was decided to use class models instead of XML Schema and to restructure according to the modules discussed in Section 8.4. Reasons for that were mainly limitations in expressive power of version 2.0, which required considerable extensions to the model, thus motivating the switch to a proper tool chain (Eclipse Modeling Framework, cf. Chapters 14 and 15) and modular design. As depicted in Fig. 8.7, this lead to the first milestone (M1) of USDL version 3.0 which was documented by detailed specifications available at the community site `www.internet-of-services.com` and released in September 2009. M1 only featured first versions of the Service (originally called "Core"), Functional, Interaction, Participants, and Foundation modules. This milestone also saw the rebranding of USDL to *Unified*, instead of Universal, Service Description Language. Since it is not the goal to replace all existing service description languages but to provide a unified means of description capabilities suited to capture aspects of services required for the universe of discourse outlined in Section 8.2.

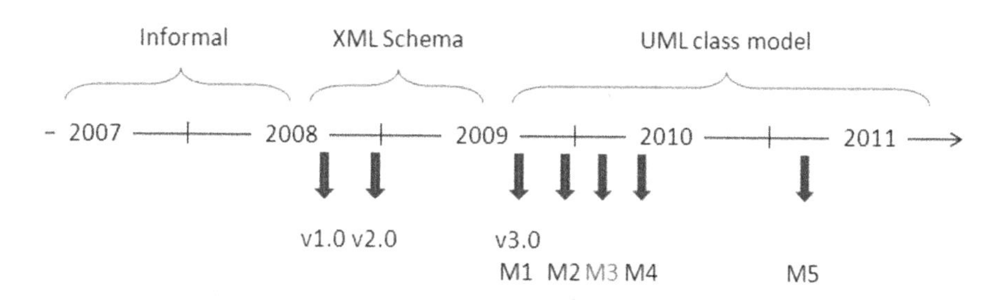

Fig. 8.7: The evolution of USDL.

The milestone M2 particularly added the Pricing Module and was released in December 2009. M3 has only been an unreleased intermediate version and has been superseded by milestone M4. Besides the constant smaller improvements that happened in every milestone, M4 particularly added a first version of the Service Level Module and Legal Module (a complete specification was published in April 2010).

Both modules have been significantly extended and revised in M5 (released March 2011). A further change worth mentioning is the introduction of a separate Technical Module which was basically split from the Functional Module.

Version 3 has been built and evaluated in a collaborative and interdisciplinary way by SAP Research. That means about a dozen researchers at SAP Research have contributed to USDL by bringing in their expertise from different backgrounds (computer scientists, incl. security and SLA experts, business economists, legal scientists, etc.) and are spread over different locations around the world. The modeling has been managed by a central governance body who coordinated the different contributions. All this has been carried out in the context of several publicly funded research projects under the Internet of Services theme, where services from various domains including cloud computing, service marketplaces and business networks, have been investigated for access, repurposing, and trading in large settings. The projects included most prominently the TEXO project [25] within the THESEUS research program initiated by the German Federal Ministry of Economy and Technology. Further projects contributed as well, including German Federal Ministry of Education and Research projects (e.g., Premium Services[4]), EU DG INFSO projects (e.g., FAST,[5] RESERVOIR,[6] MASTER,[7] ServFace,[8] SHAPE,[9] SLA@SOI,[10] SOA4ALL[11]), and the Australian Smart Services CRC.[12]

A wide variety of use cases and perspectives were investigated within the projects for forms of services and an alignment of business and technical aspects not available through previous service description efforts. The kinds of services covered include: purely human/professional (e.g., project management and consultancy), transactional (e.g., purchase order requisition), informational (e.g., spatial and demography look-ups), software component (e.g., software widgets for download), digital media (e.g., video clip players), platform (e.g., middleware services such as message store-forward) and infrastructure (e.g., CPU and storage services). Use cases from the corporate world provided insights into commercial management and arrangements of services such as cost center ownership and provisioning, releasing and dependencies in complex IT and business landscapes. Use cases involving service marketplaces procuring services as complex as those from SAP's portfolio and ecosystem provided additional insights into structures for service bundling including price-competitive of both professional and automated forms of services. Use cases from cloud computing/IT virtualization helped frame platform and infrastructure services into USDL and extended service dependency graph with hosting

[4] http://premiumservices.research-events.com/joomla/

[5] http://fast-fp7project.morfeo-project.org/

[6] http://www.reservoir-fp7.eu/

[7] http://www.master-fp7.eu/

[8] http://www.servface.eu/

[9] http://www.shape-project.eu/

[10] http://sla-at-soi.eu/

[11] http://www.soa4all.eu/

[12] http://www.smartservicescrc.com.au/

requirements for services. Use cases from business networks showed that service versioning/provisioning capabilities need to extend beyond service providers to intermediaries and outsourced players such as brokers, aggregators and channel partners — to drive up the network effect of services.

Subsequent iterations of USDL also include the contributions and evaluation feedbacks of project partners. On the one hand, specific academic partners were given mandates to incorporate or refine particular aspects in USDL. As an example, the German Fraunhofer FOKUS institute[13] was prompted to include security aspects (cf. Chapter 12). On the other hand, industrial partners of the research projects, e.g., Siemens, evaluated USDL by case studies in their setting (cf. Chapter 18). The evaluation contains feedback to further improvements and refinements of USDL. The USDL method was and still is subject to a method engineering (ME) process (cf. Section 20.2.3), i.e., USDL milestones passed several times through the ME phases of requirements engineering, method design, and method implementation. The underlying assumptions of the USDL ME process are justified by exploring the requirements of potential USDL users (cf. Chapter 20). Part IV of the book provides a detailed documentation of the evaluation efforts.

Finally, the scope of input is broadened even wider by approaching a standardization body. As a first step to standardization, a W3C Incubator group has been founded including additional players of academia and industry.[14]

8.7 Running Example

The following B2B scenario acts as running example for the remaining chapters of Part II. The chapters use one or more of the services below and show how the corresponding module would capture the corresponding information.

Consider a German medium-sized company which is world-market leader in manufacturing gas turbines. The company relies heavily on exports, usually to Europe and the US. However, the company has just received an order by a Russian company so it is in need to ship to Siberia for the first time. Therefore, the manufacturing company is in need to use a new, in an ideal case all-inclusive, service bundle for shipping, forwarding, clearance, etc. In this scenario, both the consignor (sender, i.e., German medium-sized manufacturer) and the consignee (receiver, i.e., Russian company) would classify as 1PLs.[15] Consequently, the manufacturing company consults a logistics service market place for discovering a suitable service

[13] `http://www.fokus.fraunhofer.de/en/fokus`

[14] `http://www.w3.org/2005/Incubator/usdl/`

[15] 1PL — A first-party logistics provider is an organization or natural person that needs to have cargo, freight, goods, produce or merchandize transported from a point A to a point B. The term first-party logistics provider stands both for the cargo sender and for the cargo receiver.

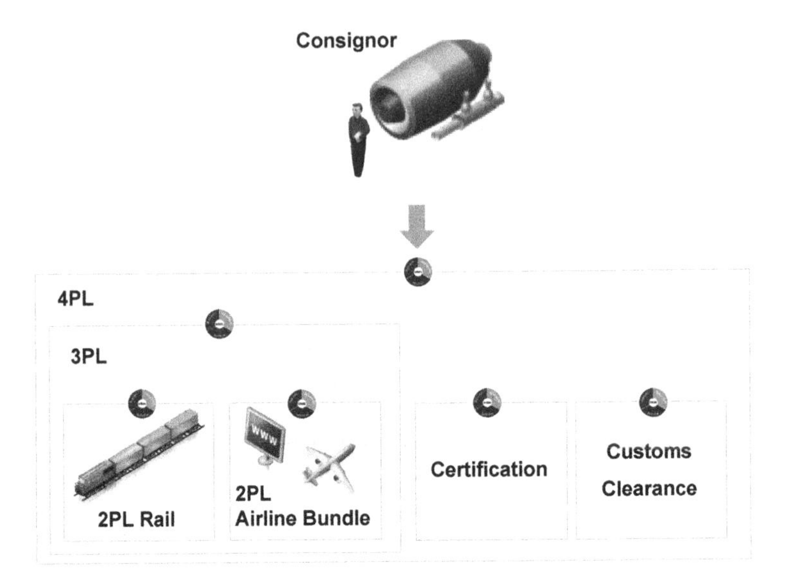

Fig. 8.8: Logistics scenario as running example for Part II.

bundle. The logistics marketplace features a range of 2PL,[16] and 3PL services,[17] accompanying services such as customs clearance, insurances, container rental, as well as 4PL services,[18] typically bundling the aforementioned services. In particular, the consignor requires a Gost-R certification[19] for exports to Russia as well as a customs clearance. Upon close inspection, the consignor decides to use the following 4PL and its bundled services (cf. Fig. 8.8):

1. 4PL
    - Name: Lead Logistics
    - Type: Composition
    - Capability: Composition of customs clearance, certification and 3PL

[16] 2PL — A second-party logistics provider is an asset-based carrier, which actually owns the means of transportation. 2PLs are: (*i*) shipping lines, which own, lease, or charter their ships, (*ii*) airlines, which own, lease, or charter their planes, (*iii*) truck companies, which own, or lease their trucks, or (*iv*) rail companies, which own their trains, warehouse owners.

[17] 3PL — A third-party logistics provider is a firm that provides a one stop shop service to its customers of outsourced logistics services for part, or all of their supply chain management functions. 3PLs typically specialize in integrated operation, warehousing and transportation services that can be scaled and customized based on market conditions and the demands and delivery service requirements for their products and materials.

[18] 4PL — A fourth-party logistics provider or lead logistics provider is a consulting firm specialized in logistics, transportation, and supply chain management. A 4PL is an independent, singularly accountable, non-asset based integrator of a client's supply and demand chains.

[19] `http://www.gost-r.info/`

2. Customs Clearance
   - Name: Customs Clearance
   - Type: Human
   - Capability: Brokering of customs
3. Certification
   - Name: Certification
   - Type: Human
   - Capability: Grants required Gost-R certification
4. 3PL
   - Name: Freight Forwarding
   - Type: Composition
   - Capability: Composition of 2PLs, acts as freight forwarding
5. 2PL Rail
   - Name: 2PL Rail
   - Type: Human
   - Capability: Transportation of goods via train
6. 2PL Airline Bundle
   - Name: 2PL Airline International Economy and Ship Manager
   - Type: Bundle
   - Capability: Bundle of 2PL Airline and 2PL Airline Manager
7. 2PL Airline
   - Name: 2PL Service International Economy
   - Type: Human
   - Capability: Transportation of goods via plane
8. 2PL Airline Manager
   - Name: 2PL Airline Manager
   - Type: Automated
   - Capability: Technical interface (Web services and frontend tool) to look up rates, kick-off and track shipments

8.8 Discussion

The chapter provided an overview of both the design and construction of USDL. The design is rooted in service concept formation requirements derived from the universe of discourse. Before the remaining chapters in this part will deep dive into the individual modules of USDL, we would like to stress benefits and limitations of USDL below.

Unification Part I highlighted that there are different strands of service description efforts — each of which concerned with specific aspects. For example, WSDL comprises detailed information about the provided functionality in form of operations with messages and data types along with details on the technical accessibility, while — in the realm of service descriptions — BPEL is primarily used to describe the observable behavior of complex services. Because of this, it is hard to obtain a unified view on the functional, technical, and behavioral aspects of services from the current description standards, which hampers the communication on specific aspects of interest. In contrast, USDL provides a novel unification of business, operational and technical aspects. As an example, USDL proposes a unified description model for the functional, technical, and behavioral aspects of services. The aim is to provide a well structured model that resides on top of existing as well as upcoming standards.

Interdisciplinary and collaborative The unification makes USDL inherently interdisciplinary. Several disciplines such as computer science, legal science, or business economics have to be involved for the specification of USDL. This can only be achieved by interdisciplinary and collaborative modeling what requires significant amount of resources, man years, as well as a governance body.

Thorough Evaluation As demonstrated by Part IV, USDL follows a method engineering (ME) process, viz., USDL milestones passed several times through the ME phases of requirements engineering, method design, and method implementation. The underlying assumptions of the USDL ME process are justified by case studies, exploring the requirements of potential USDL users, as well as a theoretical evaluation.

Tool Support It was decided to base USDL on well-established grounds with respect to software engineering. On the one hand, a wide-spread, mature, and well-known modeling environment was chosen with the Eclipse Modeling Framework. On the other hand, the decision also facilitated the creation of tools, such as the USDL Editor (cf. Section 15.2.1) because of EMF's model-driven engineering capabilities. A whole tool set was developed around USDL, including editors, repositories, and marketplaces, as discussed in Chapter 15, in order to help bootstrapping USDL.

The features above can be considered benefits and even "unique selling points" of USDL since they are not given for most of the approaches discussed in Part I. However, USDL also exhibits limitations:

Missing Formal Foundation There is no Formal Foundation (cf. Section 8.3.1.6) of USDL required especially for the design of a sound interaction module. Here, future versions should draw from the Semantic Web Services fundamentals as presented in Chapter 6.

Limitation to Offering Phase USDL captures only *master data* of a service stemming from the offering phase (cf. service lifecycle depicted in Fig. 4.3 on page

86). This includes information that a service provider can describe *before* a service is ever consumed. USDL is currently not able to capture *transactional data*, i.e., information that is created as soon as the service is used. As an example, consider pricing information. The price plan can be captured by USDL since it can be specified before the service is consumed. However, a concrete price, e.g., "consumer A has to pay 5 Euros for consuming service B on 12th of September 2011," cannot be captured by USDL since it available only after the service is consumed. Future versions of USDL might extend to further phases of the life-cycle and should draw from the service system approaches in Chapter 4.

Complexity The fact that USDL involves multiple disciplines is both an advantage and poses a challenge. As the remaining chapters in this part will show, background in all involved disciplines is required in order to comprehensively describe a service with USDL. This becomes apparent in the Pricing Module (cf. Chapter 9) and Legal Module (cf. Chapter 10), especially. The USDL light editor (cf. Section 15.2.2) is a potential remedy since it reduces complexity by the progressive disclosure paradigm. Future research should also investigate graphical representations of the language.

References

1. A. Barros, M. Allgaier, A. Charfi, M. Heller, U. Kylau, B. Schmeling, and M. Stollberg. Diversified service provisioning in global business networks. In P. Kellenberger, editor, *Proceedings 2011 Annual SRII Global Conference SRII 2011, 30 March - 2 April 2011, San Jose, California, USA*. IEEE Computer Society Conference Publishing Services (CPS), 2011.
2. A. Barros, C. Baumann, A. Charfi, J. Finzen, M. Flügge, S. Heinzl, T. Kiemes, U. Kylau, F. Marienfeld, N. May, O. Müller, F. Novelli, D. Oberle, J. Pattberg, P. Robinson, B. Schmeling, W. Theilmann, and H. Witteborg. Unified Service Description Language (USDL) — Service Level (SLA) Module. Technical Report Version 3.0, Milestone M5, SAP Research, May 2011. Available at `www.internet-of-services.com`.
3. A. Barros, C. Baumann, A. Charfi, M. Flügge, S. Heinzl, T. Kiemes, U. Kylau, F. Marienfeld, N. May, O. Müller, F. Novelli, D. Oberle, J. Pattberg, P. Robinson, B. Schmeling, W. Theilmann, H. Witteborg, J. Finzen, A. Horch, and M. Kintz. Unified Service Description Language (USDL) — Foundation. Technical Report Version 3.0, Milestone M5, SAP Research, May 2011. Available at `www.internet-of-services.com`.
4. A. Barros, C. Baumann, A. Charfi, M. Flügge, S. Heinzl, T. Kiemes, U. Kylau, F. Marienfeld, N. May, O. Müller, F. Novelli, D. Oberle, J. Pattberg, P. Robinson, B. Schmeling, W. Theilmann, H. Witteborg, J. Finzen, A. Horch, and M. Kintz. Unified Service Description Language (USDL) — Service Module. Technical Report Version 3.0, Milestone M5, SAP Research, May 2011. Available at `www.internet-of-services.com`.
5. A. Barros, C. Baumann, A. Charfi, S. Heinzl, T. Kiemes, U. Kylau, N. May, O. Müller, F. Novelli, D. Oberle, P. Robinson, B. Schmeling, W. Theilmann, and H. Witteborg. Unified Service Description Language (USDL) — Functional Module. Technical Report Version 3.0, Milestone M5, SAP Research, May 2011. Available at `www.internet-of-services.com`.
6. A. Barros, C. Baumann, A. Charfi, S. Heinzl, T. Kiemes, U. Kylau, N. May, O. Müller, F. Novelli, D. Oberle, P. Robinson, B. Schmeling, W. Theilmann, and H. Witteborg. Unified Service Description Language (USDL) — Interaction Module. Technical Report Version 3.0, Milestone M5, SAP Research, May 2011. Available at `www.internet-of-services.com`.

7. A. Barros, C. Baumann, A. Charfi, S. Heinzl, T. Kiemes, U. Kylau, N. May, O. Müller, F. Novelli, D. Oberle, P. Robinson, B. Schmeling, W. Theilmann, and H. Witteborg. Unified Service Description Language (USDL) — Participants Module. Technical Report Version 3.0, Milestone M5, SAP Research, May 2011. Available at `www.internet-of-services.com`.
8. A. Barros, C. Baumann, A. Charfi, S. Heinzl, T. Kiemes, U. Kylau, N. May, O. Müller, F. Novelli, D. Oberle, P. Robinson, B. Schmeling, W. Theilmann, and H. Witteborg. Unified Service Description Language (USDL) — Pricing Module. Technical Report Version 3.0, Milestone M5, SAP Research, May 2011. Available at `www.internet-of-services.com`.
9. A. Barros, C. Baumann, A. Charfi, S. Heinzl, T. Kiemes, U. Kylau, N. May, O. Müller, F. Novelli, D. Oberle, P. Robinson, B. Schmeling, W. Theilmann, and H. Witteborg. Unified Service Description Language (USDL) — Technical Module. Technical Report Version 3.0, Milestone M5, SAP Research, May 2011. Available at `www.internet-of-services.com`.
10. A. Barros and U. Kylau. Service delivery framework — an architectural strategy for next-generation service delivery in business network. In P. Kellenberger, editor, *Proceedings 2011 Annual SRII Global Conference SRII 2011, 30 March - 2 April 2011, San Jose, California, USA*, pages 47–58. IEEE Computer Society Conference Publishing Services (CPS), 2011.
11. A. P. Barros and A. H. M. ter Hofstede. Towards the construction of workflow-suitable conceptual modelling techniques. *Inf. Syst. J.*, 8(4):313–337, 1998.
12. C. Baumann, U. Kylau, and D. Oberle. Unified Service Description Language (USDL) — Legal Module. Technical Report Version 3.0, Milestone M5, SAP Research, May 2011. Available at `www.internet-of-services.com`.
13. M. Bunge. *Treatise on Basic Philosophy: Volume 3: Ontology I: The Furniture of the World.* Reidel, Boston, MA, USA, 1977.
14. J. Cardoso, A. P. Barros, N. May, and U. Kylau. Towards a Unified Service Description Language for the Internet of Services: Requirements and First Developments. In *2010 IEEE International Conference on Services Computing, SCC 2010, Miami, Florida, USA, July 5-10, 2010*, pages 602–609. IEEE Computer Society, 2010.
15. J. Cardoso, M. Winkler, and K. Voigt. A service description language for the internet of services. In R. Alt, K.-P. Fähnrich, and B. Franczyk, editors, *Proceedings First International Symposium on Services Science (ISSS'2009)*, volume 5 of *Leipziger Beiträge zur Wirtschaftsinformatik*, Berlin, Germany, 2009. Logos.
16. K. A. Dhanesha, A. Hartman, and A. N. Jain. A model for designing generic services. In *2009 IEEE International Conference on Services Computing (SCC 2009), 21-25 September 2009, Bangalore, India*, pages 435–442. IEEE Computer Society, 2009.
17. E. Falkenberg, W. Hesse, P. Lindgreen, B. Nilsson, J. Oei, C. Rolland, R. Stamper, F. V. Assche, A. Verrijn-Stuart, and K. Voss. Frisco — a framework of information system concepts — the frisco report. Technical report, IFIP WG 8.1 Task Group FRISCO, 1998.
18. A. R. Hevner, S. T. March, J. Park, and S. Ram. Design science in information systems research. *MIS Quarterly*, 28(1):75–105, 2004.
19. M. Lankhorst. *Enterprise Architecture at Work.* The Enterprise Engineering Series. Springer Berlin Heidelberg, Berlin Heidelberg, 2nd edition.
20. M. M. Lankhorst, H. A. Proper, and H. Jonkers. The architecture of the archimate language. In T. A. Halpin, J. Krogstie, S. Nurcan, E. Proper, R. Schmidt, P. Soffer, and R. Ukor, editors, *Enterprise, Business-Process and Information Systems Modeling, 10th International Workshop, BPMDS 2009, and 14th International Conference, EMMSAD 2009, held at CAiSE 2009, Amsterdam, The Netherlands, June 8-9, 2009. Proceedings*, volume 29 of *Lecture Notes in Business Information Processing*, pages 367–380. Springer, 2009.
21. F. Leymann, C. Fehling, R. Mietzner, A. Nowak, and S. Dustdar. Moving applications to the cloud: An approach based on application model enrichment. *International Journal of Cooperative Information Systems (IJCIS)*, 20(3), 2011.
22. O. Raabe, R. Wacker, C. Funk, D. Oberle, and C. Baumann. Lawful Service Engineering – Formalisierung des Rechts im Internet der Dienste. In E. Schweighofer, A. Geist, and I. Staufer, editors, *Globale Sicherheit und proaktiver Staat – Die Rolle der Rechtsinformatik*, volume 266 of *books@ocg.at*, pages 643–650, Wien, Österreich, 2010. Österreichische Computer Gesellschaft (OCG).

23. J. J. O'Sullivan. *Towards a Precise Understanding of Service Properties*. PhD thesis, Queensland University of Technology, 2006.
24. D. Steinberg, F. Budinsky, M. Paternostro, and E. Merks. *EMF: Eclipse Modeling Framework*. Addison-Wesley Longman, 2nd revised edition, 2008.
25. O. Terzidis, A. Fasse, B. Flügge, M. Heller, K. Kadner, D. Oberle, and T. Sandfuchs. Texo: Wie THESEUS das Internet der Dienste gestaltet — Perspektiven der Verwertung. In L. Heuser and W. Wahlster, editors, *Internet der Dienste*, acatech diskutiert, pages 141–161. Springer, 2011.
26. G. Tselentis, J. Domingue, A. Galis, A. Gavras, D. Hausheer, S. Krco, V. Lotz, and T. Zahariadis, editors. *Towards the Future Internet — A European Research Perspective*, Amsterdam, The Netherlands, 2009. IOS Press.
27. J. van Griethuysen. Concepts and terminology for the conceptual schema and the information base. Technical report, ISO/TC97/SC5/WG3, 1982.

Chapter 9
Service Pricing

Tom Kiemes, Francesco Novelli, and Daniel Oberle

Abstract The on-line discovery, trading and consumption of services as envisioned in the Internet of Services demand an advanced support of the business aspects. Pricing plays a fundamental role among such business facets on both sides of a service marketplace. This chapter discusses the scientific background of the USDL Pricing Module as a comprehensive, applicable, executable, and non-proprietary endeavor. As such, the chapter elicits what makes modeling and engineering the price of a service transaction less straightforward than in the case of a product sale. In addition, we review the state of the art with regard to price meta-models and give a detailed explanation of the most differentiating design choices in the meta-model we propose, including the interdependencies with other USDL modules.

9.1 Introduction

The USDL Pricing Module covers the range of concepts needed to model price structures in the context of service provisioning. It allows a hierarchical representation of price charges and provides the necessary elements for the specification of segmented pricing, complementing the other USDL modules with these fundamental business aspects.

As a matter of fact, given the vital role of pricing in any commercial transaction, it is striking to see how modeling price structures remains a barely touched topic in the field of service engineering. Current service description languages and standards typically lack comprehensive price modeling capabilities (see Part I of the book for

Tom Kiemes, Daniel Oberle
SAP Research Karlsruhe, Vincenz-Priessnitz-Str. 1, 76131 Karlsruhe, Germany,
e-mail: tom.kiemes@sap.com, e-mail: d.oberle@sap.com

Francesco Novelli
SAP Research Darmstadt, Bleichstrasse 8, 64283 Darmstadt, Germany,
e-mail: francesco.novelli@sap.com

an in-depth review of these approaches) while dedicated price modeling attempts either address solely the more limited case of product pricing or target too high a level of abstraction, hampering the possibility to be readily turned into executable artifacts. The USDL effort avoids these shortcomings.

We advocate that a *comprehensive*, *applicable* and *executable* pricing meta-model — such as the one we contribute — is crucial to realize the vision of the Internet of Services (cf. Chapter 1), for it represents a key enabler of effective and efficient service discovery, trade and consumption. On the supply side of a service market it does namely allow providers to implement advanced market-segmentation tactics. On the demand side it gives consumers the opportunity to automate selection and matching based on the actual value of the service offerings, i.e., the difference between the sought-after performance and the costs it shall generate for the consumer.

In this chapter, we first delineate what makes modeling and engineering the price of a service transaction less straightforward than in the case of a product sale (Section 9.2) and review the state of the art with regard to price meta-models (Section 9.3). Subsequently, we give a detailed explanation of the most differentiating design choices in the meta-model we propose, including the interdependencies with other USDL modules (Section 9.4), and highlight how the module responds to the requirements described in Chapter 8 (Section 9.5). Eventually, we apply the meta-model to a real price structure from the example introduced in Chapter 8 (Section 9.6) and conclude (Section 9.7).

9.2 The Idiosyncrasies of Pricing Services

As a first step, we consider the distinctive nature of services and the pricing issues it raises. A service is an perishable entity by definition, for it cannot be stored or inventoried by either providers or consumers. Therefore, the time of purchase — when the consumer *commits* to pay in exchange of the provisioned service (not to be confused with the time of payment) — always precedes the time of production [13]. That represents a key difference from the selling of goods, which can instead occur either before production (e.g., in a make-to-order scenario for a customized item) or afterwards (in the case of commodities). This is equally true for simple and complex, manual and automated services alike. Calling a plumber to request a job, for instance, we accept his hourly rate. Likewise, with a commercial Software-as-a-Service application, a contract including a detailed description of price charges and binding to the respective payment terms (e.g., a subscription) must be signed beforehand by the consumer of the software functionality. This is to say, the pricing necessarily takes place *in advance* of service production.

If the pricing is to precede production — and thus consumption — an important consequence to be underlined is that the total sum a consumer is charged could remain uncertain until after the performance of the service has actually taken place. The hourly rate of a legal studio may be known but the due total will be as uncertain as the length and outcome of the litigation in which it provides consulting. The same uncertainty hampers the predictability of the costs incurred by the service provider, for they might depend on the customer's specific usage profile or on external factors. Advanced price structures encompass mechanisms to mitigate this uncertainty on both sides of the transaction: for example, metered proxies that link charges to usage, aligning prices with the vendor's costs [17], or price caps and flat rates to hedge the consumer's risk [4].

In addition to that, increasingly significant segments of the service industry — namely those related to media, telecommunication and software — have a particular cost structure with high initial fixed costs and very low marginal ones. Given perishability, the crucial problem for a service provider is then matching demand and supply to avoid losing revenues from either unused or insufficient capacity (see [14] for an example of such a scenario), and the price structure becomes a key tool to manipulate demand appropriately. This is achieved by embedding in it price discrimination techniques, such as multi-part-tariffs, bundling, and versioning, alongside more common volume- or time-based discounting policies.

A fundamental challenge for a pricing meta-model is representing the complex market-segmentation rules within such price structures, i.e., those rules determining when and how different consumers are charged different prices. Given the extreme variety of service offerings targeted by USDL, all segmented-pricing practices (as listed in [12]) were considered relevant for our purpose and, therefore, their impact on the conceptual design of a price structure was assessed. However, some of them do gain importance in the light of the increasing range of IT services available. For example, while the customization of a product or of a manually-operated service may require a costly re-deployment of resources, it is instead a relatively easy step for an IT provider to design a set of variants of an electronic service responding to diverse user needs, and price them accordingly [16]. Moreover, while a manufacturing company is only rarely involved in the consumption phase, the contrary is true for a service company.[1]

We conclude this section with a significant example to anchor in the praxis what have been presented so far. Price plans offered by mobile network operators can be seen as an archetype of complex pricing practices in a service market. The same mobile communication service is typically offered under two classes of plans the consumer might choose from: contract-based plans, where a fixed fee is charged each month until renewal or termination, and pay-as-you-go plans, charging a lump sum upon purchase in exchange for a limited traffic allowance, subsequently rechargeable. Within each category a multitude of different plans is offered, with varying features, allowances and, of course, fees.[2] A single contract-based plan may itself be

[1] We assume the inseparability of production and consumption be a distinctive characteristic of services, as commonly found in the service literature. For a discussion about the validity of this point of view see [10, pages 21–23], and, consequently, several price drivers may relate to the consumer's specific usage of the service. The price meta-model must be able to express these relationships between consumers' usage levels and price charges.

[2] Vodafone, for instance, offers some 60 plans on its UK website: http://shop.vodafone.co.uk/shop/mobile-price-plans/all-plans

a composition of fees when, for example, once the monthly traffic allowance given by the monthly charge is exceeded, a pay-per-use component (e.g., by the minute for voice services) is due. Moreover, several price levels may apply depending on other conditions, such as the time of the call (cheaper rates outside of business hours) or the called network (cheaper rates to the company's own network). The same hierarchical decomposition reflects analogous pricing habits of many IT purveyors, such as web-service and cloud computing providers. Similar mechanisms apply for professional services as well: legal, accounting and consulting services are often offered with hourly rates which change with parameters such as the type and size of the project, or the length and significance of the client's contractual commitment.

9.3 Related Approaches

It is our belief that only a formalization of pricing schemes which meets specific requirements can effectively support the trade of services on the Internet. First, such a pricing meta-model should be *applicable*, that is, declarative in nature in order to be directly usable as a component in a software system and to prevent re-modeling a pricing scheme from scratch for every trading platform. Second, the meta-model ought to be *comprehensive* in terms of being able to capture every relevant piece of information used by service providers in describing their prices. Third, price determination should be *executable* on the basis of pricing model instances, rather than hard coded in calculation routines for every possible use case. Finally, the formalization should be expressed in a *non-proprietary* language.

In the world of IT there are three categories of artifacts where a more ore less explicit formalization of pricing has taken place: *Standard Business Software Applications*, *Standalone Billing Engines*, and *Dedicated Pricing Meta-Models*. We now discuss those approaches and how they rate with regard to each of the aforementioned requirements. Table 9.1 summarizes the analysis of these approaches on the basis of the given requirements.

Table 9.1: Price-Modeling approaches and requirements — a check mark is given as soon as one of a kind meets the criterion.

	Applicable	Comprehensive	Executable	Non-Proprietary
Standard Business Software Applications	-	-	✓	-
Standalone Billing Engines	-	-	✓	✓
Dedicated Pricing Meta-Models	✓	✓	-	✓

9.3.1 Standard Business Software Applications

Standard business applications such as Customer Relationship Management or Enterprise Resource Planning provide some degree of price management, usually in the form of price lists, pricing rules and pricing engines. The comprehensiveness of such applications in a context of service trade depends on the particular solution considered. Horizontal solutions are designed to be generic and address pricing mainly from a standard product-centered perspective, but vertical solution for various segments of the service industry may also be offered by some vendors. Of course, in large enterprise deployments these applications are further tailored to the customer's own requirements through customization but we are considering only the standard solution. Pricing models are typically embodied in the source code or in the data model of the application and are therefore hard to extract for reuse. Since a software application is a compiled artifact, it is indeed executable by definition. Most available options falling into this category are proprietary.

9.3.2 Standalone Billing Engines

A billing engine is a standalone software component which can execute a billing process on the basis of some input data to be provided in a predefined format. Standalone billing engines share many characteristics with software business applications. They are executable, but focus on product pricing and do not feature declarative pricing models which could be easily reused. Once again, pricing models are instead implicitly represented in the source code or in the database schemas. Both commercial engines, such as *HighDeal*,[3] and open source ones, such as *jBilling*,[4] are available.

9.3.3 Dedicated Pricing Meta-Models

Formalized meta-models which specifically address pricing can be found as part of XML-based electronic product catalogs, such as RosettaNet[5] or BMEcat [6]. Kelkar *et al.* [7] review such approaches and — concluding that none of them is comprehensive enough to adequately cover the whole pricing domain — develop their own consolidated pricing model. However, we believe that, being product-focused, even their effort is still not comprehensive enough in the context of service pricing, lacking mechanisms to incorporate dependencies, e.g., contractual ones.

---

[3] `http://www.highdeal.com/`

[4] `http://www.jbilling.com/`

[5] `http://www.rosettanet.org`

The pricing part of the GoodRelations ontology [5], based on [7], is not comprehensive enough by admission of the authors themselves, who declare it unable to represent all possible price plans. Another formalization attempt is O'Sullivan *et al.* [15] who, among non-functional service properties, investigate also pricing aspects. Their effort, based on Object Role Modeling, is comprehensive indeed, but not executable due to the mere referencing of conditions via URLs. In [18], Toma presents a syntactic translation of [15] in the proprietary Web Service Modeling Language (WSML) as non-normative extension to the WSMO framework. In [3], the Serviguration ontology [1] is also extended with pricing aspects in that they do incorporate and relate the class *price model* to other entities of the ontology. The structure of a price plan is, however, not declaratively modeled — the price model itself is merely a mathematical formula included as a string in the ontology.

Some pricing meta-models are standardized, e.g., xCBL.[6] Therefore, the requirement of being *non-proprietary* is met. To the best of our knowledge, there is no generic mechanism for computing the price determination on the basis of instances of these pricing meta-models, so they classify as non-executable. In contrast to business software applications and standalone billing engines they are reusable due to their declarative nature.

9.4 Content

In this section, we discuss the pricing meta-model we propose, as embodied by the USDL Pricing Module. The model consists of a static and a dynamic part. The static part is a conceptual modeling language which captures pricing-related concepts and relations between them. The dynamic part encompasses formal rules to declaratively express context-dependencies — a fundamental prerequisite to support practices of segmented-pricing.

9.4.1 Modeling of the Static Information

The static information in our model was assembled drawing from several sources, namely the business literature on pricing management ([11], [13], [17], [9]) and the state of the art in related IT modeling efforts (as detailed in Section 9.3). A requirements-engineering analysis of dozens of real price structures from diverse segments of the service industry was also conducted to reveal the peculiarities of service pricing. Based on the aforementioned sources, we assembled a glossary of concepts rich enough to cover the whole service pricing domain. The glossary was then formalized, whereby each glossary item is represented by a class in the model. Relevant verbs which build relations between different classes are represented by

---

[6] `http://www.xcbl.org/`

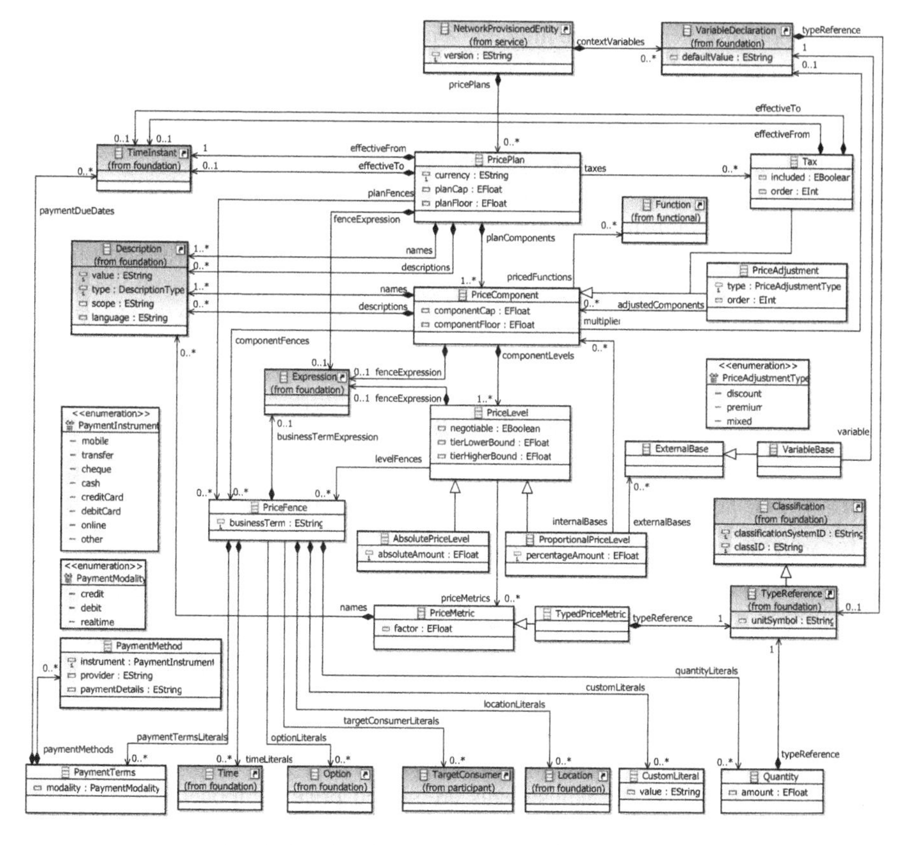

Fig. 9.1: Class Diagram of the Pricing Model.

references. Properties of a class as well as verbs which do not refer to another class are represented by attributes. The resulting meta-model was subsequently embedded in the larger context of USDL. In doing so, we were able to reuse service concepts already defined in other USDL modules, for instance, by linking a price plan to a service description.

An overview of the meta-model can be found in Figure 9.1, exemplarily depicted as a class diagram.[7] In the following we introduce the most important elements of the USDL Pricing Module.

PricePlan A price plan is a self-contained price structure attached to a service or bundle of services. The same service or bundle may be offered under different price plans the customer can choose from, as it is often the case, for instance, with the telecommunication services offered by network operators.

PriceComponent A price plan may comprise several fees to be added up in order to get the due total. Each addend is a price component. For example, in the case of

[7] In Chapter 14 other possible representations of USDL are discussed.

a transportation service there might be a price component relative to the volume of the freight, one for its weight, one for the delivery terms, and another for the specific insurance to be tied to the shipped good.

PriceLevel The monetary values a price component may assume are specified by price levels and each price component must have at least one price level. In case of multiple price levels for the same price component, rules will need to be defined to establish which one applies. There are two types of price level: an absolute monetary amount (AbsolutePriceLevels) — which represents a fixed amount per measurement unit — and a percentage amount (ProportionalPriceLevels) — an amount proportional to a certain base, i.e., another price component or an external monetary quantity. The latter would be employed, for instance, in describing value-based pricing in a B2B context, where the seller's service fee is a percentage (the ProportionalPriceLevel) of the value-added created by the buyer's own commercial endeavor (the ExternalBase). For the afore-mentioned volume price component there could be an absolute price level defined for each kilogram up to a certain weight and a second, lower price level for weights beyond that threshold (that would be one way to represent volume-discounting).

PriceMetric The price metric represents the smallest entity which is priced by the associated price level. For example, as mentioned above, one kilogram. The PriceMetric is used if the price is charged merely for the existence or availability of the priced entity, i.e., an SMS in a pre-paid allowance. The TypedPriceMetric is used if the price depends on an attribute of the performed service function, such as the number of transported kilograms.

PriceAdjustments Price adjustments are signed price components with a specific marketing semantic. For example, discounts for the shipment of small packages or surcharges for overnight delivery. When adjusting a price component, the charge is to be added or subtracted from it. In case of multiple adjustments to the same price component, these must be explicitly ordered using the dedicated attribute.

Tax A tax item is a price component, for it also represents a charge that must be added to get the (gross) total, and it too can take on a proportional amount — an *ad-valorem* tax (e.g., the VAT) — as well as an absolute one per unit of measurement (for example a specific rate duty). When the *included* attribute is true the tax amount is already included in the amounts throughout the price plan (i.e., the tax amount must be subtracted in order to obtain the net price charges), otherwise the charges throughout the price plan are to be considered net of taxes (i.e., the tax amount must be added in order to obtain the gross price charges). Multiple tax items can be associated to a price plan, but must then be ordered to avoid incongruities in applying successive taxation steps.

Function A concept from the USDL Functional Module representing the priced entity or event leading to the enforcement of a price charge. A price component always prices at least one function.

PriceFence A price fence defines a condition stating the validity of the linked price element, be it the whole price plan, a price component, or a price level. It represents the key concept to operationalize segmented-pricing mechanisms with the USDL Pricing Module, as it will be thoroughly explained in the following subsection.

9.4.2 Modeling of the Context-Dependent Dynamic Information

The price fence is the concept giving our meta-model the modeling power to represent market segmentation practices conducted through the design of the price structure [12], whereby individual customers or customers groups are charged different prices. This price differentiation can rely on the consumers' identities, on their decisions, and on the context in which these decisions are taken or in which the performance and delivery of the service takes place. For example, how the consumer decides to configure the service, the time in which the service requests occurs, etc.

Technically the price fence is a conditional expression assessing whether a particular price element — a price plan, price component, or price level — applies. The parameters entering the evaluation in the price fence may be configuration or contractual choices made by the customer prior to consumption, or some measured aspects of the service usage itself. Price fences depend on contextual information that we can generically categorize as follows:

Service Consumer Attributes of the requesting or consuming agent, such as age, residence, or employment status.
Contract Contractual choices made prior to consumption to configure the service performance and define the terms of usage.
Service Event The service event, which occurs when the respective service function from the USDL Functional Module is executed.

The business entity represented by the businessTerm attribute of the price fence will be evaluated and compared against a certain value (or set of values) — the literals. Several are made available to account for the different dimensions of the service provision process. An expression language (e.g., XPath or OCL) is required to fully specify the semantics of the condition statement. In [8], for instance, we utilize the Semantic Web Rule Language (SWRL) in conjunction with the Web Ontology Language (OWL) to perform automated reasoning over the price fences. Thereby, we show that price determination can be automated based on a valid price plan, price fence rules and usage data.

9.4.3 Interconnection of USDL Modules with Pricing Module

We now examine the relation to other USDL modules. Many classes from the Foundation Module are referenced by the Pricing Module, namely Time, TimeInstant, VariableDeclaration, TypeReference, Location, Option, Expression and Description. The TargetConsumer from the Participant Module is used as a contextual parameter influencing the price (as described in the previous subsection, 9.4.2). However, there is another semantically very important interconnection — that between the Function class from the Functional Module and price components in the Pricing Module. It links the price information to the priced entity and thus allows an algorithm to calculate the price based on the executed function and the corresponding pricing elements.

In the class diagram of Fig. 9.1 elements referenced from other USDL modules are highlighted using different color tones.

9.4.4 Extending the USDL Pricing Module

The meta-model is designed to be generic enough to represent most possible pricing schemes and should not require particular adaptation to specific domains. However, USDL can be expanded via an integrated extension mechanism (cf. Chapter 17) to support different domains and such extensions can be leveraged by the Pricing Module as well.

Moreover, besides the typed literals we provide to represent values of a specific type in a price fence expression, the CustomLiteral element can be used as a generic container for literals whose type does not fall within the predefined ones, to execute further, more exotic segmentation tactics.

9.5 Requirements Capabilities of the Pricing Module

As a non-functional aspect of service provision, the Pricing Module mainly addresses requirements of service concept formation (cf. Section 8.3). Nonetheless, it offers some particular contributions to the fulfillment of generic requirements as well. All the requirements the Pricing Module responds to and the elements involved are recapitulated in Table 9.2.

The module allows the representation of virtually any price structure that could be encountered in the service industry, from simple single-tariff ones to complex price schedules involving multiple price plans. Segmenting conditions, which dictate when certain price elements apply, are also part of the module capabilities. Moreover, alternative possibilities are given to model the same set of charges. For instance, a discounted price may be expressed as a different (lower) price level of a certain price component or as a price adjustment to be subtracted from it. The two possibilities are equivalent in terms of the total amount the price will take on for the buyer, but may differ in terms of the chosen marketing campaign and the specific communication goals it aims at. The modeled concepts cover all the requirements pricing practitioners could express with regard to the representation of advanced pricing models.

Table 9.2: Requirements Capabilities and the Pricing Module.

Requirement	Elements meeting the requirement
Generic language requirements	
Extensibility	CustomLiteral
Service concept formation requirements	
Organizational Embedding	• businessTerm (PriceFence) • PriceFence literals • ExternalBase
Cognitive Sufficiency	• Function (Functional Module) • Expression (Foundation Module) • Description (Foundation Module) • TimeInstant (Foundation Module) • VariableDeclaration (Foundation Module) • Classification (Foundation Module) • TypedReference (Foundation Module) • Location (Foundation Module) • Option (Foundation Module) • Time (Foundation Module) • NetworkedProvisionedEntity (Service Module)
Deployment Symmetry	Whole module
Execution Resilience	Whole module

From the point of view of *Extensibility* (cf. Section 8.3.1.4), the contribution of the Pricing Module is bounded to the segmentation possibilities. We provide a series of typed literals that can be used to represent values of a specific type in a price fence expression and cover the most common price-segmenting practices (e.g., segmentation by time, by location, by consumer). To allow more exotic segmentation possibilities the CustomLiteral element can be used as a generic container for literals whose type does not fall within the predefined ones.

There are three ways in which the Pricing Module can meet the requirement for *Organizational Embedding* (cf. Section 8.3.2.1). First of all, the price plan has a strong alignment to the notion of business policies in organizations. So far, price plans are contained with other considerations in business policy documents. Moreover, price plans as realized in USDL target different offers of services. The associations between price plans and service offers are also highly relevant to an organization's service portfolio management. Second of all, any organizationally-relevant aspect of the business constraining the validity of a price element through a price fence can be referenced as a business term and a corresponding set of literals. The business term specifies the variable or container of the business aspect to be evaluated against some realization set, whose elements are modeled by literals of a given type. Thirdly, monetary values other than the price charges themselves can be referenced employing the ExternalBase element. These values are used as a base for the calculation of proportional price charges (e.g., for a value-based pricing schedule that encompasses the payment by the service consumer of a percentage of his monthly revenues).

The Pricing Module supports coarse- to fine-grained pricing elements which can be referenced by coarse- to fine-grained elements in other USDL modules thus supporting *Cognitive Sufficiency* (cf. Section 8.3.2.2) (for more on the inter-module references see Section 9.4.3). At the same time, elements of the Pricing Module are referenced by other modules.

The Pricing Module greatly contributes to the *Deployment Symmetry* (cf. Section 8.3.2.4) capability of USDL in a business environment. As a service is extended, redeployed or brokered, any intermediary has the possibility to either incrementally modify the pricing by means of additional price elements or to replace the current pricing altogether.

Lastly, the pricing as a prominent non-functional property of a service addresses the requirement of *Execution Resilience* (cf. Section 8.3.2.5). For example, "cooling off" periods can be represented by price fences should there be uncertainty about customer satisfaction concerning the fulfillment of a service.

9.6 Example

We now show how to model an existing price structure with the USDL Pricing Module. Before we continue, two remarks need to be made. First, the comprehensiveness of the module is such that, given space limitations, we cannot provide an example covering the whole range of modeled concepts, instead, we focus on the most commonly found price elements: the hierarchical composition of price charges, discounts and surcharges, and consumer and contract dependencies.[8] Secondly, please bear in mind that, because of the Pricing Module's flexible design, several concrete pricing models could represent the same description of a price structure expressed in natural language, tabular, or graphical form. What we present below is therefore *one* possibility.

The price structure we are going to model is taken from the example introduced in Chapter 8, where a German manufacturer of gas turbines relies on logistic providers to have a lot of turbines delivered to Siberia. In particular, we take the perspective of the second-party logistics provider (2PL) which employs airplanes to transport the turbines along a particular leg of the delivery (from Germany to Russia). We utilize different price plans to price the delivery to different countries offered by 2PL.

Listing 9.1: PricePlan Eurasian Shipment

```
<pricePlan>
  <currency>EUR</currency>
  <names>
    <description>
      <value>Eurasian Shipment</value>
      <type>name</type>
```

[8] Additional examples can be found in the specification *USDL 3.0 module: Pricing M5* [2].

```
      <language>en</language>
    </description>
  </names>
  <effectiveFrom>
    <absolutePointInTime>
      <value>2011-03-31T08:00:00.000+02:00</value>
    </absolutePointInTime>
  </effectiveFrom>
...
</pricePlan>
```

We introduce a price component related to the volume of the good to be delivered, whereby each cubic decimeter is priced 3 euros. Please note the connection between the variable var1 of the priced function func1 and the PriceLevel with its PriceMetric, since these explicit connections are necessary for an algorithm to be able to actually calculate the price for the service by multiplying the value of the variable var1 by the amount of the PriceLevel.

Listing 9.2: PriceComponent Volume

```
<priceComponent pricedFunctions="func1" xmi:id="pc_volume">
  <names>
    <description>
      <value>Volume</value>
      <type>name</type>
      <language>en</language>
    </description>
  </names>
  <componentLevels>
    <absolutePriceLevel>
      <absoluteAmount>3.0</absoluteAmount>
      <priceMetrics>
        <typedPriceMetric typeReference="type1">
          <factor>1.0</factor>
        </typedPriceMetric>
      </priceMetrics>
    </absolutePriceLevel>
  </componentLevels>
</priceComponent>

<function xmi:id="func1" affectedContextVariables="var1">
  <names>
    <description>
      <value>FlightDelivery</value>
      <type>name</type>
      <language>en</language>
    </description>
  </names>
</function>

<variableDeclaration xmi:id="var1" typeReference="type1">
  <names>
    <description>
      <value>FreightVolume</value>
      <type>name</type>
      <language>en</language>
    </description>
  </names>
</variableDeclaration>

<typeReference xmi:id="type1">
  <unitSymbol>cubic_decimeter</unitSymbol>
</typeReference>
```

2PL offers this transportation service under a tiered pricing scheme, to differentiate charges based on the usage profiles of the customers. In the given example up to 1000 cubic decimeters are delivered for 3 euros per cubic decimeter, while exceeding cubic decimeters for 2.50 euros each. In USDL that can be realized by simply adding another PriceLevel to the price component and bounding the two price levels appropriately to create the two tiers.

Listing 9.3: Tiered Pricing

```
<priceComponent>
...
  <componentLevels>
    <absolutePriceLevel>
      <absoluteAmount>3.0</absoluteAmount>
      <tierHigherBound>1000</tierHigherBound>
      <priceMetrics>
        <typedPriceMetric typeReference="type1">
          <factor>1.0</factor>
        </typedPriceMetric>
      </priceMetrics>
    </absolutePriceLevel>
    <absolutePriceLevel>
      <absoluteAmount>2.5</absoluteAmount>
      <tierLowerBound>1001</tierLowerBound>
      <priceMetrics>
        <typedPriceMetric typeReference="type1">
          <factor>1.0</factor>
        </typedPriceMetric>
      </priceMetrics>
    </absolutePriceLevel>
  </componentLevels>
</priceComponent>
```

As mentioned in Section 9.4, multiple price components can be used, and they can also relate to each other as adjustments. 2PL incorporates a surcharge into its price plans. The surcharge is to be paid when the customer ships an item categorized as fragile, and amounts to 5% of the total expressed by the volume price component. Therefore a new price component with a ProportionalPriceLevel is introduced into the model instance and references the volume price component as its base.

Listing 9.4: Tiered Pricing

```
<pricePlan>
  <planComponents>
  ...
    <priceAdjustment pricedFunctions="func1" componentFences="fence1">
      <names>
        <description>
          <value>fragile</value>
          <type>name</type>
          <language>en</language>
        </description>
      </names>
      <type>Premium</type>
      <componentLevels>
        <proportionalPriceLevel internalBases="pc_volume">
          <percentageAmount>0.05</percentageAmount>
        </proportionalPriceLevel>
      </componentLevels>
    </priceComponent>
  </planComponents>
</pricePlan>
```

```
<priceFence xmi:id="fence1">
  <businessTerm>fragileFreight</businessTerm>
  <businessTermExpression>
    <expression>
      <value>fragileFreight IS true</value>
      <languageID>RulesLanguage</languageID>
    </expression>
  </businessTermExpression>
</priceFence>
```

9.7 Conclusion

The on-line discovery, trading and consumption of services as envisioned in the Internet of Services demand an advanced support of the business aspects of transactions alongside the technical requirements to be met. Among those business facets, pricing plays a fundamental role on both sides of a service marketplace. A pricing meta-model suitable for such a context was missing and the USDL effort therefore includes a novel attempt embodied in its Pricing Module. The module and its foundations were thoroughly described in this chapter. The USDL Pricing Module is comprehensive, applicable, executable, and part of a non-proprietary endeavor, and it can thus be a fundamental enabler for the commercial success of any service platform.

References

1. Z. Baida, H. Akkermans, and J. Gordijn. Serviguration: towards online configurability of real-world services. In *Proc. of the 5th ICEC*, pages 111–118, 2003.
2. A. Barros, C. Baumann, A. Charfi, S. Heinzl, T. Kiemes, U. Kylau, N. May, O. Müller, F. Novelli, D. Oberle, P. Robinson, B. Schmeling, W. Theilmann, and H. Witteborg. Unified Service Description Language (USDL) — Pricing Module. Technical Report Version 3.0, Milestone M5, SAP Research, May 2011. Available at www.internet-of-services.com.
3. B. de Miranda, Z. Baida, and J. Gordijn. Modelling pricing for configuring e-service bundles. In *BLED 2006 Proceedings*. AIS Electronic Library (AISeL), 2006. http://aisel.aisnet.org/bled2006/48.
4. A. Faruqui and L. Wood. Quantifying the benefits of dynamic pricing in the mass market. Edison Electric Institute, January 2008.
5. M. Hepp. Goodrelations: An ontology for describing products and services offers on the web. In A. Gangemi and J. Euzenat, editors, *Knowledge Engineering: Practice and Patterns, 16th International Conference, EKAW 2008, Acitrezza, Italy, September 29 - October 2, 2008. Proceedings*, volume 5268 of *Lecture Notes in Computer Science*, pages 329–346. Springer, 2008.
6. C. Hümpel and V. Schmitz. BMEcat - an XML standard for electronic product data interchange. In *Proceedings of the 1st German Conference on XML*, pages 1–11, 2000.
7. O. Kelkar, J. Leukel, and V. Schmitz. Price modeling in standards for electronic product catalogs based on XML. In *WWW '02: Proceedings*, pages 366–375, 2002.
8. T. Kiemes, D. Oberle, and F. Novelli. Towards a reusable and executable pricing model in the internet of services. In G. Kotsis, D. Taniar, E. Pardede, I. Saleh, and I. Khalil, editors,

The 12th International Conference on Information Integration and Web-based Applications & Services (iiWAS2010), volume 272 of *books@ocg.at*, pages 720–727, Wien Österreich, 2010. Österreichische Computer Gesellschaft (OCG).
9. S. Lehmann and P. Buxmann. Pricing Strategies of Software Vendors. *Business & Information Systems Engineering*, 1(6):452–462, 2009.
10. C. Lovelock and E. Gummesson. Whither services marketing? *Journal of Service Research*, 7(1):20–41, 2004.
11. T. T. Nagle and J. E. Hogan. *The Strategies and Tactics of Pricing*. Prentice Hall, 4 edition, 2005.
12. T. T. Nagle and J. E. Hogan. Segmented Pricing: Using Price Fences to Segment Markets and Capture Value. *SPG Insights*, 2006.
13. I. C. Ng. *The Pricing and Revenue Management of Services: A Strategic Approach*. Routledge, 1st edition, 2007.
14. F. Novelli. A Simulation Study of the Interdependence of Scalability and Cannibalization in the Software Industry. In *Proceedings of the 23rd European Modeling and Simulation Symposium*, Rome, September 2011.
15. J. O'Sullivan, D. Edmond, and A. H. M. ter Hofstede. The price of services. In *Service-Oriented Computing — ICSOC 2005, Third International Conference, Amsterdam, The Netherlands, December 12-15, 2005, Proceedings*, volume 3826 of *LNCS*, pages 564–569. Springer, 2005.
16. C. Shapiro and H. R. Varian. Versioning: The Smart Way to Sell Information. *Harvard Business Review*, pages 106–110, November, December 1998.
17. A. A. Stern. The Strategic Value of Price Structure. *Journal of Business Strategy*, 7(2):22–31, 1986.
18. I. Toma. *Modeling and ranking semantic web services based on non-functional properties*. PhD thesis, Faculty of Mathematics, Computer Science and Physics of the University of Innsbruck, 2010.

Chapter 10
Service Licensing

Christian Baumann and Maria Niedziella

Abstract Service marketplaces and service networks promote tradeable services on the Internet. With such business transactions the need for legal certainty and legal compliance arises. Two crucial aspects can be highlighted in this context: you need to know about *what* you are talking and under *which jurisdiction* a transaction is arranged. Both aspects are difficult to address in a machine-processable description for services. The subject matter — the *what* — is difficult to grasp: the service notion in USDL encompasses technical Web services to conventional business services. Furthermore, machine-processable legal attributes need to comply to the statutes of the respective country of the transaction in order to achieve enforceable legal expressions. Current models apply approaches which do not address the aforementioned aspects. We argue to overcome the difficulties by modeling legal description capabilities on the basis of the statutes. This chapter covers the modeling of licensing aspects according to two different jurisdictions. We use the copyright acts of Germany and the USA to illustrate the approach. The resulting model incorporates the terms of the statutes, for example, *Work*, which can be mapped to any licensable service artifact. Although certain notions such as *Work* have a common semantic understanding this approach requires a management of the different variants. Variants are discussed in Chapter 17.

Christian Baumann
Berkeley Center for Law & Technology, University of California, Berkeley, School of Law, 376 Boalt Hall Berkeley, 94720 CA, USA, e-mail: baumann@berkeley.edu
Center for Applied Legal Studies (ZAR), Karlsruhe Institute of Technology, Germany, e-mail: ch.baumann@kit.edu

Maria Niedziella
SAP Research Karlsruhe, Vincenz-Priessnitz-Str. 1, 76131 Karlsruhe, Germany, e-mail: maria.niedziella@sap.com

10.1 Introduction

Software-as-a-service, cloud computing, service marketplaces, and service networks are some examples for the emerging ubiquity of services as discussed in Chapter 1. With the advent of business within these structures the need for more legal certainty arises. Participants need to know about the terms of usage of a particular service, for example, liability, privacy, or copyright; however, this information is rarely provided in a machine-processable manner but rather as informal text. Such informal representations of legal clauses are not accessible, for example, to search engines — one might think of a specific search for services which can be re-sold — or an semi-automated analysis for legal consequences such as if a transfer of copyright is legally allowed.

While most of the functional and non-functional attributes can be used to describe a service for international business purposes — take, for example, the description capabilities for *Price* — the legal aspects are bound to the national context. For instance, companies such as the Internet marketplace Amazon have separate general terms and conditions for its branches in the different countries.[1] The legal terms notably differ although Amazon basically offers the same products.

The modeling of legal aspects has to account for the different national statutes. This makes it difficult to identify common concepts for the Legal Module. The extraction of generic legal terms from dictionaries, for example, is hardly an option, since a legal description model has to apply to the legal practice. The legal practice, however, does not consider generic regulations, but it rather refers to the specific norms of the specific jurisdiction. For instance, most countries have *private international laws* to determine which national law should be applied in a conflict of laws situation.[2] Although European directives and international treaties aim for harmonizing national laws, they are not binding for private entities in the different countries. These statutes are merely a mandate for the governments to implement corresponding national laws. The implementations might differ in their embodiment. Moreover, there are even more substantial differences in the common and the civil law system.[3]

As a consequence, the Unified Service Description Language (USDL) has to account for national statutory provisions in order to provide binding legal description capabilities.

USDL is an abstract and broad description mechanism for services on the Internet. Services might deliver functionality, content, and other works — basically intellectual subject matter. This intellectual property might be subject to copyright. However, certain combinations, unknown works, and new scenarios cannot be known in advance. We, therefore, apply the approach to build a generic copyright model by

---

[1] Cf. `www.amazon.com`, `www.amazon.de`, `www.amazon.fr` or other domain regions.

[2] For more information on the scope of private international law cf. [24, pp. 16-24].

[3] In common law, also known as case law, law is *developed* by judges through decisions of courts. In civil law the law is *developed* and *codified* through the legislative branch — judges apply the written statutes and follow them.

formalizing licensing aspects of the statutes itself [6]. This model is as abstract as the statutes and can be specialized for a specific domain, scenario, or artifact. Moreover, by adhering to the statutory text the specialization of the model is merely a modification of facts and legal consequences which are set by the law. Consider, for instance, the legal term *Work*. It represents a legal concept which is protected by copyright. A *Song* which is provided by a service might be considered as *Work* and, therefore, enjoy copyright. Thus, when subsuming the artifact *Song* as *Work* in order to define the corresponding usage rights, we have some level of compliance to the law [7].

Our methodology to model legal non-functional properties adheres to the principle of considering jurisdictions and the respective statutory text. To illustrate the methodology this chapter focuses on the German Copyright Act (GCA) and the Copyright Act of the United States (UCA). The advantage of copyright is, that most countries have established copyright statutes, and international treaties have promoted international harmonization of the copyright laws. Moreover, copyright is one of the most discussed legal areas in the digitized environment and important for a lot of industries on the Internet.

The remainder of this chapter is structured as follows. In Section 10.2 our approach to formalize copyrights based on the statutes is set into contrast to related work. In Section 10.3 the German module is depicted before it is applied to the example of the book (cf. Section 10.4). Differences and common concepts in relation to the U.S. module are also explained. Section 10.5 concludes the chapter and provides an outlook for future work.

10.2 Related Work

There are many approaches to the formal representation of legal clauses (rights and obligations). Because we specifically consider copyright for this chapter, we discuss in this section approaches with the capability to represent copyrights. We position our methodology, viz., considering the jurisdiction and the statutory text, to the efforts by means of what we argue is essential for legal certainty and legal compliance in USDL. The distinguishing criteria that set our model apart from related efforts are as follows.

First, we distinguish between approaches that generically express rights (and obligations) or specifically copyright. Our model follows the requirement to specifically express one field of law. This supports legal certainty, since we relate to constituted rights.

Second, we argue that by using the statutory text as a modeling basis we achieve a certain level of legal compliance, since we comply with the applicable law of a country. Generic approaches often originate from an economical background and have to be separately verified by lawyers or courts[4] for the specific jurisdiction.

[4] For instance, the *GNU General Public License version 2* (GPLv2) was in Germany approved as effective according to German law in 2004, cf. [25]. The license was developed according to the

Even if certain clauses of a generic approach are valid, they might have, depending on the jurisdiction, a different legal implication or scope.[5]

Finally, the ubiquity of the notion of services on the Internet requires a comprehensive approach. We cannot predict which copyrighted material might be described in USDL. This could vary from accessing the service itself (its program code) to an aesthetic picture, or a to date unknown work type. Moreover, the construction of the model has to account for multiple scenarios. A scenario could cover commercial or non-commercial usage, it might involve granting usage rights to several unknown users, a specific group of users, or only one known user. Therefore, we split this criterion into the works and scenarios which are covered.

The criteria are depicted as columns in Table 10.1. The following efforts are listed per row.

Table 10.1: Efforts to the formal expression of copyright.

	based on copyright	**jurisdictions considered**	**variety of works covered**	**multiple scenarios covered**
ccREL	yes	yes	no — specialized for audio, video, pictures, text, interactive content	no — mainly to provide at no cost to the public
ODRL	no	no	yes	yes
ODRL-S	partly	no	no — specialized for (technical) services	yes
XrML	partly	no	yes	yes
METSR	no	no	no — mainly for digitalized content in libraries	no — mainly for research/ libraries
IPROnto	partly	no	yes	yes
O'Sullivan	no	no	n/a	n/a

10.2.1 Creative Commons Rights Expression Language

The *Creative Commons Rights Expression Language* (ccREL) [1] focuses on the representation of the Creative Commons[6] license sets in a machine-processable lan-

U.S. legal system. The license text is officially only available in English because the Free Software Foundation (FSF) wants to avoid validation costs and the risk of introducing errors which could have extensive consequences.

[5] For instance, the GPL allows for copying, distribution, and adaptation according to German copyright law. However, the right to make publicly available (§ 19 a GCA) is not specifically included in the GPL. In Germany this right is crucial for distribution on the Internet. [35, Hören, Teil Klauselwerke, IT-Verträge, note 211]

[6] `http://www.creativecommons.com`

guage. ccREL is based on RDF [33] and implements its model in RDFa [34] and XMP [2, 3, 4]. The specification recommends a respective implementation depending on the character — if it is web-based or free-floating — of the artifact [1, p. 10].

The aim of the Creative Commons project is to provide widely applicable license sets for works. The focus of the license sets is, however, on sharing with everyone on the Internet at no charge. Even when a work is not provided to the *public domain* but rather made available with the idea of "some rights reserved," there is no consideration of using it in complex business situations.

Since the Creative Commons do not specify exclusive rights in their license sets, it is possible to license a work under additional licenses. This scenario is addressed through *CCPlus* [11], an additional option to describe more permissions than expressible through the original license sets. This enables rights holder to provide machine-processable information about commercial licensing options of their works. However, the implementation is realized by means of a simple link to some other resource on the Internet — there are no further expression capabilities.

The Creative Commons project provides its license sets as international (*unported*) versions. Since they are fairly simple a certain level of international applicability is achieved. However, the license sets are still ported to more than 70 different jurisdictions in order to account for the requirement to comply with national laws or to express subtle differences. [10]

In summary, ccREL provides a machine-processable formalism for the license sets of the Creative Commons project. The license sets address a specific field of law — copyright — and jurisdictions are considered. The approach is based on the narrow scenario to provide works at no costs to the public. The combinations are limited to six license sets and they are optimized for audio, video, pictures, text, and interactive content.

10.2.2 Open Digital Rights Language

The *Open Digital Rights Language*[7] (ODRL) is an XML based policy expression language. It provides a common vocabulary which allows to build specific profiles. ODRL is intended for publishing, distribution, and consumption of digital content. In addition to copyright concepts it introduces parameters for areas such as functionality or privacy. [22]

ODRL wants to provide a comprehensive expression language for artifacts on the Internet. This is similar to USDL, however, ODRL focuses on (generic) rights expression. The current draft of version 2.0 incorporates the semantics of the (unported) version of the Creative Commons license sets [29], hence, it emphasizes its generic approach to rights expression on the Internet.

---

[7] `http://odrl.net`

In summary, the ODRL model does not specifically address copyright. It does not consider jurisdictions but rather builds on a generic approach. By definition of its profile mechanism it is extendable to different scenarios and a variety of works.

10.2.3 Open Digital Rights Language Services

The *Open Digital Rights Language Services* (ODRL-S) is a profile which is based on ODRL version 1.1 [28]. ODRL-S specializes in the artifact *service* and addresses its licensing issues [16, 17].

The meaning of the term *service* in ODRL-S emphasizes on the technical implementation of (Web) services, thus, it is narrower than understood in USDL. Based on this meaning the model defines specific attributes which are typical for the usage of a service.

The WIPO[8] copyright treaty serves as the legal source for the analysis of copyright aspects of the artifact *service* [15, pp. 8, 49]. This treaty is, however, not binding for private entities in different countries. The ODRL-S profile has the same characteristics as the aforementioned ODRL.

10.2.4 Extensible Rights Markup Language

The *Extensible Rights Markup Language* (XrML) is the successor of the Digital Property Rights Language (DPRL) which was developed by Xerox Palo Alto Research Center. It is XML based and currently available in version 2.0. XrML is a language to express rights and business rules for using, duplicating, and distributing content. [8]

DPRL/XrML was designed to provide a comprehensive vocabulary for multiple scenarios. This, eventually, lead to the adoption by the MPEG-21 standard and made it the official part five in the ISO/IEC 21000 international standard [23]. Version 2.0 is designed for usage in any medium or type of resource and on any digital entity. Due to the integration into ISO/IEC 21000, XrML is widely used for Digital Rights Management (DRM). However, the language was not specifically designed on the basis of national statutes or jurisdictions.

The international standard ISO/IEC 21000-5 is also adapted in the OASIS Web Services Security (WSS) token profile for rights expression language [27].

[8] World Intellectual Property Organization

10.2.5 METSRights

METSRights (METSR)[9] is an extension to the Metadata Encoding and Transmission Standard [13] (METS). METSR is based on XML and specialized for digitized materials in academic institutions and libraries. It is intended to serve for informational purposes such as identification of ownership and contractual requirements for re-use of the materials. [9, pp. 6-7]

METSR is not used in complex business scenarios. It is neither designed to specifically represent copyrights, nor is it based on statutes.

10.2.6 Intellectual Property Rights Ontology

The model of the *Intellectual Property Rights Ontology*[10] (IPROnto) [12, 19, 20, 21] comprises a static and dynamic view on intellectual property rights. The static part defines legal concepts and their relations. The processes such as *creation of a work* build the dynamic model. The concepts in the static model are used to define the processes in the dynamic model.

IPROnto follows a generic approach by incorporating the WIPO copyright framework in its model [12, p. 115]. The model considers a variety of works and multiple scenarios.

10.2.7 O'Sullivan

O'Sullivan developed a comprehensive model for "a domain independent taxonomy that is capable of representing the non-functional properties of conventional, electronic, and Web services." [30, p. 19]. In his work he also considers the notion "intellectual property rights" [30, pp. 119-122], however, this is limited to registered intellectual property rights such as trademarks, designs, or patents. Copyrights are not included.

The model includes country of origin and the area where the granted intellectual property right applies [30, p. 119]. The description formalism is not capable of describing a usage right based on a specific law or jurisdiction.

---

[9] METSR is available as an XSD schema at `http://cosimo.stanford.edu/sdr/metsrights.xsd`, accessed 6 April 2011. The schema is heavily documented, there is no other documentation available at this point.

[10] IPROnto is aligned to the top level ontologies DOLCE [18] and SUMO [31].

10.2.8 Conclusion

Some of the presented works can be related to rights expression languages (REL) with the main purpose of machine-actionable control over digital content. They are designed to be applicable for a variety of works and scenarios, and it has been tried to realize generic concepts. We, in contrast, argue that an generic approach does not provide legal certainty.

Only the Creative Commons consider jurisdictions. Its application within complex business scenarios, however, is limited due to the simple expression capabilities of ccREL. Furthermore, there are only six different license sets available and the licensing model of the Creative Commons is based on the assumption to provide works at no cost to the public.

O'Sullivan's approach pursuits a comprehensive service description formalism, however, the expression of copyrights is not supported. Our approach is specialized in representing the rights and obligations granted by law, that is, copyright in our discourse. For the representation of other concepts, for example, price or functionality, we rely on the remaining USDL modules.

We also argue that by adhering to the statutes and jurisdictions we are able to foster legal compliance and legal certainty — only ccREL has a similar claim.

10.3 Modeling Copyright in USDL

Intellectual property (IP) is influenced by the principle of territoriality. Territoriality means that IP aspects have to comply with the statutes of the country where an IP issue arises [14, § 7 note 12]. In this section we use the German Copyright Act (GCA)[11] and the Copyright Act of the United States (UCA)[12] to illustrated our methodology [6]. The GCA is part of the civil law tradition, the UCA, on the other hand, is part of the common law tradition.

The modeling of copyrights does not comprise the representation of the complete copyright acts. In fact, only the relevant legal concepts which are required to address licensing are important. These concepts build the core of our model; they are integrated with other modules in USDL.

Throughout Section 10.3.1 the model according to the GCA is depicted. Legal terms which are used in the statutes, such as Work, UsageRight, or UsageType, are discussed. However, some notions which are not defined in the GCA, License, for example, have to be introduced for completeness. In Section 10.3.2 we describe the integration with concepts of other USDL modules. This integration allows us to utilize expert knowledge from other domains. Following, in Section 10.3.3 the overlaps and differences between the model of the GCA and the UCA are illustrated. The

[11] In German: *Urhebergesetz.*

[12] The Copyright Law of the United States of America and Related Laws are contained in Title 17 of the United States Code (17 U.S.C.).

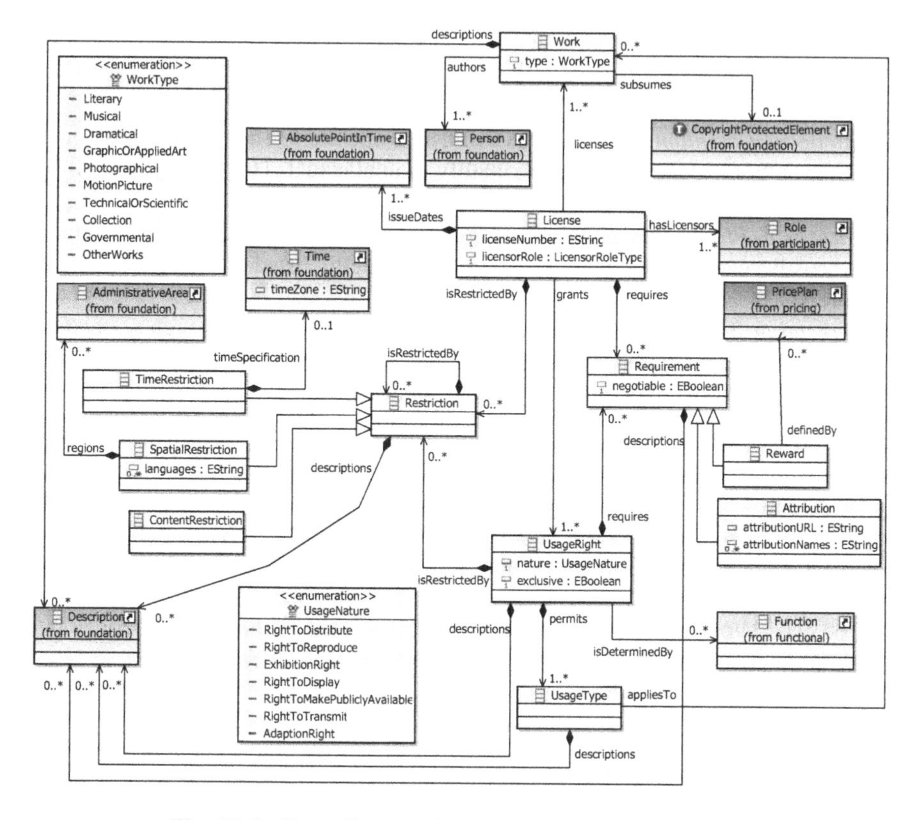

Fig. 10.1: Class diagram for the German Copyright Act.

results of this comparison are also used as an example for the management of variants in Chapter 17. This section concludes with an analysis on how the requirements put forth in the overview in Chapter 8 are met (Section 10.3.4).

10.3.1 The Module According to the German Copyright Act

The class diagram for the GCA is presented in Figure 10.1. It shows classes, relations, and the integration with classes of other USDL modules.

10.3.1.1 Work

The class Work is of central importance for the model. It is the subject matter which can be licensed. For instance, a service, a composite services, or the service output can qualify as *work* as long as it fulfills the following requirements:

- *Personal creation* [§ 2 Sec. 2 GCA]: The work has to be the product of some human action [14, § 2 note 8]. The author or creator [§ 7 GCA] can only be a natural person [14, § 7 note 2].
- *Intellectual creation* [§ 2 Sec. 2 GCA]: The work has to show or express some of the author's ideas or feelings to qualify as intellectual. Then it is individual or original. [14, § 2 note 18]
- Classify as *literature, science, or art* [§ 2 Sec. 1 GCA]: The GCA provides a non-exhaustive catalog. Computer programs are explicitly listed as literature. [14, § 2 note 3]
- *Perception*: The work has to be perceptible in some manner. Although a concrete physical materialization is not required, a mere imagination without an expression is not protected by the GCA. [14, § 2 note 13]

Novelty, however, is not a requirement. This means that the same work could (theoretically) be created by two different authors. As long as the works were created independently, for example, the authors did not know from the work of each other, each creation enjoys protection on its own. However, it is important that only the intellectual content of the work and not the embodiment itself is protected. [14, § 2 note 11]

The class Work is connected to a copyright protected element of the service by the relation subsumes. This relation reflects the legal practice of subsuming facts under abstract notions of the law. Thus, the model allows us to link elements of a service to copyright law. It is not in the scope of USDL to assess if the artifact which shall be licensed meets the requirements of § 2 Secs. 1, 2 GCA as outlined above.

The non-exhaustive enumeration of typical work types in § 2 Sec. 1 GCA is modeled as an attribute of Work. We also include the work types Governmental and Collection because they are explicitly mentioned in §§ 4, 5 GCA. Furthermore, we introduce the (non-legal) work type OtherWorks to account for to date unknown or not listed work types.

When a work is protected by the GCA, the creator gains *exclusive* moral and economic rights. Moral rights [§§ 12-14 GCA] include, for example, claim of authorship; economic rights [§§ 15-24 GCA] grant the creator exclusivity to exploit his creation economically. According to § 29 Sec. 1 GCA moral and economic rights are not transferable inter vivos — the rights are inseparable from the creator. However, the *usage rights* which are derived from economic rights of the creator are transferable.

10.3.1.2 UsageRight and UsageType

The rights that can be granted according to the GCA are *usage rights* [§ 31 GCA]. Usage rights are derived from the economic rights in §§ 15-24 GCA. In the Legal Module the usage rights are differentiated by the attribute nature, which enumerates the underlying economic rights.

According to § 31 Sec. 1 GCA a usage right might be divided into several *usage types*. A usage type defines a specific, well-defined, economic manner of how to use a work [26, § 31 note 17]. For instance, the *Right to Distribute* [§ 17 GCA] contains various distribution possibilities, such as to distribute a literary story as a hardcover book, an electronic book (eBook), or as an audio book; the audio book can be distributed on a compact disk or over the Internet as an MP3 file. The different manners of distribution are derived from the economic right of *Right to Distribute*. In the model, therefore, a UsageRight permits one or more UsageType.

Not all potential scenarios or usage types which might evolve through technological developments can be predicted and codified in the law [cf. § 31a GCA]. Therefore, the class UsageType does not have an enumeration of usage types. When USDL is applied to a specific scenario the adequate usage types should be initiated via an informal description or by introducing specializations of the class UsageType.

Usage rights can be either exclusive or non-exclusive [§ 31 Sec. 1 GCA]. A non-exclusive usage right permits the licensee to use the work in the agreed manner [26, § 31 notes 14-15]. An exclusive usage right allows the licensee to use the work in the agreed manner *and* gives him the right to prohibit others using the work in *that* manner [26, § 31 note 13]. If the author approves it, a usage right can be transferred and an exclusive licensee might grant corresponding (exclusive or non-exclusive) usage rights to third parties [26, § 31 note 9]. A creator can grant a non-exclusive usage right to several users [14, § 31 notes 50-55]. The class UsageRight has the attribute exclusive to indicate if a usage right is exclusive or non-exclusive.

A usage right can be restricted and different requirements might be attached to it. This is represented by the relations isRestrictedBy and requires to the corresponding classes.

10.3.1.3 TimeRestriction, SpatialRestriction, and ContentRestriction

According to § 31 Sec. 1 GCA the licensor is allowed to restrict usage rights in three different dimensions: temporal, spatial, and content.

A *temporal restriction* is concerned with the question if the usage right is *still* valid, as opposed to if it *exists* at all [26, Vor § 28 note 88]. The restriction can, for example, be defined as a period of time; it can also be defined in a quantitative manner, since this does also address the question for the temporal validity [26, Vor § 28 note 89].

A *spatial restriction* narrows a usage right to a specific economic territory or geographical area [26, Vor § 28 note 90]. Language regions are also considered as a spatial restriction [14, § 31 note 30].

A *content restriction* relates to the substance of a work — its content. Typically, these restrictions correspond to the economic rights in §§ 15 ff. GCA. However, you can restrict the substance of the work by additional content restrictions as long as they describe a sufficiently distinguishable, consistent, independent, technical, economic manner [26, Vor § 28 notes 91, 87]. This can be indicated, for example, by the distribution channel or the appearance of the work (cf. example above for

Right to Distribute). A restriction to a group of people is *not* a content restriction unless it is necessary because of the content itself [14, § 31 note 36].

Restrictions can be restricted by other restrictions. More information about the restrictions can be found at [5, Secs. 3.5-3.8].

10.3.1.4 Requirement

An author is entitled to demand a consideration when someone uses his work. This consideration can be regarded as the licensee's obligation which he has to fulfill in order to make the execution of a usage right valid — a requirement. In our model we distinguish between two different kinds of requirements: monetary and moral obligations.

A principle of copyright law is to incentivize creators to publish their works through monetary compensation. This is derived from the economic exploitation rights of the author [14, § 15 note 4]. The right to demand a reward is anchored in § 32 GCA. Therefore, we introduce the class Reward which can be related to other remunerations in USDL, for example, a PricePlan (cf. Chapter 9).

The moral rights are distinct from the economic rights. They are concerned with the author's relationship with the work even after it leaves his possession or ownership. The author of a work always maintains the moral rights to his work even when he grants exclusive usage rights to a third party — they are not transferable. The moral rights include the right of attribution, the right to have a work published anonymously or pseudonymously, and the right to the integrity of the work. The author has, however, the liberty to exercise those rights or not. In our model we consider the moral right to attribute [§ 12 GCA]. Further information on the class Requirement can be found at [5, Sec. 3.9].

10.3.1.5 License

The notion *license* is not defined by the GCA. In practice, it has different meanings in its usage, for instance, *license* sometimes relates to rights in the Patent Law or it does only comprise non-exclusive rights [26, Vor § 28 note 49]. In our model we introduce the class License to provide a "container" for one or more UsageRight. This class allows a licensor to relate all usage rights under which he is willing to grant the usage of his work. License can be restricted via the relation isRestrictedBy and can be provided under certain requirements via the relation requires. A restriction or requirement which relates to the class License applies to all usage rights.

The licensor might be the author or an exclusive licensee who is permitted to grant usage rights to third parties (attribute licensorRole). According to the GCA an author can only be a natural person, thus, a company as a licensor can only be an exclusive licensee. A specific license is identified by the attribute licenseNumber and the relation hasDate to the class AbsolutePointInTime in the Foundation Module. The same license number can be issued on different dates.

USDL provides description capabilities for services, however, it is not designed to manage transactional data. Therefore, License does *not* represent a contract but rather the usage rights under which a rights holder is willing to grant the usage of his work. The rights holder is related to License via the relation hasLicensor.

10.3.2 Integration with Other USDL Modules

The other USDL modules provide concepts which are also important for the description of license terms. By re-using relevant concepts in the Legal Module we utilize expert knowledge and reduce modeling complexity within the module. The present Legal Module is integrated with the Foundation, Pricing, Functional, and Participants Modules.

The Foundation Module (cf. Chapter 13) provides basic concepts of USDL of which we integrate the class Person, AdministrativeArea, Time, AbsolutePointInTime, and Description. Furthermore, the relation subsumes from the legal class Work to the interface CopyrightProtectedElement represents the legal process of subsuming a subject matter under a legal term. This allows our model to account for multiple or unknown artifacts which fulfill the requirements of a work according to the GCA. The interface potentially implements any identifiable element which is described by USDL, for example, a (copyrightable) service which shall be licensed.

The Pricing Module (cf. Chapter 9) provides a model to express pricing strategies. In the Legal Module pricing is considered as a reward for the rights holder for allowing other entities the usage of his work. Therefore, the class PricePlan represents *one* possibility to describe a Reward.

The class Function in the Functional Module (cf. Chapter 11) describes details about the usage of an artifact. This determines the actual usage right and its usage types (cf. Section 10.3 UsageRight and UsageType). For instance, when USDL is adopted by a domain which defines multiple usage rights and different usage types for a usage right, take, for example, different distribution channels of a written story (cf. aforementioned Section 10.3 for an example), the class Function provides information about the requested service and determines the concrete usage right/usage type combination for the license.

Finally, the class Role in the Participants Module (cf. Chapter 13) is used to describe a licensor. This also means that the licensor *has* to be specified in the Participants Module: he is either the Provider of the service or has to be listed as a Stakeholder.

10.3.3 Comparison to the U.S. Legal Module

Modeling of legal aspects such as licenses depends on the legislation of the country and, consequently, different modules for the jurisdictions are required. Nevertheless, copyright has overlaps across jurisdictions, since international treaties aim for harmonizing the different copyright acts. Germany and the United States (U.S.) are members of the major international copyright treaty: the Berne Convention[13] (RBC). The Berne Convention requires the signatories, for example, to recognize the copyright of works of authors from other signatory countries in the same way as it recognizes the copyright of its own nationals (Art. 5 RBC). As can be seen from the example, such a regulation does not regulate in detail, it is also not binding for private entities in the different countries. However, a country has to implement the requirements set by a treaty in its national law in order to remain a signatory. Being a member state has advantages for international relationships, for instance, the trade.

In Section 10.3.1 we demonstrate how the GCA was used to model classes and relations for granting usage rights according to the German law. The current USDL module for the UCA is depicted in Figure 10.2. In the following we discuss differences of classes and relations.

Class Work: different attributes.
According to the UCA a work can be registered [17 U.S.C. sec. 408 (a)]. The registration is not a condition of copyright protection, however, it is a prerequisite to civil infringement actions [cf. 17 U.S.C. secs. 411, 412] — it serves as *prima facie evidence*. The GCA does not consider the registration of a work; a registration is even not possible in Germany.

Class UsageRight: different attributes.
Class UsageType: only available in the German Legal Module.
The GCA distinguishes the terms *usage right* and *usage type*. Usage types are used to clearly differentiate the economic usage of a work. Since the UCA is unaware of this differentiation, the U.S. Legal Module does not provide a separate class for usage types but an enumeration of the rights according to 17 U.S.C. sec. 106. The corresponding attribute of the class UsageRight has been named type:UsageType to reflect its semantic proximity to the class UsageType in the German module. In the German module the class UsageRight has the attribute nature:UsageNature, which enumerates the (semantically similar) economic rights provided by the GCA. These economic rights are similar to the attribute type:UsageType in the U.S. Legal Module, however, we use different names for the attributes since you cannot transfer economic rights in Germany but in the U.S.

Class CopyrightTransfer: only available in the U.S. Legal Module.
In U.S. common law copyrights can be transferred. This option is not available in

[13] Convention for the Protection of Literary and Artistic Work.

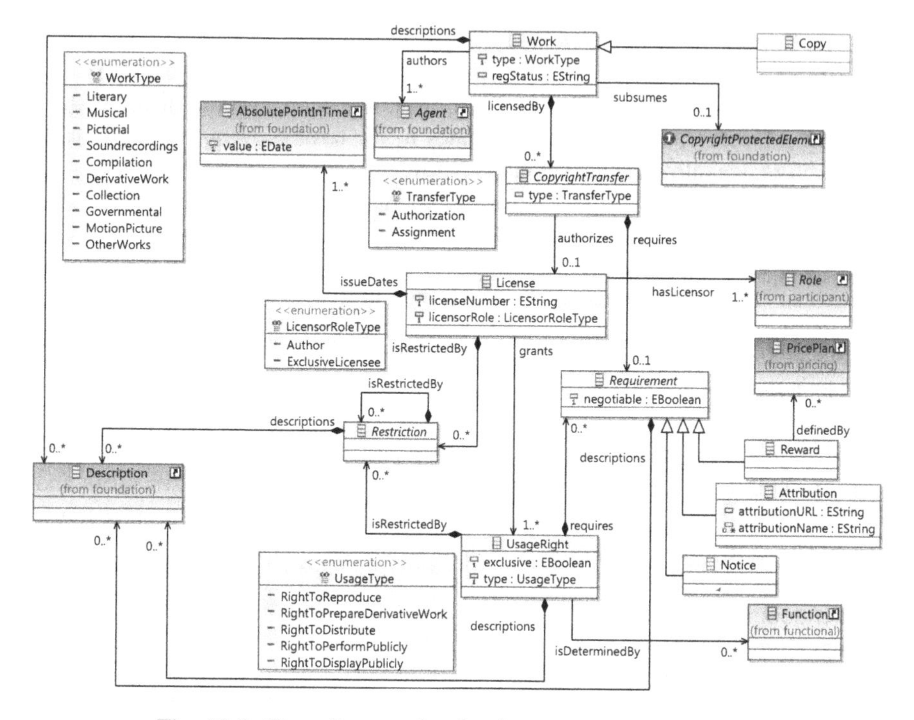

Fig. 10.2: Class diagram for the Copyright Act of the U.S.

German law. According to the GCA you can only grant usage rights, the actual copyright always remains with the original author.[14] Therefore, the U.S. Legal Module contains the class CopyrightTransfer to specify whether the copyright is assigned or transferred. The attribute type:TransferType allows an Assignment (usage right is assigned) or an Authorization (copyright is transferred).

Sub-classes of Class Restriction: not available in the U.S. Legal Module.
There is no distinction between *spatial*, *time*, and *content* restrictions in the UCA. Therefore, the corresponding classes are (currently) not available in the U.S. Legal Module.

Class Notice: only available in the U.S. Legal Module.
The legal transfer of exclusive copyrights in the U.S. is not valid unless that transfer is in writing and signed by the owner or a legal representative [17 U.S.C. sec. 204]. The transfer of copyright ownership may be recorded in the Copyright Office [17 U.S.C. sec. 205]. Consequently, the class Notice is introduced as a sub-class of Requirement.

[14] Only in the case of the death of an author you can transfer copyrights, § 29 Sec. 1 GCA.

- Class Copy: only available in the U.S. Legal Module.
  17 U.S.C. sec. 101 regards a *copy* as the material object, other than a phono record, in which the work is first fixed, thus, it is more specific than the class Work. Moreover, 17 U.S.C. often relates to the term copy in the statute text. Therefore, we introduce the sub-class Copy.

  Relation authors: relating different classes.
  German law only considers natural persons as authors of a work. Hence, the German Legal Module introduces the relation authors which points to the class Person of the Foundation Module. U.S. common law considers natural and legal persons as authors of a work. Therefore, authors points to the class Agent of the Foundation Module.

  Relations licensedBy (U.S.) vs. licenses (German).
  In the U.S. Legal Module the relation licensedBy links the classes Work and CopyrightTransfer. A work is licensed by a specific type of copyright transfer. The CopyrightTransfer then authorizes the License. The German module represents the relation between the classes Work and License via the relation licenses. In German law it is not an option to transfer the complete copyright ownership but only to grant usage rights. In U.S. common law, however, it is required to express if a license contains transferred or assigned copyrights.

Merging the U.S. and the German variants, which are derived from the statutes, into one module is not an option since further countries might require additional conceptualizations. Moreover, classes with the same name might differ in their specific conception. We also omit a generalization of the content of the two modules to a universal module in order to describe one (generic) copyright regime for the Internet. This would not support legal compliance and legal certainty as we argue in Section 10.1.

Although there are major differences between the two modules, there is also overlap, for example, the meaning of the class Work is indeed equal. The approach to variants in USDL is addressed in Chapter 17.

One has also to appreciate that the GCA and the UCA are part of different legal traditions and, therefore, very different. The differences between other countries, for instance, Germany and Austria might not be as major.

10.3.4 Design of the Legal Aspects in USDL

As mentioned throughout this chapter, the modeling of legal aspects adheres to the text of the statutes. Moreover, jurisdictions are considered and a high integration with other modules is an objective. This integration is intrinsic since the Legal Module aims to describe terms for the subject matter — the service.

Table 10.2: How are generic USDL language requirements addressed when modeling legal aspects in the Legal Module.

Requirement	Addressed how?
Conceptualization	Modeling of legal aspects adheres to the statutes of a country. Therefore, the level of abstraction is similar to the statutes. This allows for an abstract (conceptual) level. The Legal Module is restricted to the field of law, viz., copyright in our discourse. However, within this field of law the conceptualization of legal aspects for computational measure is high.
Modularity	The methodology in modeling legal aspects re-uses concepts from other base modules and, thus, accounts for the modularity of USDL.
Comprehensibility	Legal aspects should be represented in a narrative (license) text, common deed, and machine-processable form. The machine-processable form is represented by the current model, however, in order to attach textual narratives the class Artifact in the Foundation Module has to be utilized.

Table 10.3: How are service concept formation requirements in USDL addressed when modeling legal aspects in the Legal Module.

Requirement	Addressed how?
Organizational Embedding	Copyright and licensing issues have a strong alignment to the notion of business policies in organizations and statutes of jurisdictions that services operate in. So far, licenses are contained with other considerations in business policy documents. Moreover, licenses as realized in USDL target different offers of services. The associations between licenses and service offers are also highly relevant to an organization's service portfolio management.
Cognitive Sufficiency	The Legal Module supports coarse- to fine-grained modeling elements which can be referenced by coarse- to fine-grained elements in other USDL modules. The terms may also be described in a textual form (common deed/ license text) which may be provided externally. Moreover, the license can be used for various artifacts as long as they classify as work. At the same time, elements of the Legal Module are referenced by other modules.
Deployment Symmetry	As a service is extended, redeployed or brokered, any intermediary has the possibility to either incrementally modify the copyrights without overriding the previous legal constraints of providers.
Execution Resilience	Legal issues as a prominent non-functional property of a service address the requirement of *Execution Resilience*. For example, non disclaimers as put in place by service providers will not be overridden when third parties extend and deliver services elsewhere.

Another main character of the module is driven by the difficulty to predict the actual artifacts which shall be described. This leads to an abstract model (high conceptualization). Also the extensibility and comprehensibility of the module is influenced by this.

The requirements put forth in the overview in Chapter 8 are listed in Table 10.2 and 10.3.

10.4 Example — 2PL Airline Manager

Chapter 8 outlines a case example. In this example a German medium-sized company decides to use a fourth-party logistics provider (4PL) who offers a complete service bundle for the export to Russia. The service bundle includes a Web service and fronted tool, the *2PL Airline Manager*, which is provided by a second-party logistics provider (2PL). The *2PL Airline Manager* is used to look up rates, to kick-off shipments as well as to track shipments.

10.4.1 Terms of Usage

In addition to the description of the case example in Chapter 8 we consider the following information. The 2PL company of the *2PL Airline Manager* resides in Germany and provides its Web service and fronted tool to German customers amongst others. The *2PL Airline Manager* was developed and programmed by a 2PL employee. 2PL wants to allow its usage at no cost under the following conditions:[15]

(a) 2PL grants you a personal, non-exclusive, non-transferable, royalty-free license to use the 2PL Web service and to install the frontend tool only on your internal computer systems located in Germany, along with any modifications and upgrades thereof, if any, and any related manuals, documents, or other items (hereinafter collectively called "Materials") provided by or on behalf of 2PL, solely for the purpose of connecting with 2PL servers to look up rates, kick-off shipments and track shipments tendered to 2PL by you using the interface, and no other purpose.
(b) You acknowledge that you do not remove the name of 2PL from the *2PL Airline Manager* or its Materials.
(c) You may not copy, modify, adapt or reproduce the *2PL Airline Manager* or its Materials.

[15] To emphasize the applicability of our model we use an adapted version of an excerpt of a FedEx end-user license agreement which can be found at `http://images.fedex.com/us/software/pdf/license.pdf` (accessed 31 July 2011).

(d) You may not translate, decompile, reverse engineer or disassemble the *2PL Airline Manager* or its Materials in any event, except to the extend this limitation is prohibited by law.
(e) This license is effective until terminated by either you or 2PL. It will terminate independently without notice if you fail to comply with any provision of this license or any instructions regarding the *2PL Airline Manager* or its Materials.

These terms describe a restrictive usage of the *2PL Airline Manager* for an arbitrary user group, this means, the German medium-sized company as well as the 4PL can use the *2PL Airline Manager* as long as they comply with the terms. The description of the service bundle of 4PL should also include the terms of 2PL, since 1PL might make use of the *2PL Airline Manager* and, therefore, needs to know the conditions.

The terms (a)-(e) alone are not a contract and do not include transactional data, and, therefore, can be described in USDL. 2PL could define multiple licenses in USDL, for example, granting a broader usage scope[16] for the payment of a royalty.

10.4.2 Terms of Usage Transferred to USDL

In the following we apply the German Legal Module to the *2PL Airline Manager*. However, this is not a one-to-one mapping of the written terms above. The conditions (a)-(e) rather serve as instructions for the objectives of 2PL.

The *2PL Airline Manager* is implemented as a Web service and a frontend tool — basically computer programs. Computer programs are subject to the copyright law [§ 2 Sec. 1 No. 1 GCA] if they meet the requirements of § 2 Sec. 2 GCA.[17] We assume the *2PL Airline Manager* fulfills the requirements to qualify as such a work. In our model the *2PL Airline Manager* is subsumed under Work via the relation subsumes and classified as Type: Literary as defined in the GCA.[18] The other Materials mentioned in condition (a) might be classified accordingly.

Listing 10.1: Work

```
<work>
 <type> literary </type>

 <descriptions>
   <description>
    <type> name </type>
    <value> Web Service </value>
   </description>
   <description>
    <type> name </type>
    <value> Frontend Tool </value>
   </description>
 </descriptions>
```

[16] For instance, the usage of the *software package* to manage logistics provider other than 2PL.

[17] Cf. Section 10.3.1.1.

[18] The classification of a computer program as *literary work* results from the fact that it is written text. The requirements of § 2 Sec. 2 GCA do not particularly ask for poetry or such work.

```
<subsumes xsi:type='service:Service>
 <nature> Semi-Automated </nature>
 <exposedResource> <!-- urn of service --> </exposedResource>
 </subsumes>

<authors xsi:type='foundation:Person'>
 <firstName> <!-- first name of the 2PL employee --> </firstName>
 <lastName> <!-- last name of the 2PL employee --> </lastName>
 </authors>
</work>
```

The *2PL Airline Manager* was created by an employee of 2PL. The employee is the creator and, therefore, owns the copyrights. However, the employer is entitled to enact all proprietary powers [§ 69b GCA] and to limit the author's attribution rights to the effect that the attribution name might be the employer's (company) name [26, § 69b note 15]. According to condition (b) of the terms the attribution shall be given to *2PL*; this is a none negotiable Requirement (negotiable: false).

Listing 10.2: Requirement: Attribution

```
<requirement xsi:type='Attribution'>
 <negotiable> false </negotiable>
 <attributionNames> 2PL </attributionNames>
 <attributionURL> http://www.2PL.com </attributionURL>
</requirement>
```

Condition (a) lists details about the rights which shall be granted, however, it does not specify the exact usage rights. These rights are determined by the intended functionality and the manner of usage of the the *2PL Airline Manager*:[19] since it requires some installation on the customer's computer system it needs to be reproduced. Moreover, because the terms address an arbitrary user group it needs to be available to the public. Therefore, the two usage rights with the attribute nature: RightToMakePubliclyAvailable [§§ 15 Sec. 2, 19a GCA] and nature: RightToReproduce [§§ 16, 69 c GCA] are described in USDL. Moreover, condition (a) clearly states that the rights may only be non-exclusive (exclusive: false).

A UsageRight permits a certain UsageType which specifies the exact manner of usage. The manner is clearly stated in condition (a) as the purpose of *connecting to 2PL servers* to:

- *look up rates*
- *kick-off shipments*
- *track shipments*

2PL permits for every UsageRight the same manner. The UsageType can be described further via the class Description in the Foundation Module.

Listing 10.3: RightToReproduce and the corresponding manner (UsageType)

```
<usageRight>
 <nature> RightToReproduce </nature>
 <exclusive> false </exclusive>
 <permits>
```

[19] Cf. relation isDeterminedBy from the class UsageRight to the class Function in the Functional Module.

```
    <usageType>
     <descriptions>

     <!-- short description -->
      <description>
       <value> Connecting to 2PL servers is only allowed for the purpose to look
             up rates, kick-off shipments, and track shipments. </value>
       <type> freetextShort </type>
       <language> en </language>
      </description>

     <!-- keywords  -->
      <description>
       <value> rate look up </value>
       <type> keyword  </type>
      </description>
      <description>
       <value> shipment kick-off </value>
       <type> keyword </type>
      </description>
      <description>
       <value> shipment tracking </value>
       <type> keyword </type>
      </description>

    <descriptions>
   </usageType>
  </permits>
</usageRight>
```

Condition (a) lists several restrictions which restrict the UsageRight and, therefore, the UsageType. Content wise the usage right to reproduce is restricted to the installation on the internal computer system of the customer and to be used only with shipments tendered to 2PL by the customer using the interface. Furthermore, the usage is spatially restricted to computer systems in Germany only. The TimeRestriction, however, is "not limited."

Listing 10.4: Restrictions.

```
<restriction xsi:type='ContentRestriction'>
 <content> The usage is restricted to the installation on the internal computer
       system of the customer and to be used only with shipments tendered to 2
     PL by the customer using the interface.
 </content>
</restriction>

<restriction xsi:type='SpatialRestriction'>
 <regions xsi:type='foundation:AdministrativeArea'>
  <value> DE </value>
  <type> country </type>
 </regions>
</restriction>

<restriction xsi:type='TimeRestriction'>
 <timeSpecification xsi:type='foundation:DurationInterval'>
   <start xsi:type='foundation:AbsolutePointInTime'>
      <value> 2011-03-31T08:00:00.000+02:00 </value>
   </start>
   <intervalDuration>
       <value> 9999 </value>
       <type> year </type>
   </intervalDuration>
 </timeSpecification>
</restriction>
```

Condition (c) and (d) outline usage rights which are not permitted. This information is not required to be represented in USDL since in copyright law all rights are reserved by default. Each usage right one wants to utilize needs to be specifically obtained, thus, as long as a usage right is absent the work is not allowed to be used in the particular manner. Also the restriction of *non-transferable* in condition (a) and the information on the effectiveness of the license in condition (e) are negligible since this is the default of the law.

Listing 10.5 shows the snippets of Listing 10.1 to 10.4 in one exemplary license. The requirements and restrictions from Listing 10.2 and 10.4 are attached to License, since 2PL does not differentiate on a usage right level.

Listing 10.5: License *2PL Airline Manager*

```
<license>
 <licenseNumber> 298-37299-4789-G48J </licenseNumber>
 <issueDates xsi:type='foundation:AbsolutePointInTime'>
  <value>2011-03-31T08:00:00.000+02:00</value>
 </issueDates>

 <licensorRole> exclusiveLicensee </licensorRole>
 <hasLicensors xsi: type='participants:provider'>
  <enactingAgent> <!-- reference to provider element under service --> </
      enactingAgent>
 </hasLicensors>

 <licenses>
  <work>
   <!-- Listing 1 -->
   ...
 </licenses>

 <grants>
  <usageRight>
   <!-- Listing 3 -->
   ...
 </grants>

 <requires>
  <requirement xsi:type='Attribution'>
   <!-- Listing 2 -->
   ...
 </requires>

 <isResticted By>
  <restriction xsi:type='ContentRestriction'>
   <!-- Listing 4 -->
   ...
 </isRestrictedBy>

</license>
```

The information provided in Listing 10.5 represents the intention of terms (a)-(e), however, a written license text should also be attached via the class Artifact. The expression in USDL represents a machine-processable form of the license text, other aspects which are not present such as the AdaptionRight are not permitted.

10.5 Conclusion

This chapter presented our approach with modeling legal aspects in USDL. It is based on the requirement to comply with national laws by using the legal text as a modeling basis. Modeling of legal aspects in USDL is particularly difficult, because the subject matter "service" is imprecise for a legal analysis. However, when the subject matter is unknown it is challenging to define machine-processable attributes on a more precise granularity than *rights* and *obligations*. We, in contrast, argue to achieve a certain level of legal compliance and legal certainty in the area of services by complying with statutes. Additionally, the conceptualization (level of abstraction) is guaranteed because the law is abstract for a specific field of law.

To illustrate our methodology we used the German Copyright Act for modeling the Legal Module. We compared it to the Copyright Act of the United States to point out differences or common concepts. The M5 specification of the German Legal Module can be found at [5], the U.S. Legal Module is not part of this milestone, but will be integrated.

We already demonstrated the major hurdle when modeling according to the law: jurisdictions. Although not all concepts of the jurisdictions are disjunct, local peculiarities have to be accounted for. This demands for a management of variants. Chapter 17 illustrates the approach to variant management in USDL.

In future work it is planned to include description capabilities for contract clauses in the Legal Module. Such standard terms comprise aspects of private law such as liability. These clauses also have to comply with national laws.

Another advantage of modeling legal aspects according to the laws is that the notions of the legal text are used. This supports the analysis for legal consequences such as if a transfer of copyright is legally allowed. Approaches to the automated analysis of legal consequences which are based on the formal description of services can be found at [32, 7].

References

1. H. Abelson, B. Adida, M. Linksvayer, and N. Yergler. ccREL: The Creative Commons Rights Expression Language: Version 1.0, 3 March 2008.
2. Adobe Systems Incorporated. XMP Specification Part 1: Data Model, Serialization, and Core Properties, July 2010.
3. Adobe Systems Incorporated. XMP Specification Part 2: Additional Properties, July 2010.
4. Adobe Systems Incorporated. XMP Specification Part 3: Storage in Files, July 2010.
5. C. Baumann, U. Kylau, and D. Oberle. Unified Service Description Language (USDL) — Legal Module. Technical Report Version 3.0, Milestone M5, SAP Research, May 2011. Available at `www.internet-of-services.com`.
6. C. Baumann and C. Loës. Formalizing Copyright for the Internet of Services. In G. Kotsis, D. Taniar, E. Pardede, I. Saleh, and I. Khalil, editors, *Proceedings of the 12th International Conference on Information Integration and Web-based Applications & Services (iiWAS2010)*, volume 272 of *books@ocg.at*, pages 712–719, New York, NY, USA, 2010. ACM.
7. C. Baumann, P. Peitz, O. Raabe, and R. Wacker. Compliance for Service Based Systems Through Formalization of Law. In J. Filipe and J. Cordeiro, editors, *Proceedings of the 6th*

International Conference on Web Information Systems and Technology, volume 2, pages 367–371, Valencia, Spain, 2010. INSTICC Press.
8. Contentguard. XrML 2.0 Technical Overview: Version 1.0, 8 March 2002.
9. K. Coyle. Rights Expression Languages: A Report for the Library of Congress, 2004.
10. Creative Commons. CC Affiliate Network. Accessed 6 April 2011.
11. Creative Commons. CCPlus. Accessed 5 April 2011.
12. J. Delgado, I. Gallego, S. Llorente, and R. García. IPROnto: An Ontology for Digital Rights Management. In D. Bourcier, editor, *Legal knowledge and information systems*, volume 106 of *Frontiers in Artificial Intelligence and Applications*, pages 111–120, Amsterdam, 2003. IOS Press.
13. Digital Library Federation. Metadata Encoding and Transmission Standard: Primer and Reference Manual, 2010.
14. T. Dreier and G. Schulze. *Urheberrechtsgesetz: Urheberrechtswahrnehmungsgesetz, Kunsturhebergesetz: Kommentar*. C. H. Beck, München, 3 edition, 2008.
15. G. R. Gangadharan. *Service Licensing*. PhD thesis, University of Trento, Trento, Italy, 2008.
16. G. R. Gangadharan, V. D'Andrea, R. Ianella, and M. Weiss. ODRL Service Licensing Profile (ODRL-S). In R. Grimm, B. H. Hass, and J. Nützel, editors, *Virtual goods*, pages 73–90, New York, 2008. Nova Science Publishers.
17. G. R. Gangadharan, M. Weiss, V. D'Andrea, and R. Iannella. Service License Composition and Compatibility Analysis. In B. J. Krämer, K.-J. Lin, and P. Narasimhan, editors, *Service-Oriented Computing - ICSOC 2007*, volume 4749 of *Lecture Notes in Computer Science*, pages 257–269, Berlin Heidelberg, 2007. Springer.
18. A. Gangemi, N. Guarino, C. Masolo, A. Oltramari, and L. Schneider. Sweetening ontologies with dolce. In A. Gómez-Pérez and V. R. Benjamins, editors, *Knowledge Engineering and Knowledge Management. Ontologies and the Semantic Web, 13th International Conference, EKAW 2002, Siguenza, Spain, October 1-4, 2002, Proceedings*, volume 2473 of *Lecture Notes in Computer Science*, pages 166–181. Springer, 2002.
19. R. García. *A Semantic Web approach to Digital Rights Management*. PhD thesis, Universitat Pompeu Fabra, Barcelona, 2005.
20. R. García, R. Gil, and J. Delgado. Intellectual Property Rights Management Using a Semantic Web Information System. In R. Meersman, T. Zahir, W. van der Aalst, C. Bussler, A. Gal, V. Cahill, S. Vinoski, W. Vogels, T. Catarci, and K. Sycara, editors, *On the Move to Meaningful Internet Systems 2004: CoopIS, DOA, and ODBASE*, volume 3290 of *Lecture Notes in Computer Science*, pages 689–704, Berlin Heidelberg, 2004. Springer.
21. R. Gil, R. García, and J. Delgado. An interoperable framework for IPR using web ontologies. In M. Biasiotti, E. Francesconi, M.-T. Sagri, and J. Lehman, editors, *Legal Ontologies and Artificial Intelligence Techniques*, volume 4 of *IAAIL Workshop Series*, pages 135–148. Wolf Legal Publishers, 2005.
22. IPR Systems Pty Ltd. Open Digital Rights Language (ODRL), 8 August 2002.
23. ISO/IEC 21000-5:2004. Information technology – Multimedia framework (MPEG-21) – Part 5: Rights Expression Language, 1 April 2004.
24. J. Kropholler. *Internationales Privatrecht: Einschliesslich der Grundbegriffe des Internationalen Zivilverfahrensrechts*. Mohr Lehrbuch. Mohr Siebeck, Tübingen, 6 edition, 2006.
25. LG München I. Wirksamkeit der GNU General Public Licence (GPL) nach deutschem Recht, 19 May 2004.
26. U. Loewenheim, G. Schricker, and A. Dietz, editors. *Urheberrecht: Kommentar*. C. H. Beck, München, 4 edition, 2010.
27. OASIS. Web Services Security Rights Expression Language (REL), 1 February 2006.
28. ODRL Initiative. ODRL Service (ODRL-S) Profile (Working Draft: 2 April 2008). Accessed 6 April 2011.
29. ODRL Initiative. ODRL V2.0 - Common Vocabulary (Working Draft: 18 February 2011). Accessed 6 April 2011.
30. J. J. O'Sullivan. *Towards a Precise Understanding of Service Properties*. PhD thesis, Queensland University of Technology, 2006.

31. A. Pease, I. Niles, and J. Li. Origins of the IEEE Standard Upper Ontology. In *Working Notes of the AAAI-2002 Workshop on Ontologies and the Semantic Web, Edmonton, Canada, July 28-August 1, 2002*, 2002.
32. O. Raabe, R. Wacker, C. Funk, D. Oberle, and C. Baumann. Lawful Service Engineering – Formalisierung des Rechts im Internet der Dienste. In E. Schweighofer, A. Geist, and I. Staufer, editors, *Globale Sicherheit und proaktiver Staat – Die Rolle der Rechtsinformatik*, volume 266 of *books@ocg.at*, pages 643–650, Wien, Österreich, 2010. Österreichische Computer Gesellschaft (OCG).
33. W3C. RDF/XML Syntax Specification (Revised), 10 February 2004.
34. W3C. RDFa in XHTML: Syntax and Processing, 14 October 2008.
35. F. v. Westphalen, G. Thüsing, and F. Drettmann, editors. *Vertragsrecht und AGB-Klauselwerke*. C. H. Beck, München, 27 edition, 2010.

Chapter 11
Service Functionality and Behavior

Uwe Kylau, Michael Stollberg, Ingo Weber, and Alistair Barros

Abstract One of the most essential parts of every service description language is to provide suitable means for describing the following three aspects of services: (1) *what* the service does, i.e., which functionality it provides, (2) *where* the service resides, i.e., where it can be accessed and via which means it can be consumed, and (3) *how* the service behaves, i.e., how to interact with the service in order to properly consume it. These are subject to various existing and well established standards. In order to capture these aspects in an all-embracing manner, USDL defines three separate modules — the Functional, the Technical, and the Interaction Module — that each cover one aspect and together provide a holistic description of the functionality and behavior of services. The modules are commonly designed to provide a unifying description structure that abstracts from details and allows for the re-use and integration of existing as well as upcoming standards, thereby maintaining flexibility and extensibility of USDL. This chapter introduces the background and underlying design principles, and presents the USDL modules for functional, technical, and behavioral service descriptions in detail.

Uwe Kylau
SAP Research Brisbane, Building A4 Level 7, 52 Merivale Street, South Brisbane QLD 4101, Australia, e-mail: uwe.kylau@sap.com

Michael Stollberg
SAP Research Dresden, Chemnitzer Strasse 48, 01187 Dresden, Germany,
e-mail: michael.stollberg@sap.com

Ingo Weber
The University of New South Wales, School of Computer Science & Engineering, K17, The University of New South Wales, Sydney, NSW 2052, Australia,
e-mail: ingo.weber@cse.unsw.edu.au

Alistair Barros
Queensland University of Technology, GPO Box 2434, Brisbane, QLD 4001, Australia,
e-mail: alistair.barros@qut.edu.au

11.1 Introduction

The overall purpose of a service description language is to facilitate a proper communication among the parties that are involved in the provisioning and consumption of services. In order to create business value through services, service providers, consumers, and intermediaries (e.g., service brokers) need to successfully communicate on the usage of these services. Several aspects of substantially different nature are relevant in this communication, ranging from the provided features over technical details to legal and financial aspects. To allow for a common understanding — which is the crucial pre-requisite for successful communication — a service description language defines a system of symbols governed by grammatical rules which associate particular sets of symbols with a meaning. The aim of USDL is to provide an all-embracing service description language that covers all relevant aspects for communicating about business services on the basis of a structured description model with a clearly defined meaning, while preserving flexibility and extensibility for the re-use and integration of existing service description standards as well as upcoming ones.

Among the various aspects that are relevant in the context of services, the following ones relate to the basic facets of a service and hence are essential for facilitating a proper communication: (1) *what* the service does, i.e., what functionality and features it provides that create value for the involved parties when performed, (2) *where* the service resides, i.e., where to access it and by which means it can be consumed, and (3) *how* the service behaves, i.e., how the service acts when it is performed and how to interact with the service in order to properly consume it (*cf.* [9],[30]).

These aspects are subject to various existing and well established standards, such as WSDL [7], WS-BPEL [23] (or in short BPEL), etc., which constitute the core of the technology stack used in many current service-based systems. However, despite their wide adoption the existing standards commonly expose some crucial drawbacks. Firstly, they mostly provide structural descriptions on a low, technical level, which allows for automated processing but limits the common understanding and communication about services on higher levels of abstraction. Secondly, the three aspects mentioned earlier are commonly spread across several service descriptions standards; e.g., WSDL comprises detailed information about the provided functionality in form of operations with messages and data types along with details on the technical accessibility, while — in the realm of service descriptions — BPEL is primarily used to describe the observable behavior of complex services. Because of this, it is hard to obtain a unified view on the functional, technical, and behavioral aspects of services from the current description standards, which hampers the communication on specific aspects of interest.

In order to overcome these drawbacks, USDL proposes a unified description model for the functional, technical, and behavioral aspects of services. The aim is not to re-invent the wheel by introducing yet another description model for these essential aspects — but to provide a well structured model that resides on top of existing as well as upcoming standards, and facilitates the integration, interoperability, and exchange among them. For this, USDL defines three modules that each cover one of the core aspects mentioned above.

1. The *Functional Module* defines generic constructs for describing what a service does, i.e., the features and functionality provided. This covers what is commonly referred to the 'functional description' of a service.
2. The *Technical Module* encompasses generic constructs for describing the 'technical interface' of a service, i.e., how to technically access and consume the service.
3. The *Interaction Module* defines generic constructs for describing the order and other constraints over how individual parts of a service are performed and how the involved parties interact. For instance, it allows modeling that some service requires a customer to request a quotation before goods can be ordered.

With the three modules presented in this chapter, USDL defines a description model for the concepts that are common among existing standards, with clearly defined semantics, while abstracting from technical details that are already well covered by these standards. As the main goal of USDL is to provide a meta-model that captures the information required to provision, deliver and consume a service, the three modules need to address all aspects required to understand what a given service offers, how to technically access it, and how to interact with it while respecting any constraints on the execution order. However, even though the information has to be provided, it does not have to be contained entirely inside USDL itself; most of the information may be present in existing artifacts (such as WSDL documents), and can simply be referenced by the respective USDL model element. Accordingly, the three modules merely define the basic constructs in a generic manner so that existing as well as upcoming standards can be re-used and integrated, thus maintaining the foundation principle of flexibility and extensibility of USDL. Finally, addressing each of the aspect in a separate module follows the principle of modularity.

The remainder of this chapter is structured as follows. Section 11.2 reflects on the background and introduces the design of the USDL modules for describing functionality, technical interfaces, and behavior of services. Sections 11.3, 11.4, and 11.5 present the Functional, Technical, and Interaction Module of USDL in detail. Section 11.6 concludes the chapter.

11.2 Modeling Functionality, Technical Interfaces, and Behavior in USDL

USDL's modules for the functional, technical, and behavioral aspects of services define generic constructs for modeling the core aspects of services in a structured manner, while being extensible by re-using and integrating existing as well as upcoming standards. The following discusses relevant background about state of the art related to the three aspects, and motivates the focus of modeling and the constructs selected. The second part of the section then outlines the general design and relationship of the three USDL modules which are presented in detail in subsequent sections.

11.2.1 State of the Art and its Influence on the Module Design

11.2.1.1 On Functional Modeling

Functional modeling is concerned with describing what the service does. In early service frameworks, functionality has mostly been considered as the set of operations, respectively the inputs and outputs of a service (e.g., [21]). USDL takes a broader understanding of this, considering the functionality of a service to be what it achieves for the beneficiaries involved (e.g., customers and providers), i.e., its value proposition.

In order to describe the business value of a service, a means for describing the functionality on a high level of abstraction is needed. Yet, enough detail needs to be presented to enable potential consumers of the service to understand what it does, and that well enough for them to make a decision on its suitability for the consumption scenario. While no widely adopted standards for this exist, a common concept that appears to be suitable is *capability modeling* as used for example in SoaML (OMG) or the SOA Reference Model (OASIS) (see Chapter 5). [22] discusses the concept of capabilities as expressing the ability to perform a course of action that achieves a result. When viewed on a whole, this constitutes the functionality offered as the service. Other scholars have extended this by adding the notion of commitment, meaning that the entity offering the service is not only capable of doing so, but also committed to do so upon request (cf. Chapter 4 and [12]).

Originally developed as AI techniques for formal software specification, capability modeling usually takes a *black-box view*, meaning that only the external visible functionality is described but not how it is realized internally [13]. Other approaches, commonly termed *function modeling* (or *component modeling*), stem from structured systems analysis and design techniques wherein the principle of functional decomposition has been introduced [8]. They address modeling of functionality with a more detailed formalization of the (software) components that realize this functionality, thus generating a white-box or grey-box view depending on the level of detail provided about the components/processes. For example, drilling down to the lowest level of atomic functions performed in an organization including information about the roles and systems involved in the process can be regarded as a true white-box view (sometimes also called glass-box).

Recent works, especially in the context of Semantic Web Services, adopt the notion of capabilities (e.g., [11] and Chapters 6 and 7) and formally describe the provided functionality in terms of *preconditions* that specify the conditions under which a service can be consumed, and *effects* that describe the results of a regular execution (e.g., [26] and Chapters 6 and 7). It should be pointed out that capability modeling does not have to be limited to a flat, single-layer model. In fact, there are approaches that propose quite diverse hierarchical models (cf. [16]), which also capture interconnections between individual capabilities in terms of inputs, outputs and exceptions. In borrowing the principles of decomposition from function modeling, these capability models slightly extend their scope from black-box to grey-box and thus share similarities with function modeling. The difference is that they describe functionality from a business point of view, which does not go beyond a certain level of detail, as opposed to the IT systems point of view (taken by function modeling) that usually covers all aspects of realization/implementation.

For functional modeling in USDL, a mix of capability modeling and function modeling has been chosen. In particular, service functionality is modeled as a set of hierarchical functions, which, at the top-most level, are externally visible as capabilities. As outlined previously, functions are the building blocks of rendering a capability and have a number of inherent characteristics, some of which are similar to concepts of technical interfaces. Functions produce outcome, e.g., something is created, transformed, delivered or destroyed. Functions are performed by some actor (agent), who, in doing so, usually operates on one or more objects (resources), consuming and producing some of the objects, while others are maybe affected as a side effect. It is furthermore common that actors use resources as tools to perform an action.

The reason for choosing this model is that while most service consumers might not care how functions offered as capabilities are structured internally, such information is of interest to intermediaries aggregating, re-purposing and enriching services. For example, a service broker that provides payment and billing facilities may support fine-grained payment models, e.g., collecting multiple apportions during service execution. Integrating such payment models correctly, requires detailed knowledge about the structure of a capability. In some cases it is even necessary to describe conditions that have to hold before an action can be started, as well as the effects that set in once the action is completed.

11.2.1.2 On Modeling Technical Interfaces

The second of the core aspects refers to describing how to access a service, which is necessary to provide service consumers with the relevant details on how to technically invoke the service in order to consume its functionality. This refers to the set of concrete technologies through which the service can be accessed, which is commonly referred to as the *technical interface* of a service. Description methods for the technical interface are already well covered by the existing standards, which form the heart of service-based IT systems, at least for automated services that offer IT-supported means of access [9].

The currently most prominent technologies for technical interfaces of automated services are SOAP-based interfaces as, e.g., supported by WSDL [7] and, lately emerging, RESTful Web service interface as, e.g., supported by WADL [15] or structured according to ODATA [20] (see also Chapter 5). These have emerged from the domain of distributed computing, and expose certain similarities of concepts:

they consist of operations that have typed input and output parameters, and faults / exceptions that may occur during the execution.

USDL supports the modeling of such technical interfaces by defining an abstract notion of a technical interface which can be extended by referencing respective modeling constructs of existing standards. This allows providing the detailed information for the technical access and consumption of services, also covering the case where a business service exposes several technical interfaces, e.g., a SOAP-based interface and a proprietary EDI interface (such an SAP RFC). Similar to constructs that have been defined in other higher-level service description efforts (e.g., by the 'grounding' concept of OWL-S [19]; cf. also Chapter 7), this provides a means for describing technical interfaces in an explicit and well structured manner while allowing for the re-use and integration of existing as well as emerging standards that capture technical interface descriptions in sufficient detail and widely employed as the technical foundation of service-based systems.

In addition, the generic constructs in USDL allow integrating description standards of other types of technical interfaces, e.g., for services that can be accessed by email, fax, telephone as well as manual services that cannot be accessed (requested or delivered) by any technical means. Take, for example the service provided by a hair dresser. Although you can request it via phone, it can only be delivered through manual labour involving direct human interaction. With this, the support for modeling technical interfaces USDL allows the re-use and integration of existing standards while remaining extensible for emerging standards as well as for the broader understanding of business services that is not limited to automated services.

11.2.1.3 On Behavioral Modeling

In addition to the knowledge about the provided functionality and the technical interface, it might be necessary to know how to interact with the service in order to properly consume its functionality. This is commonly referred to as *behavior modeling* of services, which is concerned with the external behavior of the service, i.e., constraints on the order in which individual functions are performed and when and how it is necessary to interact with the service, respectively with the actors performing it [6]. The artifact resulting from behavior modeling is referred to as behavioral interface, business protocol, or public view on a process. Behavior modeling is particularly important for properly consuming complex or long-running services that offer several functions, as common in business contexts.

Interaction structures are well known in the domain of workflow and business process modeling (e.g., BPEL [23] or BPMN [24]), and have been applied in choreography languages to define collaborations (e.g., WS-CDL [17] or WSCI [1]). Within the spectrum of different modeling approaches, two general styles of representing control flow can be identified. On the one hand, there are graph-based models that capture ordering as a set of relationships between vertices and edges (examples include BPMN and Petri Nets). On the other hand, there are hierarchical, block-structured languages (e.g., WS-CDL and WSCI) that capture ordering through the semantics of the blocks. BPEL is a hybrid, offering both block-structured and graph-based modeling.

For reasons of simplicity and traceability, USDL supports behavioral modeling by a basic version of the block-structured approach which allows capturing sequences. This appears to be sufficient for services that do not actually have an existing formal description of their externally observable behavior and hence do need such a description (such as, e.g., manual services). To not exclude complex behavioral interaction protocols, providers get the possibility to attach the definition of their protocol(s) in a formalism of their choice and potentially have an abstraction of these protocol(s) mapped to the USDL model. Phases may be introduced as necessary, as they usually only make sense for long running process-based services. With this, USDL provides a means for describing the externally observable behavior of services in a basic manner, which can — analogously to the other models — be extended with more expressive languages in case this is desirable.

11.2.2 Design and Relationship of USDL Modules

The split of modeling of functionality and behavior in three modules in USDL was motivated primarily by the requirement of *Modularity* (cf. Section 8.3.1.3). Only the Functional Module is mandatory in each USDL service description (there can be no service without value-creating functionality). The other two modules are optional, with the Technical Module applying mostly to automated services and the Interaction Module being only relevant for complex, longer-running services.

Even though Functional, Technical and Interaction Modules are distinct in USDL, all three form a close unity and together provide a description model for three important aspects: *what*, *where* and *how*. They are modules where almost all the general principles and modeling requirements of USDL are applied. This is due to the fact that they cover the very core of service description, overlapping with many other description languages. Accordingly, all three modules are concise and try not to replicate existing standards, e.g., for describing technical service interfaces. They do, however, fill gaps between capabilities of existing description languages and identified description requirements.

In the attempt to find a language suitable for a wide range of services, the fact that existing standards do not always provide the right means for the description purposes of USDL has to be addressed with proper *Conceptualization* and *Expressive Power*. For instance, a non-technical user should be able to define a simple interaction protocol in USDL.

Overall, one of the goals of USDL is to enable provisioning, delivery and consumption of services, which means the combination of all modules needs to provide sufficient information for potential consumers to make their decision and understand how to consume services (*Comprehensibility*, cf. Section 8.3.1.5).

Table 11.1 lists all requirements that were considered during modeling of functional, technical and interaction aspects. The table also outlines how these requirements were implemented in the three modules. Please note that for the remainder of this section requirements introduced in Section 8.3 are given without reference.

Table 11.1: USDL requirements addressed by the Functional, Technical and Interaction Module (cf. Chapter 8).

Requirement	**Requirement addressed how**
Generic language requirements	
Conceptualization	Functional, Technical and Interaction modules are described on an abstract level and where relevant allow for language specific artifacts, e.g., BPEL processes, to be integrated not subsumed.
Expressive Power	Functional, Technical and Interaction modules are sufficiently expressive to support description of manual and automated services in varying degrees of detail.
Comprehensibility	Support of different views of functional, technical and interaction aspects through black-box, gray-box, white/glass-box.
Modularity	Functional Module does not require elements of Technical Module or Interaction Module to be present.
Extensibility	Technical Module and Interaction Module allow introduction of alternative interface and behavioral models.
Service concept formation requirements	
Organizational Embedding	Service capabilities can be traced from business to technical levels (through cyclic associations on Functions).
Cognitive Sufficiency	Functional, Technical and Interaction modules allow referencing of external (e.g., organizational) artifacts and support interconnecting artifacts to core service elements.
Service Information Hiding	Functional Module descriptions can be limited to externally visible capabilities (black-box); Technical Module and Interaction Module — by definition — only capture externally observable aspects.
Deployment Symmetry	Different degrees of detail (black-box, gray-box, white-box) can be developed for Functional, Technical, and Interaction modules supporting different stakeholder extensions.
Execution Resilience	Functional Module supports definition of (conceptual) faults that can be mapped to technical fault messages (Technical Module) and exception handling procedures (Interaction Module).

The interconnection between USDL and existing standards, respectively the interconnection of *Services* and *Service Dependencies and Composition* (technical and behavioral aspects of interaction; cf. universe of discourse in Section 8.2), is what mainly characterizes the structure of the modules. Driven by the requirement of *Conceptualization*, functionality — as modeled in the Functional Module — is understood in an abstract, conceptual way and thereby freed from any concerns of access. This has the advantage that multiple technical means of access and behavioral protocols can be defined for a single piece of functionality. *Cognitive Sufficiency* across all three modules is instilled through cross-referencing. Specifically, concepts in the Technical Module and Interaction Module reference concepts in the Functional Module. Both modules describe concrete means of access to functionality either directly or through references to external descriptions of such means, e.g., in other service artifacts.

With respect to the requirement for *Organizational Embedding*, the Function has a cyclic association which supports traceability of service functions from the business to technical levels. Interaction protocols can be defined at each level. Technical interfaces can be defined for the technical level.

Allowing existing description standards, and their models, to be integrated into USDL and interconnected with core functional concepts requires the three modules to be flexible. Only a small number of concepts define the necessary structural skeleton for describing functional, technical and interaction aspects (Function, Interface and InteractionProtocol, respectively), which is in accordance with the requirement of *Conceptualization*. The aforementioned integration of existing description models is enabled by a few additional concepts included in the modules by default. These concepts represent alternative concrete implementations of the skeleton. New implementations can be introduced as well, using the extension mechanisms proposed in Chapter 17. Addressing the requirement of *Extensibility*, an extension may define a complete description model (for technical or behavioral aspects) or augment the currently non-exhaustive integration capabilities to cover additional external description standards.

Separating *Services* and the two aspects of *Service Dependencies and Composition* (technical and behavioral, as previously mentioned) into three distinct, but interconnected, modules presents one of the unique properties of USDL. Current service description languages do not address this the same way, i.e., they do not combine all three characteristics.[1]

Some languages, e.g., WSDL [7], separate functionality from concerns of access, but then do not cover all aspects of the latter (e.g., only technical). Other languages cover all three, but do not clearly separate. For example, OWL-S [19] separates technical aspects into *groundings*, whereas functional and behavioral aspects are captured together in *process models*.

Some of the other principal commonalities and differences with existing service description models were outlined in Section 11.2.1. Among them is the use of capability modeling and functional decomposition as a combined approach for modeling functional aspects in USDL. It was chosen to meet the requirement of *Service Information Hiding*. In particular, this approach facilitates description of functionality on different degrees of visibility, exposing extensive functional descriptions (glass/white-box) or hiding pieces of information as necessary.

This concludes the overview of the design of the three modules. For a full formal specification please refer to [2], [4] and [3], respectively.

[1] (1) combining functional, technical and interaction aspects; (2) addressed separately and modular; (3) interconnected to form a whole.

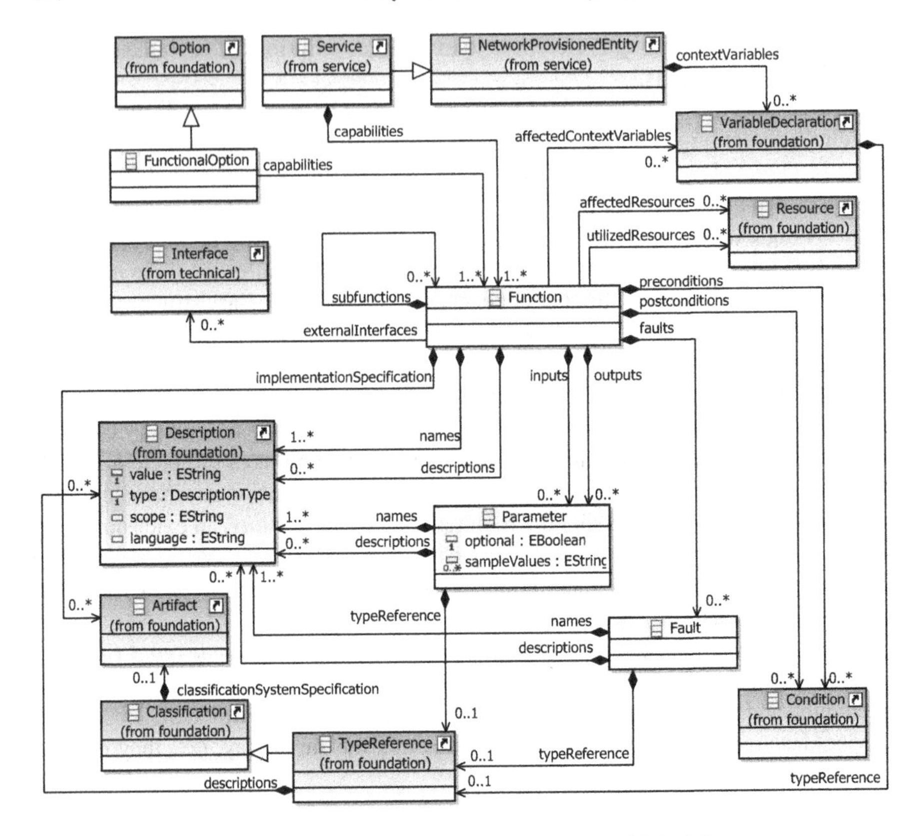

Fig. 11.1: Class diagram of the Functional Module.

11.3 The Functional Module

As outlined in previous sections, modeling of functionality is one of the fundamental components of a service description language. The Functional Module of USDL defines a concise, yet powerful, meta-model that explicitly deals with the value-creating functionality provided by a service, abstracting from technical implementation. As USDL is intended to support a wide range of services, it provides modelers with the flexibility to express different degrees of detail and complexity (from black-box to white-box views), while being designed to remain comprehensible. Following these principles, the Functional Module combines key concepts of capability modeling and functional decomposition. The central construct is the class Function, which is understood as representing a course of action that can be decomposed into sub-functions. The following explains the model in detail. Figure 11.1 depicts the class diagram of the module.

11.3.1 Overview and Main Constructs

The course of action modeled as a Function is usually performed by a single actor,[2] which is specified as the provider of the service (see Chapter 13). However, there are also cases where more than one actor is involved, e.g., if the "leading" actor decides to outsource parts of the activity to other parties. The course of action may be defined by providing an informal textual description using the features names and descriptions. Alternatively, modelers have the ability to recursively decompose a function into sub-functions (feature subfunctions), i.e., lower-level building blocks that together make up the higher-level function, but are not themselves exposed for isolated external consumption. A third variant of describing functionality is to link the USDL element to an external vocabulary, e.g., a different specification language, classification system or ontology. This can be achieved through referencing an actual formal specification of the function via feature implementationSpecifications, or by specifying a Description of type concept as part of feature descriptions. Depending on the level of detail of textual description and decomposition, or of the referenced vocabulary, a black-box, gray-box or white-box view of functionality is produced.

Apart from describing and structuring the course of action, there are other aspects that characterize a function. For example, a Function may feature one or more input and output Parameters, as well as one or more Faults. The latter is used to capture information about exceptions that may arise while the course of action is performed. Both elements are named and may be typed via feature typeReference, essentially being a pointer into an external type system. It was decided not to incorporate elements for the definition of types into USDL, because there exist sufficiently powerful type definition languages (e.g., XML Schema [28] or ontology languages — see Chapters 5 and 7). Adding such elements to USDL would have replicated these efforts and potentially jeopardize the comprehensibility of the model.

Other features of Function comprise required pre-conditions and produced post-conditions (effects), as well as references to context variables, i.e., variables held in the context of a service and are affected as part of performing the function. This completes the set of features considered by most service description languages to be part of the specification of a service function. In USDL modelers have, on top of that, the possibility to specify resources involved in the course of action defined as a service function. Two categories are distinguished: resources that are used for performing the function (utilizedResources), e.g., tools or organizational roles, and resources that are manipulated in that process (affectedResources), e.g., business objects (such bank accounts).

The concepts of capability modeling (cf. Section 11.2.1) are applied at the connection point between the Functional and Service Module. Namely, a Service is required to provide one or more functions, which in this context are defined to be the capabilities offered by the service. The notion of capability carries additional semantics that go beyond the general definition of function. In particular, it means

[2] The term *agent* is used in USDL, cf. Chapter 13.

that the actors performing the course of action are not only able to do so, but they are also willing to accept requests and establish an agreement for this to take place. Accordingly, when using a function as a capability, the course of action in that function is understood as the process of value creation[3] (as viewed from the providing side).

Capabilities are the externally-consumable portions of functionality. This means they can be made accessible through technical interfaces, a fact that is captured in the model by feature externalInterfaces. This feature may reference a number of Interface objects in the Technical Module, each of them describing an interface through which the capability can be requested/invoked. The exact part of the interface responsible for providing access is identified in the Technical Module by referencing back to the capability or its sub-functions (cf. Section 11.4). The idea is that externalInterfaces is only used with capabilities, i.e., a sub-function cannot reference a technical interface object. This is due to the fact that a whole technical interface is used to implement a capability, and hence logically belongs to the top-level function.

As capabilities are defined in the scope of a Service object, the default case is that all capabilities are made available upon contracting the service. Alternatively, a subset of capabilities may be selected and a USDL service description is derived specifically for a particular consumer contract. In order for a provider to indicate what the valid subsets are, s/he can use the option concept of USDL and define a FunctionalOption that references the respective Function objects.

11.3.2 Illustrative Example

The object diagram in Figure 11.2 shows an example of how some of the elements of the Functional Module are used to describe service functionality. It refers back to the example service introduced in Section 8.7, which is an automated service used by clients to manage shipment processes provided by an airline. One of the functions that is offered as a capability lets clients initiate the shipment of goods by booking certain capacities on a specific flight (function F1, "Request Booking"). They send a request for booking, in this case assumed to be corresponding to a type defined by a fictitious IATA[4] specification. The request furthermore has to include a client-specific X.509 certificate containing a client key, which was issued as part of an initial registration step prior to using the service. The registration process also creates a client account, and both items (account and client key) are necessary preconditions (subsumed in C1) for calling this particular function through the offered WSDL interface (interface I1).

[3] Technical services may offer functionality that is only loosely connected to the value proposition of a service, or even completely disconnected from it. For instance, deprecated functions may still be present, but new consumers may be discouraged from using them. It is recommended to simply ignore such functionality in USDL, i.e., to not provide model elements for them.

[4] International Air Transport Association

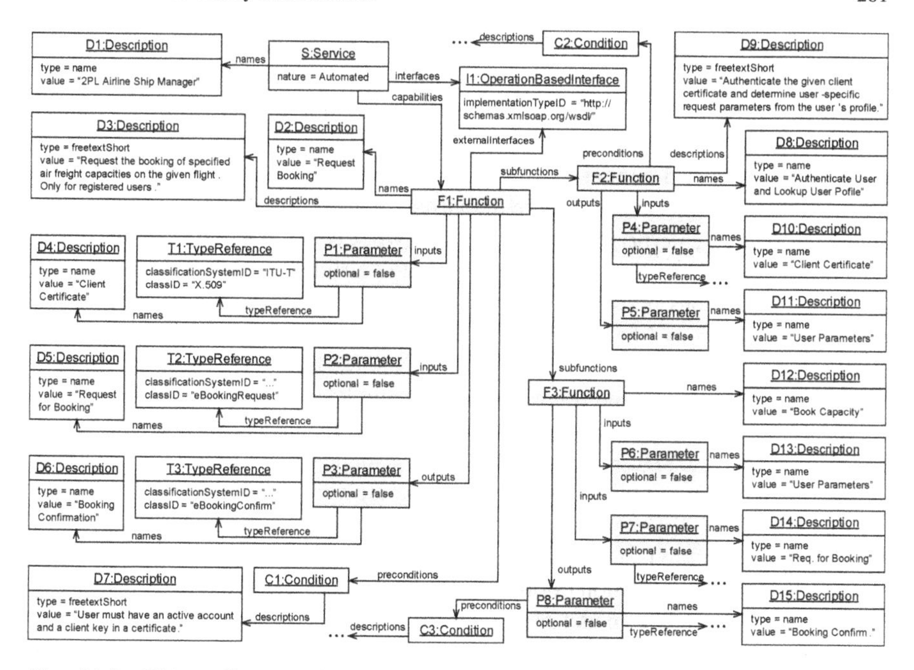

Fig. 11.2: Object diagram depicting functional elements of example service 2PL Airline Manager.

The function itself is realized by two component functions (F2 and F3). The first one takes the client certificate and uses it to authenticate the client. It furthermore looks up the client's account profile and extracts additional parameters from it. These parameters are then passed to the second function, which performs the actual booking according to the information in the booking request and the received parameters. If the request contains no errors and there is enough capacity on the selected flight, a booking is made and a booking confirmation is returned. Both functions have a precondition (C2 and C3, only indicated in Figure 11.2) that is identical to the precondition of their "parent" (F1).

In this example one of the core concepts of USDL, which also constitutes a limitation, can be observed. It concerns how USDL objects, i.e., class instances, are understood and how their scope of existence and visibility is defined. For example, take the first sub-function of "Request Booking," which is "Authenticate User and Lookup User Profile." One could easily imagine that exactly the same function is used in other higher-level functions, e.g., authentication is also required for "Change Booking." Maybe it behaves slightly different, because of the changed context (request vs. change), but this could be captured in the implementation. When mapping this to USDL, however, there would be two different Function objects describing what in reality is one and the same function. The same applies to other elements of USDL, including Parameter and Fault, as seen, for example, with the two objects called "User Parameters" in Figure 11.2 (P5 and P6).

The main reason for this is to be found in the notion of description, which in USDL is clearly distinguished from the concept of representation. Most USDL objects describe entities (e.g., Functions), they are not understood as 1:1 model representations (with a few notable exceptions[5]). In a sense, they can be seen as views onto real-world entities — and possible model representations of these. To illustrate this, consider a case where the airline manages its business and IT operations through an enterprise architecture framework (e.g., TOGAF [14], Zachman Framework [31]).[6] As part of these activities the airline created a capability map showing all functions performed within the organization, including the externally-consumable functions (the capabilities), and their relationships. In this map there exists exactly one record with name "Authenticate User and Lookup User Profile." It has a unique identifier and represents the intangible code routine performed by a computer system, which is used as a building block for many higher-level functions. The two (or more) USDL Function objects (in a single service description) created to describe the code routine constitute two separate views onto this routine and its record in the capability map. Each view exists locally within the scope of its parent function and therefore is not visible to objects outside this scope.

Even though this definition of object scope introduces modeling redundancies in the form of independent model objects, there is still the possibility to capture the information that several objects actually describe one and the same entity. This is achieved in this case by linking both views (Function objects) to the record held in the capability map. For this purpose a Description object of type "concept" is added to each Function object (using feature descriptions). The description references the unique identifier of the record in the map, which thereby enables viewers of the USDL document to infer that both Function objects refer to the same entity. In fact, using such "concept links" is a generic method to align elements of USDL service descriptions with artifacts capturing organizational models (cf. *Organizational Embedding*, Section 8.3.2.1).

There is another effect of this limitation to the Functional Module. Function objects that are sub-functions of another Function object cannot be referenced as capabilities. The reason is that, because they only exist in the scope of their parent function, they are not visible to a Service object. In case a sub-function describes a function that is actually consumable by itself, a new top-level Function object has to be defined and can then be referenced as a capability (by a service or a functional option).

11.4 The Technical Module

The modeling of technical aspects in the Technical Module is inspired by what was outlined in Section 11.2.1 as the two main styles of describing technical interfaces.

[5] In fact, all major core elements, e.g., Service, Agent, or Resource, are modeled in a representative manner, i.e., defined once and then referenced.

[6] For a comprehensive overview on Enterprise Architecture see [18].

On the one hand, there are those service description languages that originated from distributed systems computing and promote a style centered around the notion of methods (operations) with inputs and outputs. On the other hand, Web-based APIs of recent years follow a style centered around the notion of resources (as in REST). In order to be compatible with both and future styles, the Technical Module introduces an extensible model built around the abstract class Interface. The intention is that, per style of interface description, one concrete class implementing Interface is introduced. Surveying all available interface technologies is not an immediate task of USDL, and it is moreover impossible to foresee every new style of interface possibly being developed in future. Therefore, the Technical Module comes with only two concrete classes. In particular, there is one class for interfaces based on the concept of operations and one class for interfaces based on the concept of resources. Other interface classes may be introduced through extensions to USDL — see Chapter 17.

11.4.1 Overview and Main Constructs

Figure 11.3 depicts the class diagram of the Technical Module. Note that the model refrains from capturing any sort of interface structure, because, more than any other aspect of services, technical interfaces are well covered by specialized, best-of-class description languages. These languages are connected to the technologies employed for service implementation, and over the years many such technologies have been developed. Most service providers have their technology and language of choice (including tool suite), and have produced many interface descriptions. Replicating the information stored in these artifacts would introduce unnecessary issues of keeping USDL synchronized with them. Hence the chosen approach omits interface structure, e.g., which operation an input parameter belongs to. The Technical Module captures only information that is of interest to parties forming a decision about a service (in order to make USDL self sufficient for this purpose), and information that connects interface elements with concepts in other USDL modules, e.g., functional concepts.

With the decision not to replicate interface description languages, the main purpose of the Technical Module is the integration of different description artifacts. In line with the idea of Semantic Web Services (cf. Chapter 7, [27]), it "annotates" elements defined in interface descriptions with elements of USDL, especially those elements describing conceptual functionality which are exposed by the interface in question. Thus, the two concrete interface classes of the Technical Module mostly perform the function of a container for "interface element" objects (e.g., objects of classes OperationBasedInterfaceElement and OperationBasedInterfaceElement). These objects essentially are connectors in the form of a typed double pointer. Firstly, they reference a particular element in an external interface description, e.g., an operation or parameter definition. The typing of the connector ensures that it is clear what kind of interface element is referenced. A link

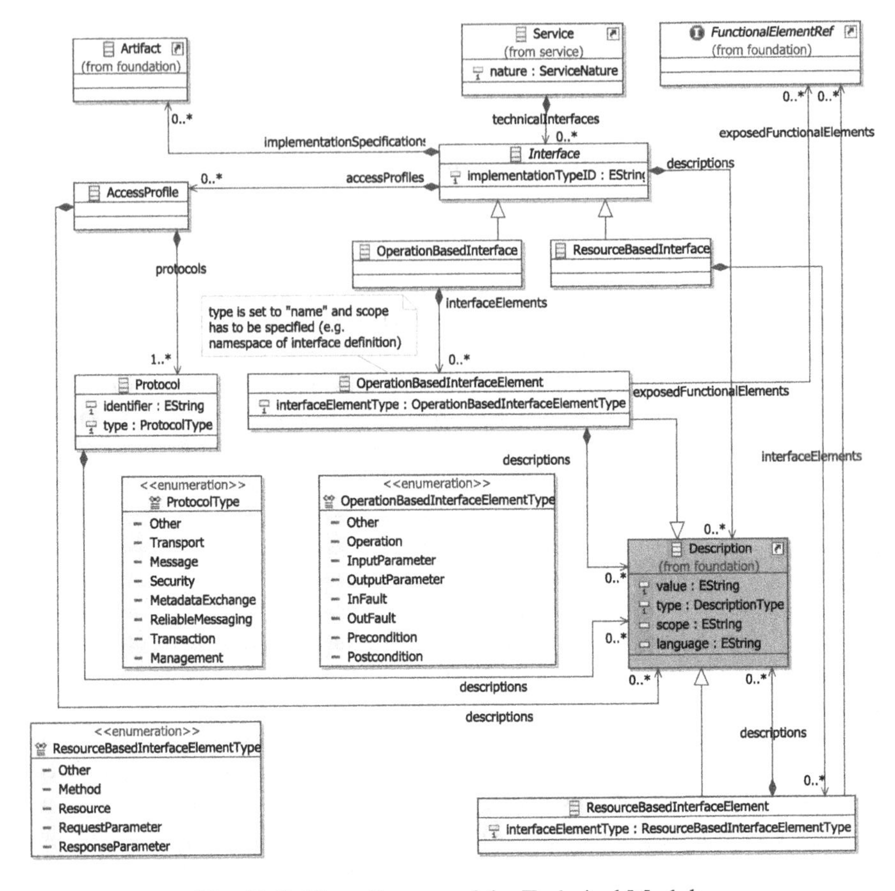

Fig. 11.3: Class diagram of the Technical Module.

to the actual interface description artifact (e.g., a WSDL) that should contain such definitions is provided through implementationSpecifications. Secondly, connectors reference a USDL object implementing the interface FunctionalElementRef (defined in the Foundation of USDL, cf. Chapter 13). In the current version of USDL [2], Function, Parameter and Fault implement this interface.

Aside from connecting interfaces with conceptual functionality, the Technical Module also captures information about the technologies used for communicating with the service, e.g., transport and messaging protocols.[7] Similar to the concept of bindings in a WDSL interface, there may be several alternative sets of co-acting protocols. These are called AccessProfiles in USDL.

[7] In the context of the Technical Module, the word *protocol* refers to such *communication protocols*. These are distinct from the *interaction protocols* or *business protocols* covered in the Interaction Module.

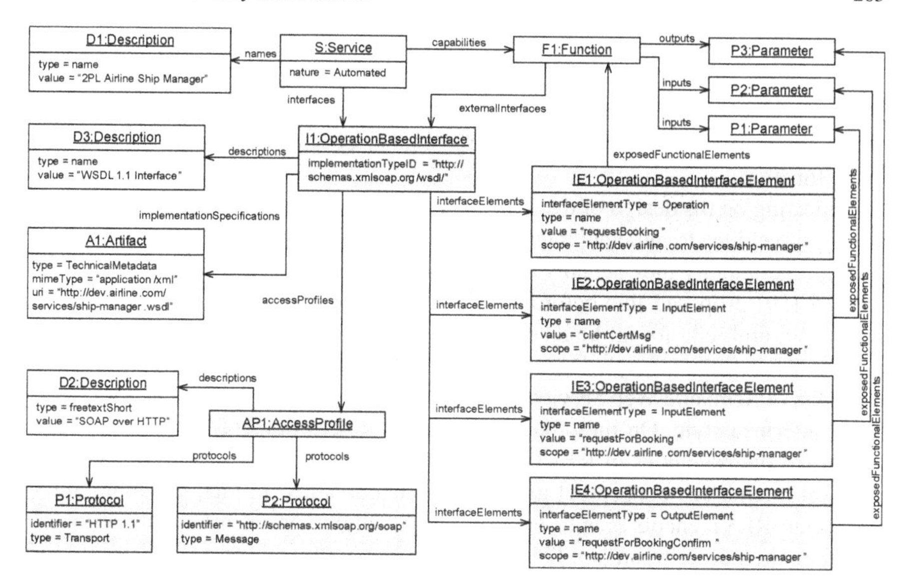

Fig. 11.4: Object diagram depicting technical elements of example service 2PL Airline Manager.

Overall, the Technical Module provides a lot of flexibility. For instance, it allows to specify nothing but a single link to a description artifact, e.g., WSDL file. In such a minimalist case, however, the annotations linking technical with conceptual elements would be missing. On the other hand, it can accommodate new interface styles through extensions. Note that an extension may provide a complete interface description language in its own right, meaning it is not necessarily bound to the design decision made for the Technical Module, i.e., to reference and not (re-)model a language. Refer to Chapter 17 for more information about extension mechanisms in USDL.

11.4.2 Illustrative Example

The object diagram in Figure 11.4 depicts an excerpt of a possible interface description for the *2PL Airline Manager* service (see Section 8.7). As indicated by feature implementationTypeID on the operation-based interface object (I1), it is a WSDL 1.1 interface that provides its WSDL file at the URL given in the artifact object under implementationSpecifications (A1). The figure shows one access profile grouping two protocols, HTTP and SOAP, which means there is a "SOAP over HTTP" binding in the WSDL interface. In addition, four interface element objects are listed, one WSDL operation ("requestBooking," IE1), two WSDL messages of type InputParameter (IE2 and IE3), and one WSDL message of type OutputParameter

(IE4). They reference the function "Request Booking" (F1, see Figure 11.2) and its three parameters (two inputs, P1 and P2, and one output, P3) via feature exposedFunctionalElements, respectively. This suggests that the WSDL operation and the three WSDL messages comprise a number of elements that provide access to the function.

Reflecting on the design of the Technical Module, it should be pointed out that it is not a novel idea to annotate interfaces with references to conceptual descriptions (of the functionality they actually expose). Other languages such SAWSDL [10] or OWL-S [19] proposed this before. However, what is new in USDL is the aim to provide an umbrella for all kinds of technical interface descriptions and, more important, for all different styles of structuring interfaces (and services). Approaches in the realm of Semantic Web Services, due to their heritage, only cover the operation-based interface style. On the other hand, formal descriptions of services built according to the principles of REST are still in their infancy (e.g., WADL [15]) and have yet to embrace conceptual modeling of functionality. In fact, the community promoting REST encourages to employ a lightweight approach towards services, including their description (see [25]). Hence, it is doubtful whether REST technologies will ever be natively extended with semantic descriptions. With USDL there now exists one possibility for a non-intrusive link between the two.

11.5 The Interaction Module

The Interaction Module captures behavioral aspects of services that govern the interaction protocols (also called behavioral interfaces, business protocols, or public view on a process) to be followed by the different parties engaged in value co-creation. Accordingly, the central concept of the module is the class InteractionProtocol, which groups the set of mandatory and optional interactions taking place between the parties. This section introduces the module.

Modeling of behavioral aspect is greatly influenced by the underlying question to what extent USDL should allow the modeling of flow structures and which style of modeling is most appropriate. As explained in Section 11.2, there exist two major, fundamental styles for modeling the structure of processes: hierarchic block structures and directed graphs. Adding to this question is the fact that, as with technical interfaces, describing the externally observable behavior of a service has received considerable attention. Thus, a number of description/modeling languages are available and actively used. In consequence, the principles outlined for the Technical Module also apply here: existing languages and description artifacts should not need to be replaced, but should be augmented with additional information linking them to other elements of the service. This caters in particular for scenarios where existing methods for service/system description are used. Providing the augmented information for service interaction protocols is the main purpose of the Interaction Module.

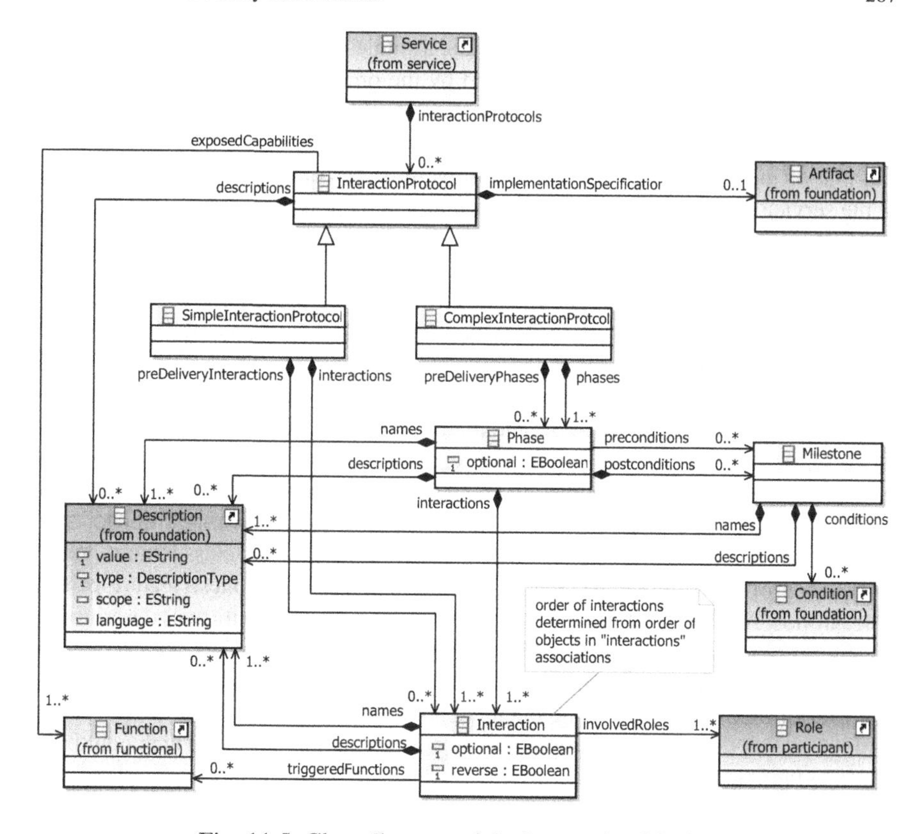

Fig. 11.5: Class diagram of the Interaction Module.

11.5.1 Overview and Main Constructs

Figure 11.5 depicts the class diagram of the Interaction Module, which employs a basic version of the block-structured approach: only sequences of (possibly optional) activities are supported. This approach was chosen for reasons of simplicity and traceability, with the goal of enabling non-technical users to model the externally observable behavior of services where no USDL-external model exists. Similar modeling approaches can be found in the literature [5, 29], some of which have been shown to be comprehensible and usable by non-technical users. These are mostly manual services with no technical footprint, or automated services with simple interaction protocols. They often have a textual description of how customers should interact with them, and therefore would need such a description, in order to be monitored and tracked, e.g., by a service marketplace.

According to the design decisions made, two types of classes capturing interaction protocols are introduced. They both extending a common super-type InteractionProtocol, which essentially represents a nameless container. The class

SimpleInteractionProtocol is used to define a single sequence of interactions, with the order of interactions being determined by the order in which they are listed in feature interactions. An Interaction is the basic building block of interaction protocols. It models an act of communication between the consumer (user) of the service and one or more other parties. These parties must be roles defined in the context of the service (feature involvedRoles), and an interaction is by default directed towards them. It may also run in the reverse direction, i.e., from a role towards the customer, or be skipped entirely (indicated by features and reverse and optional, respectively). The main sequence of interaction may be preceded by a second sequence that is situated during the time of service matchmaking and offering (feature preDeliveryInteractions).

In case of long-running services it may be desirable to split the overall sequence of interactions into groups marking the major phases of service delivery. Breaking down service execution into a manageable number of phases may make it easier for consumers to track the progress and understand what achievements have already been made, e.g., during the application for a visa. The class ComplexInteractionProtocol is used to define such protocols as a set of phases. The class Phase holds a sequence of interactions and may yield one or more milestones (feature postconditions). A Milestone, in turn, is defined as the (formal or informal) description of state of objects that are affected by the service. It thus describes achievements, and phases may require certain milestones to be reached before they can start (feature preconditions). At this point the model deviates slightly from the simple path, as this construct allows to define an arbitrary order among phases (which could be translated into a directed graph).

Just like the other two modules that are presented in this chapter, the Interaction Module leaves room for flexibility. The only mandatory requirement is that modelers have to indicate which capabilities of the service are being rendered when running through a protocol (feature exposedCapabilities). This constitutes the main link to the Functional Module. Fine-grained links may be provided on the level of interactions by referencing one or more ("sub-")functions that are triggered after an interaction successfully took place (triggeredFunctions). Modelers with existing complex interaction protocols have the possibility to attach their protocol definition(s) in a formalism of their choice (feature implementationSpecifications on abstract class InteractionProtocol). They may even have an abstraction of these protocol(s) mapped to the model in USDL, which would allow them to annotate a protocol with references to conceptual functionality and service participants. However, unlike "interface element" objects in the Technical Module, Interaction objects do not refer to an external artifact by default. In order to establish the link to a specific element in a behavioural interface description, a corresponding Description object has to be added to feature names. This object has to specify the unique name of the element in the external description artifact and has to have its scope set to the unique identifier of a namespace in that artifact.

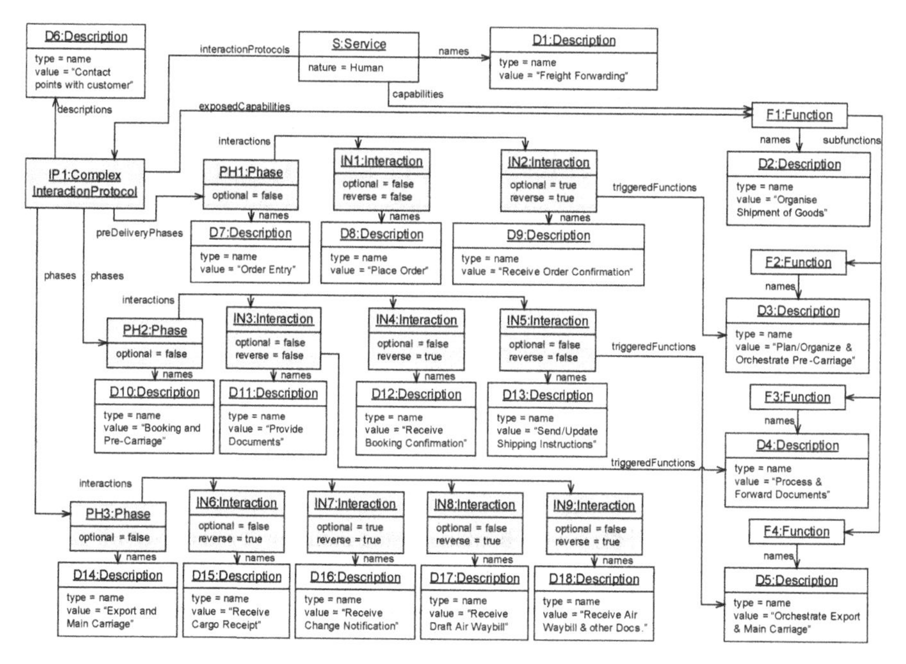

Fig. 11.6: Object diagram depicting interaction elements of example service Freight Forwarding.

11.5.2 Illustrative Example

The object diagram in Figure 11.6 illustrates an example of an interaction protocol. It belongs to the USDL description of example service "Freight Forwarding" provided by the 3PL (see Section 8.7). Contrary to service *2PL Airline Manager* which was used to explain the other two modules, this service is a manual service. There is no technical interface and no defined interaction protocol. Even so, providing a freight forwarding service actually comprises a substantial number of interactions between the consumer (shipper, aka. cargo owner) and the provider (freight forwarder). Furthermore, the overall time that it takes to export and/or import, the process managed by the forwarder, may range from a few days to several weeks, meaning this service can be considered long-running.

In making the effort of describing the service in USDL, the 3PL decides to capture the interaction protocol that shippers should follow. For illustration purposes please refer to Figure 11.7, which presents an alternative model of the interaction protocol in BPMN notation (cf. [24]). The 3PL creates a ComplexInteractionProtocol object (IP1) that consists of three phases. The first phase "Order Entry" (PH1) takes place before the actual service is provided and is initiated by the shipper upon sending of a forwarding order (IN1). Being a pre-delivery phase, it may end with the order being rejected (not shown in Figure 11.6), whereupon no service is delivered. On the other hand, if the order is accepted, a confirmation is returned to the shipper

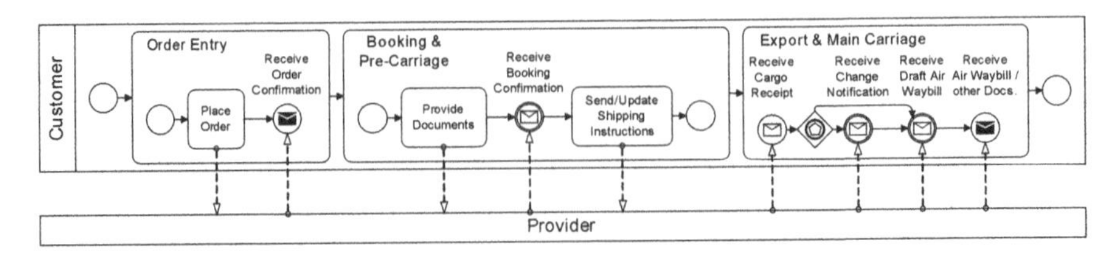

Fig. 11.7: BPMN diagram depicting the interaction protocol of example service Freight Forwarding.

(optional and reverse interaction IN2) and the phase concludes with freight forwarding capabilities being rendered. In particular, the forwarder commences planning activities and organization of pre-carriage (pick-up of the goods and transport to an export facility, e.g., port), which is modeled as (sub-)function F2 and triggered by IN2.

The second phase (PH2) covers booking of the main carrier and executing pre-carriage. It starts with the arrival of a number of documents that the shipper has to provide (IN3), e.g., invoice, packing list, or certificate of origin. These documents are then checked by the 3PL for completeness and correctness (triggered function F3) (and some will later be forwarded to the carriers). Booking of the main carrier (e.g., airline) is performed as part of function F2 and a booking confirmation is returned to the shipper once booking was successful (IN4). The shipper then sends final, binding shipping instructions (IN5). Alternatively, it may update or confirm instructions sent earlier (e.g., in IN1 or IN3). With the final instructions received, the 3PL is now able to orchestrate the main export steps (function F4 triggered by IN5). The first interaction of this main phase (PH3) occurs when the cargo has arrived at the export handling facility and the 3PL receives a cargo receipt. It forwards this receipt to the shipper (IN6). Before creating and sending the draft house bill of lading (ocean transport), or in this case air waybill (air transport), the 3PL performs another round of checks. If anything has changed with the shipment (e.g., different flight), it sends a notification to the shipper (optional interaction IN7). Assuming there is no intervention (not shown here), the shipper receives the draft air waybill indicating that the cargo is about to be loaded and ready to go on its main journey (IN8). Some time thereafter the 3PL sends the the final air waybill (as received by the airline) together with a few other documents (e.g., dispatch note and export confirmation) to the shipper (IN9). The export is finished at this point. Please note that import and invoicing is not shown here for brevity.

11.6 Conclusion

This chapter has presented the USDL constructs for modeling the functional, technical, and behavioral aspects of services. These are essential for facilitating an appropriate communication between service providers, consumers, and intermediaries for the proper provisioning and consumption of services, and are subject to various existing standards which form the core of many existing platforms for service-based systems. USDL introduces three separate modules for this, which define basic modeling constructs and foster the re-use and integration of existing as well as upcoming standards, therewith maintaining the foundation principles of conceptualization, flexibility and extensibility of USDL.

In this chapter the relevance of functional, technical, and behavioral modeling as well as the applied design principles have been outlined. The three USDL Modules have been presented in detail along with illustrative examples. The first aspect covered here is concerned with functional modeling, i.e., with describing what a service does. In USDL, this is captured in the Functional Module, which defines generic constructs for describing the abstract functionality in terms of the functions that a service provides, following previous works on capability modeling. The second aspect is concerned with modeling the technical interfaces of services, which describes how to technically access and consume services. For this, the Technical Module of USDL defines the notion of an abstract interface, which can be connected to existing as well as upcoming interface description standards. This provides a structured means to explicitly connect the technical accessibility with the abstract functionality while remaining extensible. The third aspect addresses behavioral modeling, i.e., the externally visible behavior of a service that needs to be known in order to properly consume the provided functionality. Behavioral protocols are especially relevant for services with more complex interaction patterns. For this, the Interaction Module of USDL provides basic constructs for behavior modeling that can be extended with more expressive behavioral languages if desirable.

In summary, USDL provides an integrated meta-model for functional, technical, and behavioral aspects of services, which in turn constitute the core of technology platforms for service-based systems. In accordance with general design principles of USDL, the respective modules define a skeleton of basic constructs in a generic manner and foresee the re-use and integration of existing as well as upcoming standards. As outlined in the illustrative examples, the intended usage of the presented USDL modules is to extend the core skeleton with the specific notions and standard languages that are relevant for a particular application scenario. This enables a tailored description that only uses the actually needed technology stack while providing a holistic view on the services by associating external artifacts in this stack with core USDL constructs.

References

1. A. Arkin, S. Askary, S. Fordin, W. Jekeli, K. Kawaguchi, D. Orchard, S. Pogliani, K. Riemer, S. Struble, P. Takacsi-Nagy, I. Trickovic, and S. Zimek. Web Service Choreography Interface (WSCI) 1.0. Note 8 August 2002, W3C, 2002. Online at: `www.w3.org/TR/wsci`.
2. A. Barros, C. Baumann, A. Charfi, S. Heinzl, T. Kiemes, U. Kylau, N. May, O. Müller, F. Novelli, D. Oberle, P. Robinson, B. Schmeling, W. Theilmann, and H. Witteborg. Unified Service

Description Language (USDL) — Functional Module. Technical Report Version 3.0, Milestone M5, SAP Research, May 2011. Available at www.internet-of-services.com.
3. A. Barros, C. Baumann, A. Charfi, S. Heinzl, T. Kiemes, U. Kylau, N. May, O. Müller, F. Novelli, D. Oberle, P. Robinson, B. Schmeling, W. Theilmann, and H. Witteborg. Unified Service Description Language (USDL) — Interaction Module. Technical Report Version 3.0, Milestone M5, SAP Research, May 2011. Available at www.internet-of-services.com.
4. A. Barros, C. Baumann, A. Charfi, S. Heinzl, T. Kiemes, U. Kylau, N. May, O. Müller, F. Novelli, D. Oberle, P. Robinson, B. Schmeling, W. Theilmann, and H. Witteborg. Unified Service Description Language (USDL) — Technical Module. Technical Report Version 3.0, Milestone M5, SAP Research, May 2011. Available at www.internet-of-services.com.
5. J. Becker, D. Pfeiffer, and M. Räckers. PICTURE - a new approach for domain-specific process modelling. In *CAiSE'07 Forum, Proceedings of the CAiSE'07 Forum at the 19th International Conference on Advanced Information Systems Engineering, Trondheim, Norway, 11-15 June 2007*, 2007.
6. B. Benatallah, F. Casati, and F. Toumani. Web Service Conversation Modeling, A Cornerstone for E-Business Automation. *IEEE Internet Computing*, 8(1):46–54, 2004.
7. R. Chinnici, J.-J. Moreau, A. Ryman, and S. Weerawarana. Web Services Description Language (WSDL) Version 2.0 Part 1: Core Language. Recommendation 26 June 2007, W3C, 2007. Online at: www.w3.org/TR/wsdl20.
8. T. DeMarco. *Structured Analysis and System Specification*. Prentice Hall, 1979.
9. T. Erl. *Service-Oriented Architecture: Concepts, Technology, and Design*. Prentice Hall International, 2005.
10. D. Farrell and H. Lausen. Semantic Annotations for WSDL and XML Schema (SAWSDL). W3C Recommendation August 2007, W3C, August 2007. Online at: http://www.w3.org/TR/sawsdl/.
11. D. Fensel, H. Lausen, A. Polleres, J. de Bruijn, M. Stollberg, D. Roman, and J. Domigue. *Enabling Semantic Web Services. The Web Service Modeling Ontology*. Springer, Berlin, Heidelberg, 2006.
12. R. Ferrario and N. Guarino. Towards an Ontological Foundation for Services Science. In *Future Internet - FIS 2008, First Future Internet Symposium, FIS 2008, Vienna, Austria, September 29-30, 2008, Revised Selected Papers*, pp. 152-169, 2008.
13. J. D. Gannon, J. M. Purtilo, and M. V. Zelkowitz. *Software Specification: A Comparison of Formal Methods*. Ablex Publishing Co., 1994.
14. T. O. Group. *TOGAF™ Version 9 Enterprise Edition*. Van Haren Publishing, February 2009.
15. M. Hadley. Web Application Description Language. W3C Member Submission 31 August 2009, W3C, 2009. Online at: http://www.w3.org/Submission/wadl/.
16. U. Homann. A Business-Oriented Foundation for Service Orientation. MSDN, Microsoft Corporation, 2006.
17. N. Kavantzas, D. Burdett, G. Ritzinger, T. Fletcher, Y. Lafon, and C. Barreto (eds.). Web Services Choreography Description Language Version 1.0. Candidate Recommendation 9 November 2005, W3C, 2005. Online at: http://www.w3.org/TR/ws-cdl-10/.
18. M. e. a. Lankhorst. *Enterprise Architecture at Work*. Springer, Berlin, Heidelberg, 2009.
19. D. Martin. OWL-S: Semantic Markup for Web Services. W3C Member Submission 22 November 2004, W3C, 2004. Online at: http://www.w3.org/Submission/OWL-S/.
20. Microsoft. Open Data Protocol (OData) Specification, Version v20101230, 2010.
21. E. Newcomer. *Understanding Web Services: XML, WSDL, SOAP, and UDDI*. Addison-Wesley Professional, 2002.
22. P. Oaks, A. Hofstede, and D. Edmond. Capabilities: Describing What Services Can Do. In *Service-Oriented Computing - ICSOC 2003, First International Conference, Trento, Italy, December 15-18, 2003, Proceedings*, pp. 1-16, 2003.
23. OASIS. *Web Services Business Process Execution Language Version 2.0*, Apr. 2007. Online at: http://www.ibm.com/developerworks/webservices/library/ws-bpel/.
24. OMG. Business Process Model and Notation, V1.1, 2008.

25. C. Pautasso, O. Zimmermann, and F. Leyman. RESTful Web Services vs. 'Big' Web Services: Making the Right Architectural Decision. In *In Proceedings of 17th International World Wide Web Conference (WWW 2008)*, pages 805–814, April 2008.
26. M. Stollberg, U. Keller, H. Lausen, and S. Heymans. Two-phase Web Service Discovery based on Rich Functional Descriptions. In *Proc. 4th European Semantic Web Conference (ESWC 2007), Innsbruck, Austria*, 2007.
27. R. Studer, S. Grimm, and A. Abecker. *Semantic Web Services. Concepts, Technologies, and Applications*. Springer, 2007.
28. H. S. Thompson, D. Beech, M. Maloney, and N. Mendelsohn. XML Schema Part 1: Structures Second Edition. W3C Recommendation 28 October 2004, W3C, October 2004. Online at: `http://www.w3.org/TR/xmlschema-1/`.
29. I. Weber, H. young Paik, B. Benatallah, C. Vorwerk, Z. Gong, L. Zheng, and S. Kim. Managing long-tail processes using FormSys. In *ICSOC'10: 8th International Conference on Service Oriented Computing, Demo Track*, San Francisco, CA, Dec. 2010.
30. S. Weerawarana, F. Curbera, F. Leymann, T. Storey, and D. F. Ferguson. *Web Services Platform Architecture: SOAP, WSDL, WS-Policy, WS-Addressing, WS-BPEL, WS-Reliable Messaging, and More*. Prentice Hall PTR, 2005.
31. J. A. Zachman. The Zachman Framework for Enterprise Architecture™: A Primer for Enterprise Engineering and Manufacturing.

Chapter 12
Service Levels, Security, and Trust

Florian Marienfeld, Edzard Höfig, Michele Bezzi, Matthias Flügge, Jonas Pattberg, Gabriel Serme, Achim D. Brucker, Philip Robinson, Stephen Dawson, and Wolfgang Theilmann

Abstract This chapter covers the scientific background for the Service Level Module of the Unified Service Description Language (USDL). In addition to general service level concepts, we expand on two specific service level fields: security and trust. For that end we first review the state of the art in service level modeling, then we explain the design of the Service Level Module and position it among the rest of USDL. For security, two possible perspectives, a high level business view and a low level engineering approach, are introduced. With regards to trust, USDL is suitable to specify how a service can be rated by its consumers and to ensure that ratings of competing services are comparable, and hence to determine trustworthiness. Additionally, we present a description of non-security-related elements that can be exploited for trust estimation.

Florian Marienfeld, Edzard Höfig, Matthias Flügge, Jonas Pattberg
Fraunhofer FOKUS, Kaiserin-Augusta-Allee 31, 10589 Berlin, Germany,
e-mail: firstname.lastname@fokus.fraunhofer.de

Michele Bezzi, Gabriel Serme
SAP Research Sophia-Antipolis, 805, Avenue du Dr. Maurice Donat, 06250 Mougins, France,
e-mail: michele.bezzi@sap.com, e-mail: gabriel.serme@sap.com

Achim D. Brucker, Wolfgang Theilmann
SAP Research Karlsruhe, Vincenz-Priessnitz-Str. 1, 76131 Karlsruhe, Germany,
e-mail: achim.brucker@sap.com, e-mail: wolfgang.theilmann@sap.com

Philip Robinson, Stephen Dawson
SAP Research Belfast, Concourse, Queen's Road, Queen's Island, Titanic Quarter, BT3 9DT Belfast, United Kingdom,
e-mail: philip.robinson@sap.com, e-mail: stephen.dawson@sap.com

12.1 Introduction

The USDL Service Level Module captures concepts concerned with guarantees regarding quality of service operation, which are claimed/requested by different actors involved in the provisioning, delivery and consumption of a service. Given the role of service levels as a vital component of any commercial transaction, it is striking to see how poorly service levels are supported in commercial offerings and also to see the lack of a systematic approach to these in the research arena. Most approaches are missing formal semantics, leave fine-grained content unspecified, lack flexibility and are tuned to specific scenarios and domains. Based on the research results of the SLA@SOI project [21], the USDL effort tries to avoid these shortcomings.

We advocate that a comprehensive, applicable and executable service level metamodel such as the one we contribute is crucial to realize the vision of the Internet of Services (IoS), for it represents a key enabler of effective and efficient service discovery, trade and consumption. It basically gives both, service providers and customers dependability on the quality levels a service comes with and the related obligations of the involved stakeholders.

In our view, two special service level topics deserve to be discussed in more detail in this chapter: security related concepts are actually realized as part of the service level module; trust related elements in contrast, are technically located in other modules, but are conceptually most closely related to SLA. Therefore, they are covered here, too.

There are a number of security focused service description languages, which express the security properties of services and service message exchanges in a standardized way. However, they are often not sufficient to address new scenarios where security is becoming a key aspect in business decisions. The USDL security elements we present in this part of the chapter are a description of security and trust aspects of a service, making the bridge between the IT level and the business level.

As for other USDL modules, the security part provides a minimal set of elements to allow for simple and fast description of security features of services. More technical descriptions can be integrated in the USDL, using appropriate references to standard security description languages.

The question of trust is closely related to security, yet subtly different. When concerned with security, one assumes that chosen business partners, i.e., providers, are both competent and benevolent and considers threats emanating from third parties. With many principals unknown to each other, however, this assumption does not hold, and some prediction has to be made, regarding the reliability of potential partners. The means to perform this calculation and to track and communicate trust information are quite different from those that serve to secure operation against interference from external entities. Nevertheless, trust has a dependency on security, since a smart choice about the provider is worthless if no appropriate security mechanisms are in place. In the traditional world of business services a human user can partly assess the trustworthiness relying on cues like brand of the provider, "word of mouth" recommendations [7] and perceived quality of the website [5]. This does not scale to the level of an Internet of Services, where services are automatically composed and delivered with limited human intervention, and explicit trust and security properties are becoming a key for a broad adoption of service technology [17].

The structure of this chapter is the following: In the subsequent Section 12.2 we review related approaches. In Section 12.3 we position the Service Level Module among the other modules and layout its content. Section 12.4 introduces the USDL representation of security properties. Trust awareness is covered in Section 12.5. Conclusions are drawn in the Section 12.6.

12.2 State of the Art

The state of the art in SLA specification and management motivates the features of USDL. USDL is a synthesis and generalization of existing specifications, capturing the essential elements of an SLA specification. In doing so it satisfies key requirements for SLA management and enables new business-oriented aspects of service management (e.g., security and privacy) to be encoded in SLA documents. SLA Management includes the specification of machine and human readable documents, the configuration of systems based on the content of these documents and the monitoring of parameters expressed in these documents in order to achieve compliance. Failure to achieve SLA compliance can lead to losses in efficiency, performance, reputation and business opportunities. This section discusses the state of the art in SLA specification, monitoring, negotiation and enforcement as well as security models. A comprehensive solution to SLA management addresses each of these areas. For each area of SLA management the requirements and challenges are described, such that existing approaches can be compared to the SLA principles of the USDL Service Level Module.

12.2.1 SLA Specification

SLA specification can be viewed as both a process and a document. It is the process of a service consumer initially declaring and agreeing to specific service requirements, when entering a contract with a service operator or provider. The consumer hence specifies *Service Level Objectives (SLOs)*, before or after the availability and capability of services and service providers are known. In the case this is done beforehand, SLOs are used for service discovery. If SLOs are specified when the service and provider are known, this is the process of initiating or requesting the provisioning of a specific service on their behalf. From the provider's perspective it is the process of declaring their service capabilities and quality guarantees in a form of advertisement. This is also known as a *Service Level Agreement Template (SLAT)*, and acts as the baseline for contractual agreement with customers, potentially in different classes. The specified documents all have the same minimal requirements for structure and content:

- Definition of individuals, organizations and roles involved in the agreement. The roles are typically consumer/customer and provider/operator, but can also include a third-party broker, an intermediate actor in the SLA management process.
- Functional description of the service's purpose and capabilities. In the case of a technical, IT service, such as a Web service, the functional description refers to the set of operations, methods and parameters. For example, the Web-Services Description Language (WSDL) provides a standard specification for SOAP-based Web services.
- Costs to the consumer for receiving the service. The units for costs are defined by relating financial costs to utility functions of the resources consumed by the service. For example costs can be defined per requests, per volume of storage used, per user or on a fixed-term or unlimited basis.
- Guarantee or Quality of Service (QoS) terms define the non-functional properties of the service. These properties include availability, performance, response time, reliability and security.
- Compensation terms define what the consumer can rightfully demand from the service provider in return, should the functional or guarantee terms not be fulfilled.

The distinction between a SLA document and a configuration document for a service infrastructure is becoming increasingly fuzzy. The contents of a SLA are inevitably translated into concrete configuration directives that are used to guide the provisioning of resources, deployment of software and tuning of settings to enable effective operation of the service. Effective operation means that all functional and non-functional terms in the SLA can be satisfied without exploding costs for the provider. Providers need to keep their costs down so that they can offer an attractive service deal to consumers without sacrificing their profit targets. Given that this becomes a more complex problem for human operators to deal with manually and rapidly, the following new requirements arise for SLA specifications in a world of service management automation:

1. SLA specifications need to be both human and machine readable, given that people need to define, exchange and agree terms, while algorithms are required to parse, extract and analyze terms, and translate terms and analysis results to configuration directives in an efficient and accurate manner.
2. Traceability to organizational Key Performance Indicators (KPIs) such as performance, availability and cost.
3. All terms must be associated with concrete metrics that enable them to be monitored and audited.
4. Sufficiently flexible for application in different service operation domains and contexts.
5. Support for the entire SLA life cycle, including negotiation, provisioning, monitoring and decommissioning.

There are various specification languages in existence, each with different motives but similar concepts. The Open Grid Forums's WebService-Agreement [1] and

IBM's Web Service Level Agreement (WSLA [12]) are the most well-known languages for expressing SLAs. WSLA is a comprehensive specification for describing SLAs for Web services, providing a template for defining concrete metrics. WSLA however does not support the entire life cycle of SLAs, as negotiation of terms is not supported for flexibility. WS-Agreement superseded WSLA in order to address these points. While WS-Agreement is very flexible, given its generality, and provides a comprehensive schema that enables human and machine interpretation, it still requires more concrete metrics and explicit traceability to KPIs. Another point to mention is that WSLA and WS-Agreement were developed for WSDL-type Web services and hence do not have the semantics included for dealing with non-WSDL services.

There are other specifications existing that are not tied to WSDL. SLAng [9] for example is specified in the Object Management Group's (OMG) Meta Object Facility (MOF) and, thus, has a degree of language independence with mappings to XML and Human-Usable Textual Notation (HUTN). SLAng also places greater emphasis on semantics, providing formal notions of SLA compatibility, monitorability and constrained service behaviour. It is, however, targeted at electronic services and provides only a limited set of domain-specific QoS constraints. Another language, viz., CC-Pi [2] is more generic, offering a theoretical framework for mapping SLAs to service constraints. The CC-Pi model is, however, tightly-coupled to the mechanics of negotiation, and does not address common constructs such as agreement party details or service interfaces. The SLA* approach from SLA@SOI [8], in contrast, is a complete abstract SLA syntax which has been designed to be independent of underlying technologies. It is decoupled both from particular notions of service, and from particular modes of expression, and can be extended to diverse scenarios without sacrificing formality or semantics.

12.2.2 SLA Monitoring

The formal specification of SLAs enables monitoring the status and compliance of services with the organizational KPIs of the parties involved. As stated in the previous section, in order to monitor aspects of a service's operation, it must be possible to measure that aspect using concrete metrics. There are critical areas of service and resource management that depend on this information and delivering it to the right people and systems in a timely manner. These are discussed in the following and are the set of standard activities defined for IT Service Management (ITSM) in the Information Technology Infrastructure Library (ITIL).[1]

Capacity planning and management involves the allocation of resources in order to avoid over-spending, wastage and under-provisioning. However, there is still the over-arching objective of satisfying the obligations and guarantees stated in SLAs. Resources include people, hardware, software licenses, software in-

[1] `http://www.itil-officialsite.com`

stances and materials. SLA monitoring is the timely update of information about resource capabilities, availabilities and performance during operation. In order for services to be delivered efficiently in an on-demand manner, there is a need for mixing historical, predictive and live information about resources for dynamic replanning and provisioning of resources. In the case of people this includes the assignment of tasks and access to resources through justifiable provisioning of users and assignment of privileges.

Availability analysis and management is the set of activities done to maximize the likelihood that resources will be available when required, as well as recoverable from unsafe or fault states. Availability is both an explicit and implicit objective of SLA management. As an explicit objective the guarantees of availability and recovery are agreed in the SLA. As an implicit objective an analyst or management system needs to monitor the resource to identify when associated services are required. This also has relations to capacity planning and management, as the sizing and number of resources will change the availability of services.

Operations management is the coordination tasks and processes amongst resources to ensure that the KPIs of an organization are met, along with the objectives in the SLAs the organization has with customers and partners. SLA monitoring should provide feedback about the current load on different resources and the criticality of service request to be handled. Operations management usually includes optimization of handling service requests and assigning tasks based on multiple objectives. These multiple objectives are derived from the set of objectives in multiple SLAs, such that conflicts and contentions will arise in an environment that allows concurrent services and service users.

Incident management is the handling of inevitable failures and unexpected events that arise during service operation. Capacity planning has to take incident management into account, as redundant resources and back-up resources usually need to be deployed to for executing contingency plans. Incident management is also related to availability analysis and management, as effective incident management increases the likelihood that resources and services will be available even if experiencing known and unexpected incidents. Operations management also extends to incident management, as the coordination of resources during contingency and recovery operations might be more critical than during normal operation. SLA monitoring is hence critical for identifying and characterizing incidents and deviations from SLA obligations and guarantees, such that effective incident response actions can be executed.

SLA monitoring is more than blanket monitoring of every possible operation, property and behavior of resources, although this might be necessary to some extent. SLA monitoring has to be purposeful and driven by measurable indicators derived from business objectives. Even if there is extensive monitoring of all resources, a system of filtering and routing information to meaningful endpoints or sinks is necessary. Failure to meet this requirements results in monitoring information consumers being overwhelmed, the processing of irrelevant or redundant data, and the introduction of unnecessary communications and processing bottlenecks.

Figure 12.1 shows a conceptual architecture for SLA monitoring for the purpose of discussing the state of the art as opposed to proposing a blueprint.

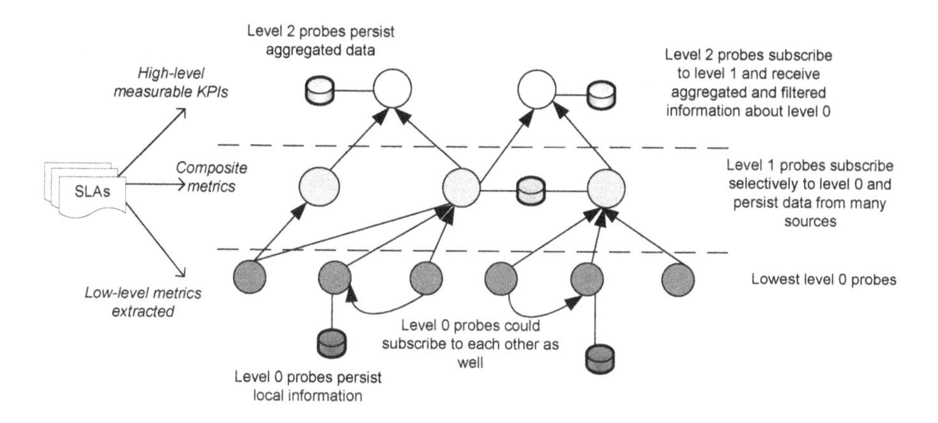

Fig. 12.1: Conceptual architecture for SLA monitoring.

The typical components in a SLA monitoring solution are collections of probes or sensors that gather localized resource information and publish it to a set of information subscribers. The assignment of subscribers to channels that the probes publish on is then based on the types of services, location of resources, connectivity and the respective metrics associated with services. This assures that the collection and distribution of monitoring data is traceable to specific KPIs and operational contexts. Higher-level probes subscribe to lower-level probes and hence act as data aggregators, filters and transformers, based on the information needs of the SLA management system at the time of deployment. The needs for monitoring are determined by analyzing the existing SLAs. Very low-level metrics such as CPU, memory and network utilization and availability can be mapped directly to localized probes that gather raw resource performance information. Higher-level metrics that described collections of resources, functions (e.g., predictions and trends) require higher-level probes to subscribe to lower level probes. These are referred to as level 1 probes in Figure 12.1. The information required to configure monitors is extracted from SLA specifications.

Monitored data can be persisted at each level in order to have different levels of granularity for historical data. Determining when to capture, aggregate, archive and delete data will differ from solution to solution. Furthermore, the selection of where to persist data is dependent on the local storage capabilities of the respective monitored resources, as well as the overall storage architecture and network topology. Centralized persistence has the advantage of simplicity and faster queries, as data is stored in one location. The disadvantage is the single point of failure that can cause all historical data to be damaged or lost. The loss of historical data can be problematic for optimizing the way in which SLAs are enforced and the availability of audit data when payment is due or disputes arise.

12.2.3 SLA Negotiation and Enforcement

SLA negotiation is the process of a service provider and consumer reaching consensus on the terms to be included in a SLA document. Negotiation is complete when all parties agree to the terms. This process can be automated but is often done as manual exchanges of proposals/tenders and offers. A generic protocol for negotiation is known as the *alternating offers-based protocol* [24], which is shown in Figure 12.2.

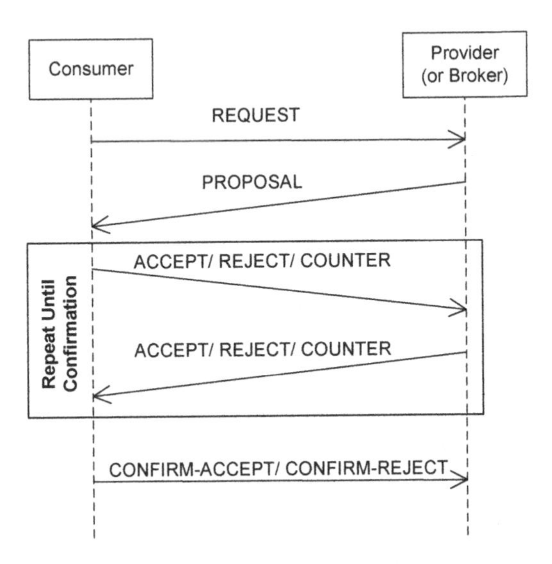

Fig. 12.2: Alternating offers-based protocol adapted from Venugopal, Chu and Buyya [24].

The protocol begins with the consumer (or initiator) sending a request to the provider or broker, who take the role of responder. The provider issues a proposal of how they can satisfy the request. For example, the request might have stated "*provide service X with a guarantee of less than 3 ms response time for 1000 concurrent users.*" The proposal would use the same functional specification and quality metrics although the provider might not be capable of exactly matching the request. The provider might offer a proposal such as "*Can provide service X with a guarantee of less than 3 ms response time if there are less than 750 concurrent users.*" The consumer can accept, reject or counter (i.e., update the request) the proposal, to which the provider can do the same until a confirmed state is reached. The confirmed state can be either of acceptance or rejection. This plain protocol assumes that both parties will adhere to the protocol and that the provider will typically have counter offers. However, the style of negotiation varies based on the number of parties involved and the set of options available. Three styles of SLA negotiation are as follows:

- **Boolean**: there are no alternatives offered by providers. Consumers either accept or reject a provider's offer without requesting alternatives. This style of negotiation is typically brokered, as the consumer considers the offerings of multiple providers registered with the same broker. Consumers select providers that have offers that best fit their requirements. The proposal is that of a single provider and counters are alternative providers. It can be the case that the provider chooses to maintain a very generic, one-size-fits-all policy for services to avoid any liability. However such a style of negotiation is not attractive for critical services where the consumers need to know what to explicitly expect from the provider.
- **Template**: in this case a provider has several options in the form of templates. Examples of this are the Amazon EC2 services[2] that offer T-shirt sizes of small, medium, large and x-large, which all have predefined service guarantees and technical specifications. This style has issues for over and under-provisioning, as the capabilities of the provider might change over time. They might need to dynamically update the specifications of their templates and offerings based on their current and planned prediction.
- **Scalar**: the most complex style of negotiation is where fine-grained adjustments are permitted on a dynamic basis. The provider does not counter with a suite of templates but makes adjustments in their guarantees and obligations, which then have an impact on how resources are sized and configured. Elastichosts[3] are a provider of infrastructure services similar to Amazon, but allow customers to state the exact amount of capacity (memory, storage and CPU) they require.

Given that the style of negotiation varies from business domain to business domain, the Service Level Module must be sufficiently configurable to support any of these styles. Enforcement of SLAs is the translation of the terms in the agreement to concrete configuration directives. There are three possibilities that exist for handling the translation of SLAs to directives, each having their advantages and disadvantages for building a complete solution.

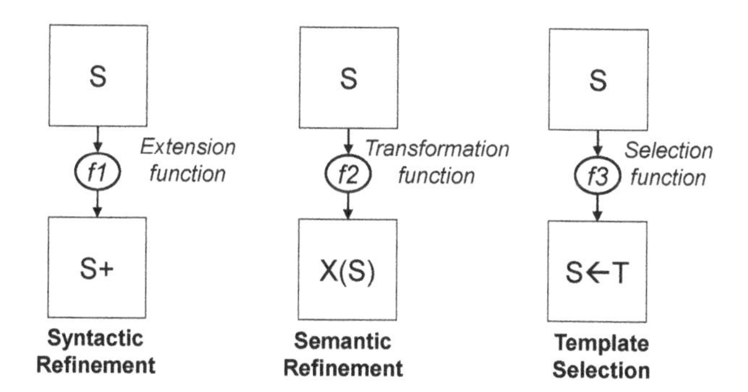

Fig. 12.3: Different approaches to SLA translation to configuration directives

[2] `http://aws.amazon.com/ec2`

[3] `http://www.elastichosts.com`

The advantages and disadvantages of these different approaches are discussed below, as it is important for any user of the Service Level Module to know which approach should be used and the consequences of that selection.

1. **Syntactic Refinement**: In this case there are homogeneous modeling semantics for SLA terms and configuration directives. It is only a case of adding more details (e.g., missing parameter values) to the specification without changing the semantics or schema of the specification. Syntactic refinement is hence transforming a SLA model S to a configuration $S+$. For example, S could be WSLA or WS-Agreement specification with many null fields, where every $S+$ is more information provided to those fields.

    - *Advantages*: simpler process for moving through the SLA transformation process as there are less processing and transformation logic involved. This also implies higher scalability, easier rollback, consistency checks and simulation.
    - *Disadvantages*: there is the initial agreement that all management components use the same semantics. Secondly, some of the human readability would have to be sacrificed in order to have a specification that is already at the level of configuration directives without the ability to separate concerns.

2. **Transformation (Semantic Refinement)**: In this case there is no assumption of homogeneous specification templates, such that the SLAs can be specified in any language. In order to obtain configuration directives, the specification would then have to be transformed into lower level semantics using a transformation function. The transformation function must be provably correct for deriving a specification with different syntactical properties. There is also need to add annotations to the initial specification in order to supply sufficient information for concrete configuration directives. In Figure 12.3 this is illustrated as performing a transformation function $X(.)$ on the specification S, such that $X(S)$ is produced, which could be a totally different format for specifying configuration directives.

    - *Advantages*: lower coupling and dependencies between components that handle SLA specifications and configuration directives. Existing control and management components do not have to be changed in order to configure resources based on SLA specifications. There is also better support for human and machine readability, as S could be intended for humans and $X(S)$ is compiled for machines.
    - *Disadvantages*: transformation can be quite complex and hence could introduce errors that take a long time to debug; additional model annotations would have to be introduced in order to perform this automatically.

3. **Template Selection**: In this case there are no assumptions of homogeneous models or templates, but there is a logic implemented for mapping higher level models to lower level deployable component templates, in such a way

that selected templates are understood as being conformant with a higher-level model specification. Figure 12.3 illustrates this by showing the template selection $S \leftarrow T$, where the template T is selected given the specification S. The set of templates are discrete and predefined.

- *Advantages*: even lower coupling and dependencies between models; selection logic is relatively easy to encode and tolerates manual interaction. Higher flexibility gained in how the SLA is specified. Templates can be made to be platform independent, such that configuration directives can be further compiled for different management systems.
- *Disadvantages*: carries the additional overhead of creating templates and loses the dynamic property of the above methodologies.

This coverage of the state of the art in SLA management shows that there is still a gap in the areas of SLA specifications that are flexible for multiple domains and not restricted to IT-centric services. Secondly, there should not be an assumption of what level of monitoring is going to be associated with the SLAs specified in the language. Finally, the language needs to be sufficiently flexible and expressive to support different forms of translation into configuration directives, without sacrificing human-readability and machine processing that enable advanced analytics and automation in service management.

12.3 The Service Level Module

12.3.1 Position within USDL

The Service Level Module is an integral part of USDL. The module ties together all the functional and non-functional guarantees that are stated on top of a core service description (as described in the Service Module, cf. Chapter 13). Furthermore, all guarantees are clearly linked to its related (and probably obligated) stakeholder (as expressed in the Participants Module, cf. Chapter 13). Last there is a strong semantic linkage to the Legal Module (cf. Chapter 10). While the Legal Module describes the constraints and aspects of licensing, the Service Level Module complements this with the specific conditions that are to be guaranteed.

The Service Level Module directly responds to the main requirements for USDL. It satisfies a clear *Conceptualization* (see Section 8.3.1.1) as it realizes the core principles of guarantee, obligated party, affected service elements and negotiable parameters on an abstract level. It comes with means for *Extensibility* (see Section 8.3.1.4) that allows for incorporating arbitrary, domain specific type/term systems. It also supports *Comprehensibility* (see Section 8.3.1.5) as it allows description of service levels in human-readable, semi-structured, and fully structured ways. *Organizational Embedding* (see Section 8.3.2.1) is achieved by the association to participants. And last, it supports *Deployment Symmetry* (see Section 8.3.2.4) as it allows mutual, symmetric obligation relationship between arbitrary roles in the service value creation chain.

12.3.2 Construction Rationale

Service Level Agreements, as considered in the research community, specify all the conditions under which services are to be delivered. In that sense the whole specification of USDL can be considered as a language to describe SLAs.

However, within USDL several key perspectives have got high priority and likewise shall get high visibility. For this reason the actual Service Level Module is just one module among others (such as pricing or legal).

The Service Level Module is intentionally kept completely generic, as it does not specify how concrete service levels on concrete aspects (such as legal, pricing or security) shall be specified. Instead, its main purpose is twofold. First, it shall provide a proper glue between other USDL concepts. For example it specifies to which elements of a function a certain service level shall apply and who is the related stakeholder. Second, it should allow for incorporation of arbitrary attribute and expression languages. One particular attribute language is specified further below in this chapter and deals with security aspects. Other languages could be integrated as well, e.g., the UML profile for Modeling and Analysis of Real-Time and Embedded Systems (MARTE) or the full SLA model from SLA@SOI.

The Service Level Module intentionally does not specify any concrete attribute type systems. This is done for three reasons. First, there is no commonly agreeable type system that applies to all kinds of domains. Secondly, even some common core for such a type system can easily grow to extensive size and therefore contradicts our ambition to keep the Service Level Module as lean as possible. Third, there are already established type systems for different domains, and we wanted to keep the Service Level Module neutral and fully extensible towards these type systems. However, for pragmatic reasons, the Service Level Module also comes with a base extension (not part of the core USDL language) which provides common notions for frequently used metrics such as reliability, security, location, time, performance, and availability.

12.3.3 Module Overview

The main concepts of the module (see Figure 12.4) are the ServiceLevel (specifying either a state or an action), the ServiceLevelExpression, and the ServiceLevelAttributes. Furthermore, the module contains important references to other modules' concepts such as the ObligatedParty (Participant Module), the VariableDeclaration (Foundation Module), and the relatesTo reference (Foundation Module).

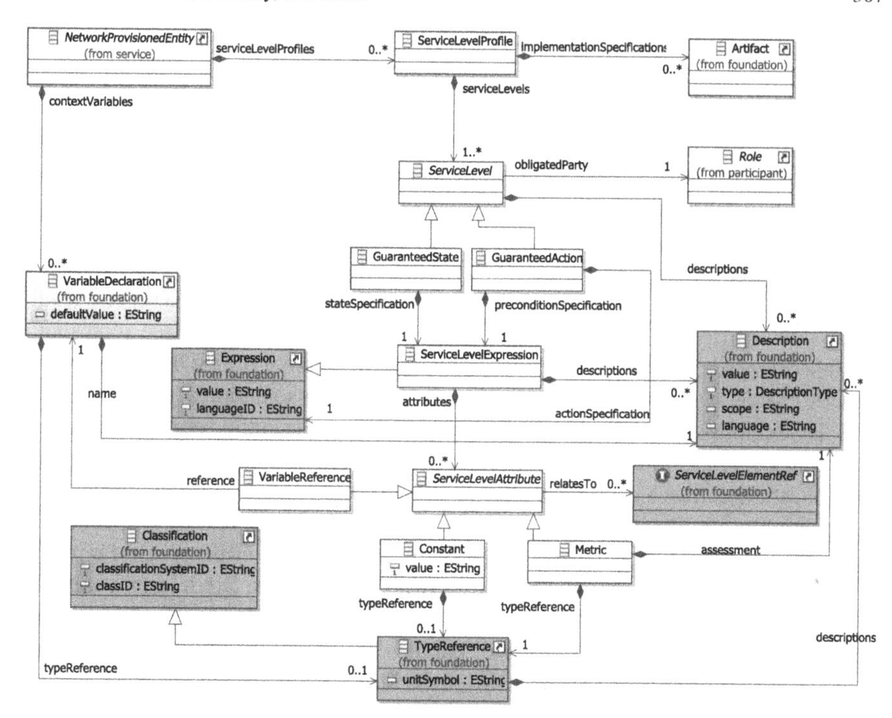

Fig. 12.4: Overview of the Service Level Module.

The entry point to specify the service levels of a given service is realized via ServiceLevelProfile. A set of service level specifications are combined into one profile and are offered, negotiated, or agreed upon as a whole. Different profiles can be used to specify different options of how service levels may be specified and grouped (e.g., as gold, silver, bronze profile). A ServiceLevelProfile resembles the concept of a Service Level Agreement Template as for example specified in WS-Agreement.

A ServiceLevel specifies a single service level objective as it characterizes an offered, negotiated or agreed service. ServiceLevels are defined by the parties participating in service provisioning, delivery, and consumption and express assertions that are claimed or expected to hold during these activities. Such assertions are always attributed to a single party, which is obligated to enforce the service level. From the viewpoint of the party defining the ServiceLevel two cases are distinguished. Either the defining party obligates itself to ensure the ServiceLevel, i.e., it claims that the assertion will hold, or the defining party expects the obligated party to ensure the ServiceLevel, i.e., it requires the other party to enforce the assertion. A ServiceLevel can be either a GuaranteedState (specifies a single state that must be maintained within the lifetime of any service instance, to which the respective service level profile applies) or a GuaranteedAction (specifies a self-contained activity that must be performed, if and only if during the lifetime of any service instance to which the respective service level profile applies a specific precondition is fulfilled).

A ServiceLevelExpression specifies an expression that is evaluated in the context of a service level state or action. For this purpose it may reference a set of ServiceLevelAttributes (constants, metrics or variable references) and define relationships between these attributes, e.g., Boolean or arithmetic operands. Typically, it resolves to a Boolean value that indicates whether a GuaranteedState is met or whether the precondition to a GuaranteedAction is fulfilled.

A ServiceLevelAttribute specifies a single attribute that is part of a service level expression. Attributes can take various concrete forms, of which three (constant, variable reference and metric) are defined in the core version of USDL. A ServiceLevelAttribute has a scope, i.e., it exists in reference to something to which it applies. By default, all attributes are defined in relation to the entire service (including its overall context). Alternatively, a scope that covers only parts may be specified. A Constant specifies a single ServiceLevelAttribute which is constant during service operation, i.e., during the lifetime of any service instance. A Metric specifies a single ServiceLevelAttribute which refers to the observation (measure) of a property of the service at service runtime. It may change over the lifetime of a service instance. Last, a VariableReference allows for referencing a variable declared in the global context of a service or service bundle. VariableReferences are used, for instance, as part of service level expressions.

In Listing 12.1 (cf. appendix of this chapter) we provide a more detailed example of how the Service Level Module can be used. It is related to a 3PL logistics provider as specified in the running example introduced in Chapter 8. The XML snippet shows a service level specification, where

1. Customers can specify their expected delivery duration
2. A provider-obligated term guarantees the delivery within the specified duration (ServiceLevel nb 1)
3. A provider-obligated term guarantees that goods are maintained at -5 degrees Celsius (ServiceLevel nb 2)
4. A provider-obligated action regulates penalty payments for delayed deliveries (ServiceLevel nb 3)

12.4 USDL and Security

The implementation of security measures for electronic services, such as the proper authentication of consumers and the encryption of service data, can become a complex and error-prone effort. Service providers do not always possess sufficient knowledge and experience on technical security mechanisms and standards. The same applies to the definition of service security characteristics as part of a formal service description. Such technical issues may be far out of the core competencies on which a service provider should be able to concentrate in a globalized and competitive market.

With regard to the security issues mentioned above, this means that platforms on which business services are developed, hosted and traded will provide — as part of the platform infrastructure — shared security services to handle the management of identities, the access control to services, data protection and related tasks. Business services strongly vary in their security requirements. General information services, such as rental car availability information, and transaction oriented services, such as the booking of a rental car in combination with a debit advice, are possibly traded on one and the same platform. Hence, security services provided by the platform are not statically bound to business services. Rather they can be integrated with and bound to business services on demand and in a flexible manner (Security-as-a-Service).

Ideally, a business service provider is enabled to implement service security measures in a declarative manner, i.e., by specifying the desired security characteristics as part of a service model or service description. In this case the underlying platform (operated by a third party) takes care of fulfilling the declared security requirements by including the required security services, e.g., by enforcing user authentication or by ensuring non-repudiation of the service usage. In order to anticipate the potentially limited technical security knowledge of business service providers, the description of security characteristics should be supported on an abstract, non-technical level. The abstract description of business service security characteristics also enables non-technicians on the consumer side to express their security demands and to find business services that comply with these demands.

The USDL (v3 Milestone 5 specification) elements that serve to model these claims and requirements are depicted in Figure 12.5. The key building blocks are the classes SecurityMetric and SecurityAttribute, which are explained in detail here. They are special cases of the class ServiceLevelAttribute and both entail properties of the enumeration types SecurityGoal, SecurityRequirementLevel and RealizationLevel. The former two are straightforward elements that can be used as detailed below. RealizationLevel refers to the OSI reference model [3] for layered networks and specifies at what layer in the communication stack the security element is targeted at. These classes, that were introduced due to security considerations were placed in the module Service Level Base Extension. On the one hand, non-functional security properties are clearly part of the agreed service level. On the other hand, they are considered a domain specific extension and not part of USDL since they are not inherently part of all domains of business services, but only geared towards domains involving extensive network communication.

The SecurityAttribute and SecurityMetric elements support the description of service security characteristics in two alternative, mutually exclusive ways, they mainly differ in the abstraction level. SecurityAttribute specifies security characteristics in terms of security goals and security requirement levels, but it does not provide reference to specific technical security measures. Such SecurityGoals are integrity, confidentiality, identification, authentication, authorization, privacy and accountability. A SecurityRequirementLevel is used to indicate the desired implementation degree of a security goal, i.e., the required level of protection/security with respect to the security goal. The granularity of the security requirement levels is inspired by the "authentication assurance levels" [23] as developed by the Euro-

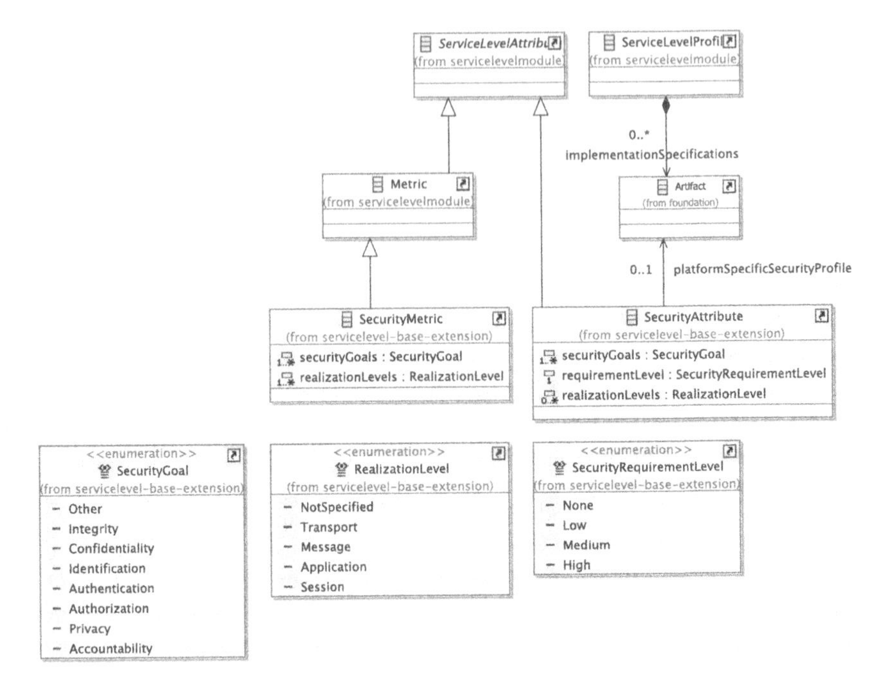

Fig. 12.5: Model elements introduced for security infrastructure.

pean STORK[4] project. On the other hand, SecurityMetric describes the high level security goal, but also the specific security mechanism used to address this goal, possibly including values of parameters and/or pointers to concrete security policies written in a standard policy language, such as WS-Security, P3P, XACML.

USDL allows one to browse services, and select them according to their capabilities and features, and security features may be important criteria for such a choice. Thus, USDL elements for security can express claims and/or requirements about security properties, with information on protections that are enforced by the service provider. For example, a service provider may claim that the customer data remain confidential, but at the same time it may require that the consumer should support message encryption to send input data in a confidential way.

In the next subsections we will describe the two approaches in more detail.

12.4.1 SecurityAttribute

Defining security characteristics in terms of security goals and requirement levels enables actors (on the provider as well as on the consumer side) who are not "techni-

[4] `https://www.eid-stork.eu`

cal security experts," i.e., who are not familiar with, e.g., WS-Security policies [11] or XACML [16] statements, to express their security demands. These requirements are then interpreted in the context of the particular platform that provides access to the service, using a platform-specific security profile (additional service metadata). The platform may undertake the task of implementing the security goals at the desired requirement levels by generating appropriate technical security policies (e.g., WS-SecurityPolicy artifacts) and by involving suitable platform services that are handling authorization, authentication, encryption etc.

The platform operator may create a "platform-specific security profile" to specify the technical details on how a security goal is realized on a given platform. The platform-specific security profile maps security goals that are defined at certain security requirement levels to concrete security mechanisms and technical standards that are supported by the given platform. For example, a platform operator may map the security goal "Authentication" at the security requirement level "low" to a formal WS-SecurityPolicy statement which specifies that a "UsernameToken" with a "password digest" and a "creation time stamp" is required to be authenticated. The same security goal with the security requirement level "high" could, e.g., be mapped to a WS-SecurityPolicy statement demanding an "X.509-Token."

The security mechanisms as well as the mapping of security goals and requirement levels to security mechanisms and technical standards are likely to vary from platform to platform, depending on the application domain and on the technical security services and standards supported by the platform. For this reason, USDL does not prescribe the form or structure of a platform-specific security profile. It is simply referenced.

In summary, the specification of security requirement levels for security goals enables service providers to express business service security characteristics on a an abstract rather than on a technical level. The same applies to service consumers that may search for appropriate business services based on these abstract characteristics.

The following short example further refines the 2PL Airline Manager (cf. Example 8 in Section 8.7), and illustrates the utilization of security attributes. The "2PL Airline Manager" provides two interfaces for *looking up rates* and *kicking-off shipments*. It was decided, that the operation "looking up rates" should be public, and therefore does not require any security characteristics; resulting in the USDL specification as outlined in Listing 12.2 on page 324: lines 28 to 48. Whereas the kick-off shipment requires medium Authentication and high Encryption, as it has to be known, who initiated the shipment, and the data of the operation should be kept confidential.

The service provider of the 2PL Airline Manager utilizes the "platform-specific security profile," defined by the platform operator of the logistics marketplace. At first the service provider has to select the "platform-specific security profile," which is in this case an ontology.[5] Then, the service provider defines the desired security requirements by associating the adequate security goals with the correspondent

[5] Accessible via `http://ontology.logistics_service.org/security/lso_profile123`

security requirement levels (see Figure 12.6). The USDL serialization is also illustrated in Listing 12.2: lines 50 to 73.

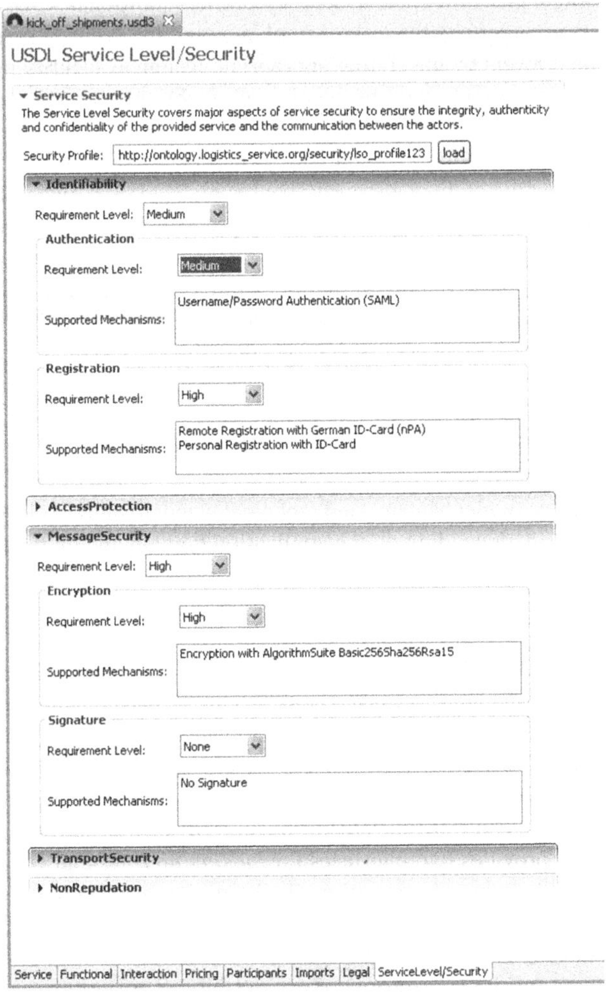

kick_off_shipments.usdl3

USDL Service Level/Security

Service Security

The Service Level Security covers major aspects of service security to ensure the integrity, authenticity and confidentiality of the provided service and the communication between the actors.

Security Profile: http://ontology.logistics_service.org/security/lso_profile123 load

Identifiability

Requirement Level: Medium

Authentication

Requirement Level: Medium

Supported Mechanisms: Username/Password Authentication (SAML)

Registration

Requirement Level: High

Supported Mechanisms: Remote Registration with German ID-Card (nPA)
Personal Registration with ID-Card

AccessProtection

MessageSecurity

Requirement Level: High

Encryption

Requirement Level: High

Supported Mechanisms: Encryption with AlgorithmSuite Basic256Sha256Rsa15

Signature

Requirement Level: None

Supported Mechanisms: No Signature

TransportSecurity

NonRepudation

Service | Functional | Interaction | Pricing | Participants | Imports | Legal | ServiceLevel/Security

Fig. 12.6: Specifying security via SecurityAttributes with the USDL Editor (cf. Section 15.2.1).

12.4.2 SecurityMetric

Similar to the SecurityAttribute element, the SecurityMetric element specifies both a SecurityGoal and a RealizationLevel, but does not feature a Security-RequirementLevel. Note that this is an alternative way to describe security requirements/claims, not intended to be used in combination with SecurityAttribute. Here, we do not need to express the requirement level, from low to high as in this approach we define more concrete security properties with link to technical artifacts. Security is then expressed in terms of concrete actions rather than an abstract level. Unlike SecurityAttribute, which only defines an abstract requirement level, the security metric enables a direct mapping with technical artifacts. A service provider can then specify some claims in terms of mechanism, such as internal procedure to erase Personally Identifiable Information after a certain amount of time to cover a privacy SecurityGoal. Also, the specifications to communicate with the service provider are no longer platform-specific, but rather described and decided when one operates service elicitation.

Having security described with technical artifacts allows actors that understand security protocols and standards to express their security demands in such terms. For example, during the matchmaking phase, a service consumer might restrict search to services that support data confidentiality at the application level through usage of a specific encryption algorithm. Prior to publishing the service, the provider defines which algorithms are accepted, such as AES and RSA for symmetric and asymmetric cryptography, and puts these capabilities in USDL. The consumer is then able to check which service is compatible with his requirements.

SecurityMetric defines what the service provider claims or requires in terms of technical artifacts. As this element is mostly for automated services, we can foresee usage of this element for manual or semi-automated services, such as making sure in a parcel shipment service that the warehouse clerk sets the seal on the box and then signs the registry.

Listing 12.3 on page 325 is an instantiation of the model shown in Figure 12.5. It provides information for an automated service on how identity is managed to certify authentication and provide message encryption to avoid leakage of data. The example is based on the example of the 2PL Airline Manager (cf. Section 8.7) where we try to specify more concrete mechanisms than previous section.

Instead of declaring the security level in an abstract way, we observe that the snippet gives us details about security mechanisms and goals. It automatically links security requirements with two external security policies, as outlined in Listing 12.3: lines 2 to 26. The first one is set to express encryption of a SOAP Body and the second adds support for SAML token for identity propagation, i.e., authentication. From line 35 to 38, we specify a protocol that is not an implementation specification. In our case, we introduce the HTTP Authentication scheme to express future usage of BASIC or DIGEST authentication. Then, lines 39 to 62 is the first block that expresses a concrete security mechanism. With the goal to describe authentication security at the session level, the block lists security mechanisms accepted by the service provider. Technologies can be linked coming from various sources, such as the reference to SAML Token coming from a WS-SecurityPolicy or from protocol documentation such as the HTTP Auth link. USDL provides expression logic to decide and restrict application of security mechanisms. The second block in lines 63 to 75 indicates the usage of message confidentiality and contains a summary of what is used. The service consumer is able to quickly understand the profile used by the service, and in case of further and detailed information he can consult directly the referenced security policy.

12.5 Trustworthiness of Service Providers

In contrast to security measures, which are targeted at third party threats, trust considerations are concerned with the risk emanating from business partners, in particular, unknown service providers. Trust calculation in this field evaluates which of the suitable providers is most likely to actually deliver what was promised [7]. This definition of trust is in line with Marsh's dissertation thesis [15], viz., the earliest transfer of the concept from humanities to computer sciences. He defines trust as the confidence towards a decision that entails obvious risks.

In an Internet of Services environment, where automatic, manual, and hybrid services are traded, there are three conceivable sources that can be considered to estimate the amount of a trust a consumer should put into a particular offering. The actual calculation of a trust score depends highly on the domain and the consumers preferences, but in any domain and for any user, the following categories apply.

The first source is provider and service information supplied by the provider himself. This is precisely the data covered in a USDL service description. It can safely be called *objective*, since any piece of information can, at least in theory, be verified by a neutral third party. This type of source is dealt with in the following section.

As opposed to the objective data of the service description, trust can be based on subjective data, i.e., feedback of other users. The connection of USDL with this reputation-based trust is discussed in the subsequent Section 12.5.2.

The third possibility for service seekers to judge trustworthiness is run time data, i.e., the information that the service platform collects about providers, offerings and invocations as they are delivered over time. This material, however, is out of scope of the static service description which is the purpose of USDL.

12.5.1 Trust Directly Based in Service Description

The USDL service description may offer various cues for confidence in a service offering. In Chapter 13, the class Certification is introduced (see also Figure 12.7). By asking neutral third parties to issue certificates about agents or a resources a provider can establish trustworthiness. Straightforward examples for such are TrustedShops[6] and organic food.[7] Ratings issued by agencies can also be modeled using the Certification class, such as Skytrax's World Airline Star Rating[8] or the stars of the European Hotelstars Union.[9]

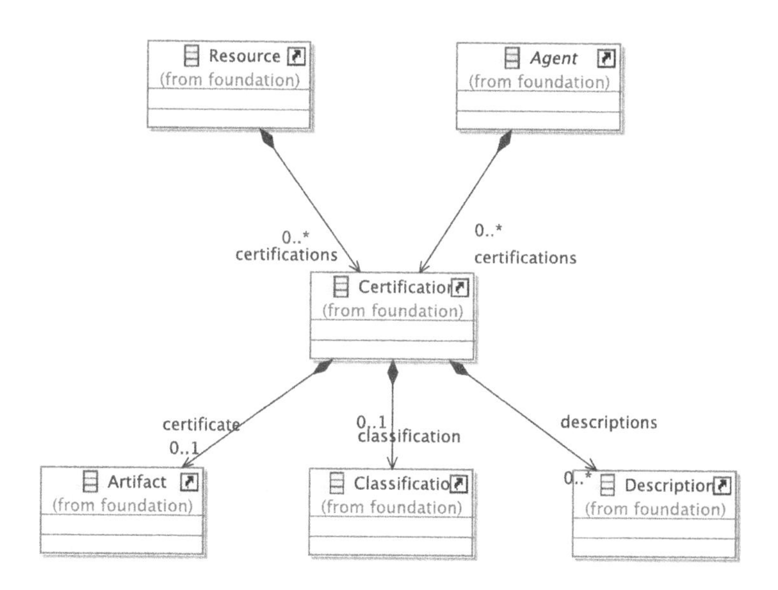

Fig. 12.7: Class Certification.

Inside the framework of USDL, there is no way to ensure that only rightful certifications are claimed. Nevertheless, it is reasonable to trust the claim of the certification as much as one trusts the neutral party that supposedly issued it: to make sure that no one falsely shows certificates is of vital business interest to the respective rating agencies and USDL makes it easy for them to scan for abuses. For example, TrustedShops can easily check if everyone referencing them actually shows in their files. Since we are dealing with an electronic market place and formalized service descriptions, this is much easier than in the case of physical shops that hang up physical documents.

Apart from these explicit trust cues, there is a range of USDL elements that implicitly induce some amount of trustworthiness. An example for this is the physical location of a provider. In the logistics domain, a courier company looking for an ocean carrier may not strictly require that it be based in a certain country, but it may find EU based carriers more trustworthy than American competitors. Likewise the attribute yearOfFounding in class Organization can contribute to the trust calcu-

[6] `http://www.trustedshops.de`

[7] `http://www.bio-siegel.de`

[8] `http://www.airlinequality.com/StarRanking/ranking.htm`

[9] `http://www.hotelstars.eu`

lation. Beyond these there are various pieces of information contained in the USDL description, that can be exploited for trust calculation. However, that calculation depends highly on the domain and on consumer preferences and is therefore out of the scope of this book. In the THESEUS/TEXO project, we realized a sample trust prediction for a car rental scenario [10] based among on USDL data. Similar to certification, all this information is provider supplied and hence questionable. So reasonable trust preferences put more weight on such certain elements; namely those that are likely to be audited by a relevant actor, who has a natural incentive. The location, for example, is most likely to be checked by the service market place operator.

On a conceptual level, a clear distinction can be made between two classes of USDL items: on the one hand, items that imply how well an offering matches the demand of a service seeker. On the other hand, items that assures confidence in the provider. An example of the former is the price, the latter could be a certificate. In practice, however, the line is blurry, certain elements can easily appeal to both the liking and the trusting of a service seeker. Imagine a traveler looking for a hotel: A two-star cuisine may not be a requirement, yet its presence can be interpreted as a sign of overall reliability. In other circumstances one piece of information can even contribute in a contradictory way to preference and reliability. To illustrate this, let us consider a 3PL courier company (cf. running example from Chapter 8) that seeks an airline service. Given their functional parameters such as origin, destination, dates, weight, etc. there might be ten airlines offering that particular service at a particular price. Now, while a lower price is quite to the liking of the seeker, a fee lower than half the mean price may be a cue to distrust that provider.

12.5.2 Using the Service Description to Harness User Ratings

A different approach to evaluate the trustworthiness of a provider is to consider his reputation, i.e., the reported experience of other users. This field is currently still under investigation and a recent overview about reputation in service-oriented environments is provided in [13]. The essential idea is that of wisdom of crowds [22]: the more users report on a given provider, the better his reputation predicts his behavior. Additionally, and in contrast to the static data mentioned in the previous section, a reputation system makes white-washing difficult, i.e., the cost of building up a trusted profile can stop malicious providers from cheating with fresh accounts [4].

The actual reputation can obviously not be part of a static USDL service description, since it is highly dynamic. Nevertheless, USDL can be used to remedy some of the problems often encountered in reputation systems.

In many domains a simple measure of how happy a consumer was with a service is too undifferentiated. This also applies if the rating system is used for other purposes than trust calculation. For instance, providers want specific aspects of their services rated to guide their innovation process. As these two were the main applications for feedback information foreseen by the USDL meta-modelers, the elements discussed here cater for the needs of both. Consequently, a framework for evaluating specific aspects of a service or provider is needed. Moreover, domain specific scales may be necessary, wherever a simple five star scale is insufficient. On the other hand, there is a mechanism that ensures that comparable services can be rated in a consistent way. Otherwise, providers would make their offering ratable only in those categories where they excel.

In order to meet the requirement of *Extensibility* (cf. Section 8.3.1.4) a complete feedback meta-model was developed in THESEUS/TEXO [14], that allows for the definition of a hierarchy of rating aspects on a corresponding scale. Since that meta-model is not specific to the realm of services, but could also be used for products, it was not incorporated in USDL. Instead, instances of the feedback meta-model can be referenced using the USDL class Artifact.

The interlinking of feedback with relevant USDL classes is depicted in Figure 12.8. Essentially, the relation feedbackModels adds 0 to n feedback model Artifacts to NetworkProvisionedEntity, AbstractService and Agent. ServiceBundle and Service inherit the link from NetworkProvisionedEntity and they represent the core objects to be rated. By means of the super class Agent, USDL users can describe how Persons and Organizations can be rated. The class AbstractService deserves a more detailed discussion.

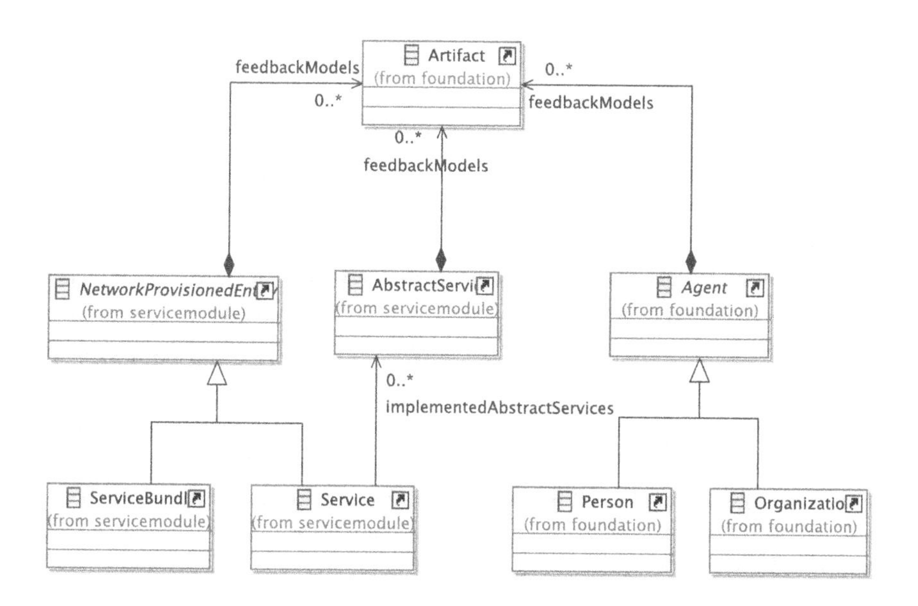

Fig. 12.8: Model elements introduced for a feedback infrastructure.

The feedback models attached to AbstractService enable consist rating schemes across similar services across multiple providers. A domain knowledgeable authority such as the service market place platform operator can define AbstractServices and specify how instances of these are to be evaluated. Concrete services claim to implement a given AbstractService in order to be suggested to seekers of this kind of offering. By that they automatically are subject to the rating scheme attached to the abstract one, and hence all competing services are guaranteed to sharing the most important evaluation aspects for this kind of offering. Moreover, providers can attach further rating models inside their USDL description to get a more detailed feedback on their performance.

To illustrate this, let us consider an example. We chose a domain where most principals are unknown to each other, and hence must base their trust on other consumers' rating. This diverges from the running example where principals are most likely to have some past interaction and do not have to rely on third party opinions. Consider a service market that trades automotive services. The platform host would reasonably create an AbstractService "car repair" and link it up with a feedback model that covers at least "quality of repair" and "speed of repair." Thus, all providers that want to be taken into account when a car repair is wanted, must reference this AbstractService and can therefore automatically be evaluated in these two most relevant categories. This in turn leads to a consistent reputation landscape of car repair services, which is suitable for service seekers to base their trust on.

12.6 Conclusion

In this chapter we presented an overview of existing approaches to model service levels with an extra glimpse on security specific languages. Building on that, we described how USDL addresses existing gaps, followed by a discussion of how the Service Level Module in constructed and how it relates to other USDL elements. Again, special care was taken to discuss how security properties are covered in USDL. Namely, USDL allows service providers to specify security offerings and state their security requirements.

Subsequently, we explained the USDL meta-model elements that are related to security and trust in detail. Most notably the classes SecurityAttribute and SecurityMetric for specifying security goals either on a high level or on a technical level, respectively. The relation feedbackModel was introduced explicitly to facilitate trust calculation. It associates services with rating schemes. Additionally, we surveyed preexisting USDL elements that offer cues for trust estimation.

We decided that trust and security elements do not form a USDL module of their own, since they do not represent a group of functional elements. They rather form part of the Service Level Module which host most non-functional properties.

Based on the foundations laid out in this chapter, we see several lines of future work: First, the discussed security properties are on a merely technical level and are not well suited for describing security properties on a more abstract, i.e., business level. On a business level, users usually do not want to specify properties such as Confidentiality or Authorization. Instead, they want to specify more abstract properties such as "comply to the following regulations" or "it is only allowed to share this data between the following parties." Supporting such high-level specifications requires, on the one hand, to link the security extension with the Legal Module (see Chapter 10). This allows for an extensive support of legal compliance regulations such as the Sarbanes-Oxley Act [20] in the financial industry or HIPAA [6] in the health care industry. Supporting such legal compliance regulations is particular challenging as they combine legal requirements based on abstract concepts with technical security and privacy aspects. On the other hand, this requires a process for mapping high-level requirements to technical realizations of those requirements and, thus, allow business experts and security experts to work together for providing secure, trustworthy, and compliant applications on top of the Internet of Services. Second, we plan to provide extensions that serve domain specific needs, e.g., that allow for describing advanced access control or privacy needs in the health care domain. The reader interested in a detailed description of such requirements is, e.g., referred to the documents describing the security requirements England's National Programme for Information Technology (NPfIT) of the National Health Service (NHS) [18, 19]. Third, we plan to empirically evaluate the concepts introduced here, i.e., to investigate how well they help to describe and trade business services.

References

1. A. Andrieux, K. Czajkowski, A. Dan, et al. Web services agreement specification (ws-agreement). Technical report, OpenGridForum, 2007. http://www.ogf.org/documents/GFD.107.pdf.
2. M. G. Buscemi and U. Montanari. Cc-pi: A constraint-based language for specifying service level agreements. In *Programming Languages and Systems, 16th European Symposium on Programming*, pages 18–32. Springer, 2007.
3. J. Day and H. Zimmermann. The OSI Reference Model. *Proceedings of the IEEE*, 71(12):1334–1340, Dec. 1983.
4. M. Feldman and J. Chuang. The Evolution of Cooperation under Cheap Pseudonyms. In *7th IEEE International Conference on E-Commerce Technology (CEC)*, pages 284–291, München, jul 2005. IEEE Computer Society.
5. D. Gregg and S. Walczak. The relationship between website quality, trust and price premiums at online auctions. *Electronic Commerce Research*, 10:1–25, 2010.
6. HIPAA. Health Insurance Portability and Accountability Act of 1996. http://www.cms.hhs.gov/HIPAAGenInfo/, 1996.
7. A. Jøsang, R. Ismail, and C. Boyd. A survey of trust and reputation systems for online service provision. *Decision Support Systems*, 43(2):618–644, 2007.
8. K. Kearney, F. Torelli, and C. Kotsokalis. SLA*: An Abstract Syntax for Service Level Agreements. In *Proceedings of IEEE Grid2010 conference; Service Level Agreements in Grids Workshop*, Brussels, 2010.
9. D. D. Lamanna, J. Skene, and W. Emmerich. Slang: A language for defining service level agreements. In *Future Trends in Distributed Computing Systems*, pages 100–106. IEEE Computer Society, 2003.
10. E. Lapi, E. Höfig, and F. Marienfeld. THESEUS/TEXO Consortium: TRICE Deliverable E4 – Demonstrator Based on TEXO Platform, Feb. 2011.
11. K. Lawrence, C. Kaler, and al. Ws-securitypolicy 1.3. http://docs.oasis-open.org/ws-sx/ws-securitypolicy/v1.3/os/ws-securitypolicy-1.3-spec-os.html, 2009.

12. H. Ludwig, A. Keller, A. Dan, et al. Web service level agreement (wsla) language specification. Technical report, IBM Research, 2003. http://www.research.ibm.com/wsla/WSLASpecV1-20030128.pdf.
13. Z. Malik and A. Bouguettaya. *Trust Management for Service-Oriented Environments*. Springer US, 2009.
14. F. Marienfeld, E. Höfig, and E. Lapi. THESEUS/TEXO Consortium: TRICE Deliverable E2 – Extension of USDL with Trust and Quality Criteria, Feb. 2011.
15. S. P. Marsh. *Formalising Trust as a Computational Concept*. PhD Thesis, University of Stirling, 1994.
16. T. Moses. eXtensible Access Control Markup Language(XACML) Version 2.0. http://docs.oasis-open.org/xacml/2.0/access_control-xacml-2.0-core-spec-os.pdfl, 2005.
17. L. Nixon, D. Lambert, A. Filipowska, and E. Simperl. Future of the Internet of Services for Industry: the ServiceWeb 3.0 Roadmap. *Future Internet Assembly (FIA 2009)*, 2009.
18. D. of Health. The Care Record Guarantee. Our Guarantee for NHS Care Records in England. Technical report, Department of Health, 2009.
19. D. of Health. Information Governance (IG) Concepts, 2010. http://www.connectingforhealth.nhs.uk/systemsandservices/infogov/.
20. P. Sarbanes, G. Oxley, et al. Sarbanes-Oxley Act of 2002. 107th Congress Report, House of Representatives, 2nd Session, 107–610, 2002.
21. F. I. P. SLA@SOI. Empowering the service industry with sla-aware infrastructures. http://sla-at-soi.eu.
22. J. Surowiecki. *The Wisdom of Crowds*. Doubleday, 2004.
23. the STORK-eid Consortium. STORK Deliverable D2.1 - Framework Mapping of Technical/Organisational Issues to a Quality Scheme. https://www.eid-stork.eu/index.php?option=com_processes&Itemid=&act=streamDocument&did=579, 2011.
24. S. Venugopal, X. Chu, and R. Buyya. A Negotiation Mechanism for Advance Resource Reservations Using the Alternate Offers Protocol. In *16th International Workshop on Quality of Service*, pages 40–49, june 2008.

Listings

Listing 12.1: USDL Service Level Agreement sample

```
1  <identifiableElement xsi:type="service:Service">
2
3    <contextVariables>
4      <!-- variable for expected duration of delivery in weeks,
5           default: 1 week -->
6      <variableDeclaration xsi:id="varExpDelDuration">
7        <name>
8          <value> expectedDeliveryDuration </value>
9          <type> name </type>
10       </name>
11       <defaultValue> 1 </defaultValue>
12       <typeReference>
13         <classificationSystemID> http://www.internet-of-services.com/
14   serviceTypes </classificationSystemID>
15         <classID> duration_week </classID>
16         <unitSymbol> wk </unitSymbol>
17         <descriptions>
18           <description>
19             <value> duration in weeks </value>
20             <type> freetextShort </type>
```

```
21            <language> en </language>
22          </description>
23        </descriptions>
24      </typeReference>
25    </variableDeclaration>
26  </contextVariables>
27
28  <serviceLevelProfiles>
29    <serviceLevelProfile>
30
31      <serviceLevels>
32
33        <!-- service level #1 -->
34        <serviceLevel xsi:type="servicelevel:GuaranteedState">
35          <!-- provider is obligated -->
36          <obligatedParty> prov321 </obligatedParty>
37
38          <descriptions>
39            <description>
40              <value> Delivery duration must not longer than what has been
41                      specified by the customer. </value>
42              <type> freetextShort </type>
43              <language> en </language>
44            </description>
45          </descriptions>
46
47          <stateSpecification>
48
49            <!-- measured delivery duration is less than or equal to what is
50                 set in the service context (cust. input -->
51            <value> metric["m_delDuration"] <= variable["v_expDelDur"] </value>
52            <languageID> urn:example:expression_language </languageID>
53
54            <attributes>
55              <!-- reference to the variable -->
56              <attribute xsi:id="v_expDelDur"
57                         xsi:type="servicelevel:VariableReference">
58                <reference> varExpDelDuration </reference>
59              </attribute>
60              <!-- metric for measuring delivery duration -->
61              <attribute xsi:id="m_delDuration" xsi:type="servicelevel:Metric">
62                <typeReference>
63                  <classificationSystemID>
64  http://www.internet-of-services.com/serviceTypes
65                  </classificationSystemID>
66                  <classID> duration_day </classID>
67                  <unitSymbol> d </unitSymbol>
68                  <descriptions>
69                    <description>
70                      <value> duration in days </value>
71                      <type> freetextShort </type>
72                      <language> en </language>
73                    </description>
74                  </descriptions>
75                </typeReference>
76                <assessment>
77                  <value> delivery duration as measured by receiving party
78                  </value>
79                  <type> freetextLong </type>
80                  <language> en </language>
81                </assessment>
82              </attribute>
83            </attributes>
84
85          </stateSpecification>
86        </serviceLevel>
87
```

```
88          <!-- service level #2 -->
89          <serviceLevel xsi:type="servicelevel:GuaranteedState">
90            <!-- provider is obligated -->
91            <obligatedParty> prov321 </obligatedParty>
92
93            <descriptions>
94              <description>
95                <value> The temperature of the goods is maintained at minus 5
96                    degrees. </value>
97                <type> freetextShort </type>
98                <language> en </language>
99              </description>
100           </descriptions>
101
102           <stateSpecification>
103
104             <!-- measured temperature is approximately equal to minus 5
105                 degrees Celsius -->
106             <value> metric["m_temp"] ~= constant["c5"] </value>
107             <languageID> urn:example:expression_language </languageID>
108
109             <attributes>
110               <!-- temperature threshold (modeled as constant) -->
111               <attribute xsi:id="c5" xsi:type="servicelevel:Constant">
112                 <value> -5 </value>
113                 <typeReference>
114                   <classificationSystemID>
115                     http://www.internet-of-services.com/serviceTypes
116                   </classificationSystemID>
117                   <classID> temperature_celsius </classID>
118                   <unitSymbol> dC </unitSymbol>
119                   <descriptions>
120                     <description>
121                       <value> temperature in degrees Celsius </value>
122                       <type> freetextShort </type>
123                       <language> en </language>
124                     </description>
125                   </descriptions>
126                 </typeReference>
127               </attribute>
128               <!-- metric for measuring temperature, related to goods (input to
129                   service function "Transport") -->
130               <attribute xsi:id="m_temp" xsi:type="servicelevel:Metric">
131                 <relatesTo> paramGoods </relatesTo>
132                 <typeReference>
133                   <classificationSystemID>
134                     http://www.internet-of-services.com/serviceTypes
135                   </classificationSystemID>
136                   <classID> temperature_celsius </classID>
137                   <unitSymbol> dC </unitSymbol>
138                   <descriptions>
139                     <description>
140                       <value> temperature in degrees Celsius </value>
141                       <type> freetextShort </type>
142                       <language> en </language>
143                     </description>
144                   </descriptions>
145                 </typeReference>
146                 <assessment>
147                   <value> temperature of goods as constantly measured
148                       by provider </value>
149                   <type> freetextLong </type>
150                   <language> en </language>
151                 </assessment>
152               </attribute>
153             </attributes>
154
```

```
155        </stateSpecification>
156      </serviceLevel>
157
158      <!-- service level #3 -->
159      <serviceLevel xsi:type="servicelevel:GuaranteedAction">
160        <!-- provider is obligated -->
161        <obligatedParty> prov321 </obligatedParty>
162
163        <descriptions>
164          <description>
165            <value> Penalty payment of EUR100 per full day of delayed
                   delivery
166            </value>
167            <type> freetextShort </type>
168            <language> en </language>
169          </description>
170        </descriptions>
171
172        <preconditionSpecification>
173          <!-- measured delivery delay is greater or equal than 1 -->
174          <value> metric["m_delDelay"] >= constant["c6"] </value>
175          <languageID> urn:example:expression_language </languageID>
176
177          <attributes>
178            <!-- delivery delay threshold (modeled as constant) -->
179            <attribute xsi:id="c6" xsi:type="servicelevel:Constant">
180              <value> 1 </value>
181              <typeReference> <!-- same type description as specified in
182                    metric below --> </typeReference>
183            </attribute>
184            <!-- metric for measuring delay in delivery -->
185            <attribute xsi:id="m_delDelay" xsi:type="servicelevel:Metric">
186              <typeReference>
187                <classificationSystemID>
188                      http://www.internet-of-services.com/serviceTypes
189                </classificationSystemID>
190                <classID> duration_day </classID>
191                <unitSymbol> d </unitSymbol>
192                <descriptions>
193                  <description>
194                    <value> duration in days </value>
195                    <type> freetextShort </type>
196                    <language> en </language>
197                  </description>
198                </descriptions>
199              </typeReference>
200              <assessment>
201                <value>delayed delivery as measured by receiving party </value
                      >
202                <type> freetextLong </type>
203                <language> en </language>
204              </assessment>
205            </attribute>
206          </attributes>
207        <preconditionSpecification>
208
209        <actionSpecification>
210          <value> <!-- credit the customer with (EUR 100
211              * floor(metric["m_delDelay"])) --> </value>
212          <languageID> urn:example:action_language </languageID>
213        </actionSpecification>
214
215      </serviceLevel>
216    </serviceLevels>
217
218  </serviceLevelProfile>
219 </serviceLevelProfiles>
```

```
220
221
222 </identifiableElement>
```

Listing 12.2: USDL SecurityAttribute

```
1
2  <!-- ... -->
3  <serviceLevelProfile>
4    <implementationSpecifications>
5      <implementationSpecification xsi:id="lso_profile123">
6        <type> TechnicalMetadata </type>
7        <mimeType> application/xml </mimeType>
8        <uri> http://ontology.logistics_service.org/security/lso_profile123 </uri
           >
9        <descriptions>
10         <description>
11           <value> security profile 123 </value>
12           <type> name </type>
13           <language> en </language>
14         </description>
15       </descriptions>
16     </implementationSpecification>
17   </implementationSpecifications>
18 <!-- ... -->
19   <serviceLevels>
20     <serviceLevel xsi:type="servicelevel:GuaranteedState">
21        <!-- ... -->
22       <stateSpecification>
23         <!-- ... -->
24         <value> attribute["sec1"] applies </value>
25         <value> attribute["sec2"] applies </value>
26         <!-- ... -->
27         <attributes>
28           <!-- look_up_rates has no security restrictions -->
29           <attribute xsi:id="sec1" xsi:type="slbaseext:SecurityAttribute">
30             <relatesTo>airline_mgmt:look_up_rates </relatesTo>
31             <securityGoals>
32               <securityGoal>Authentication </securityGoal>
33             </securityGoals>
34             <requirementLevel>none</requirementLevel>
35             <platformSpecificSecurityProfile>
36               lso_profile123
37             </platformSpecificSecurityProfile>
38           </attribute>
39           <attribute xsi:id="sec2" xsi:type="slbaseext:SecurityAttribute">
40             <relatesTo>airline_mgmt:look_up_rates </relatesTo>
41             <securityGoals>
42               <securityGoal> Confidentiality </securityGoal>
43             </securityGoals>
44             <requirementLevel>none</requirementLevel>
45             <platformSpecificSecurityProfile>
46               lso_profile123
47             </platformSpecificSecurityProfile>
48           </attribute>
49           <!-- ... -->
50           <!-- kick_off_shipments requires medium Authentication
51                and high Confidentiality
52             -->
53           <attribute xsi:id="sec3" xsi:type="slbaseext:SecurityAttribute">
54             <relatesTo>airline_mgmt:kick_off_shipments </relatesTo>
55             <securityGoals>
56               <securityGoal> Authentication </securityGoal>
57             </securityGoals>
58             <requirementLevel>medium</requirementLevel>
59             <platformSpecificSecurityProfile>
60               lso_profile123
```

```
61            </platformSpecificSecurityProfile>
62          </attribute>
63          <attribute xsi:id="sec4" xsi:type="slbaseext:SecurityAttribute">
64            <relatesTo>airline_mgmt:kick_off_shipments</relatesTo>
65            <securityGoals>
66              <securityGoal>Confidentiality</securityGoal>
67            </securityGoals>
68            <requirementLevel>high</requirementLevel>
69            <platformSpecificSecurityProfile>
70              lso_profile123
71            </platformSpecificSecurityProfile>
72          </attribute>
73        </attributes>
74          <!-- ... -->
75      </stateSpecification>
76    </serviceLevel>
77  <serviceLevels>
78 <serviceLevelProfile>
79 <!-- ... -->
```

Listing 12.3: USDL Security Metric

```
1  <serviceLevelProfile>
2    <implementationSpecifications>
3      <implementationSpecification xsi:id="lso_Wssp1.2_EncryptBody">
4        <type> TechnicalMetadata </type>
5        <mimeType> application/xml </mimeType>
6        <uri> http://logistics_service.org/security/lso-Wssp1.2-EncryptBody.xml
              </uri>
7        <descriptions>
8          <description>
9            <value> WS-SecurityPolicy to express that the entire body of a soap
10           message has to be encrypted </value>
11           <language> en </language>
12         </description>
13       </descriptions>
14     </implementationSpecification>
15     <implementationSpecification xsi:id="lso_Wssp1.2_SupportSAMLToken">
16       <type> TechnicalMetadata </type>
17       <mimeType> application/xml </mimeType>
18       <uri> http://logistics_service.org/security/lso-Wssp1.2-SupportSAMLToken.
              xml </uri>
19       <descriptions>
20         <description>
21           <value> WS-SecurityPolicy to express the support of SAML token for
                 Identity </value>
22           <language> en </language>
23         </description>
24       </descriptions>
25     </implementationSpecification>
26   </implementationSpecifications>
27   <!-- -->
28   <serviceLevels xsi:type="servicelevel:GuaranteedState" obligatedParty="//
          @Roles.0">
29     <stateSpecification>
30       <!-- ... -->
31       <attributes>
32         <!-- kick_off_shipments requires medium and
33               high Confidentiality
34           -->
35         <attribute xsi:id="sec:httpauth" xsi:type="servicelevel:GenericConstant
               ">
36           <!-- HTTP AUTH (BASIC or DIGEST) -->
37           <value>http://www.ietf.org/rfc/rfc2617.txt</value>
38         </attribute>
39         <attribute xsi:id="sec5" xsi:type="slbaseext:SecurityMetric">
40           <relatesTo>airline_mgmt:kick_off_shipments</relatesTo>
```

```
41          <securityGoals>
42            <securityGoal> Authentication </securityGoal>
43          </securityGoals>
44          <RealizationLevel> Session </RealizationLevel>
45          <securityMechanisms>
46            <securityMechanism xsi:type="sec:httpauth">
47              <value>HTTP AUTH</value>
48            </securityMechanism>
49            <securityMechanism xsi:type="lso_Wssp1.2_SupportSAMLToken">
50              <value>SAML</value>
51            </securityMechanism>
52          </securityMechanisms>
53          <expressionSpecification>
54            <description>
55          Authentication is verified through HTTP headers
56          or token in SOAP messages
57        </description>
58            <expression>
59          sec:httpauth OR lso_Wssp1.2_SupportSAMLToken
60        </expression>
61          </expressionSpecification>
62        </attribute>
63        <attribute xsi:id="sec6" xsi:type="slbaseext:SecurityMetric">
64          <relatesTo>airline_mgmt:kick_off_shipments</relatesTo>
65          <securityGoals>
66            <securityGoal>Confidentiality</securityGoal>
67          </securityGoals>
68          <RealizationLevel> Message </RealizationLevel>
69          <securityMechanisms>
70            <securityMechanism xsi:type="lso_Wssp1.2_EncryptBody">
71              <type>Encryption</type>
72              <value>Basic256Sha256Rsa15</value>
73            </securityMechanism>
74          </securityMechanisms>
75        </attribute>
76      </attributes>
77      <!-- ... -->
78    </stateSpecification>
79    <!-- ... -->
80  </serviceLevels>
81 </serviceLevelProfile>
```

Chapter 13
Modeling Foundations

Steffen Heinzl, Uwe Kylau, and Norman May

Abstract One of the basic purposes of USDL is to provide a "commercial envelope" around a service for exposition in Web-based service networks (i.e., the Internet of Services). Besides the description of pricing, licensing, functionality, behavior, service levels, and security aspects, there remain the following fundamental facets: the interweaving of all the aspects, the modeling of basic concepts such as service, composite service, or bundles, and the information about participants in the delivery of the service. These concepts are reflected in the modules introduced in this chapter: the Service Module and the Participants Module. These two modules are complemented by the Foundation which covers aspects that are used in multiple other modules. This chapter presents the module design taking into account the influence of the state of the art and several of the general language requirements as well as the service concept formation requirements from Chapter 8.

13.1 Introduction

USDL is a service description language to be used in multi-party service networks in the Internet (frequently dubbed the *Internet of Services*). In particular, USDL is targeted as an integral building block for supporting the value creation processes taking place in such ecosystems — by providing the means to encapsulate a service as self-contained asset which is exposed in novel ways (cf. Chapter 1). The inher-

Steffen Heinzl
SAP Research Darmstadt, Bleichstrasse 8, 64283 Darmstadt, Germany,
e-mail: steffen.heinzl@sap.com

Uwe Kylau
SAP Research Brisbane, Building A4 Level 7, 52 Merivale Street, South Brisbane QLD 4101, Australia, e-mail: uwe.kylau@sap.com

Norman May
SAP AG, Dietmar-Hopp-Allee 16, 69190 Walldorf, Germany, e-mail: norman.may@sap.com

ent business nature of this context has to be addressed in the foundational parts of USDL. In order to make a decision, consuming participants want to know (1) *who* they are dealing with, i.e., who is involved in providing and delivering a service, and (2) *what* exactly they are paying for.

The second part has several aspects. First and foremost, it is about what a service does, i.e., its functionalities and features. This is covered in a dedicated chapter (see Chapter 11). Second, the it concerns how the service is embedded into the network that makes up a service ecosystem. Relationships to participants are subject of the *who* aspect, but relationships and dependencies to other services, and resources in general, have to be covered as well. This comprises, in particular, the issue of combining services into new service offerings, an act that may take either the form of composition or bundling of services. Both types have received considerable attention by existing standards and prior research. Composition of services is addressed by many different approaches. There are, for example, process-driven orchestration languages such as WS-BPEL [18] (BPEL hereafter) or BPMN [19], choreography languages such as WS-CDL [14], and data-driven mash-up techniques. The third aspects concerns the classification (or categorization) as a means of positioning a service in the overall service space by associating it with well-defined meaning. Typical types of classifications are, for instance, service type hierarchies, which are often industry-specific. The United Nations Standard Product and Services Code (UNSPSC[1]) or eCl@ss[2] (cf. also [9]) are examples of global, generic type hierarchies. One particular type of classifying a service is the *service nature* in USDL. It concerns the involvement of human or machine actors in the process of providing and delivering a service. Another categorization often used relates to the granularity or complexity of a service, i.e., does it provide a single operation or a complex set of features. It is easy to imagine that there are many more of these category-like characteristics, all of which could play a role in describing a service.

Besides (1) and (2), USDL defines a unifying frame that is referential and extensible, and avoids to replicate. The frame ties together foundational parts of service description with all other aspects required for a comprehensive description discussed in Chapters 9 to 12. The foundational parts provide the core information about a service, including its general properties, involved participants and dependencies. As USDL does not aim to re-invent standards for composing services or building classification hierarchies, it supplies mechanisms that allow such standards to be integrated into the language and connected to other description aspects. USDL captures the foundational parts in three distinct modules.

1. The *Service Module* captures the non-functional portions of the *what* aspect, alongside general information about a service. The module also covers dependencies to other entities, except for those that relate to the *who* aspect which are subject of the next module.
2. The *Participant Module* defines a number of abstract and concrete roles, which were identified in the context of the Internet of Services.

[1] `http://www.unspsc.org/`

[2] `http://www.eclass.de/`

3. The *Foundation* (module) of USDL contains elements that are of general nature and re-used throughout the other modules of USDL. This includes elements that describe entities independent of the concept of service, e.g., organization and person.

The remainder of this chapter is structured as follows. Section 13.2 reflects on state-of-the-art background and introduces the design of the three USDL modules describing foundational parts of services. Sections 13.3, 13.4, and 13.5 present the Service and Participant Modules, and Foundation of USDL in detail. Section 13.6 concludes the chapter.

13.2 Modeling Foundational Service Aspects in USDL

Foundational service aspects constitute the backbone of service description in USDL and cover general information about a service, its relationships to participants, as well as structural and commercial dependencies. USDL defines three modules for these aspects and takes a generic, in some parts abstract, modeling approach, which ensures enough flexibility to address a wide variety of services. The following section discusses relevant background for some of the aspects mentioned, in order to motivate the focus of modeling and the constructs selected. The section outlines the design and relationship of the three USDL modules which are presented in detail in subsequent sections.

13.2.1 State of the Art and its Influence on the Module Design

13.2.1.1 On Service Composition and Bundling

Building service-based applications through composition has been addressed by two major areas: (business) process composition and mash-ups. Business process composition deals with representing steps of a process as single software components. The software components are ordered and data dependencies are established, in order to define the composition. The components can be reused to rapidly change existing processes or build new ones and are often conceptually based on the idea of a service-oriented architecture. Two wide-spread standards used to represent and execute business processes are BPEL [18] and BPMN [19]. Building of mash-ups often follows more informal approaches. The representation of mash-ups is not standardized in a formal way such as, e.g., BPEL. Applications are not only built from consuming reusable software components, but also by aggregating data sources (including data transformation and making the results accessible). One popular example for creating data mash-ups is Yahoo Pipes [8]. While BPEL and BPMN often rely on Web services (i.e., SOAP, WSDL, etc.) as the technical implementation of the SOA paradigm, mash-ups often use a number of different technologies and protocols (including HTTP accessible (quasi-)REST-based interfaces, RSS [15], ATOM [21], OData [17], RDF [16], etc.), which allow representing and consuming data.

Beside the widespread BPEL and BPMN standard, there's also the Service Component Architecture (SCA) [1],[3] a set of specifications that describe how to build systems and applications out of reusable components around the SOA paradigm. SCA components among others support a number of different interfaces for local as well as remote consumption. Furthermore, composition of components is done in general-purpose programming languages, such as Java and C#.

While the Interaction Module focuses on the behavioral aspect of compositions, the modeling foundation reflects the structural aspects of composition to capture how services are embedded into a business network. Section 8.3.2.3 already explains that USDL addresses the requirement of information hiding by hiding the internal structure of the composition (including language-specific composition details). So, USDL just grasps enough information to show which services interact with each other, short of knowing how this is implemented. Basically, the embedded composition information in USDL shows which services and thus which partners interact in the ecosystem.

To further emphasize the importance of trading in business networks, also bundling is possible in USDL. Bundling allows providers to combine services to increase sales numbers. The services are not necessarily connected on a technical level and can even be completely independent of each other.

13.2.1.2 On Service Participants

As already explained in Chapter 1, the participants in service networks and ecosystems go beyond the traditional SOA roles (provider, broker, and consumer) especially with respect to monetization and fulfillment. [5] A service has to be described, offered and found on a service marketplace for monetization. A service broker may provide marketplace functionality and additionally coordinate the communication with a set of services that fulfill a larger task, such as founding a new company and thus take part in service delivery. To communicate with a service, data translations might be necessary being performed by a service gateway. To add value to existing services, service aggregators may combine existing services to new services, host these with a service hoster (e.g., offering a cloud infrastructure) and sell these. In some scenarios, a service may even be provided by another legal entity, viz., a business owner, for example, a national subsidiary of a multi-national organization, or the legislative authorities force the provider to adhere to certain conditions to protect the environment.

To address all of these roles, USDL introduces — beside the provider and consumer — the general concepts of intermediaries and stakeholders, and specifically the business owner. Intermediaries comprise all the participants that are part of

[3] http://oasis-opencsa.org/sca

the service delivery, e.g., service hoster, service broker, etc. Stakeholders reflect all additional participants beside providers, consumer, and intermediaries, that have 'stakes' in the service delivery, such as the legislative authorities.

13.2.1.3 On Time and Location

The conceptualization of location and time has been designed to specifically cater for the needs (and challenges) of service description and is aligned with concepts in the domain of geographic information systems and international time standards. Specifications that were consulted include the Geographic Markup Language (GML, ISO 19136) [12] of the Open Geospatial Consortium,[4] as well as ISO 8601 [11] and ISO 19108 [10] (a standard associated with GML), for the exchange of time-related information. Whereas GML and ISO 19108 were deemed too complex to be referenced in USDL (the required tooling support was considered to be too much overhead), ISO 8601 covers only a part of what was identified as relevant. As a consequence, both set of concepts (time and location) were re-modeled in USDL, even though in a way that allows mapping between the different languages.

13.2.1.4 On Organizational Modeling

USDL was modeled taking into account the requirement of *Organizational Embedding* as explained in Section 8.3.2.1. Since services cannot be described independent of organizational details, such as the actors that are part of the fulfillment of the service, the resources needed, etc., USDL takes on the concepts of agents and roles. Hence, relationships of a service to resources, stakeholders, business services inside the organization are included in USDL.

Among other works, USDL draws from the Organization Ontology [20] introducing concepts, such as Organization, Role and Agent. The ontology focuses on these concepts to describe the structure of organizations. Further, Bottazzi [6] defines an ontology of organizations based on DOLCE. In this ontology the main concepts are organizations, norms, agents and roles. The norms define the main concepts of the organization and the behavior of the agents. Agents and norms are connected by roles.

13.2.2 Design and Relationship of USDL Modules

Capturing foundational service concepts in three distinct modules follows the requirement of *Modularity* (cf. Section 8.3.1.3). The Service Module and Participant Module are each concerned with a specific aspect, which is intended to help mod-

[4] `http://www.opengeospatial.org/`

elers focus on that aspect when defining a service description step-by-step. Moreover, separating core service and participant aspects follows proven principles of role modeling and resource allocation, as employed in, e.g., work-flow and process management systems [22]. Such systems define roles and participants independent of the subject of activity (the work-flow or process). Roles are an abstract notion and represent a set of capabilities or permissions. They are assigned to portions of activity, e.g., specific tasks, at design time or deployment time, and are later resolved according to defined rules. This enables systems to manage allocation of individual participants and resources more dynamically by taking into account the context of a specific case or process instance during role resolution.

Elements in the Foundation fall in one of the following three categories:

1. Elements independent of services, e.g., Organization, Person, or Resource.
2. Elements re-used across modules, e.g., Description, Artifact, or Classification.
3. Connecting interfaces: These elements are interfaces that are used to connect elements in one module with many different elements in other modules. A particular example is the link that connects the element Work in the Legal Module (cf. Chapter 10) with the element represented (or subsumed) by Work. This could be an entire Service defined in the Service Module or just an output parameter of a single service function defined in the Functional Module (cf. Chapter 11). In order not to introduce many module dependencies in the referencing module (the Legal Module in this case), an interface is added to the Foundation which is then inherited by the elements that can be referenced. This way there are only dependencies on the Foundation.

Table 13.1 lists the complete set of requirements that were considered during modeling of core service, participant and foundation aspects. The table also briefly outlines how these requirements were implemented in the three modules. A more in-depth explanation is given in the remainder of this section.

The main purpose of the foundational modules of USDL is twofold. On the one hand, they capture information about *Service Agents and Networks* (participants) and *Service Dependencies and Composition* (cf. Section 8.2). On the other hand, the Service Module provides the anchor for linking the different aspects of service description, including the aspects of *Services*, *Service Delivery*, and *Service Consumption*, all of which are covered by other USDL modules. Addressing the requirement of *Cognitive Sufficiency*, this anchoring represents the structural backbone of USDL's "unifying" character. For reasons of getting a quick entire overview when inspecting a service description, all links to other modules (except for the Legal Module) are directed outwards, i.e., a relationship exists from a core service element to a single ("anchoring") element. These elements act as containers for the information captured by their respective module. The Legal Module (cf. Chapter 10) connects to core service elements in the reverse way (via a connecting interface in the Foundation). This method was chosen, because of the way that laws, in this case copyright laws, are written. In copyright law there is a clear distinction between the *work* that can be protected as intellectual property and an entity such as service that

Table 13.1: USDL requirements (cf. Section 8.3) addressed by the Service and Participant Module, as well as the USDL Foundation.

Requirement	Requirement addressed how
Generic Language Requirements	
Conceptualization	Service Module views service as business asset, i.e., independent of particular implementation; definition of types (services, roles, etc.) is kept on higher (abstract) level
Modularity	Service Module and Participant Module are separated along conventional lines; re-use is promoted through moving elements to USDL Foundation
Extensibility	For business networks, the Participant Module supports future specialization in the area of intermediaries.
Service Concept Formation Requirements	
Organizational Embedding	Services are captured on technical as well as business level. The Service Module with its dependency structure allows tracing services back to the organizational structures, resources, and services on business level. The Service Module furthermore allows referencing of external (e.g., organizational) artifacts for service composition.
Cognitive Sufficiency	Service Module provides the anchor for interconnecting different service aspects and external description artifacts. A coarse-grained dependency structure of services referring to other services may be refined on-demand by obtaining the relevant information from other modules.
Service Information Hiding	Service Module permits to publish or hide structure of a service, e.g., in terms of its composition. Furthermore, language-specific artifacts, such as process models can be hidden.
Deployment Symmetry	Service and Participant Module both define structures for services with its participants that are common for most application contexts and service ecosystems. Participants may add information corresponding to their roles to an existing USDL description.

is an embodiment of this work. This does not rule out that both are congruent, and it is common that the entire service is protected and licensed.

Cognitive Sufficiency, i.e., understanding how different aspects of a service relate to each other, is complemented by *Comprehensibility*. This requirement primarily calls for easy to understand representations of the structured USDL model. On a coarse level services can have dependencies to other services. These dependencies may be refined — if needed — by obtaining the relevant information from other modules, e.g., by resolving the function of the service and in a next step the price of the function, etc. Beside the dependency structure, this requirement also motivates supplementing USDL with additional information. This includes structured and unstructured materials that provide further descriptive value, e.g., implementation guides, demonstration videos or reviews by other users. A similar need for connections to external description artifacts exists for the requirement of *Organizational Embedding*. Therefore, the USDL Foundation introduces elements to provide references to external artifacts in a generic way. References can be on the level of entire artifacts, as well as individual artifact elements.

All three modules presented here are mandatory in USDL for the simple reason that there can be no service description without defining a service object that refers at least one participant identified as the provider. Together with the Functional Module (cf. Chapter 11) they form the minimum set of modules that need to be supported in a USDL deployment (cf. Chapter 17). Deployment is the process that comprises the definition of a USDL variant and configuration/adaptation of USDL tools to be used in a particular application context. The structures of the service with its participants are common for different ecosystems into which services can be deployed. Thus, they build the foundation to achieve *Deployment Symmetry* for different application domains. A cloud hoster may for example add information about himself to the intermediaries. He may also make additions to the Pricing Module (see Chapter 9), e.g., adding a certain percentage of the original price for the hosting.

Having this in mind, USDL is specifically designed to be used for a wide range of services. In consequence, it has to satisfy the description requirements of many service types; requirements that potentially are very different. This problem is addressed by a generic and open approach, i.e., an approach that leaves a lot of constraints unspecified. The main reason is that the alternative would have been to introduce specializations of concepts such as Service, Role or Resource, in order to cater for the specific requirements of each concrete type. Done properly, this would have led to the definition of tens or hundreds of individual types, given the fact that the services domain is so vast and can be categorized along so many dimensions. However, it would have been a futile undertaking to find a single set of specializations that is comprehensive and satisfying, and would be accepted as an adequate decomposition. This is not only because of the large number of possible dimensions, but also because many of them overlap. Thus, instead of unnecessarily enlarging the USDL core model (the model presented in this part of the book) with countless subtypes, the intention is that domain-specific USDL variants introduce concrete requirements particular to their domain by means of extensions to the generic USDL. Alternatively, it is possible to just indicate a type through a generic classification mechanism available in the Foundation.

Altogether this approach offers the best basis for introducing USDL as a truly "unifying" service description, as it maintains a generic level, but allows fine-grained scoping of concrete variants for different application domains. Please refer to Chapter 17 for more information about variant management and extensibility. In addition, Chapter 16 discusses governance processes to ensure correct use of these mechanisms during USDL deployment and maintenance.

Even though domain-specific types are the subject of USDL extensions, there are a few important general type dimensions for the categorization of services that have to be addressed in the core service description language. This is due to the fact that these type dimensions influence the handling of a service in almost every service ecosystem. The two dimensions considered in USDL are *(a)* ServiceNature and *(b)* different types of composition structures. As mentioned earlier, ServiceNature refers to the way a service is rendered (manual labor, semi-automated, automated), which usually also influences how it can be requested and accessed. Composition structures, introduced earlier in this section, are concerned with functional (re-)use or commercial packaging and imply dependencies between services. Following the requirement of *Service Information Hiding*, they are captured in USDL on an abstract level that omits any detail regarding how individual services are composed with each other. The only information available is information about which services are part of a composite or bundle and, in the case of composites, what type of functional composition was used. Facilities to reference external artifacts that describe the composition in detail ensure that *Organizational Embedding* can be achieved. Through the dependency mechanism in the Service Module, services can be traced back to the organizational structure, resources, and services at business level.

Since composition details as well as information for *Organizational Embedding*, such as dependencies to other services and resources, are abstracted from the technical details of concrete languages (such as BPEL) and from implementation details, the general language requirement of *Conceptualization* is fulfilled. Furthermore, to additionally allow for *Extensibility*, the Intermediary role is abstract. Thus, USDL is not limited to a predefined set of roles, but different types of intermediaries can partake in the fulfillment of the service (e.g., a Service Hoster, a Service Broker, etc.) as already described in Section 13.2.1.2.

This concludes the overview of the design of the three modules. For a full formal specification please refer to [2], [4] and [3], respectively.

13.3 The Service Module

As previously mentioned, the Service Module forms the central backbone of USDL where all other description aspects are tied together into a "unified" model of services. According to the design principles introduced in Section 13.2, the elements in the Service Module are defined on a generic level without formalizing any details specific to a single application domain. They are also conceptual, meaning they do not suggest of preclude a particular implementation. Figure 13.1 depicts the class diagram of the module. Please note that the diagram does not show the entire model, as connections to other modules are omitted. Please refer to Chapter 8 (Fig. 8.6 on page 215) for a complete overview about module dependencies and other chapters in this part of the book for details on these dependencies.

13.3.1 Overview and Main Constructs

Given the prominent and generic position of service composition, it was a natural choice to architect the core structure with that dimension in mind. Correspondingly, the module's core structure consists of three service concepts that are unified under the umbrella of the abstract element NetworkProvisionedEntity. As suggested by the name, this element subsumes all entities that are subject of provisioning (and delivery and consumption) in Web-based service networks (cf. Chapter 1). It con-

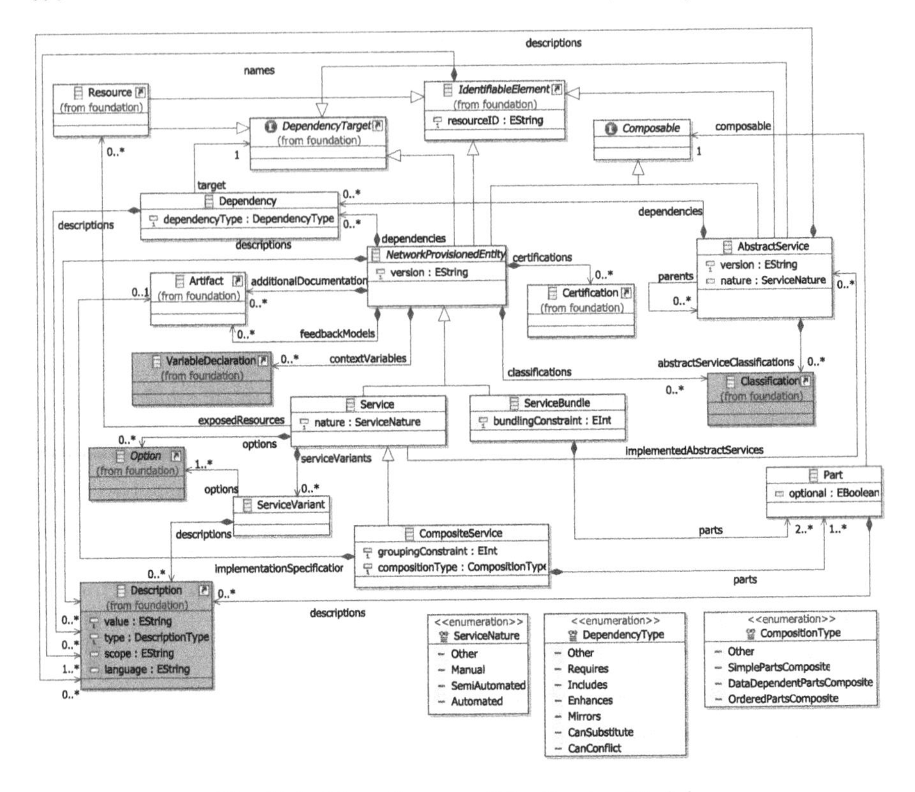

Fig. 13.1: Class diagram of the Service Module.

stitutes the least common denominator among the three concrete service concepts, namely Service, CompositeService and ServiceBundle. All three concepts manifest the following commonalities.

- They each describe an identifiable entity (represented by the class Identifiable-Element), which means it has a unique identifier and at least one name.
- The entity described may specify a version, which indicates its evolution.
- The entity described may be classified/categorized in a defined classification system (classifications), and may hold certifications.
- They may provide additional information about the entity described, either as informal descriptions or through links to external artifacts (additional-Documentation).
- The entity described may manifest one or more dependencies to other services or resources.
- They may link to a number of feedbackModels. These are artifacts that define a model for collecting feedback about the entity described. Please see Chapter 12 for more information about service feedback.

- The entity described may declare one or more variables which are held in its context (contextVariables), either permanently or for the lifetime of a single invocation.

Besides the three concrete service concepts there also exists the concept of AbstractService which is used to represent classes of services, i.e., groups of services that comply with a number of predefined description properties. Services developed to operate in the context of particular functional or organizational structures can be abstracted and carried through into new deployments where they can be concretely instantiated. An AbstractService is itself an IdentifiableElement that has a version, can be classified (abstractServiceClassification) and supplemented with additional descriptions. On top of that, it may contain nearly an entire service description, i.e., it has references to almost all service aspects (not shown in Figure 13.1). This is the template defined by the AbstractService. Concrete instances have to adhere to it completely. AbstractServices may be composed hierarchically by overwriting and extending other abstract services that act as their parents.

The basic definition of the most elemental concrete service concept, Service, is that of a distinct set of capabilities intended to be rendered to a customer, partially or as a whole. It exposes the set of capabilities through one or more technical interfaces, under well-defined service levels and according to particular pricing schemes and license agreements. Services are also of a specific nature meaning that they can either be performed manually (by a human), automated (e.g., by computers without human interaction), or semi-automated. Finally, some services act as facades to access and manage resources (type of resources indicated by exposedResources).

Because the service model by default allows for flexible definition of consumable functionality and delivery constraints, it is sometimes desirable to clearly define what combinations of capabilities, price plans, etc. are available. This is possible by giving a number of alternative options, e.g., functional options or pricing options, which essentially represent subsets of the overall set of service aspects. In order to furthermore capture dependencies between different kinds of options, one or more ServiceVariant objects can be defined that group options into predefined packages, e.g., a certain set of capabilities with a certain service level profile.

The second concrete service concept, CompositeService, is itself a Service in the sense that it provides capabilities to consumers. Unlike Service, it uses and aggregates other Services, ServiceBundles or AbstractServices available in the service ecosystem (subsumed under the interface Composable). The composition of these components (or parts) may be arbitrary, i.e., does not follow a particular order (SimplePartsComposite), or it may be governed at its highest level by a process (OrderedPartsComposite). Components may also be composed loosely on the basis of data producer/consumer relationships (DataDependentPartsComposite). Details of the composition are not captured in USDL, but may be provided through referencing a formal specification of the composition (implementationSpecification). The only pieces of information available are: what the components are, how many of them have to be included, and which components are mandatory (feature optional of Part). In the case of dynamic composition, it is possible that only some of the referenced parts are included in the final composite. A groupingConstraint indicates whether it has to be one (1), all (-1), or some (the number given) part(s).

Part represents the concept of a slot within a composition structure. Using this abstraction layer on top of the actual Service, ServiceBundle or AbstractService (component of a composition structure) allows for fine-grained configuration of a composition, e.g., by making parts optional. It should be pointed out that general composition constraints (of a CompositeService or ServiceBundle) always take precedence over part configurations. For example, if the composition constraint is set to all (-1), it is not possible to omit optional parts. In general, it is envisioned that part configurations refine overall composition constraints and are not defined in a conflicting way.

The third concrete service concept is ServiceBundle. In contrast to CompositeServices, which provide new functionality by combining the capabilities of their parts, ServiceBundles aggregate Services, ServiceBundles or further AbstractServices for commercial reasons. In other words, bundling is defined as offering several services for sale as one combined package. It is motivated by, e.g., competitive pricing or combining a low-selling service with a highly demanded service in order to drive up sales figures. The degree of collective delivery/consumption of the bundled items can be limited through a bundlingConstraint (similar to groupingConstraint of CompositeService).

The element Dependency captures information about the relationships that exist between a Service or ServiceBundle and another Service, ServiceBundle, AbstractService or Resource. This element expresses various generic types of dependencies such as Requires, Includes and canSubstitute, amongst others. Each DependencyType has particular semantics. For example, a dependency of type Requires implies that the operation of the Service or ServiceBundle depends on the operation of the referenced Service, ServiceBundle, AbstractService, or Resource; consequence is that the referenced object has to be made available by the consumer (i.e., has to be ordered separately). Please refer to [2] for an explanation of all DependencyTypes.

13.3.2 Illustrative Example

Listing 13.1 depicts an excerpt of the service description for service *2PL Airline Manager* (cf. Section 8.7). It shows general service properties and references to other service aspects (covered in other USDL modules). It is an automated service (nature). It is available in its first major version (1.0) that was published on March 31st, 2011, at 8am UCT+2h (feature publicationTime). There is one long description with information about the service, which is supplemented by an external document about certification guidelines for client implementations (feature additionalDocumentation).

Listing 13.1: Core properties of example service 2PL Airline Manager.

```
<service>
  <resourceID>http://usdlbook.org/services/2pl-ship-manager</resourceID>
  <version>1.0</version>
  <nature>Automated</nature>
  <names>
    <description>
      <value>2PL Airline Manager</value>
      <type>name</type>
    </description>
  </names>
  <publicationTime xsi:type='foundation:AbsolutePointInTime'>
      <value>2011-03-31T08:00:00.000+02:00</value>
  </publicationTime>
  <descriptions>
    <description>
      <value>2PL Airline's Manager Software makes business shipping
            easier and more efficient. ...</value>
      <type>freetextLong</type>
      <language>en</language>
    </description>
  </descriptions>

  <additionalDocumentation>
    <artifact>
      <type>UsersManual</type>
      <mimeType>application/pdf</mimeType>
      <uri>https://www.2pl.com/downloads/pdf/CertificationGuidelines.pdf</uri>
      <descriptions>
        <description>
          <value>Guidelines for certifying a client implementation.</value>
          <type>freetextShort</type>
          <language>en</language>
        </description>
      </descriptions>
    </artifact>
  </additionalDocumentation>

  <provider> <!-- Participant Module --> </provider>
  <pricePlans> <!-- Pricing Module --> </pricePlans>
  <serviceLevelProfiles> <!-- Service Level Module --> </serviceLevelProfiles>
  <capabilities> <!-- Functional Module --> </capabilities>
  <technicalInterfaces> <!-- Technical Module --> </technicalInterfaces>
  <interactionProtocols> <!-- Interaction Module --> </interactionProtocols>

</service>
```

The second example depicted in Listing 13.2 and shows how 2PL Airline bundles services into an *International Economy Bundle*. The bundle consists of the automated *2PL Airline Manager* service and the manual *International Economy* airfreight service. Both services have to be included as indicated by bundlingConstraint (set to -1).

Listing 13.2: Bundling of 2PL Airline example services.

```
<!-- 2PL Airline Manager -->
<service>
  <resourceID>http://usdlbook.org/services/2pl-ship-manager</resourceID>
  ...
</service>

<!-- 2PL Airline International Economy -->
<service>
  <resourceID>http://usdlbook.org/services/2pl-intl-economy</resourceID>
  <nature>Manual</nature>
```

```
11    ...
12  </service>
13
14  <!-- 2PL Airline International Economy Bundle -->
15  <serviceBundle>
16    <resourceID>http://usdlbook.org/bundles/2pl-intl-economy-1</resourceID>
17    ...
18    <bundlingConstraint>-1</bundlingConstraint>
19    <parts>
20      <part>
21        <optional>false</optional>
22        <composable>http://usdlbook.org/services/2pl-ship-manager</composable>
23      </part>
24      <part>
25        <optional>false</optional>
26        <composable>http://usdlbook.org/services/2pl-intl-economy</composable>
27      </part>
28    </parts>
29  </serviceBundle>
```

13.4 The Participant Module

The Participant Module connects provisioned entities, i.e., Services or Service-Bundles, with the participants that are involved in their provisioning, delivery and consumption. In this context participants fulfill a particular role. The Section 13.2.1.2 introduced a set of general roles identified as relevant for service ecosystems. It goes beyond the set of roles usually considered in traditional service-oriented architectures (SOA), which is one of the reasons why USDL dedicates a separate module to the modeling of roles. The other reason was outlined in Section 13.2 and concerns the level of abstraction applied to modeling. Keeping the model on an abstract level ensures that it is generic enough to cover a wide range of application scenarios.

These two aspects, extended role model and high level of abstraction, formed the Participant Module. Figure 13.2 depicts the class diagram of the module.

13.4.1 Overview and Main Constructs

The central concept of the Participant Module is that of a Role. It is an abstract element that is extended by a number of concrete roles. These are the high-level roles identified earlier in this book (cf. Section 8.2.2). Each one of them can be defined either in an abstract way, i.e., only using textual descriptions, or in a concrete way by connecting it to a participant (abstract element Agent in USDL). Enabling an abstract definition allows to express certain requirements before a Role is actually resolved and an enacting participant is determined. Such requirements are, for example, that a certain role is necessary for the delivery of a service, or that the en-

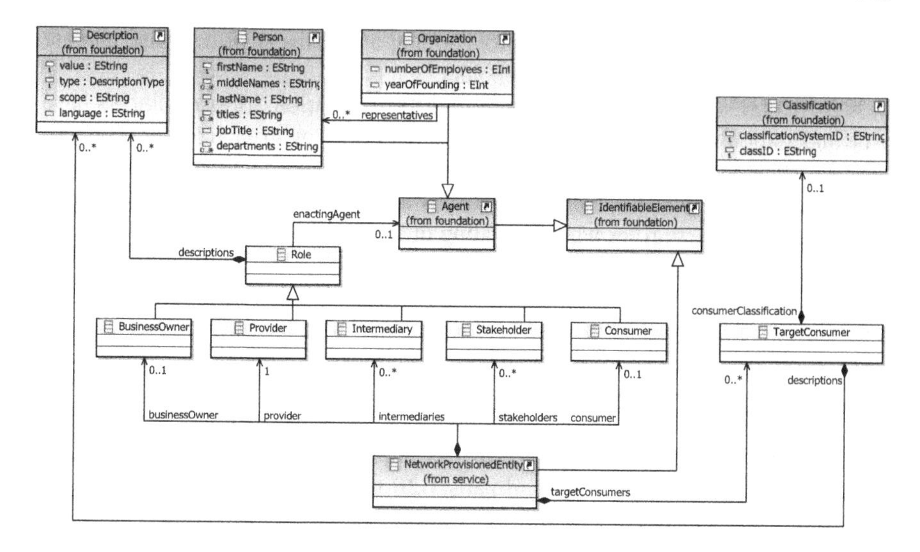

Fig. 13.2: Class diagram of the Participant Module.

acting participant has to comply with constraints and obligations. The default case, however, is that a role is defined concretely, i.e., a participant exists.

In the process of provisioning, delivery and consumption of services through service ecosystems there is one participant that holds governance and operational responsibility for a service. This participant is commonly referred to as service Provider. It controls how the service is provisioned to consumers, e.g., what are the organizational and system resources used, or how it is implemented. In most cases the provider will also act as the trading partner to consumers and define business constraints of delivery. However, there are scenarios in which this function is performed by another legal entity — here called a BusinessOwner. This could be, for example, a national subsidiary of a multi-national organization, where the subsidiary sells the service and the parent organization provides it.

Especially in diversified service ecosystems, there are often entities other than the provider that hold stakes in a service. For example, composite services and service bundles are aggregations that comprise services from different providers. Each aggregated provider performs part of the aggregation and hence becomes a Stakeholder. This is due to the fact that it is the providers, who largely control the terms of engagement with aggregators regarding the re-use or re-purposing of their services. In other words, they have a certain influence on the composite services and service bundles that re-use their services. Further examples of stakeholders include regulation bodies, such as governments or industry associations, which have the authority to prescribe certain aspects of service delivery. There are also third-party providers of delivery functions (e.g., billing or authentication) that can be orchestrated with a service enabling the outsourcing of these functions. All these participants are associated with dedicated parts of the provisioning and delivery of a service and are summarized here under the term stakeholder.

A particular subset of Stakeholders are Intermediaries. Intermediaries have to be understood as one possible set of architectural roles, with each role conceptualizing a certain behavior or mode of operation requiring a specific set of applications and tools. Participants that facilitate delivery and consumption of services act in one or several of these roles while operating in service ecosystems. The Participant Module clearly distinguishes between a pure Stakeholder and an Intermediary, in order to capture the difference of purpose associated with the stake. Stakeholders can influence delivery of services, whereas intermediaries facilitate delivery.

The *Service Hoster* is an example for an intermediary that catalogues special types of services, namely infrastructure-as-a-service and platform-as-a-service offerings (commonly termed cloud computing services). Likewise, the *Service Gateway* is a specific intermediary that provides interoperability through cataloguing and interfacing with a choice of a 3rd-party B2B gateways, which provide services such as message translation and store-forward processing. The *Service Aggregator* provides added value by packaging and combining services. In addition, the *Service Broker* cares for central service publication, discovery, and ordering. Finally, the *Service Channel Maker* is positioned at the consumer end of the service provisioning chain where services are channeled into user environments and consumed. [5]

Finally, on the consumer side, the model defines two elements: Consumer and TargetConsumer. TargetConsumer describes potential user groups of a service, which offers a means to advertise in targeted manner. The definition of the user group is done either through referencing an entry in a predefined taxonomy (classifications) or via textual descriptions. Consumer is a specific role that represents the participant consuming a service. In most cases it will be abstract, i.e., with no particular participant assigned, and will be used to specify requirements consumers must meet, e.g., comply with certain service levels. It is only in those cases where a service description becomes part of contracting that a concrete consumer can be identified.

13.4.2 Illustrative Example

An example of how to use different participant roles is shown in the service description excerpt depicted in Listing 13.3. It is the *Lead Logistics* service composition offered by 4PL (cf. Section 8.7). To be specific, it is actually a version of the service that is offered in a German logistics service ecosystem. It consists of three services: 3PL's *Freight Forwarding* composite service, *Certification* provided by 4PL and *Customs Clearance* performed by Russia's customs agency. Only 3PL's service is mandatory. The other two services are optional, as they depend on the type of cargo, i.e., not everything needs to be cleared and certified. Accordingly, the groupingConstraint of the composition is set to 1.

The participant role setup for services *Freight Forwarding* and *Customs Clearance* is simple. They each have a provider only (3PL and Russian customs, respectively). *Certification* is provided by 4PL — a Russian company, but sold in the German market by 4PL's German subsidiary. This makes the subsidiary the service's businessOwner, while the employees of the Russian parent company do the work.

With 4PL's composite service, this is a bit different. Employees of the German subsidiary are actually organizing the shipment, invoking the individual parts and interacting with them. Consequently, it is the subsidiary that is listed as the provider. However, because Lead Logistics is a composite service, there exists a relationship to the parent company. 4PL, next to 3PL and Russian customs, is a stakeholder of the service, as it provides one of its components.

The description of the different participants (agent entries) are given for completeness. Please refer to Section 13.5 for a comprehensive example of participant description.

Listing 13.3: Bundling of 4PL example service.

```
<!-- 3PL Freight Forwarding -->
<compositeService>
  <resourceID>http://usdlbook.org/services/3pl-freight-fwd</resourceID>
  <nature>Manual</nature>
  <provider>http://usdlbook.org/organizations/3pl</provider>
  ...
</compositeService>

<!-- GOST-R Certification -->
<service>
  <resourceID>http://usdlbook.org/services/4pl-gost-r</resourceID>
  <nature>Semi-Automated</nature>
  <provider>http://usdlbook.org/organizations/4pl</provider>
  <businessOwner>http://usdlbook.org/organizations/4pl-de</businessOwner>
  ...
</service>

<!-- Customs Clearance -->
<service>
  <resourceID>http://usdlbook.org/services/customs-russia</resourceID>
  <nature>Manual</nature>
  <provider>http://usdlbook.org/organizations/customs-russia</provider>
  ...
</service>

<!-- 4PL Lead Logistics -->
<compositeService>
  <resourceID>http://usdlbook.org/bundles/4pl-lead-logistics</resourceID>
  ...
  <provider>http://usdlbook.org/organizations/4pl-de</provider>
  <stakeholders>
    <stakeholder>http://usdlbook.org/organizations/3pl</stakeholder>
    <stakeholder>http://usdlbook.org/organizations/4pl</stakeholder>
    <stakeholder>http://usdlbook.org/organizations/customs-russia</stakeholder>
  <stakeholders>
  ...
  <groupingConstraint>1</groupingConstraint>
  <compositionType>OrderedPartsComposite<compositionType>
  <parts>
    <part>
      <optional>false</optional>
      <composable>http://usdlbook.org/services/3pl-freight-fwd</composable>
    </part>
    <part>
```

```
45        <optional>true </optional>
46        <composable>http :// usdlbook . org / services /4 pl−gost−r </composable>
47      </part>
48      <part>
49        <optional>true </optional>
50        <composable>http :// usdlbook . org / services / customs−russia </composable>
51      </part>
52    </parts>
53  </compositeService>
54
55  <agent>
56    <resourceID>http :// usdlbook . org / organizations /4 pl </resourceID>
57    ...
58  </agent>
59  <agent>
60    <resourceID>http :// usdlbook . org / organizations /4 pl−de</resourceID>
61    ...
62  </agent>
63  <agent>
64    <resourceID>http :// usdlbook . org / organizations / customs−russia </resourceID>
65    ...
66  </agent>
67  <agent>
68    <resourceID>http :// usdlbook . org / organizations /3 pl </resourceID>
69    ...
70  </agent>
```

13.5 The Foundation

As outlined in Section 13.2, the Foundation contains a number of different base elements, ranging from elements re-used across two or more modules to elements independent of service. This section gives an overview of these elements and presents a few examples. A detailed specification can be found in [3].

13.5.1 Overview and Main Constructs

13.5.1.1 Time and Location

Among the most universal elements, which are independent of service description, are time and location. Both concepts are used in the context of service description, for instance, to express temporal and geographical availability, i.e., the time when and location where a service can be requested and delivered. Figure 13.3 and Figure 13.4 depict time and location elements, respectively.

The Time concept in USDL offers a number of concrete types to capture temporal parameters. Four main types are distinguished, which together cover many known use cases. The more common ones are TimeInstant and TimeInterval. TimeInstants are either defined as absolute points in time, where value is a representation according to ISO 8601 [11], or relative points in time. The latter is expressed relative to a ReferenceEvent and may either use a duration (the sum of

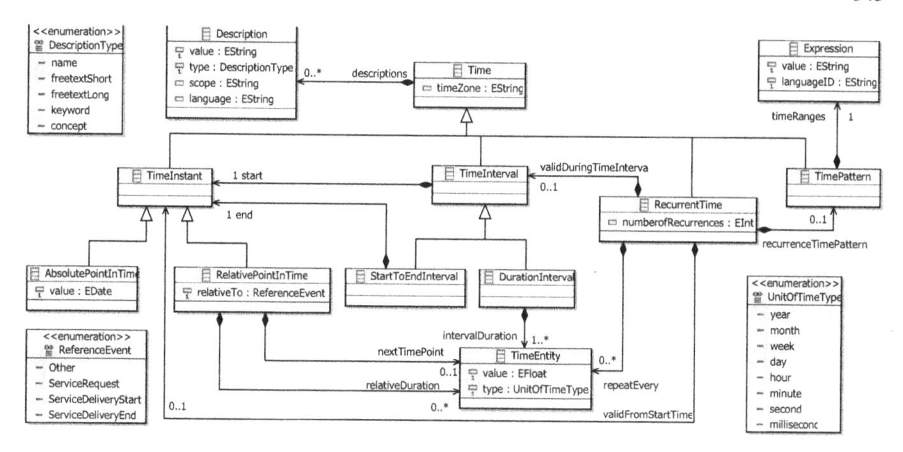

Fig. 13.3: Excerpt of the class diagram of the Foundation: time elements.

TimeEntity objects, e.g., two weeks and three days) or the next occurrence of a specific temporal entity (e.g., next April) to mark the time instant. TimeIntervals may also use durations (DurationInterval) to define a time span from a specific time instant (start). Alternatively, StartToEndInterval uses two time instants to mark the time span.

The third concrete type of time is RecurrentTime and defines a set of time instants by means of recurrence. Again, there are several ways to express recurrence. If the recurrence is periodic, a fixed duration is specified (repeatEvery). If it is not, a more complex TimePattern can be defined. The number of recurrences is either known and can be given in combination with a start time instant (validFromStartTime), or unknown, in which case an interval marks the time span during which recurrence takes place (validDuringTimeInterval).

AbsolutePointInTime, the two concrete time intervals and RecurrentTime make up a portion of the model that is equivalent to what can be expressed with ISO 8601. USDL adds to that RelativePointInTime and TimePattern. TimePattern is the fourth main type and is meant to express more elaborate temporal properties, e.g., opening times of a shop. Patterns are defined using a formal expression in an arbitrary expression language. Finally, to ensure that time properties are unambiguous around the world, each concrete type may specify a timezone identifier (e.g., +1000 or CET).

The Location concept in USDL distinguishes between real physical (geographic) locations, e.g., street address, and electronic locations, e.g., a telephone number or a URL. Both types are understood to identify one or more locations in an address space, which is why they are subsumed under one umbrella concept. For real physical locations there exist two encodings of a single address space, the global union of postal addresses and the global geographic coordinate system (limited here to WGS84,[5] for the sake of simplicity). It is possible to do a mapping from one to the

[5] World Geodetic System 1984

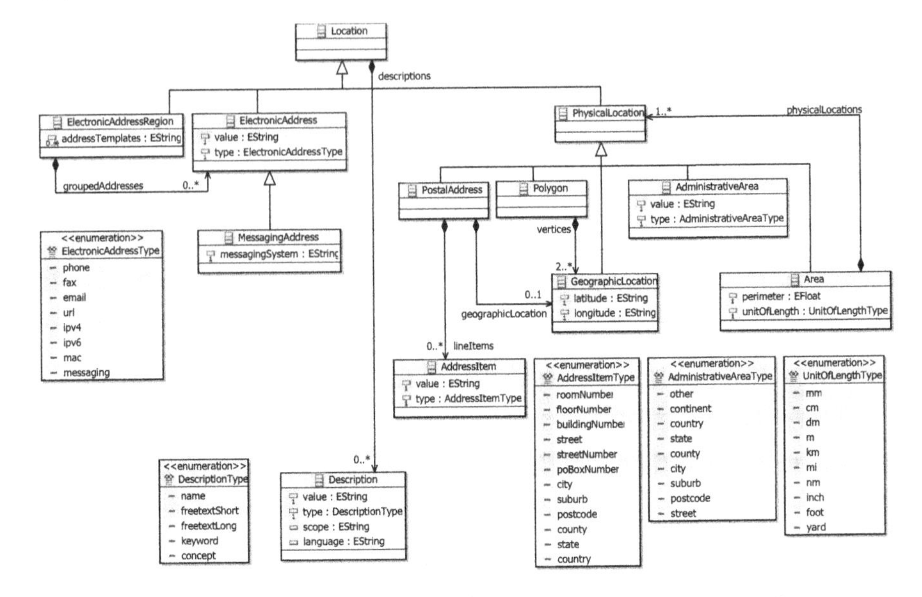

Fig. 13.4: Excerpt of the class diagram of the Foundation: location elements.

other, and they are sometimes referenced implicitly and simultaneously. For example, when talking about a country one could either think of it as a self-contained region in the global space of street addresses, or as a defined set of geo-coordinates. Please note that physical locations are limited to locations on the surface of the earth. USDL defines the following concrete types of PhysicalLocation.

- GeographicLocation: A single point on the earth's surface, defined by WGS84 coordinates.
- PostalAddress: A street address that is expressed as a set AddressItem objects. A mapping to GeographicLocation may be provided.
- Polygon: An area on the earth's surface delimited by the convex hull that is defined by 2 or more vertices (GeographicLocations).
- AdministrativeArea: An area on the earth's surface that is delimited by a political definition or claim of ownership, e.g., country, state (province, territory), city, or postal code area.
- Area: An area on the earth's surface that is defined by one or more reference locations and a perimeter around these locations (product of perimeter value and unitOfLength).

Opposite to physical locations, the realm of electronic addresses has more than just one address space and often many encodings for a single one of them. This is due to the fact that electronic addresses are created with information technology and are thus virtual by default. Because of the variety of address spaces, it is necessary to identify that space when specifying an ElectronicAddress. USDL defines a simple set of address types, which in most cases is sufficient to uniquely identify

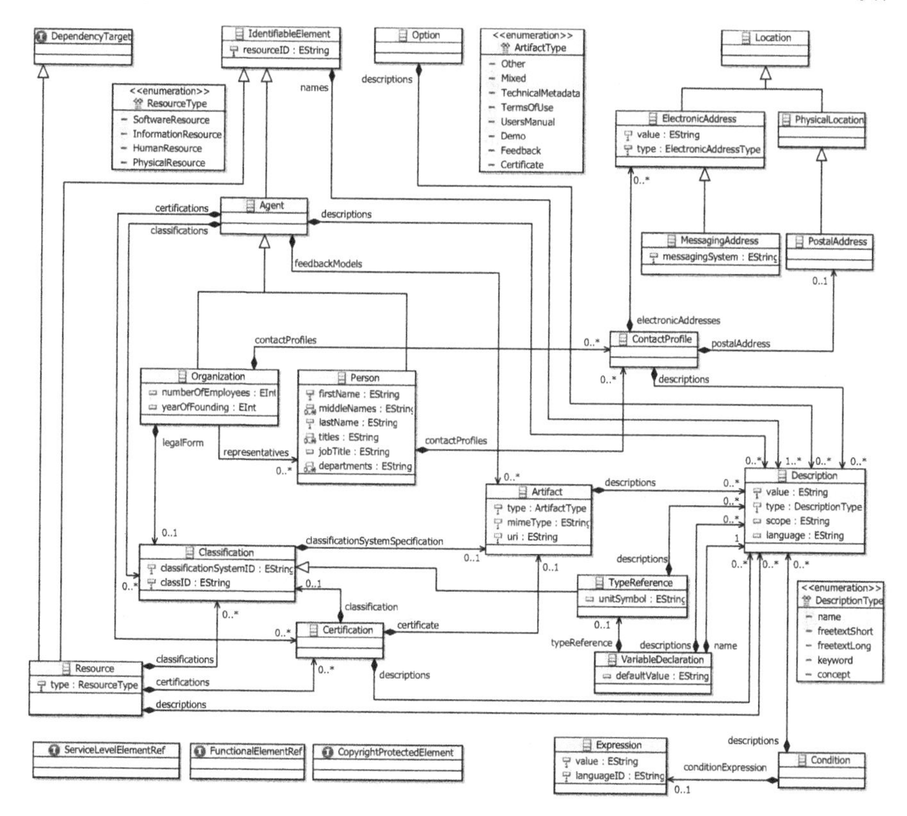

Fig. 13.5: Excerpt of the class diagram of the Foundation: generic elements, agents, resources and connecting interfaces.

a space (e.g., phone number, email, IP v4). However, in the case of instant messaging the type is not enough and an additional identifier of the messaging system has to be supplied. Lastly, electronic addresses can be aggregated into sets or regions, either through explicit reference (groupedAddresses) or through specification of templates in the form of regular expressions (addressTemplates).

13.5.1.2 Generic Description Elements

The USDL Foundation defines a number of generic description elements that are often used from other modules. They are depicted in Figure 13.5 and introduced in the following.

One of the most widely used elements is Description. It is intended to attach typed and scoped textual descriptions to other USDL elements. Typing relates to how the textual description that is given as the value should be understood, e.g., name, short free-text or keyword. Scoping can be used to limit the validity of the textual description to a specific space. It is useful, for instance, to define the context wherein a name represents a unique identifier. Another option is to specify a language identifier to indicate the language of the textual description. This enables a form of internationalization by attaching multiple Description objects of the same type and scope, but different languages, to a single USDL element instance.

Concerning the aspect of identification and naming, it should be noted that many elements in USDL provide a description of real world objects (material and immaterial). One inherent characteristic of such real world objects is that they can be distinguished from another. In order to capture their individuality, Identifiable-Element provides means to make them uniquely identifiable. The class is abstract and requires its concrete subtypes to specify one global identifier (resourceID), to be used mainly by systems and tools, and at least one name. Names are Description objects of type name and usually represent identifiers introduced by humans. There can be several names that are used synonymous, with probably some of them limited to a certain scope.

In addition to generic descriptions and identifiers, there are several elements in the Foundation that can be used to reference external resources. One of them is the element Artifact. It represents a link or pointer to a resource that can be located with a URI. The content type of the resource (type) and its encoding (mimeType) are given to ensure correct lookup and handling during resolution of the URI, while a set of descriptions may give detailed information about the resource. Another referential element is Classification. It identifies a specific entry (or class) in a classification system, e.g., a taxonomy or ontology. The classification system is uniquely identified (classificationSystemID) and a link to its formal definition or implementation can be attached using an Artifact object. A special type of classification is TypeReference, which identifies an element in a type system (a classification system of data types).

While ordinary classifications are "self-assigned" with no sure method the check whether they are true, a Certification provides a classification that was verified by an independent party. The proof that the classification in the certification system is not part of USDL itself, but can be given in form of a certified, e.g., digitally signed, artifact (the certificate).

The last of the referential elements in the Foundation of USDL is Variable-Declaration. It represents variables that are defined in the context of an entity, e.g., a service whose functions read and manipulate such variables. The represented variable has to have a single name and may be typed (using TypeReference). It may also specify a defaultValue and descriptions that give more information about the purpose and use of the variable. No assumptions are made about the lifetime of the variable. It could exist permanently, i.e., as long as the service is deployed and available, or just for the lifetime of a single service invocation.

The set of generic description elements is completed by the elements Expression, Condition and Option. Expression can be used to specify formal expressions in an arbitrary language (identified by languageID). Condition represents an expression that is used to express state. Option defines an abstract super-type that is meant to be extended in future by concrete option elements. The concept of option is introduced to capture subsets of particular service aspects, e.g., functionality or pricing, in order to limit the possible combinations of individual elements that can be contracted and consumed. For instance, a service provides three capabilities and the service provider wants to express that users can only consume them in two combinations, because two capabilities are interchangeable (two versions of a particular functionality). Supplying two options, the provider can rule out that the wrong combination is selected.

13.5.1.3 Agents and Resources

In order to describe the participants that are involved in the provisioning, delivery and consumption of services, the Foundation introduces the abstract element Agent, which is an IdentifiableElement (see Figure 13.5). The term "agent" is used to indicate that these are entities that act independently, and they are also understood in the legal sense as legal persons. USDL distinguishes two concrete types of agents: Organization and Person. Both share a number of properties. For example, they may be categorized into multiple classification systems and may hold multiple certifications, e.g., ISO9001 [13] for organizations and CISA [7] for persons. They are also entities that can receive feedback from other participants and therefore may reference a number of feedback models (cf. Chapter 12).

An Organization is an institutionalized legal person. It is mainly characterized by its name (or names), its legalForm (business type in a country-specific system of legal entities) and its contact data. The latter is grouped into profiles that may contain multiple electronic addresses and up to one postal address. Each ContactProfile is meant to represent the official contact information for a specific location or branch of the organization. Information about this location may be given as additional descriptions. Similarly, additional information about the organization itself can be attached to an Organization object via descriptions. Finally, persons can act as representatives of organizations.

A Person is an individual natural person. She or he has a firstName, lastName, and potentially many middleNames. She or he may also specify her or his dateOfBirth, jobTitle. In addition, she or he may hold a number of official titles, e.g., "Prof.," "Ph.D," or "Sir," and may be affiliated with one or more functional groups in an organization (departments). Similar to organizations, a person may have multiple contactProfiles for different purposes, for instance, one profile for work and one personal profile. A person's displayable name and aliases can be expressed with feature names, and additional information, such as biographies, can be attached using feature descriptions.

Next to active agents there is also an element to describe passive objects used in the context of service provisioning and delivery. To be specific, the element Resource captures information about types or classes of objects. These resource classes are uniquely identifiable entities (superclass IdentifiableElement) and are categorized into one of four general types (software, information, human, or physical). They can be further classified, as well as hold certifications.

13.5.1.4 Connecting Interfaces

Connecting interfaces are used as a non-intrusive method of referencing when the referenced target can be one of many USDL elements. The intent is to reduce module dependencies which would otherwise be introduced, because targeted elements are likely to be scattered across multiple modules. The following interfaces are defined (see also Figure 13.5).

- DependencyTarget is inherited by elements that describe entities upon which a service can depend (cf. Section 13.3).
- FunctionalElementRef is inherited by elements that describe functionality which may be exposed by elements of a technical interface (cf. Chapter 11).
- ServiceLevelElementRef is inherited by elements that describe aspects of a service to which service levels may apply (cf. Chapter 12).
- CopyrightProtectedElementRef is inherited by elements that describe aspects of a service to which copyright protection may apply (cf. Chapter 10).

13.5.2 *Illustrative Example*

The following illustrates the use of a few selected elements of the Foundation. More examples are given in Chapters 9 and 10, as well as Section 13.3 of this chapter.

Listing 13.4 depicts a fragment of the description of service *Lead Logistics* (see Section 8.7). It shows the part where the organizational data of the service provider, i.e., 4PL's German branch (cf. Listing 13.3), is defined. Organization is a concrete type of Agent, a fact expressed by feature xsi:type. The company has the official name "4PL GmbH" (feature names) and its business type, "GmbH," is formally specified through a Classification (feature legalForm). 4PL GmbH has 63 employees and was founded in 1993 (features numberOfEmployees and yearOfFounding). The company can be contacted at its office, which is located in Berlin at approximately 55.7237 degrees latitude and 37.6826 degrees longitude. There is one person representing the company (feature representatives; see also Listing 13.5).

Listing 13.4: Definition of 4PL organizational data: 4PL German branch.

```
1  <agent xsi:type="foundation:Organization">
2    <resourceID>http://usdlbook.org/organizations/4pl-de</resourceID>
3    <names>
4      <description>
5        <value>4PL GmbH</value>
6        <type>name</type>
7      </description>
8    </names>
9    <numberOfEmployees>63</numberOfEmployees>
10   <yearOfFounding>1993</yearOfFounding>
11   <legalForm>
12     <classificationSystemID>German Business Entity Types</
           classificationSystemID>
13     <classID>GmbH</classID>
14   </legalForm>
```

```
15    <representatives>
16      <representative>http://usdlbook.org/persons/4pl-yuri</representative>
17    </representatives>
18    <contactProfiles>
19      <contactProfile>
20        <postalAddress>
21          <lineItems>
22            <addressItem>
23              <value>Logistik Strasse</value>
24              <type>street</type>
25            <addressItem>
26            <addressItem>
27              <value>43</value>
28              <type>streetNumber</type>
29            <addressItem>
30            <addressItem>
31              <value>Berlin</value>
32              <type>city</type>
33            <addressItem>
34            <addressItem>
35              <value>10123</value>
36              <type>postcode</type>
37            <addressItem>
38            <addressItem>
39              <value>Germany</value>
40              <type>country</type>
41            <addressItem>
42          </lineItems>
43          <geographicLocation>
44            <latitude>55.723727</latitude>
45            <longitude>37.682562</longitude>
46          </geographicLocation>
47        </postalAddress>
48        <descriptions>
49          <description>
50            <value>4PL German Branch Office</value>
51            <type>freetextShort</type>
52            <language>en</language>
53          </description>
54        </descriptions>
55      </contactProfile>
56    </contactProfiles>
57  </agent>
```

The service description fragment in Listing 13.5 depicts the definition of a person, who is a representative of 4PL's German branch. It is another part of the description of service *Lead Logistics*. As indicated by feature xsi:type, Person is also a special type of Agent. The person is called "Yuri Petrovich Andropov;" a name that consists of a firstName, a middleName and a *lastName*. In addition to individual name composites, a label for display is given in feature *names*. Yuri is the sales manager and has a Ph.D. (features jobTitle and titles). He moreover holds a certified accreditation of the status 'Expert Customs Broker' from the German ministry of trade, as specified in feature certificates. A PDF copy of the certificate can be obtained from the German Website of 4PL. Lastly, a phone number and email address are listed as Yuri's business contact details (feature contactProfiles specifying two ElectronicAddress objects).

Listing 13.5: Definition of 4PL organizational data: representative of 4PL.

```
1  <agent xsi:type="foundation:Person">
2    <resourceID>http://usdlbook.org/persons/4pl-yuri</resourceID>
3    <names>
```

```
4         <description>
5           <value>Yuri P. Andropov</value>
6           <type>name</type>
7         </description>
8       </names>
9       <firstName>Yuri</firstName>
10      <middleNames>
11        <middleName>Petrovich</middleName>
12      </middleNames>
13      <lastName>Andropov</lastName>
14      <titles>
15        <title>Ph.D.</title>
16      </titles>
17      <jobTitle>Sales Manager</jobTitle>
18      <certifications>
19        <certification>
20          <certificate>
21            <type>Certificate</type>
22            <mimeType>application/pdf</mimeType>
23            <uri>http://www.4pl.de/company-data/yuria/agent-certification.pdf</uri>
24          <certificate>
25          <classification>
26            <classificationSystemID>urn:trade-office:customs-agents:de</
                  classificationSystemID>
27            <class>certified-expert<class>
28          <classification>
29          <descriptions>
30            <description>
31              <value>'Expert Customs Broker' accreditation from ministry of trade</
                    value>
32              <type>freetextShort</type>
33            <description>
34          <descriptions>
35        <certification>
36      <certifications>
37      <contactProfiles>
38        <contactProfile>
39          <electronicAddresses>
40            <electronicAddress>
41              <value>+49 030 999 1234</value>
42              <type>phone</type>
43            </electronicAddress>
44            <electronicAddress>
45              <value>yuri.a@4pl.de</value>
46              <type>email</type>
47            </electronicAddress>
48          </electronicAddresses>
49        </contactProfile>
50      </contactProfiles>
51    </agent>
```

13.6 Conclusion

This chapter has presented the modeling foundation of USDL. The design of the Service Module, Participants Module, and the Foundation have been influenced by the state of the art from the areas of service composition, of actors that partake in the fulfillment and delivery of services, of time and location modeling, and of organizational modeling. Furthermore, it has been outlined how the generic language and service concept formation requirements from Chapter 8 have been addressed. From each module, the concepts and entities have been shown and exemplified.

References

1. Service component architecture. Technical report, BEA Systems, Cape Clear Software, IBM, Interface21, IONA Technologies PLC, Oracle, Primeton Technologies Ltd, Progress Software, Red Hat Inc., Rogue Wave Software, SAP AG, Siebel Systems, Software AG, Sun Microsystems, Sybase, TIBCO Software Inc., Nov 2006.
2. A. Barros, C. Baumann, A. Charfi, M. Flügge, S. Heinzl, T. Kiemes, U. Kylau, F. Marienfeld, N. May, O. Müller, F. Novelli, D. Oberle, J. Pattberg, P. Robinson, B. Schmeling, W. Theilmann, H. Witteborg, J. Finzen, A. Horch, and M. Kintz. Unified Service Description Language (USDL) — Service Module. Technical Report Version 3.0, Milestone M5, SAP Research, May 2011. Available at www.internet-of-services.com.
3. A. Barros, C. Baumann, A. Charfi, M. Flügge, S. Heinzl, T. Kiemes, U. Kylau, F. Marienfeld, N. May, O. Müller, F. Novelli, D. Oberle, J. Pattberg, P. Robinson, B. Schmeling, W. Theilmann, H. Witteborg, J. Finzen, A. Horch, and M. Kintz. Unified Service Description Language (USDL) — Foundation. Technical Report Version 3.0, Milestone M5, SAP Research, May 2011. Available at www.internet-of-services.com.
4. A. Barros, C. Baumann, A. Charfi, S. Heinzl, T. Kiemes, U. Kylau, N. May, O. Müller, F. Novelli, D. Oberle, P. Robinson, B. Schmeling, W. Theilmann, and H. Witteborg. Unified Service Description Language (USDL) — Participants Module. Technical Report Version 3.0, Milestone M5, SAP Research, May 2011. Available at www.internet-of-services.com.
5. A. Barros and U. Kylau. Service delivery framework — an architectural strategy for next-generation service delivery in business network. In P. Kellenberger, editor, *Proceedings 2011 Annual SRII Global Conference SRII 2011, 30 March - 2 April 2011, San Jose, California, USA*, pages 47–58. IEEE Computer Society Conference Publishing Services (CPS), 2011.
6. E. Bottazzi and R. Ferrario. Preliminaries to a DOLCE ontology of organisations. *Intl. Journal Business Process Integration and Management*, 4(4):225–238, 2009.
7. D. L. Cannon. *CISA: Certified Information Systems Auditor Study Guide*. Sybex, 3rd edition, 2011.
8. J. C. Fagan. Mashing up multiple web feeds using Yahoo! pipes. *Computers in Libraries*, 27(10), 2007.
9. M. Hepp, J. Leukel, and V. Schmitz. A quantitative analysis of product categorization standards: content, coverage, and maintenance of eCl@ss, UNSPSC, eOTD, and the RosettaNet technical dictionary. *Knowl. Inf. Syst.*, 13(1):77–114, 2007.
10. International Organization for Standardization. *ISO 19108:2002, Geographic information - Temporal schema*, 2002. online at: http://www.iso.org/iso/catalogue_detail?csnumber=26013.
11. International Organization for Standardization. *ISO 8601:2004, Data elements and interchange formats - Information interchange - Representation of dates and times*, 2004. online at: http://www.iso.org/iso/catalogue_detail?csnumber=40874.
12. International Organization for Standardization. *ISO 19136:2007, Geographic information - Geography Markup Language (GML)*, 2007. online at: http://www.iso.org/iso/catalogue_detail?csnumber=32554.
13. International Organization for Standardization. *ISO 9001:2008, Quality management systems - Requirements*, 2008. online at: http://www.iso.org/iso/catalogue_detail?csnumber=46486.
14. N. Kavantzas, D. Burdett, G. Ritzinger, T. Fletcher, Y. Lafon, and C. Barreto (eds.). Web Services Choreography Description Language Version 1.0. Candidate Recommendation 9 November 2005, W3C, 2005. online at: http://www.w3.org/TR/ws-cdl-10/.

15. R. M. Lerner. At the forge: syndication with RSS. *Linux Journal*, 126(8), 2004.
16. F. Manola and E. Miller. RDF Primer. W3C Recommendation, Feb 2004. http://www.w3.org/TR/rdf-primer/.
17. Microsoft. Open Data Protocol (OData) Specification, Version v20101230, 2010. `http://www.odata.org/media/16352/[ms-odata].pdf`.
18. OASIS. *Web Services Business Process Execution Language Version 2.0*, Apr. 2007. online at: `http://www.ibm.com/developerworks/webservices/library/ws-bpel/`.
19. OMG. Business Process Model and Notation, V1.1, 2008.
20. D. Reynolds. An organization ontology. Epimorphics, `http://www.epimorphics.com/public/vocabulary/org.html`, Oct 2010.
21. R. Sayre. Atom: The standard in syndication. *IEEE Internet Computing*, 9(4):71–78, 2005.
22. A. ter Hofstede, W. van der Aalst, A. ter Hofstede, and M. Weske. Business process management: A survey. In M. Weske, editor, *Business Process Management*, volume 2678 of *Lecture Notes in Computer Science*, pages 1019–1019. Springer Berlin / Heidelberg, 2003.

Part III
USDL — Methods

The previous part of the book has described the meta-model of USDL in terms of its abstract syntax using natural language, examples, and UML class diagrams. This part presents methods, viz., approaches and tools to the representation, creation, communication, and management of actual service descriptions.

Chapter 14 deals with technical aspects of how USDL service descriptions can be read from and written to different representations for use by humans and tools. A combination of techniques for representing and exchanging USDL have been drawn from Model-Driven Engineering and Semantic Web technologies. Chapter 15 argues that fundamental tooling is required in order to apply USDL in practical settings. The chapter discusses three fundamental types of tools for USDL. First, USDL editors have been developed for expert and casual users, respectively. Second, several USDL repositories have been built to allow editors accessing and storing USDL descriptions. Third, our generic USDL marketplace allows providers to describe there services once and potentially trade them anywhere. The operation of such a service marketplace requires a governance approach that lies adjacent to the requirements of a SOA and the more general governance of IT. It also has requirements of its own, especially when it comes to the description of services with languages such as USDL. Therefore, Chapter 16 proposes four building blocks as a basis for a governance framework that is capable of supporting the operation of a service marketplace. Finally, Chapter 17 addresses the fact that different variants of USDL are required for different contexts. This is already shown by the Legal Module which requires different contents depending on the jurisdiction of a country. The issue aggravates if more and more parameters are relevant to determine the correct variant. The chapter presents one possible solution for variant management consisting of a canonical data model, a context driver mechanism, governance processes, and appropriate tooling.

Chapter 14
Representing USDL for Humans and Tools

Keith Duddy, Matthias Heinrich, Steffen Heinzl, Martin Knechtel, Carlos Pedrinaci, Benjamin Schmeling, and Virginia Smith

Abstract This chapter deals with technical aspects of how USDL service descriptions can be read from and written to different representations for use by humans and tools. A combination of techniques for representing and exchanging USDL have been drawn from Model-Driven Engineering and Semantic Web technologies. The USDL language's structural definition is specified as a MOF meta-model, but some modules were originally defined using the OWL language from the Semantic Web community and translated to the meta-model format. We begin with the important topic of serializing USDL descriptions into XML, so that they can be exchanged between editors, repositories, and other tools. The following topic is how USDL can be made available through the Semantic Web as a network of linked data, connected via URIs. Finally, consideration is given to human-readable representations of USDL descriptions, and how they can be generated, in large part, from the contents of a stored USDL model.

Keith Duddy
Queensland University of Technology, GPO Box 2434, Brisbane, QLD 4001, Australia,
e-mail: keith.duddy@qut.edu.au

Steffen Heinzl, Benjamin Schmeling
SAP Research Darmstadt, Bleichstrasse 8, 64283 Darmstadt, Germany,
e-mail: steffen.heinzl@sap.com, e-mail: benjamin.schmeling@sap.com

Matthias Heinrich, Martin Knechtel
SAP Research Dresden, Chemnitzer Strasse 48, 01187 Dresden, Germany,
e-mail: matthias.heinrich@sap.com, e-mail: martin.knechtel@sap.com

Carlos Pedrinaci
Knowledge Media Institute, The Open University, Walton Hall, Milton Keynes, MK7 6AA, UK,
e-mail: c.pedrinaci@open.ac.uk

Virginia Smith
Hewlett-Packard Company, 8000 Foothills Blvd, Roseville, CA 95747, USA,
e-mail: virginia.smith@hp.com

14.1 Introduction

The previous part of the book has described the semantics of the USDL language in terms of its abstract syntax shown as UML *class diagrams* with natural language explanations, and examples. This part presents tools and approaches to the creation, representation and communication of actual service descriptions using a variety of formats. The class diagrams show the structure of information that the USDL language expresses, with classes representing the main concepts we wish to capture about services, attributes of these classes representing properties of the concepts, and references between the classes representing links between the concepts. The underlying specification in terms of which the USDL is defined is the Ecore language from the Eclipse Modeling Framework (EMF), which is derived from the Meta-Object Facility (MOF) [19] — an Object Management Group (OMG) standard for meta-data definition which shares the same semantics as the UML class modeling language. MOF and Ecore also use a subset of the graphical language of UML to provide diagrammatic representations of meta-models. The diagrams do not show every feature of UML, MOF or Ecore, but the concepts they do show share the same semantics. Ecore has an XML representation with filenames ending in `.ecore` which captures all of the details.

For readers not familiar with EMF, the following comparison with XML may assist. The Ecore language plays the same role as the XML Schema language for defining XML schemas — it provides the basic concepts in terms of which object-oriented models are defined: XML element types are roughly equivalent to classes, XML attribute types are similar to Ecore classes' attributes, and XML ref element types are similar to Ecore references. A document type X is defined by a schema definition `X.xsd`, in the same way as a language Y is defined by the Ecore model (also called a meta-model) `Y.ecore`.

A schema document (`.xsd` file) contains element type definitions that constrain what elements may appear in an XML document that validates against the schema, in the same way that a meta-model (serialized as an `.ecore` file) defines what objects may appear in a model instance (serialized as an `.xmi` file) that conforms to the meta-model.

Figure 14.1 shows the modeling hierarchy in the EMF technical space in which the USDL is defined. It has the Ecore language at the top layer, with its concepts of packages, classes, attributes and references. The next layer down contains the USDL meta-model, which identifies concepts about services in terms of Ecore classes, attributes and references. The lowest layer shows USDL service descriptions expressed in the USDL language, as instances of the classes defined in the USDL meta-model.

As you can see, there are two possible instantiations that come as standard within EMF: Java objects which are instances of Java classes generated according to standard mappings from the Ecore meta-models; and XMI files, which conform to the XML Schemas that are also generated from Ecore meta-models. In EMF the XMI mapping is used to create a default serialization from a set of in-memory Java objects to an XML document.

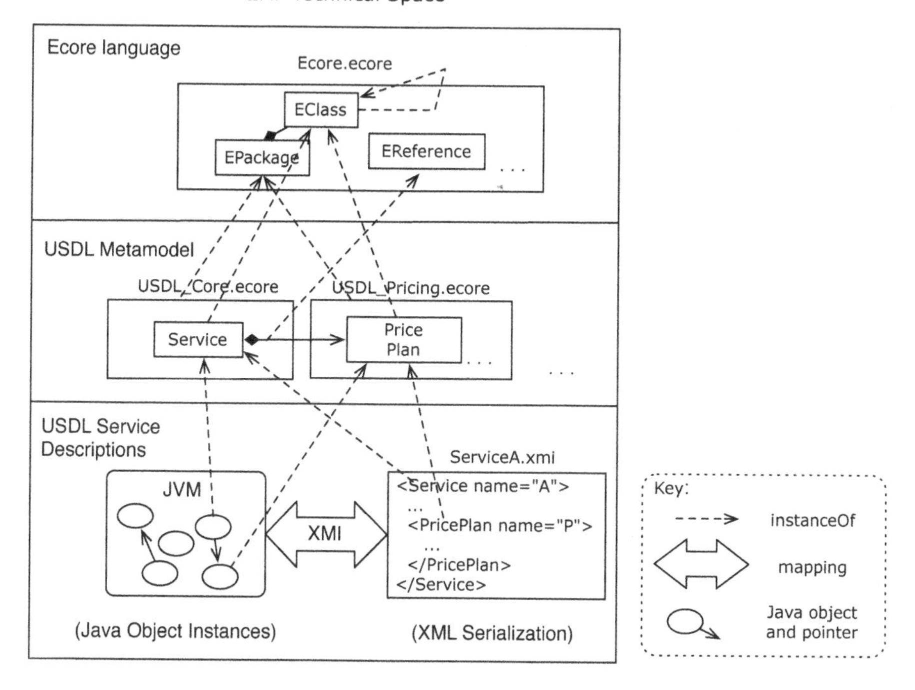

Fig. 14.1: Meta-model hierarchy for USDL.

In addition to the EMF code generators, there is a large developer community, both open-source and private, which has extended EMF with myriad tools for editing, serializing, and transforming models; for storing them in various kinds of databases; and for creating tools around them. EMF is a plugin for Eclipse, which is the world's most widely used Java developer environment, and therefore integrates the use of Ecore models in Java with hundreds of other frameworks and plugins.

The use of tools which manipulate models to produce textual syntaxes, code, editors, data stores and other models of various kinds is known generically as *Model-Driven Engineering (MDE)*. One of the earliest articulations of the concept of models as first class artifacts in software engineering was through the OMG's *Model Driven Architecture (MDA)* [26], introduced in 2000, which began by suggesting that abstract models of business functionality could be transformed into running code through a series of layered transformations: *Computation Independent Models* could be augmented by some additional information and transformed into *Platform Independent Models*, which could then be transformed using mappings to *Platform Specific Models* from which code could easily be derived by (re-)using best-practice patterns. Although MDA is only being used in this suggested form in a minority of cases, the basic concept of specifying higher-level abstractions of domain concepts as meta-models, and then using mappings to different formats, and code-generation tools to support the models, rather than writing code, is now well established. The approaches we call MDE also overlap with technologies from the *Domain Specific Language (DSL)* community, in which the concept of the domain language definition is very close to the abstraction of the meta-model, and many tools bridge the gap between textual and graphical DSLs and object-oriented models. The storage and manipulation of object models in programs which are derived from grammars representing textual syntaxes is now commonplace, and integrates tools designed for compiler construction, such as parser generators and parse-tree manipulation tools with other model transformation languages and tools.

Furthermore, there are automated mappings from Ecore to other representational formats such as non-XML textual languages, as well as Semantic Web languages such as OWL and RDF. However, it is more valuable in a Semantic Web environment to use human-guided mappings which match the concepts in the USDL to those represented in existing vocabularies so that service meta-data from USDL can be integrated into a larger web of Linked Data.

This chapter shows how model-driven techniques are used to manipulate the Ecore models of the USDL language in order to create a number of concrete representations of USDL service descriptions. This chapter is structured as follows: Section 14.2 explains the approach used to serialize USDL for interchange between tools that can place an entire service description into a single XML document, or can exchange an XML representation of one of the modules of USDL at a time. The use of the augmented XML framework, viz., the *Service Modeling Language (SML)*, is also considered for its document cross referencing, packaging and validation capabilities. This is followed in Section 14.3 by an analysis of the existing Semantic Web environment for established ontologies and vocabularies with which the USDL is well aligned. Section 14.3 then describes approaches across several Semantic Web technologies for representing USDL so that existing Linked Data approaches to modeling domain semantics can be matched and re-used, and to overcome some of the limitations of semantic representation in the object-oriented MOF specification. Finally, Section 14.4 demonstrates the use of template-based query approaches over EMF models to provide human-readable representations of USDL.

14.2 Serialization of USDL models

The approach used for the serialization of USDL models is based on the XML Metadata Interchange standard (XMI), which is defined by the Object Management Group (OMG). A few requirements are introduced that are necessary (or nice to have) for the goal of creating well-formed XML documents that describe a concrete USDL model (i.e., a service description). Based on the requirements, two ways of creating a serialization module for USDL are described. Leveraging XMI, it is possible to generate XML documents from the USDL models, serving as a concrete syntax for the USDL's abstract syntax. Finally, the use of the Service Modeling Language is considered for its capacity to cross-link models between documents, and to package a set of XML Schema and instance documents into a single XML file for distribution.

14.2.1 Model Requirements for an XML-based Concrete Syntax

When using XML as a concrete syntax, a number of requirements arise that we will categorize into two types: *technical* and *structural* requirements. Technical requirements are necessary for the serialization of our model into a single XML document, and for the ability to serialize the subset of a service description that pertains to a particular module. Structural requirements make the concrete syntax simpler and easier to read.

Technical Requirement 1:

The model to be serialized must define a single root element. The reason is that XML is a textual representation that is tree-based, whereas Ecore models are graph-based. From the root, navigation to all other elements that are part of the tree is possible.

Technical Requirement 2:

Each model class that is not the root element for a module must be contained by another model class. This implies that there is always a navigable path from the root element class to all other classes. This path must be available using Ecore containment references only. XMI already allows for XML path navigation to subclasses of any class that is already reachable from the root element by containment. Any class navigable from such subclasses is also transitively considered navigable from the root class, and its XMI serialization will form a correct nested element set.

Technical Requirement 3:

Non-containment references must be mapped by some textual linking mechanism. Otherwise already contained XML elements cannot be reused but have to be duplicated which would destroy the well-formed structure of the concrete model instance.

Different XML specifications provide different mechanisms for referencing other elements, such as XML Schema's ref mechanism or RDF's URIref mechanism. In general, there are three strategies to handle references. The first strategy identifies XML elements by their location in the XML subtree (often referred to as URI fragments), the second uses unique characteristics of an XML element such as an attribute with unique values, and the third introduces dedicated unique identifier attributes which are added to the XML element. The first two strategies have one significant disadvantage: If the identifier changes, the references in the document are broken. The third strategy on the other hand has no need to change identifiers even if the document structure or attribute values change. However, the challenge in using the third strategy is to generate unique identifiers across multiple documents.

Structural Requirement 1:

If a model consists of several modules (or packages in MOF terminology), each module should define a module root element. This allows for the definition of documents containing contents from only a single module, e.g., a USDL fragment containing only the pricing model could be serialized and exchanged.

Structural Requirement 2:

Model classes that are referenced (by non-containment references) should be top elements contained directly by the document root. The reason for this requirement is that referenced elements are reusable: once defined they can be referenced from different parts of the XML document. When enforcing this requirement, the semantics of the model should be taken into account, i.e., there may be reasons to not strictly follow this requirement.

Structural Requirement 3:

Imported elements should be part of the serialization model. There is no theoretical reason for this requirement, because elements that are referenced by URI in other documents can be located. However, it should be obvious and directly visible which documents need to be accessed to resolve all references. Unless we define an import element in the model which maps to an import schema for the document, all references will have to be checked and document dependencies calculated. This is why XML-based specifications such as WSDL and XML Schema introduce dedicated import elements.

14.2.2 The USDL Serialization Model

To allow the serialization into an XML-based syntax, a few extensions to the abstract USDL meta-model are needed. These extensions are used to address the above requirements and result in a concrete USDL meta-model.

We introduce two variants for the concrete USDL meta-model: a lightweight model that fulfills the technical requirements and a full-fledged model that also fulfills the structural requirements. The lightweight version is easily adaptable for tool builders since it directly builds upon available XMI tools. The full-fledged model also needs extensions in each module as we explain below.

14.2.2.1 Lightweight USDL Meta-model

Technical Requirement 1 has been addressed by the introduction of an additional root class in Ecore — the USDL3Document — which has a containment reference to all other classes either directly or indirectly. This class is in a Serialization Module which imports the other modules and makes references to classes in other modules.

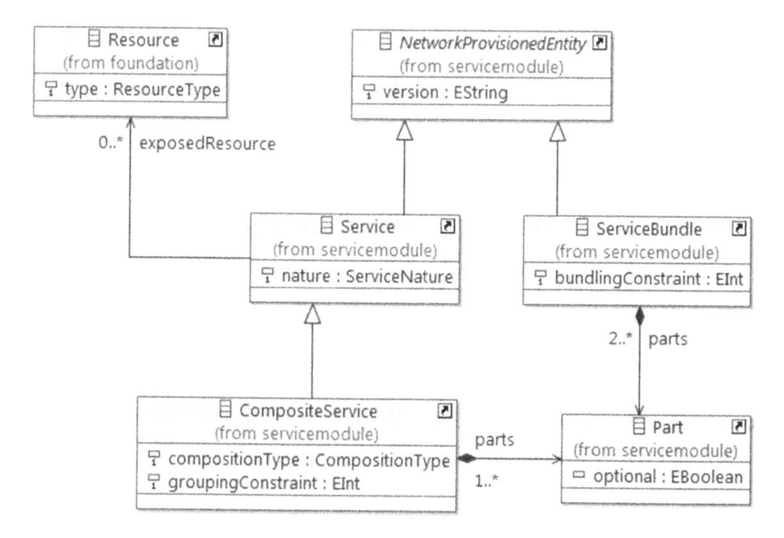

Fig. 14.2: Excerpt from the USDL meta-model.

Technical Requirement 2 can be met by defining references from the root element USDL3Document which contain either (i) all non-contained, concrete classes or (ii) all non-inheriting, non-contained classes.

The definition of the root class and its module can either be achieved manually or derived automatically by applying a model transformation to the standard USDL meta-model. Take for example the excerpt from the USDL model shown in Figure 14.2. The model shows the interrelations of a few classes. The ServiceBundle and the CompositeService classes contain the class Part, whereas the Resource class is only referenced by Service. CompositeService inherits from Service. Service andServiceBundle inherit from NetworkProvisionedEntity.

Following the approach (i) above (containing all non-contained, concrete classes), the classes that need to be contained by the newly introduced root element in this diagram are Resource, Service, Composite Service and ServiceBundle. Ap-

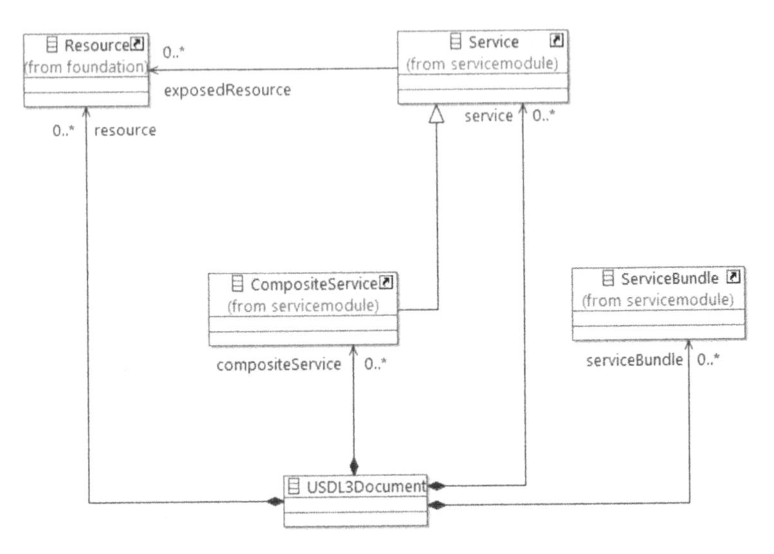

Fig. 14.3: Serialization module for the USDL excerpt.

plying the model transformation to the USDL meta-model excerpt results in the (automatically generated) Serialization module shown in Figure 14.3. The same model transformation can be applied to all USDL modules to create the lightweight serialization module for USDL.

Technical Requirement 3 is already addressed by virtue of the choice of the standard XMI mapping rules (see Section 14.2.4). One advantage of the lightweight model is apparent: The serialization module can simply be *added* to the other Ecore modules without modifying them.

14.2.2.2 Full-fledged USDL Meta-model

A full-fledged approach to USDL Serialization requires changing the standard meta-model to address the structural requirements as well. In order to meet Structural Requirement 1, a root element for each module is introduced which is used as a container for all non-contained classes of a module, e.g., a PricingElements class has been added to the Pricing Module. Technical Requirement 2 is solved in a slightly different way than the lightweight model, namely by adding containment references between the root class USDL3Document and each of the module roots. In addition, containment references are created between all classes that are not already reachable via containments from the USDL3Document root class via each module root class (and thus Structural Requirement 2 is met). For example the PricePlan class is not reachable from the module root, PricingElements, and therefore we add a containment reference from it to PricePlan. Consequently, the PricePlan is now reachable from the USDL3Document by navigating from USDL3Document to PricingElements to PricePlan.

Structural Requirement 3 has been addressed by introducing the Import class. The Import class provides a uri attribute that points to the imported USDL document. This convenience class is especially helpful for tools to preload referenced documents in order to navigate cross-document references.

The full-fledged USDL meta-model has the advantage that structural requirements are fulfilled and thus that the readability of the resulting USDL is better due to a better separation of the different modules inside the XML document (or documents — if serialized per module). A major disadvantage is that the existing USDL modules have to be extended with module root elements, such as PriceElements. Thus, the abstract syntax of USDL has to be made more complex in order to make the concrete syntax easier to understand.

14.2.3 Serialization Model

The Serialization model that is used for USDL (Version 3, Milestone 5) is a lightweight one with additional support of an Import mechanism. Figure 14.4 depicts the Serialization model in that module.

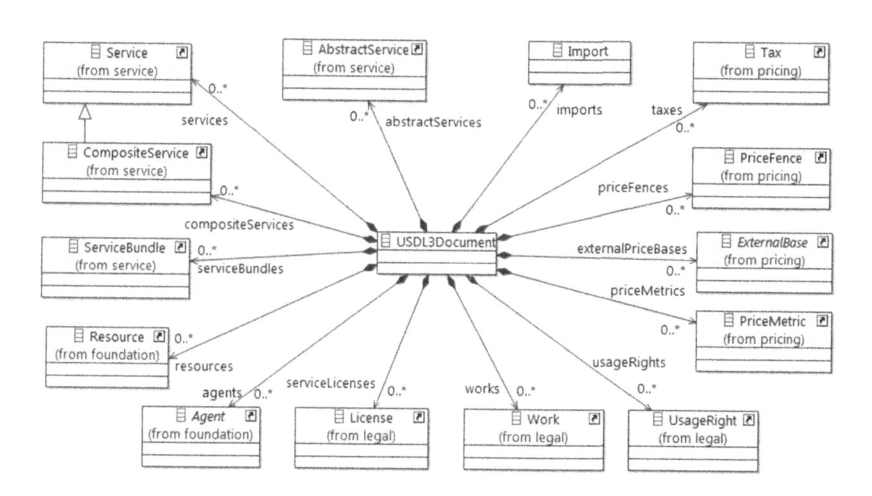

Fig. 14.4: Serialization module for USDL (Milestone 5).

14.2.4 Mapping the USDL Serialization Model to XML

After defining a Serialization Module — which meets our technical (and structural) requirements — this model needs to be mapped to XML. The mapping is based on the default Ecore serialization mechanism which itself is based on XMI. Hence, a short introduction to XMI follows.

The XML Metadata Interchange (XMI) specification (V2.0 was used for Ecore) defines a standard for exchanging any type of metadata that is compliant to the Meta Object Facility (MOF). Ecore has been built to comply to a subset the MOF standard, and thus satisfies this requirement. XMI defines XML Schema constructs for the purpose of identification, linking, object type hierarchies, and data typing. These predefined schema element patterns allow for the serialization of instances of any MOF-conformant object-oriented meta-model to XML, including standards such as UML 2.0 and BPMN 2.0.

Table 14.1: Overview of XMI Mapping from Ecore to XML

Ecore	XML
Instance of EClass	XML Element with xmi:id
Instance of EAttribute of Datatype	XML Attribute
Instance of EReference (containment)	Nesting XML Elements
Instance of EReference (non-containment)	XML Ref Attribute
Inheritance — A is a concrete subtype of B	`type` Attribute set to "A" in the XMLElement representing "B"

To reduce the effort of building tools on top of USDL, the default XMI mapping of Ecore to XML has been applied to USDL. A summary of the major parts of this mapping is shown in Table 14.1. This mapping fulfils Technical Requirement 3 since each generated element has an `xmi:id` that can be referenced. Listing 14.1 shows an example XMI serialization of a USDL service model instance, as specified by the Serialization Module and transformed USDL Modules, as documented in the USDL Version 3 Milestone 5 specification.

Listing 14.1: XMI serialization of a USDL service model instance.

```
1  <usdl3:USDL3Document  xmlns:xmi="http://www.omg.org/XMI"  ...>
2
3   <services xmi:id="Service_178"  version="1.0"  nature="Manual"  ...>
4
5    <names xmi:id="Description_389"
6        value="Lead Logistics - General Freight"  type="name" />
7
8    <provider  xmi:id="Provider_2562"  enactingAgent="Organization_432"/>
9
10  </services>
11
12  <agents  xsi:type="foundation:Organization"  xmi:id="Organization_432">
13
14   ...
```

```
15
16    </agents>
17
18  </usdl3:USDL3Document>
```

The USDL3Document is the root element. It may contain several service description which are in turn represented by a services element (named after the containment reference from the USDL3Document class to the Service class). The services element furthermore has an xmi:id attribute that allows other elements to reference the services element. The nature of the service is a simple EAttribute and therefore mapped to an XML attribute. Names are contained by the Service class and therefore realized by nesting the names element inside it. The Provider has an enacting agent. This is realized by including an agents element with type Organization and using the Organization's xmi:id.

It is common practice for Ecore modelers to apply plural nouns for representing multi-valued containment references, such as, "services," but when mapped to XML, this results in multiple nested elements, each of which is named after the reference. The resulting XML would read better with singular names — such as "service," but this would require the adaptation of the default XMI mapping.

14.2.5 Serialization and Exchange of USDL Models Using SML

The Service Modeling Language (SML) [21], which is a W3C recommendation, provides some useful tools for representing and exchanging USDL models. SML is not a domain language itself, i.e., it does not define the domain entities. However, SML does provide useful constructs for representing complex service descriptions. In addition, the corresponding SML Interchange Format (SML-IF) [20] provides a convenient and standardized way to represent and exchange self-contained USDL models.

14.2.5.1 SML Models

An SML model is a set of interrelated documents that describe a service (or other domain) model. This set of documents consists of model definition documents and model instance documents. The model definition documents contain information about the service (the service model), as well as constraints on the model that must be satisfied for the service to function properly. The model instance documents contain the data for the modeled instances.

There are two types of model definition documents: schema documents and rule documents. The model definition documents provide much of the information a model validator needs to decide whether a given model is valid. Schema documents define constraints on the structure and content of the instance documents in a model. SML uses XML Schema as the schema language and defines a set of extensions to

XML Schema to support references that may cross document boundaries. Rules are boolean expressions that additionally constrain the structure and content of documents in a model in ways that may not expressible within a given XML Schema. SML uses Schematron [13] and XPath [7] for rules. While the rules can be embedded in SML model schema documents, they can also be placed in separate documents so that the schema documents themselves are not altered.

SML schema documents are defined as a strict superset of XML Schema. All valid XML Schema documents are valid SML schema documents. SML does define an extension of XML Schema, namely the SML references. SML references are used to link from one element in a model to another element in the same model. The linked elements may reside in separate XML documents at runtime. In addition, SML references may be constrained to specific elements or element types. Extending the Ecore default mapping by defining the mapping of non-containment references to SML references instead of XML Ref attributes enhances the cognitive sufficiency of the USDL model by making the reference endpoint more explicit while still satisfying Technical Requirement 3. However, it is worth noting that, as long as the Schema extensions defined by SML are not used, then SML model documents can still be processed by currently available XML processors.

SML model instance documents are XML documents that together form a service's description (instance). They describe or support the description of the individual resources that the model portrays and must conform to the structure and constraints as defined in the model definition documents, as shown in Figure 14.5.

14.2.5.2 SML Interchange Format

The SML-IF specification defines a standard interchange format that preserves the content and interrelationships among the model documents. It also defines a constrained form of model validation to ensure interoperability when specific conditions are met and to increase the likelihood of interoperability in other cases. But, at a minimum, the SML-IF interchange format provides a well-defined standard for exchanging a set of model documents regardless of the validation process. An SML-IF document packages the set of SML model documents to be interchanged as a single XML document. Each model document appears as content in either the 'definitions' or 'instances' subsection of the SML-IF document, depending on whether the model document in question is a model definition document or a model instance document. Each model document can be represented in either of two ways, by embedding its content or by providing a reference to it.

14.2.5.3 USDL Models as SML Models

As mentioned, SML is agnostic of any domain model such as a service description model. USDL provides such a domain model. As specified previously in this chapter, USDL Version 3 Milestone 5 defines a concrete representation of the USDL

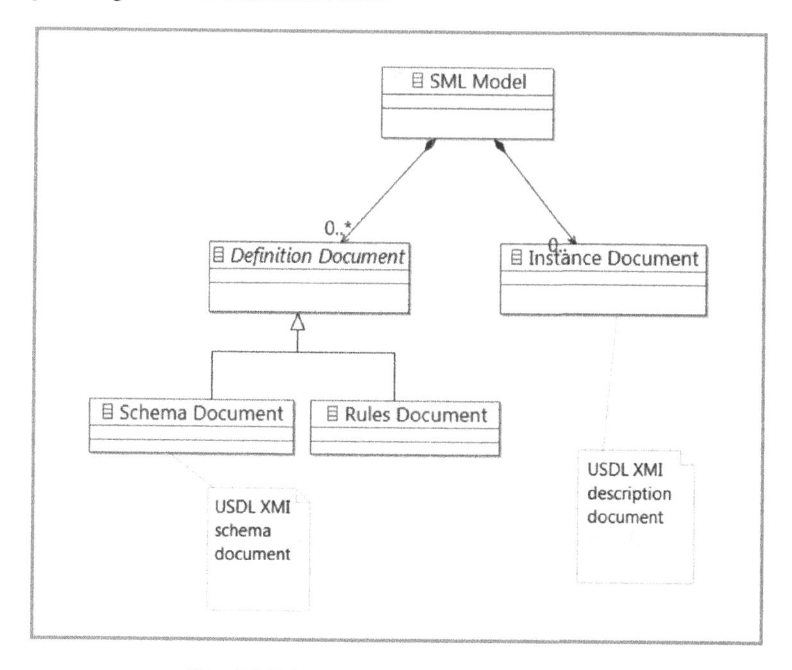

Fig. 14.5: The SML document structure.

model as an XMI schema and a specific model instance as an XMI document that validates to that schema. These documents form a valid SML model as described here and can be packaged into one SML-IF document. Note that in the case of full-fledged USDL serialization, a serialization may be in a single file, or each module may be represented in a separate XML document. In the latter case an SML-IF document provides a convenient way to package multiple documents into one deliverable XML document. The value of SML-IF is that a single USDL model made of multiple schema and instance documents can easily be passed around as a coherent whole thus satisfying the requirement of cognitive sufficiency.

The user could also manually expand the model description beyond the USDL meta-model to formalize specific business rules as SML rule documents. This can be done without altering the XMI schema for USDL. SML-IF provides a way for the USDL model to incorporate this additional model definition document. In addition, SML provides the capability to link references across schema documents which would assist in the implementation of Structural Requirement 1 and the full-fledged USDL serialization model by enabling cross document links. However, this capability would require an SML processor rather than a standard XML processor and, therefore, this capability comes at an additional cost of implementation.

14.3 Representing USDL as Linked Data

The W3C defines the Web as "an information space in which the items of interest, referred to as resources, are identified by global identifiers called Uniform Resource Identifiers (URI)" [14]. The Web is based on three main aspects, namely the identification of resources, enabling the interaction between agents (software or humans) via well-defined protocols, and formats that govern the representation of data and resources transmitted.

These principles have effectively governed the Web and still maintain the ability for extension to cope with new kinds of resources, or to enable more complex activities to be carried out. A good example is the work carried out on the Semantic Web towards providing machine interpretable semantic descriptions of resources, which could pave the way for the development of more intelligent agents. Most relevant is the use of RDF and OWL, which are based on pre-existing Web standards, to define domain-specific models of concepts in an effective and extensible manner.

The Linked Data principles were suggested[1] by Tim Berners-Lee in 2006 as a means of creating a Web of Data better suited for machine processing. These principles recommend that one should:

1. use URIs as names for things,
2. use HTTP URIs so that people can look up those names,
3. provide useful information, using the standards (RDF, SPARQL) when someone looks up a URI,
4. include links to other URIs so that they can discover more things.

Since these principles were proposed we have witnessed an outstanding growth in terms of data and vocabularies allowing people to freely expose and interlink large quantities of heterogeneous data. In fact, for raw data that can effectively be modeled in RDF, Linked Data principles are considered by well cited authors [4] as the best means for publishing to the Web.

The first principle ensures that resources are uniquely identified. The second principle ensures that their identification is such that HTTP can be used for obtaining information about the resource. The third principle establishes standard technologies for exposing data in a manner that is suitable for machine processing. Finally, the fourth principle aims to promote the interlinking of data. The same way hyperlinks connect Web documents into a single global information space, Linked Data uses hyperlinks to connect data into a single global data space. These links allow applications to navigate the Web of Data and, since the data is exposed through HTTP (see principle 2) and represented in some standard format (see principle 3), machines can obtain it, interpret it, and act accordingly in an automated manner.

USDL was originally modeled in Ecore and integrates a number of different perspectives on services (e.g., pricing, technical details, legal aspects, stakeholders in service provision, etc). In addition, the Pricing and Legal Modules were also modeled ontologically in OWL. In [15] Semantic Web tools are used to create and man-

[1] `http://www.w3.org/DesignIssues/LinkedData.html`

age instances of pricing plans for billing purposes. In work related to the USDL Legal and Service Level Modules [3] the German copyright legislation was modeled as an ontology from which licence rights models can be derived to describe the conditions for use of copyright material provided by a service. In the remainder of this section we shall thus focus on how all of USDL can adopt linked data principles, focusing mainly on the representation of USDL in RDF(S), and on the interlinking with external vocabularies and data sources.

14.3.1 Linked USDL

Creating a linked-data-ready version of USDL involves modeling USDL data and thus the USDL language (meta-model) itself in RDF(S) [5] or related standards, such as RDFa [1] and OWL [8]. USDL is composed of a number of modules some of which started out as Ecore models only, and need to be re-modeled in RDF(S)/OWL accordingly. Redescribing the whole USDL model again in the form of ontologies is beyond the scope of this book, so instead we shall focus on the main design decisions concerning the lightweight semantic representation of USDL in RDF(S)/OWL. Then some examples targeted at the reuse of existing vocabularies and instances are considered. Through this exercise it shall be seen not only that existing vocabularies cover a good part of USDL, but also that modeling USDL in this manner has a number of benefits from the use of Semantic Web tools and formalisms (e.g., temporal reasoning) and from compatibility with existing datasets.

14.3.1.1 Integrating USDL in the Web of Data Through Reuse

The fourth Linked Data principle is to include links to other data sources from the Web. These links are an essential means of generating a Web of Data as opposed to disconnected silos. Linked data simplifies data integration and interpretation, as well as enabling the discovery of related data that is not part of a USDL service description.

Most often there are three kinds of links contemplated [9]:

Relationship Links whereby entities from a data source are linked to entities from other data sources through relationships. For instance, *Service A* stored in a USDL repository can be described as being provided by *Company C* defined in an external *Companies Catalogue*. This type of link enables the reuse of data and establishes links across datasets.

Identity Links which indicate that two URIs refer to the same entity. This allows us to state, for instance, that *Company C* from the previous catalogue is actually the same as *Company X*, rated by customers in a certain social Web site. By means of these links, machines can incorporate different views about the same entity based on data coming from diverse sources.

Vocabulary Links which are established between entities and the vocabularies used to define them, thus allow machines to retrieve the definitions and interpret them. For instance, knowing that *Service A* is a ServiceBundle informs us that it has other *Services* as part of it. These links also allow us to indicate relationships (e.g., equivalence, subsumption) between concepts, and re-use relationships defined in different vocabularies. This helps integrate data from diverse datasets.

A fundamental activity in adapting USDL for the Web of Data therefore concerns the analysis of existing vocabularies and datasets, in order to i) identify reusable vocabularies to avoid reinventing the wheel and promote reuse and integration; and ii) identify possible relationships with USDL concepts and USDL data to support navigation across datasets and to simplify data integration.

14.3.2 Design Decisions

In this section we introduce some of the main decisions that have been adopted while creating Linked USDL. We first introduce general modeling decisions and we then cover choices made concerning capturing information of particular kinds, such as geospatial and temporal.

14.3.2.1 Classifications and SKOS Schemes

The main purpose of the USDL specification is to provide a schema, or type system, which defines the contents and structure of concrete service descriptions, e.g., *Service A*, its kind, e.g., *Service*, and a certain categorization of the Service kind, e.g., an *Automated Service*. When defined in terms of ontologies the USDL Ecore model could be replicated using sub-classing and meta-modeling, however, the use of a hierarchy of classes essentially establishes explicitly and *a priori* the subclasses available. Some parts of the model, however, consist of a set of enumerated values which act as classification categories, and are not expressed as class hierarchies.

The conceptual distinction introduced by a hierarchy of classes is sometimes appropriate as is the case for instance when capturing the relationship between *Service* and *Composite Service*, in which the latter has a grouping constraint which makes no sense for *Services* in general. However, in the Participants module, there are many different subtypes of Role, including BusinessOwner, Provider and Stakeholder, which do not introduce any extra data values or structural constraints. We consider that these subtypes would be better expressed as simple categories.

As a general decision for Linked USDL, we have captured the relationship between concepts via subsumption relationships whenever there was a structural and semantic difference like in the case of Service and CompositeService. For cases like the nature of services which are represented as an enumeration of values in Ecore, we use Simple Knowledge Organization System (SKOS) [18] schemes for defining the different categories. SKOS is a common lightweight data model represented in RDF, which supports the capture of knowledge organization systems such as classification schemes and taxonomies. Using SKOS, categorizations can be represented in a machine processable manner. We can easily integrate different perspectives simply by changing the definition of the concept categorized by means of a property whose range is *skos:Concept*. In cases where the USDL specification uses subclassing to introduce categories, such as the *Role* example above, we can also use SKOS, and apply simple taxonomies to the concept of agents, rather than having a fixed set of subsumed concepts. Indeed, this mechanism does not prevent users of USDL from providing their own vertical domain-specific categorizations through subsumption if they wish to.

14.3.2.2 Types as Properties

The USDL specification defines some kinds of things using a property indicating the type (e.g., dependencyType) and others by using a hierarchy of concepts (see for instance the different Roles above). In RDF(S) both these approaches could be best modeled using several properties, possibly in a hierarchy. For example, in the case of Dependency and DependencyType, where the relationship is binary, RDF properties are the natural choice as they allow capture of relationships between the properties where necessary. In this case we have therefore modeled all the DependencyTypes as properties, we have defined their range accordingly, and we have dropped the concept DependencyTarget since it becomes redundant with this modeling approach. We have thus defined a top level property dependsOn and a number of sub-properties including requires, includes, mirrors, etc.

14.3.2.3 Partonomy

Part-whole relations are very common structuring primitives of the universe, and indeed they are represented in the USDL Ecore model by containment references (black diamonds in the visual representation). For example, ServiceBundle and CompositeService have a number of constituent parts which can be either AbstractServices or NetworkProvisionedEntities. The existence of a part or parts in the model can be specified as a cardinality, or range of numbers, which if the lower bound is zero is *optional*, or if the lower bound is some other number, then that number of parts is *mandatory*. In the case of CompositeServices the Ecore has no way of directly specifying the CompositionType, and so an enumerated value in an attribute of Services specifies whether the sub-services are *data dependent*, *ordered*, or just an *aggregation*.

RDFS and OWL do not have specific construct for modeling part-whole relations but there are however a number of general purpose proposals for capturing these. The reader is referred to [23] and [24]. For the purposes of USDL we adopted the latter as it helps define both direct and transitive relations. We thus include hasPartTransitive as a transitive relation and hasPart, a sub-property of hasPart-

Transitive, as a normal property. Doing so allows us to simply capture the hasPart relations and if necessary be able to deduce the existence of transitive containment relationships automatically through reasoning over hasPartTransitive.

Additionally, as indicated earlier, USDL constrains the cardinality of parts, with the most common being single-valued, either mandatory or optional. These different kinds are captured through a hierarchy of properties refining hasPart and hasPartTransitive respectively. Notably we have included hasOptionalPart and hasOptionalPartTransitive as well as hasMandatoryPartTransitive and hasMandatoryPartTransitive. The current version does not insert the inverse relation isPartOf, but, should it be necessary, this would be an easy addition.

14.3.2.4 Agents and Roles

Given that the provisioning of services necessarily involves a number of individuals or organizations taking part, USDL provides a number of classes and relations covering this. In particular, Agent represents all the entities that can take active part in the provisioning of a Service. USDL identifies Organization, NaturalPerson, and ResourceAgent as the main kinds of Agents. This term appears in a number of vocabularies, notably in Dublin Core,[2] and FOAF [6] to name the main ones. The notion of Agent also concerns organizations which are covered in other vocabularies, for example by *gr:BusinessEntity* in GoodRelations [10].

Closely related to the notion of Agent, USDL includes the notion of Role. Role serves as the super type of all concrete USDL classes that represent roles found in a service network (e.g., service provider). Agents participating in the provisioning and delivery of a service perform distinct functions, which define their Role. Roles may either be bound to a concrete Agent or may be used as placeholders. The latter is necessary if the Service is in a stage where some Agents are yet to be determined. For example, a service description may specify that there needs to be a B2B gateway in order to deliver the service to a consumer. Which gateway provider will be chosen, however, depends on the message/interface standards supported by a consumer and the consumer's preferences.

In order to maximize reuse and integration across vocabularies we have adopted Reynolds' organization ontology [25], which covers all the core notions, provides basic modeling constructs for Roles and is already integrated with existing vocabularies such as FOAF and GoodRelations. The main design decisions in this ontology are i) capturing the relationship between Agents and Roles played in an organization through an n-ary relationship represented by the intermediate class Membership, and ii) integration with GoodRelations, FOAF and the vCard vocabulary [12]. By means of these alignments we enable the re-use of many FOAF profiles, GoodRelations descriptions, and other existing Web data.

[2] http://dublincore.org/

14.3.2.5 Geospatial Modeling

One aspect of the USDL Foundation module concerns location-related entities and relationships. In particular USDL specifies a super type for all location related entities, namely Location, and the subtypes PhysicalLocation, GeographicalPoint, PhysicalAddress, AdministrativeArea, and many others.

There has been a considerable amount of work devoted to creating ontologies and services in this area. Currently, perhaps the most reused vocabulary for geographic concepts is the W3C Basic Geo vocabulary, which facilitates the capture of GeographicalPoints on the basis of their latitude, longitude and altitude. In addition to this effort, the W3C Geo Incubator Group also devoted some effort to creating a simple and reusable vocabulary [16] for capturing some basic geometry relations, which we have adopted in Linked USDL.

There is also a range of complementary vocabularies, data sources, and services available on the Web with which it would be interesting to integrate for data reuse. It is worth noting the work by several organizations: firstly, the UK Ordnance Survey[3] as part of the data.gov.uk initiative for the public release of a large quantity of governmental data in the UK; the site Geonames.org, which comes with a large knowledge base of locations and services for accessing it; and the geospatial data set authored by the FAO.[4]

14.3.2.6 Temporal Modeling and Reasoning

USDL includes quite a few classes for representing time, including Time Instant, Time Interval, etc. Time representation and reasoning has been addressed quite often by researchers. Indeed, Semantic Web researchers already have several works on time representation. In particular, perhaps the most popular for Linked Data is OWL Time [11] which is hosted by W3C.

Time Ontology defines temporal entities and temporal relationships based on James Allen's interval temporal algebra [2]. It therefore identifies Instants, defines Intervals on the basis of beginning and end Instants and includes the typical temporal relationships between Instants and between Intervals (e.g., before, during, etc). Linked USDL supports the capture of most of the temporal aspects of USDL using OWL Time, and additionally supports the implementation of Allen's interval temporal algebra for reasoning about intervals and instants. Some issues like the precision of OWL Time (currently limited to seconds) and notions such as Recurrent Time and Time Pattern still need to be addressed.

[3] http://www.ordnancesurvey.co.uk/

[4] http://www.fao.org/countryprofiles/geoinfo.asp?lang=en

14.3.3 Services and Service Vocabularies

Modeling the central notion of Service in USDL with Linked Data did not require any particular decisions other than those mentioned above. However, we carefully reviewed the state of the art in service ontologies and vocabularies in order to identify the main alignments to be addressed. The main vocabularies used can be divided into those that address business aspects of services, such as e3Service [27] and GoodRelations, and those that tackle the technical aspects of services, for example, OWL-S [17], WSMO-Lite [28], and the Minimal Service Model [22].

In the current version of Linked USDL, we have performed the following alignments with GoodRelations since it is the most widely used vocabulary for (business) services on the Web:

- AbstractService is a subclass of *gr:Product Or Service Model* since it provides prototypical definitions of Services.
- Service and CompositeService are subclasses of *gr:Product Or Services Some Instances Placeholder* as they both identify a placeholder for instances of a service.
- The inter-service relationship enhances is equivalent to *gr:addOn.*

Possible additional alignments could be carried out with the e3 family of ontologies. However, at this stage these ontologies are not offered publicly on the Web in a resolvable manner which is a requirement for Linked Data.

14.3.4 Summary of USDL as Linked Data

This section has outlined the approach to mapping USDL to existing Linked Data and the Semantic Web resources. The structural specification of the USDL metamodel replicates a lot of vocabularies and relations specified in ontologies, for which there are meaningful and interlinked instances available on the Web through Linked Data. In fact the original form of some parts of USDL, notably the Pricing and Legal modules, were as OWL specifications, to facilitate the use of existing tools to create and manipulate service descriptions. By drawing equivalences between classes, attributes and relationships in the USDL specification with existing vocabularies, relations, and stores of data in the Linked Data Web, we can re-use both common concepts and instances of these which already exist, and use a broad range of tools for reasoning about the contents of USDL service descriptions.

14.4 USDL Documentation Generation using USDL-Doc

USDL service descriptions are a convenient way to capture various aspects of a service (e.g., technical, legal or operational aspects) in a structured manner which allows for further automatic processing. However, raw formats embedding structuring tags (e.g., XMI) typically fall short of providing a decent human-readable description. Therefore, we have implemented the USDL-Doc tool capable of transforming USDL service descriptions into HTML or PDF documents.

14.4.1 USDL-Doc Architecture

The USDL-Doc tool is a simplistic but powerful USDL editor add-on dedicated to mapping existing USDL service descriptions, stored as EMF models in XMI format, to HTML or PDF documents. Generated HTML/PDF documents may serve as service detail pages exhibited on the service marketplace, developer documentation, etc.

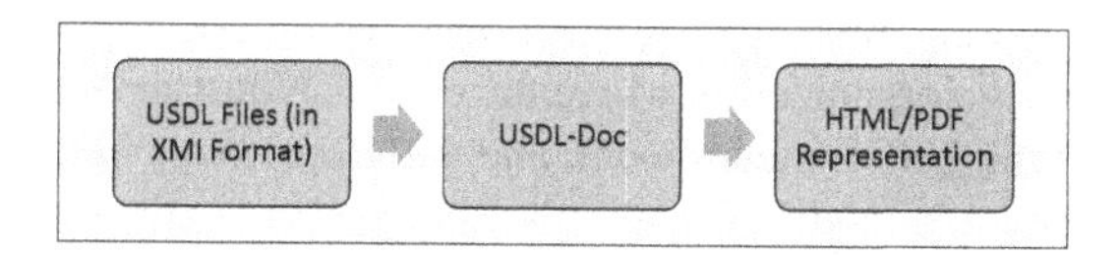

Fig. 14.6: Document Generation Workflow.

The document generation process is fully automated and follows the workflow depicted in Figure 14.6. Existing USDL service descriptions persisted in the XMI format are processed by the USDL-Doc tool which eventually produces HTML or PDF files.

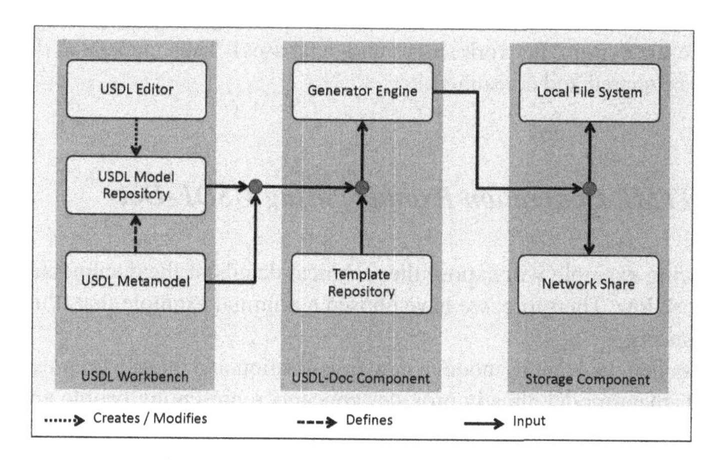

Fig. 14.7: Document Generation Platform Architecture.

In order to provide a fully automated workflow, the architecture illustrated in Figure 14.7 was devised. Essentially, there are three major building blocks: (i) the USDL workbench, (ii) the USDL-Doc component and (iii) the storage component. The USDL workbench is used by the USDL designer when creating and revising USDL descriptions by means of the USDL editor. The USDL editor creates USDL models which are compliant to the USDL meta-model. In essence, the USDL meta-model provides a fixed language vocabulary which can be combined to form arbitrary USDL models. Those models are stored in a dedicated USDL model repository which may be implemented in any form from a complex shared repository to a folder of XMI files on the local file system.

After USDL services are completely described, the USDL-Doc component may be triggered to produce self-contained HTML or PDF documents. The USDL-Doc component derives the documents taking into account the USDL service description instance itself as well as the associated USDL meta-model. Therefore, the generator engine — having access to the model and meta-model — can query the model in a declarative style. For example, a declarative query could ask for all natural persons belonging to a specific organization where the classes NaturalPerson and Organization are part of the USDL meta-model, and therefore nouns in language vocabulary implied by the meta-model. The template language to express these declarative queries is the Xpand language. Besides specifying dynamic queries, Xpand templates can also include static code blocks (e.g., HTML code). Hence, Xpand serves as a flexible code generation language targeting any kind of textual generated code (e.g., HTML, Java, C). The Xpand template language, together with the Xpand editor, is bundled in the openArchitectureWare (oAW) framework that also provides a generator engine executing the Xpand templates. In summary, the USDL-Doc component leverages the oAW template language Xpand and the oAW text generation engine to transform USDL models into HTML or PDF files.

These generated files are consequently transferred to the storage component. The storage component may put files into a variety of configured storage locations (including the file system, network shares and databases). Thus, generated documents can easily be shared and consumed.

14.4.2 HTML Generation Example using USDL-Doc

The following example will expose the technical details of the documentation generation workflow. Therefore, we have chosen a minimal example that illustrates all relevant aspects.

Let's assume we want to model a new organization and all of its representatives. The USDL meta-model already provides concepts representing people and organizations and their properties and relationships. Figure 14.8 shows a small fragment of the USDL meta-model (residing in the Participant module) depicting the classes Organization, NaturalPerson and the reference representatives. Consequently, an arbitrary USDL editor may instantiate these classes to create an object instance

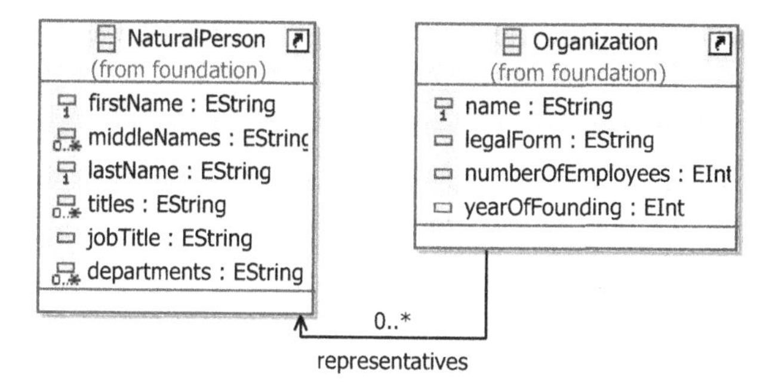

Fig. 14.8: Fragment of the USDL meta-model.

representing a new organization, its properties, and objects representing its associated people. Figure 14.9 shows such a USDL editor where the organization Lead Logistics (cf. running example in Section 8.7) is established and multiple natural persons are the associated representatives.

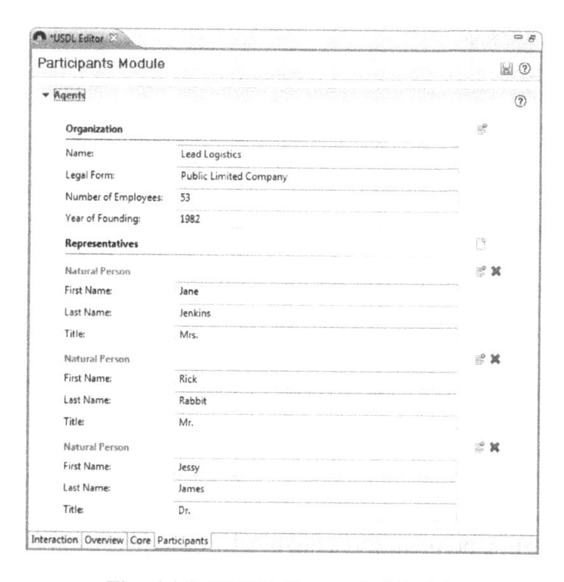

Fig. 14.9: USDL Example Model.

In order to advertise the service specified in USDL, we can deploy the service description we have created in the editor to a broker of services in a service marketplace in XMI format. We also might want to offer the information about the service on a Web page to inform potential customers. Therefore, we initiate an USDL-to-HTML conversion by invoking the USDL-Doc tool. The tool processes the USDL service description file, along with the USDL meta-model, and uses the Xpand template to produce a human-readable representation.

```
«IMPORT usdl»
...
«DEFINE expandOrganization FOR Organization»
    <html>
    <body>
    <div>
    <p>organization.name</p>
    <p>organization.legalForm</p>
    <p>organization.numberOfEmployees</p>
    <p>organization.yearOfFounding</p>
    </div>
    «EXPAND expandNaturalPerson FOREACH representatives»
    </body>
    </html>
«ENDDEFINE»

«DEFINE expandNaturalPerson FOR NaturalPerson»
    <div>
    <p>naturalPerson.firstName</p>
    <p>naturalPerson.lastName</p>
    <p>naturalPerson.title</p>
    </div>
«ENDDEFINE»
...
```

Fig. 14.10: Xpand Code Generation Template.

An example Xpand template is shown in figure 14.10. It defines the rules (e.g., `expandOrganization`, `expandNaturalPerson`) for how to map USDL model elements to HTML code. While the bold black font denotes dynamic code generation, the lighter font expresses static HTML code. To have a means to access the USDL model elements, the template has to declare the available types by importing the USDL meta-model. Once the generator engine is aware of the USDL types, various typed rules might be defined. In the example in Figure 14.10 there are two rules specified: `expandOrganization` and `expandNaturalPerson`. These rules are applied to all organizations and all natural persons defined in the USDL model. The `expandOrganization` rule actually creates a new HTML file and prints the organization properties to a div-block. Moreover, within the *expandOrganization* rule an `expandNaturalPerson` rule is called creating a dedicated div-block for each person. Note that Xpand language is capable of (i) defining declarative rules for specific types, (ii) accessing element properties using the dot-operator and (iii) navigating through the model to discover linked objects via references (e.g., representatives).

Finally, the generator engine processing the Xpand template creates an HTML page like the one depicted in Figure 14.11. The illustrated HTML generation is also feasible for PDF documents. It merely requires a PDF-specific Xpand template.

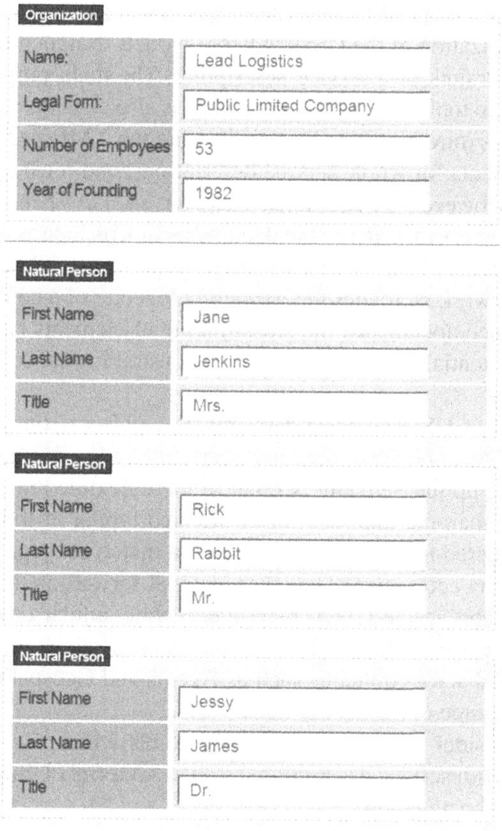

Fig. 14.11: Generated HTML Page.

14.4.3 Summary of USDL-Doc

In this section, we have demonstrated the USDL-Doc tool which transforms USDL service descriptions into HTML or PDF documents. The resulting documents may address various needs of service providers such as providing developer documentation or marketing material.

14.5 Conclusion

This chapter has explained how tools can use the USDL meta-model to make representations of concrete service descriptions for use by humans and tools.

Firstly, we can see that additional meta-model machinery is needed to allow tree-based textual serialization of the USDL's Ecore models designed for describing the structure and constraints of a service description. The main purpose is for model interchange between tools, but human readability is also considered. A simple structural containment is introduced to allow tools using the XMI specification to create valid XML documents for whole service descriptions, and for fragments from particular modules. However, an additional class for an *import* mechanism are also used to facilitate the easy tracing of the documents in which parts of a USDL model are located. The use of SML is and its interchange format SML-IF are also discussed. This framework facilitates the packaging together of a coherent set of XML Schema and instance documents, the cross-linking of elements between these documents, and the potential to validate additional constraints that cannot be expressed via XMI.

The integration of USDL service descriptions within existing vocabularies and ontologies in the Semantic Web space has been explored in Section 14.3. The technologies that make up the Semantic Web have more flexible kinds of relationships between concepts than the minimal object-oriented typing of the MOF, and some of these are considered to maximize the ability to match concepts in USDL within a larger Linked Data ecosystem. The other idea put forward is that an initial effort to match the concepts in the USDL meta-model with similar concepts in existing ontologies will allow the USDL service descriptions available in a Semantic Web context to be linked across domains such as Agents and Geospatial data, and to be manipulated by a range of reasoning tools.

Finally, we consider the use of USDL-Doc tools which use the USDL meta-model in concert with service description instances to format USDL data for human comprehension and navigation.

References

1. B. Adida, M. Birbeck, S. McCarron, and S. Pemberton. RDFa in XHTML: Syntax and Processing. `http://www.w3.org/TR/rdfa-syntax/`, October 2008.
2. J. F. Allen. Maintaining knowledge about temporal intervals. *Communications of the ACM*, 26(11):832–843, 1983.
3. C. Baumann and C. Loës. Formalizing copyright for the internet of services. In G. Kotsis, D. Taniar, E. Pardede, I. Saleh, and I. Khalil, editors, *iiWAS'2010 - The 12th International Conference on Information Integration and Web-based Applications and Services, 8-10 November 2010, Paris, France*, pages 714–721. ACM, 2010.
4. C. Bizer, T. Heath, and T. Berners-Lee. Linked data - the story so far. *Int. J. Semantic Web Inf. Syst.*, 5(3):1–22, 2009.
5. D. Brickley and R. V. Guha. RDF Vocabulary Description Language 1.0: RDF Schema, 2002. `http://www.w3.org/TR/rdf-schema`.
6. D. Brickley and L. Miller. FOAF Vocabulary Specification 0.98. http://xmlns.com/foaf/spec/, August 2010. Last Visited: July 2011.
7. J. Clark and S. DeRose. XML Path Language (XPath) Version 1.0. Recommendation, W3C, November 1999.

8. M. Dean, G. Schreiber, S. Bechhofer, F. van Harmelen, J. Hendler, I. Horrocks, D. McGuinness, P. F. Patel-Schneider, and L. A. Stein. OWL Web Ontology Language Reference. http://www.w3.org/TR/owl-ref/, February 2004. Last Visited: March 2005.
9. T. Heath and C. Bizer. *Linked Data: Evolving the Web into a Global Data Space*, volume 1 of *Synthesis Lectures on the Semantic Web: Theory and Technology*. Morgan & Claypool, 1st edition edition, 2011.
10. M. Hepp. Goodrelations: An ontology for describing products and services offers on the web. In *16th International Conference on Knowledge Engineering and Knowledge Management (EKAW2008)*, volume 5268 of *LNCS*, pages 332–347, Acitrezza, Italy, October 2008. Springer.
11. J. R. Hobbs and F. Pan. Time Ontology in OWL. Available at http://www.w3.org/TR/owl-time/, September 2006.
12. R. Iannella, H. Halpin, R. Iannella, B. Suda, and N. Walsh. Representing vcard objects in rdf. Member submission, W3C, January 2010.
13. ISO. Information technology – Document Schema Definition Languages (DSDL) – Part 3: Rule-based validation – Schematron . ISO/IEC 19757-3, June 2006.
14. I. Jacobs and N. Walsh. Architecture of the world wide web, volume one. Recommendation, W3C, December 2004.
15. T. Kiemes and D. Oberle. Generic modeling and management of price plans in the internet of services. In K.-P. Fähnrich and B. Franczyk, editors, *Informatik 2010: Service Science - Neue Perspektiven für die Informatik, Beiträge der 40. Jahrestagung der Gesellschaft für Informatik e.V. (GI), Band 1, 27.09. - 1.10.2010, Leipzig*, volume 175 of *LNI*, pages 533–538. GI, 2010.
16. J. Lieberman, R. Singh, and C. Goad. W3c geospatial vocabulary. Incubator group report, W3C, October 2007.
17. D. Martin, M. Burstein, J. Hobbs, O. Lassila, D. McDermott, S. McIlraith, S. Narayanan, M. Paolucci, B. Parsia, T. Payne, E. Sirin, N. Srinivasan, and K. Sycara. OWL-S: Semantic Markup for Web Services. Member submission, W3C, 2004. W3C Member Submission 22 November 2004.
18. A. Miles and S. Bechhofer. SKOS Simple Knowledge Organization System Reference. Recommendation, W3C, August 2009.
19. Object Management Group. Meta Object Facility (MOF) Core Specification Version 2.4. OMG Document No. formal/2008-12-10, December 2008.
20. B. Pandit, V. Popescu, and V. Smith. Service modeling language interchange format, version 1.1. Recommendation, W3C, May 2009.
21. B. Pandit, V. Popescu, and V. Smith. Service modeling language, version 1.1. Recommendation, W3C, May 2009.
22. C. Pedrinaci, D. Liu, M. Maleshkova, D. Lambert, J. Kopecký, and J. Domingue. iServe: a Linked Services Publishing Platform. In *Proceedings of Ontology Repositories and Editors for the Semantic Web at 7th ESWC*, 2010.
23. V. Presutti. Part whole design pattern.
24. A. Rector, C. Welty, N. Noy, and E. Wallace. Simple part-whole relations in owl ontologies. Editor's draft, W3C, 2005.
25. D. Reynolds. An organization ontology. http://www.epimorphics.com/public/vocabulary/org.html, October 2010.
26. Richard Soley. Model Driven Architecture. OMG Document No. formal/2000-11-05, November 2000.
27. P. van Eck, J. Gordijn, and R. Wieringa. Value-based design of collaboration processes for e-commerce. In *2004 IEEE International Conference on e-Technology, e-Commerce, and e-Services (EEE 04), 29-31 March 2004, Taipei, Taiwan*, pages 349–358. IEEE Computer Society, 2004.
28. T. Vitvar, J. Kopecky, J. Viskova, and D. Fensel. WSMO-Lite Annotations for Web Services. In M. Hauswirth, M. Koubarakis, and S. Bechhofer, editors, *Proceedings of the 5th European Semantic Web Conference*, LNCS, Berlin, Heidelberg, June 2008. Springer Verlag.

Chapter 15
Enabling USDL by Tools

Markus Heller, Benjamin Schmeling, Steffen Heinzl, Torsten Leidig, Keith Duddy, Thorsten Sandfuchs, Andreas Klein, and Matthias Allgaier

Abstract Fundamental tooling is required in order to apply USDL in practical settings. This chapter discusses three fundamental types of tools for USDL. First, *USDL editors* have been developed for expert and casual users, respectively. Second, several *USDL repositories* have been built to allow editors accessing and storing USDL descriptions. Third, our generic *USDL marketplace* allows providers to describe their services once and potentially trade them anywhere. In addition, the marketplace software can be customized to different settings and considers the idiosyncrasies of service trading as opposed to the simpler case of product trading. The chapter also presents several deployment scenarios of such tools to foster individual value chains and support new business models across organizational boundaries. We close the chapter with an application of USDL in the context of service engineering.

15.1 Introduction

In order to be of practical value to users, a set of enabling tools for USDL is required. In essence, there are three basic types of enabling USDL tools: USDL editors, USDL repositories, and USDL service marketplaces. In this chapter, we provide an overview of available tool support for USDL and discuss possible deployment scenarios as well as advanced applications.

Markus Heller, Torsten Leidig, Thorsten Sandfuchs, Andreas Klein, Matthias Allgaier
SAP Research Karlsruhe, Vincenz-Priessnitz-Str. 1, 76131 Karlsruhe, Germany,
e-mail: `firstname.lastname@sap.com`

Benjamin Schmeling, Steffen Heinzl
SAP Research Darmstadt, Bleichstrasse 8, 64283 Darmstadt, Germany,
e-mail: `firstname.lastname@sap.com`

Keith Duddy
Queensland University of Technology, GPO Box 2434, Brisbane, QLD 4001, Australia,
e-mail: `keith.duddy@qut.edu.au`

The presented tools address different phases of the service lifecycle. The service lifecycle is comprised of several phases that describe all work activities that are needed to handle services: service innovation, service offering, service matchmaking, service usage, and service feedback.

To support the service offering phase, two different editor environments are presented: the USDL editor and the USDL light editor. Both editors offer basic functionality to create, manipulate and store USDL descriptions. The USDL editor (Section 15.2) targets the user group of modeling experts and supports the full range of concepts of the USDL specification. The editor is automatically generated from the USDL meta-model by means of model-driven engineering. The USDL light editor targets casual users with a simpler and reduced user interface and allows working with a restricted well-designed set of USDL concepts.

The USDL repository (Section 15.3) supports the storage and retrieval of USDL descriptions in the service offering phase. The USDL repository provides a user interface and an application programming interface (API).

The most prominent tool for the matchmaking phase is the USDL service marketplace (Section 15.4). It provides a virtual marketplace for the publication, search and selection of services and brings service providers and service consumers together on a trading platform.

The three types of tools can be combined differently in various deployment scenarios which are exemplarily discussed in Section 15.5.

Finally, an application of USDL in a more complicated application scenario is detailed as an example of advanced tools that leverage the potential of USDL in a technical engineering context of service engineering. A framework for the integration of services into service-based enterprise systems is described in Section 15.6.

15.2 Editors

The editors for the creation and modification of USDL descriptions are the most basic tools. Both editors address a typical dilemma when tool support is provided for a basic meta-model. On the one hand, the complexity and expressive power of the modeling language needs to be supported in the editors while on the other hand the user's mental models for the editor usage need to be as simple and straightforward as possible. Both USDL editors address this dilemma with different emphasis:

- The first editor — the USDL editor — supports the full range of concepts of the USDL meta-model, and, thus, is meant for modeling experts.
- The second editor — the USDL light editor — targets casual users and offers a restricted well-designed set of USDL concepts to be able to specify basic service characteristics within a form-based editor with a simpler and reduced user interface.

Both editors have been developed by SAP Research and they are presented in the two subsequent sections.

15.2.1 USDL Editor for Experts

The USDL meta-model is the main artifact for language specification and formalizes the USDL concepts and the relationships among them. It defines the abstract syntax of the language and hence specifies the rules that instances of the meta-model must adhere to. As USDL covers business, operational and also technical aspects, there are many concepts that have been assembled into the USDL meta-model.

In order to facilitate this task, the USDL editor presented in the following supports users in defining valid (i.e., meta-model conforming) service descriptions and also in displaying such descriptions in a structured and understandable way.

15.2.1.1 Requirements and Design Choices

To understand the design choices that have been taken for this USDL editor we present the most important requirements that have driven the development. As there are implementations of a USDL editor currently available, we categorize the requirements into general requirements (that every USDL editor should support) as well as specific requirements that motivate the development of the USDL editor presented here.

General requirements for USDL editors:

Easy creation of new service descriptions from scratch Creating new USDL descriptions should be as easy as possible, e.g., by using wizards. The editor can start with a minimal description which only contains the mandatory description elements predefined by default values.

Editing and displaying already existing service descriptions Service descriptions can be displayed in a structured way and reduce complexity whenever possible or applicable, e.g., the editor should provide an easy and consistent way to instantiate relationships, i.e., links between description elements.

Assuring that descriptions conform to the USDL meta-model The editor should help the user to define models that adhere to the rules given by the meta-model. This can either be achieved by validation functionality or even by assuring that only valid models can be saved.

Abstraction from concrete syntax Although there is a concrete USDL syntax based on XML (cf. Chapter 14), it is tedious to edit USDL files in an XML editor, because it is too low level, e.g., references and id generation has to be done manually. Moreover, large XML documents tend to be unreadable for humans. This can for example be overcome by offering subsections per USDL module and only displaying information that is currently relevant for the user.

Requirements that were taken into account specifically for the USDL editor discussed here are:

Support for all USDL concepts There should be at least one editor that supports all the concepts defined in the USDL meta-model, otherwise those concepts cannot be instantiated. This requirement was especially important for being able to validate all language concepts during the development of the USDL language.

Changes in the meta-model are automatically reflected in the editor The concepts added to the USDL meta-model during the development of the language had to be validated by a set of examples, e.g., to see how a certain concept can be instantiated and how it would work for a concrete service description. However, during the development of the language the meta-model is not stable and will change frequently. Every change in the meta-model will affect the code of an editor. If changes in the code have to be done manually, there will be high development efforts.

15.2.1.2 Realization Concepts

Especially the last requirement has a direct impact on the architecture and set of frameworks to be used for the implementation of the USDL editor. As discussed in Chapter 14, the USDL meta-model has been specified by the Eclipse Modeling Framework (EMF). EMF provides Ecore as a meta-modeling language with a corresponding Ecore editor, an Ecore2Java transformation, an EMF Reflection API, Ecore2XSD transformation, and a XML/XMI Serialization for Java objects generated by the Ecore2Java transformation. All the features are supported by tools integrated in the Eclipse IDE and facilitate the automatic, model-driven, generation of user interface and application code for the editor. An overview of the architectural components and models is depicted in Figure 15.1.

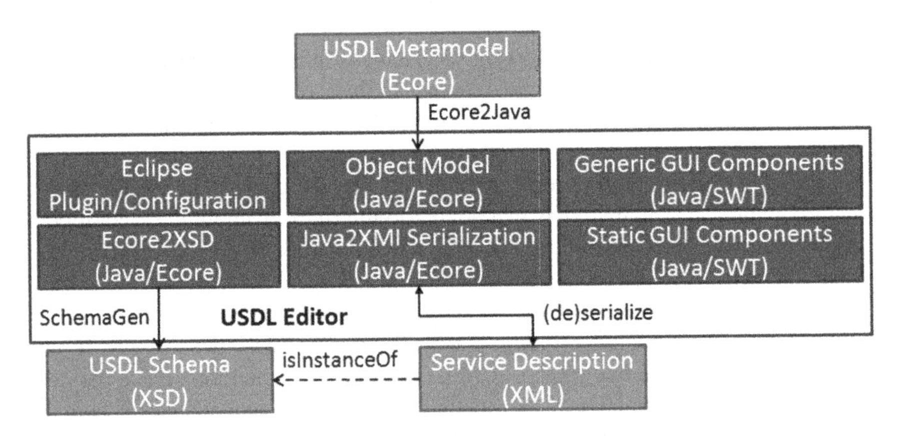

Fig. 15.1: Architecture of the USDL Editor.

The USDL Editor comprises six architectural components: The *Eclipse Plugin/-configuration* component contains the Eclipse-specific configuration artifacts. The *Object Model* is the meta-model's representation as Java Classes which is generated by the Ecore2Java transformation and is used as data model for the editor. The

Generic GUI Components are based on the EMF Reflection API and render input forms based on the attribute types and values of the object model, e.g., combo boxes in case of enumeration types. There is also a smaller part of GUIs that is static and has to be written manually (*Static GUI Components*). This includes for example the main user interface and the structuring into different tabs. The *Ecore2XSD* component is a model-to-model transformation that generates the USDL XML Schema out of the Ecore meta-model. This module is an extension of the default Ecore2XSD transformation provided by EMF. The *Java2XMI Serialization* component — also a slightly customized version of the EMF serialization — supports writing and reading of USDL descriptions from service descriptions, which can be validated against the USDL meta-model.

Whenever there is a change in the meta-model, the Ecore2Java transformation can be run which updates the Java object model. Because of the generic GUI Components, the user interface of the editor automatically changes according to the new object model. As an example consider the introduction of a new attribute to the Service class. After invocation of the Ecore2Java transformation the editor will analyze the changed Java classes at runtime (using EMF reflection API) and automatically add a new input field, e.g., a text field for attribute type String. This works because the generic GUI components always rebuild the user interface of the editor after restarting, reflecting changes immediately. A nice side effect of this approach is the uniform user interface. However, not the complete structure of the user interface can be generated automatically. That is why the Eclipse Plugin/Editor Widget component allows for the manual adaptation of the main widget, e.g., the subtabs for modules or the elements shown in each tab can be specified programmatically. Figure 15.2 shows a screenshot of the USDL editor.

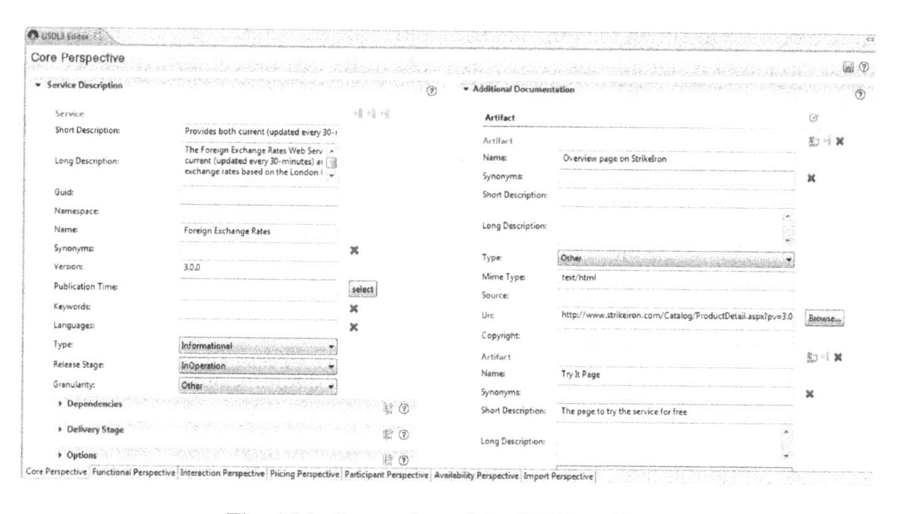

Fig. 15.2: Screenshot of the USDL Editor.

15.2.2 USDL Light Editor for Casual Users

The USDL light editor is targeted at users, who want to specify the core business information of the service, such as general header data, participant, and contact information, abstract capabilities and actions of the service, the core interaction schema as well as basic pricing, availability information. In contrast to the editor discussed in the previous section, the USDL light editor hides some of the complexities of the language in favor of lowering the entry barrier for casual users.

15.2.2.1 Requirements and Design Choices

The editor should be easy to deploy and access by a potentially large group of diverse target users in arbitrary application domains and business sectors. So we decided to implement the editor as a Web application. Another important requirement results from the fact that our target users are mainly non-technical people. Therefore, the editor should be easy to use, which means it should reduce the complexity of USDL and translate technical details into a user friendly and understandable representation. Simplification was the main goal to lower the entrance barriers for new users. Not every service needs the full breadth of USDL and we should not penalize the many simple services for the sake of the few complex ones. Specific details, e.g., complex pricing models, can be added later by the specialists using the previously presented editor.

Other important principles to achieve usability are a) focus on aspects, b) progressive disclosure of details [20], and c) utilization of context [22]. The principle of focus means that the visual appearance (color, font, visibility, layout, lines, etc.) should be used to attract the user's attention to the aspects that are really necessary to do the job in a certain situation. This could mean that only information and interaction elements are presented to the user if they are relevant to a certain aspect. For example, if the user is specifying the interaction aspects of the service, he will only be presented UI elements that are dealing with interaction behavior. Elements out of focus will be eliminated, blurred, or put to background. Also the interaction elements need to be offered in the right representation, which would include the use of graphical representations in this case. Changes of the focus or scene can be indicated by smooth animations. [18]

In addition, the principle of context suggests to limit the user activity to one specific context supporting it by the adequate operations and contextual information available inside the service description but also related information from other services and the background business systems. The editor tool must not be a monolithic isolated tool. It rather should be embedded into an overall business setting, which provides valuable information and constraints about the business context. Therefore, the editor must be open to be integrated into the business environment of the company.

Progressive disclosure [20] is another principle that helps to avoid scaring people and reduces the user interface elements to the amount that is necessary in the

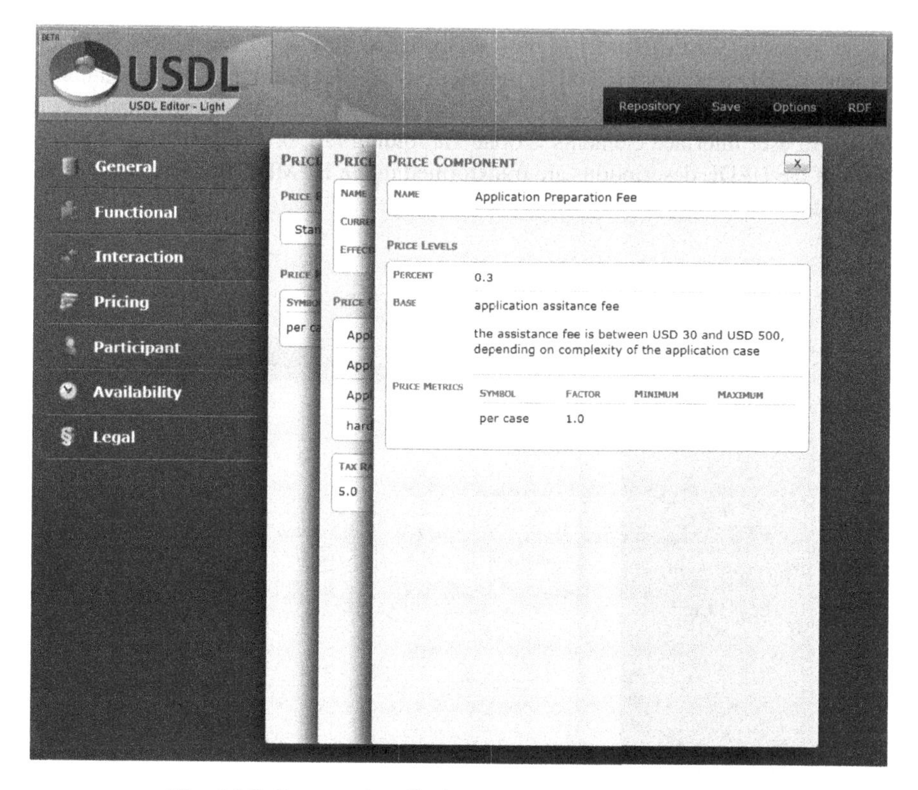

Fig. 15.3: Progressive disclosure of pricing information.

current situation. Only if the user is going to be more specific on an issue, the respective user interface elements will be shown. Figure 15.3 shows how a price plan is progressively disclosed to its price components.

15.2.2.2 Realization Concepts

The USDL light editor is realized as a single-page application (SPA) based on HTML5, SVG and JavaScript. This ensures that it can be used on major desk-top and mobile platforms that come with a Web browser. No dedicated deployment procedures are necessary. Utilizing the local store and caching capabilities of modern Web browsers, the editor also can be used online. The editor can be packaged as a WebApp for the GoogleChrome and Mozilla Firefox browser. The package size remains less than eight megabytes, which leads to acceptable loading time even if loaded from the Web server.

JQuery is used as core DOM and UI framework and we can rely on a rich set of available plug-ins, which we use for in-place editing, smooth animation, gesture-based interactions, especially on mobile devices. The editor is using the RESTful

USDL repository API in order to browse, load, and save service descriptions from various USDL repositories. USDL descriptions are delivered as XML documents according to the USDL XSD (cf. Chapter 14). Mapping of XML representations of USDL to user interface elements is done via folding [23, 6, 21]. Specific substructures of the USDL descriptions are transformed into a HTML/SVG structure that is folded into the corresponding XML part.

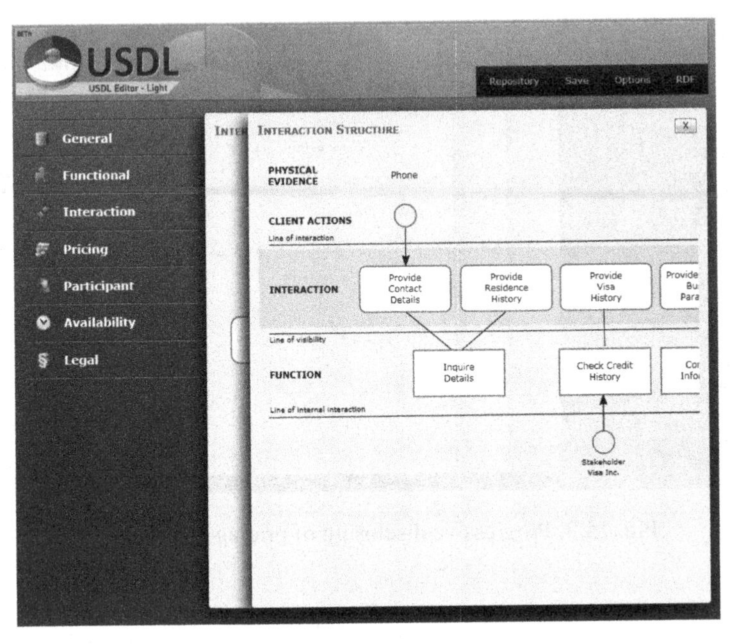

Fig. 15.4: Graphical representation of interaction model.

Folding provides an easy way to achieve a bidirectional mapping between the USDL model and the user interface representation. Changes of the HTML part will result in corresponding changes in the XML part and vice versa. Extensions are simply done by specifying HTML representations and their corresponding model parts. A set of generic folding operations is doing the actual mapping and updating. In HTML5 this also can be done for graphical representations using SVG (Figure 15.4).

We use the card stack metaphor in order to achieve focusing and progressive disclosure. Each major aspect is summarized on a card. If the user wants to navigate to specific sub-aspects, a new card describing this aspect will be opened on top of the current card. The user can jump between the cards ("look back") by moving the cards to the left or right. The UI representation on a card consists of forms containing in-place editable text fields, which are supported by pop-up tools depending on the type of the input field, e.g., rich text commands, option menus, date-time selectors. Elements of the interaction module of USDL have a graphical representation, loosely based on service blueprints, because it is more intuitive to specify behavioral aspects in a graphical notation.

15.3 Repositories

For the storage of the created USDL descriptions, a set of USDL repositories has been developed. The repositories, once deployed to many sites, allow the interchange of USDL descriptions that have been described once for the deployment into multiple repository sites. In the remainder of this section, two USDL repositories with different addressed needs are explained. Besides the implemented state within these two repositories, a standard interface and protocol for accessing USDL repositories in general is currently under discussion.

15.3.1 USDL Repository as an Enterprise Application

After creating USDL descriptions for his services, the service provider usually publishes the descriptions either for broad access through the Internet or for (internal) consumption in the corporate intranet. This is where USDL repositories come into play. A model-driven approach is suitable to construct a repository that fully covers a complex model as USDL, that provides persistence and de-/serialization of the USDL descriptions, and that offers web-based access to create, read, update and delete descriptions. Implementing the persistence layer alone means that

- a database schema reflecting the whole USDL model has to be created (either by modeling it with ER diagrams or creating it directly with SQL statements),
- SQL statements to create, read, and update service descriptions have to be written, and
- serialization into XML and deserialization from XML according to the concrete syntax format defined in Chapter 14 has to be undertaken.

Through model-to-model transformations, it would be possible to create an ER-diagram or the SQL statements could be generated from the model. However, the USDL model is structured according to object-oriented principles while the database structures are relational.

Taking this into account, it is reasonable to use an ORM (Object-relational mapping) framework, such as Hibernate[1] or EclipseLink[2] which are both JPA (Java Persistence API) providers. JPA provides an annotation-based mechanism to map Java

---

[1] Hibernate, JBoss Community, `http://www.hibernate.org/`

[2] EclipseLink, The Eclipse Foundation, `http://www.eclipse.org/eclipselink/`

objects to database tables which is indeed a challenging task because object-oriented principles cannot be directly mapped to relational data models. The most prominent examples are inheritance (which is not directly supported), attributes with complex data types (which cannot be directly mapped to columns) and polymorphic queries (also not directly supported). These and other shortcomings are compensated by the ORM frameworks. JPA allows different inheritance mapping strategies (Single Table, Joined, Table per Class), which result in a different number of tables, number of joins, and different amount of replicated data. The Java classes with the required annotations can directly be generated from the USDL meta-model. The Eclipse Modeling Framework (EMF) offers a rich tool set for this purpose. The generated Java classes are based on the Ecore class model which offers a richer reflection API than the standard class model of Java. The Ecore class model can, however, not be directly processed by JPA frameworks, such as Hibernate. Hence, Teneo[3] was used to provide this mapping. While Hibernate allows Java objects to be written to the database directly, Teneo extends the reach of Hibernate and EclipseLink to Ecore models. It is possible to create a class model in Ecore, provide the annotations (such as adding an ID annotation for all classes through a model-to-model transformation) needed by the JPA framework, and then create the Java classes that are further used in the application implementation. Besides the Java classes, also JEE 6 Data Access Objects (DAOs) and internal services for transaction handling are created which can be used to facilitate the backend access.

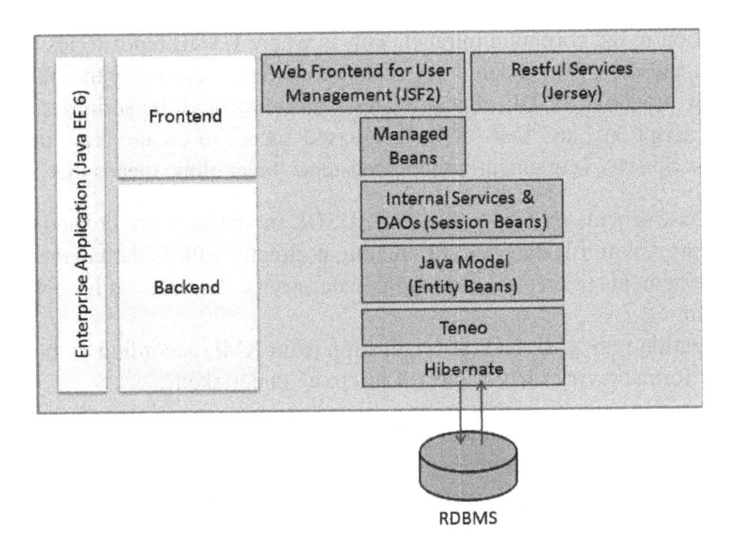

Fig. 15.5: Technology Stack of the USDL Repository.

[3] Teneo, The Eclipse Foundation, `http://wiki.eclipse.org/Teneo`

The JBoss 6 Application Server[4] was used to host the repository, providing a technology stack which is depicted in Figure 15.5. The next section shows how a remote access to a repository can be realized.

We have focused mainly on the backend of the repository, since the frontend functionality is standard functionality which can — to a certain degree — be independent of the backend application.

Remote access to the repository is enabled via Jersey[5] (REST style access to the service descriptions) and odata4j[6] (an implementation of the OData[7] protocol for Java).

15.3.2 Building a USDL Repository as a Service

A project of the Smart Services Australian Cooperative Research Centre (CRC) at Queensland University of Technology (QUT) has developed a repository generation tool called Repository as a Service (RaaS). Its first prototype implementation is documented in [14]. Unlike other model repositories, such as EMFStore [17], the Jazz repository [8], and others [9], RaaS is more concerned with storing a large number of instances of a given data model than supporting models within software engineering approaches. RaaS builds a Web service accessible repository of model objects, as specified by any EMF Ecore meta-model. The USDL was the first substantial model for which a repository service has been generated and deployed for use within service-oriented projects in the CRC and its partners. The size and complexity of USDL, and the likelihood of USDL repositories needing to contain tens or hundreds of thousands of service descriptions have driven the requirements for the structure and scalability of the RaaS tools.

15.3.2.1 High Level Design Rationale

USDL is meant to define a set of service descriptions for shared use in an enterprise or marketplace, and therefore, the implementation of a repository for USDL descriptions is necessarily a datastore with distributed access by all service providers and their potential clients, as well as service portfolio managers. The design of efficient and scalable distributed systems requires the smallest number of remote interactions as possible, conveying the largest payload possible without forcing the unnecessary duplication of data in these payloads. The current trend in services design is toward document-oriented computing (using large gained XML documents), rather than application integration using remote procedure calls (using a one call per fine-grained

[4] JBoss Application Server, JBoss Community, `http://www.jboss.org/jbossas/`

[5] Jersey, java.net, `http://jersey.java.net/`

[6] odata4j, An OData Toolkit for Java, `http://code.google.com/p/odata4j/`

[7] Open Data Protocol (OData), `http://www.odata.org/`

data element update). Our design is based on the premise that the granularity of interaction with repository services should be based on the semantics of the data being exchanged, rather than dictates of the middleware style being employed to implement the repository.

Another dichotomy in service provision which invites much polarizing debate is the choice of the W3C and OASIS standards for Web services, known as WS-*, versus the more minimal approaches to HTTP protocol reuse known as Representational State Transfer (REST). We wish to allow both styles of interaction with USDL repositories to gain the widest possible adoption of USDL.

Here is a list of requirements that formalizes some of these high-level design goals:

As a Service Our first requirement for a USDL repository is that it also be available to users as a service. We have named our implementation of a repository generation framework Repository as a Service (RaaS).

Full code generation, with maximal off-the-shelf reuse The experience of building a prototype repository for an early version of USDL by hand has led us to the conclusion that hand-coding tools to support metadata management are a wasted effort in an environment where the meta-model is undergoing continuous change. The RaaS tools must reuse code generation frameworks where possible, and use best-of-breed model transformation technologies where new code generation approaches are needed to fulfill the other requirements. As the USDL abstract syntax is specified using MOF models, and as EMF is mature and an open implementation of the MOF, with the most active developer community, we chose to use Eclipse/EMF as the base technology platform.

Multi-Granular Access The use of EMF as the in-memory Java programming framework for programmers to create, access and modify USDL metadata provides a very fine grain of manipulation of individual objects and their attributes and references using either generated accessor methods or the EMF reflective interfaces. This granularity of access should be made available to distributed clients who need to update particular attributes in a USDL service description on a regular basis. However, if the generated interfaces of all of the EMF classes in the USDL model were the only basis on which a Web service was made available, there are a number of problems that this would bring with it. Firstly, the generated WSDL which represents the EMF interfaces for the whole of USDL contains thousands of operations, grouped into hundreds of port definitions. Secondly, the latency associated with the hundreds or thousands of remote operation invocations required to populate a single full service description creates a very inefficient distributed application. The option to upload a whole service description, which has been edited locally, into a remote repository allows us to overcome this inefficiency. However, there are many use cases where some subset of a service's descriptive objects and attributes may need to be updated as a group, for example in the update of a Price Plan (which is only valid with at least one attached Price Component, which comes with at least one Price Level), or the addition of a new Service Level Agreement (which also must have many additional contained objects in order to be valid). Therefore it is a requirement that a user of the RaaS be able to choose the granularity of object group creation and update, both to match their requirements for consistent updating of a related group of model elements, and for distribution efficiency.

Multiple User Roles with Defined Rights and Access Control The USDL brings together a number of different aspects of service metadata that have to date been dispersed throughout an organization, and which are represented in diverse formats. The set of roles in an organization that are responsible for this metadata includes at least: technical, marketing, provisioning, legal and business owners. Then there are clients of the services described in USDL who need access to the descriptions via search and browse applications. We require that a set of arbitrary roles may be defined to represent the various users of a USDL repository which is deployed in a particular setting, and that they be given read or read/write access to the parts of the USDL model that they are intended to populate or query. We require that this specification be in terms of a subset of the MOF packages and classes defining the USDL abstract syntax, and that appropriate access control mechanisms be derived from the specification. A deployed USDL RaaS must also offer an administration interface which allows roles to be granted appropriate access rights and users to be assigned to roles. Furthermore, if certain roles are only permitted access to a subset of the full create/read/write capabilities then an appropriate role-based WSDL specification (or REST interface) must be generated that hides irrelevant detail from users in that role (such as create and write operations for read-only users, or even read operations for modules which are outside the purview of the role).

Reliable and performant storage We require that a repository service can safely store a set of USDL service descriptions in such a way that the data is robust and that it can be searched and retrieved efficiently. We have chosen the Teneo plugin for object-relational storage.

Rich query language support Service portfolio managers and service clients need to be able to query a repository of USDL service descriptions in a flexible manner to allow them to discover appropriate services for their needs.

15.3.2.2 A Design which Meets the Requirements

The tools we have developed using a combination of off-the-shelf and bespoke code generators and wizards are known as Repository as a Service. Fig. 15.6 shows the RaaS code generators and the architecture of the deployment of a USDL repository, which includes objects in a Java Virtual Machine which conform to the EMF types generated from the USDL Ecore model, and a persistence layer using Teneo and Hibernate, as well as a Web services layer supporting WSDL and REST services. The following subsections describe the design in detail.

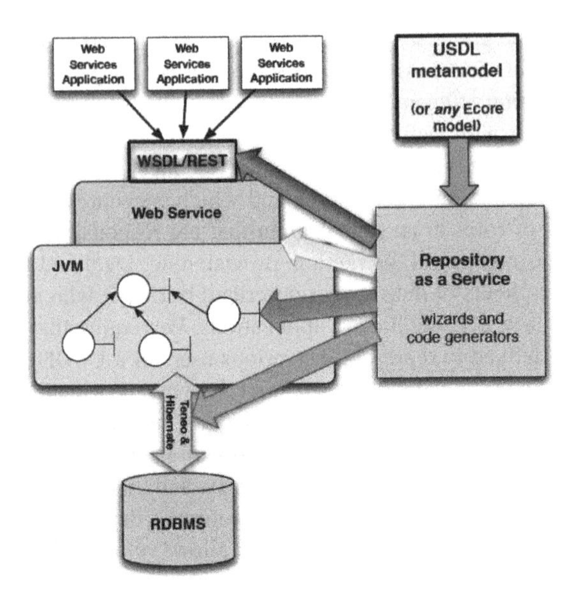

Fig. 15.6: The Architecture of the RaaS Tools.

WS-* and REST Services with Multi-granular Access

The design of the serialization of the contents of a graph of objects is the key to providing "multi-granular access." An object that contains other objects must be able to serialize its own attributes as well as the object(s) that it contains. We identify the classes in the USDL meta-model which represent roots of trees of contained objects, and use these as the candidates for read/write operations in the Web service which can be passed a complete set of connected objects making a well formed piece of a service description. We generate appropriate XML parameter schemas, and allow repository users to identify any additional meta-model classes as candidates for operation generation. For example: PricePlan is selected as a candidate, but a user can also chose PriceLevel (a contained class) as a candidate for read/write operation generation, as this is the most likely class to be frequently updated when a price changes.

Another approach which is supported is the choice to assert a containment relationship over a reference in the meta-model which is specified as non-contained. For example, in a number of places in USDL, Expressions are referred to by other classes, perhaps with the rationale that they may be shared by several instances that require them, in the same way as an Organization is defined once per USDL repository, and referenced by all the USDL model elements that need to nominate a particular Organization as a BusinessOwner, Stakeholder, etc. But in the service description editing scenario, it is unlikely that an Expression can be shared by multiple service description instances, and so if we annotate the meta-model with an asserted containment then the Web service can treat each Expression usage in this context as a contained element, and need not generate a separate interface for it.

Object Identification across Address Spaces

The first practical consideration when implementing a distributed version of the in-memory EMF object graphs that exist in a Java program is the identification of objects previously created in the USDL repository. Unlike previous generations of middleware, such as DCOM and CORBA, Web services toolkits make distributed object identification the problem of the application. Even when a service description is created as a single XML document, and uploaded into the repository, there are some things that must be provided by reference, rather than by value. For example, the Organizations, NaturalPersons and Resources specified in the Foundation Module of USDL must be defined in the repository once only, and have references to them from other USDL objects. If not, the change in contact details, locations, etc. for these entities cannot be made in a single place, and their utility as shared classes is lost.

The module structure of USDL is designed to provide a default separation of concerns, and granularity of service description population, even though the main classes in each non-core module are contained by references from the Service class in the Service Module. The module structure implies a module-at-a-time population of service descriptions by different roles in the maintenance of USDL data. However, the existence of many cross-module references in the USDL meta-model implies that each module must to be able to refer to objects that are already present in the repository.

To support the identification of each extant object in RaaS, we insert a new attribute into each class in the USDL meta-model, named raasId. This is a simple, but effective way of being able to populate references to objects already created by another user, which are not present in the address space of a client. These identifiers also allow the creation of meaningful URIs for REST access to the repository — especially where multi-valued references don't allow these to be determined from context by their type name alone.

15.3.2.3 Implementation

In this section we first introduce the set of tools which is used to create the majority of the code for the RaaS, and then explain the process in which we use them to meet our design goals.

Tool Reuse

As stated in the design goals, we use off-the-shelf tools and frameworks where possible, and model transformation and code generation approaches where no appropriate tool exists. We use the EMF generated classes, wrapped by Connected Data Objects (CDO)[8] as our Java object platform, and Teneo[9] with Hibernate[10] to generate a relational persistence layer from the EMF.

The fact that Teneo uses the Hibernate framework, gives us an implementation of the Hibernate Query Language (HQL) for use by clients which we provide through a generated query operation. Optionally, the repository may also support the use of the OMG standard Object Constraint Language (OCL), which is supported by an open-source Eclipse plug-in.

The Web services layer reuses the Java XML Schema generation tool JaxB, and Java Web Service implementations JaxWS and JaxRS. The use of JaxB in combination with EMF provides a number of options for client applications to communicate with the repository. They can populate documents that conform to the generated XML Schemas, and invoke the WSDL operations from any Web services toolkit, or they can use the EMF interfaces to create groups of interconnected objects, and then pass these to a proxy for the Web service which performs the SOAP/XML serialization and remote operation invocation for them. There are also REST URIs generated which allow XML or JSON formatted documents to be passed via HTTP GET and PUT for ease of Web browser-based repository client creation, and integration into RESTful applications. The XML generated by JaxB is flexible enough to allow a whole subgraph of serialized EMF objects to be transmitted as the parameter to a single SOAP invocation or REST interaction. This capability is matched by a deserializer in the repository that accepts XML fragments of arbitrary size and creates matching structures in the datastore when they are created, or patches them into the existing structure in the datastore when updates are performed. Finally, in order to manipulate the input meta-model to RaaS we use the Tefkat model transformation language and engine [19].

Combining the Tools and Frameworks

Firstly, we create a new version of the USDL meta-model using a Tefkat transformation to insert a unique identifier attribute "raasId" into each class in the meta-model. Then EMF and CDO code is generated from the new meta-model. A proxy is implemented in CDO to allow these identifiers to behave as de-facto references when an object referred to is not in memory. Teneo is nominated to the CDO code generator as the relational persisence layer, and its code is also generated in this step.

[8] http://wiki.eclipse.org/CDO
[9] http://wiki.eclipse.org/Teneo
[10] http://www.hibernate.org

Then, using pattern matching, we identify the "root classes" of each package in the meta-model, taking into account any asserted containment relationships. An Eclipse wizard for RaaS presents these to the repository builder as default medium-grained model entry points to the Web service generation tools. For example, in the Pricing Module there are twenty classes, but as only four are not contained by another class, it would be sufficient to generate the WSDL operations (or REST URIs) for only these classes, and allow interaction with the repository only for creation and update of whole PricePlans, PriceFences, PriceMetrics and Taxes, with all their contained classes attached to them by containment references.

Once the RaaS repository builder has chosen any additional classes for which to generate operations, the generated EMF/CDO classes are annotated for JaxB, JAX-RS and JAX-WS code generation by the wizard, and a project build is initiated. The generated Web service, including the persistence layer, is then packaged as a WAR file, and may be deployed into Tomcat or other Web services container on any server machine.

15.3.3 Conclusion

The design of USDL as a set of EMF Ecore models allows for the use of many powerful model-driven engineering and code generation tools and frameworks. It suits both the generation of a repository as an enterprise application as well as a repository as a service. However, the combination of these tools into a scalable three tiered distributed server requires a number of important additional design choices. We have seen that the selection of only a subset of meta-model classes for remote access provides a basis from which to match the semantics of the model with an appropriate granularity of distribution of related model objects. The use of additional identifiers in combination with the CDO framework allows programmers to manipulate a subset of objects, while preserving references to extant objects that are resident in the remote repository.

15.4 USDL Marketplace

Creating descriptions with USDL editors and storing the descriptions in USDL repositories happens in the offering phase. Referring back to the running example introduced in Section 8.7, the 4PL would have to capture a description of its service in USDL and store it in a local or third-party repository. A marketplace is required to enable the trading of services in the matchmaking phase. The example mentioned a logistics marketplace for enabling trade of logistics services.

However, 4PL might want to offer its service on different marketplaces and trading platforms. Currently, 4PL would have to capture its service several times conforming to the representation requirements of each individual marketplace. In order to remedy this situation, we have developed a generic standard software for USDL service marketplaces. The generic software can be customized to different settings (e.g., the logistics domain), is able to import and export USDL descriptions, and can subscribe to several USDL repositories thus enabling providers to potentially *describe once and deploy anywhere* (further discussed in Section 15.4.1).

The generic marketplace software particularly considers the idiosyncrasies of service trading as opposed to the simpler case of product trading. One such idiosyncrasy concerns the pricing of services as discussed in Section 9.2. A service is an intangible entity by definition, for it cannot be stored or inventoried by either providers or consumers. Therefore, the time of purchase — when the consumer *commits* to pay in exchange of the provisioned service (not to be confused with the time of payment) — always precedes the time of production. That represents a key difference from the selling of goods, which can instead occur either before production (e.g., in a make-to-order scenario for a customized item) or afterwards (in the case of commodities). Another idiosyncrasy and benefit of our solution is discussed in 15.4.2.

Our generic USDL service marketplace is implemented as a J2EE[11] application based on JBoss's Seam framework.[12] JBoss Seam supports REST for communicating with the USDL repository as well as a fully-fledged tool set for creating user interfaces, business logic and database persistence.

15.4.1 USDL Service Publishing and Presentation

Providers can publish USDL services into a marketplace through direct uploading or using a certificate-based subscription mechanism, which connects the providers' marketplace accounts with a USDL repository. In this case, updates on a USDL service in the repository get automatically propagated to the connected marketplaces. See details of possible deployment scenarios in Section 15.5.

The service marketplace is capable of extracting different aspects of a USDL service description and present them tailored for a specific role. Figure 15.7 shows how a USDL service description exposing the customer relevant aspects of the service such as a price plan, main description attributes and references to the provider's main contact. The USDL description is parsed and rendered in human readable format (cf. also Section 14.4), exposing, e.g., price fences in a tabular design. The given USDL description is exposed with a link for further programmatic access to the description.

---

[11] `http://java.sun.com/j2ee/docs.html`

[12] `http://seamframework.org/`

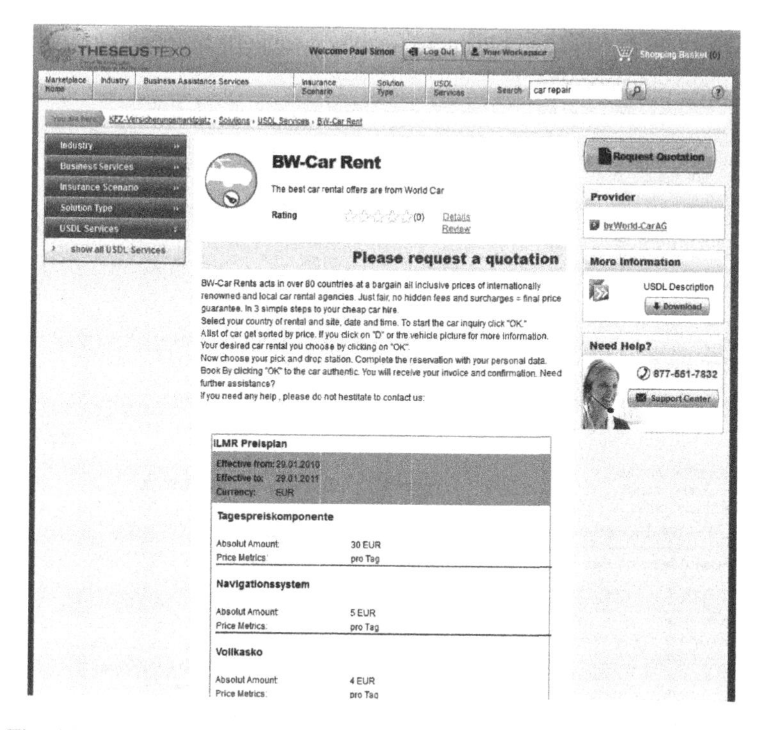

Fig. 15.7: Visualized USDL service description targeted at the consumer.

15.4.2 Business Scenarios — Enhanced User Guidance through Abstract Services

Another central functionality of the service marketplace is the support of matchmaking and bundling options for the involved roles in the concept called "business scenarios."

A business scenario enables the customer to find a non-trivial combination of services to fulfill a complex business goal. Complex business goal cannot be achieved by a single service or a simple service bundle, but can be solved through logically combining different services usually coming from different services providers. Such business scenarios have to be understood by marketplace customers and the marketplace itself, therefore they are supported with a textual and a formal description. The textual description is written by a business expert, targeted at specific customers. The formal description of a "service plan" is the corresponding technical representation and also created by the business expert.

A service plan is an abstract solution to a problem and consists of abstract services and abstract service modules. An abstract service module contains different abstract services, or other abstract service modules combined with the logical operators AND, OR, or XOR. A built-in SAT-solver[13] supports the expert in configuring service plans consistently.

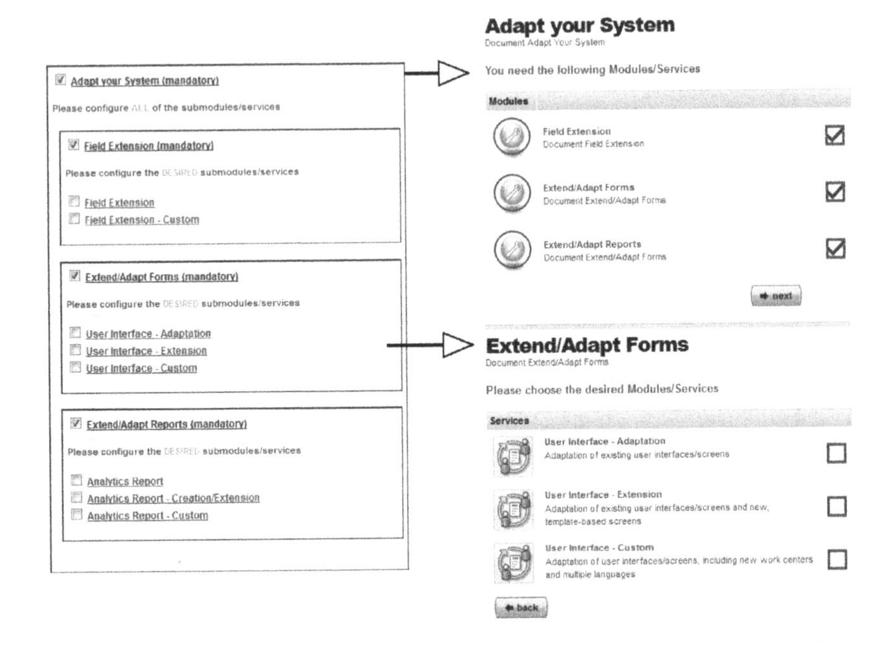

Fig. 15.8: Guided service plan procedure.

When a consumer instantiates a business scenario, the marketplace system generates a guided procedure out of the service plan model and supports the user in fulfilling the scenario on an abstract level. The contained abstract service modules are displayed successively and the consumer can choose exactly one, multiple or all abstract services or abstract service modules from each module depending on the logical operator defined by the business expert. Figure 15.8 shows the logical model behind a service plan on the left side. On the right side two guided procedure steps are visualized.

After configuring the service plan on an abstract level, the marketplace system generates a proposal with concrete services described in USDL that fulfill the user's service plan. The proposal is a best bet, based on parameters such as minimal price, service rating, provider rating and geographical closeness. Furthermore, the importance of the different factors can be weighted by the user as shown in Figure 15.9.

This split approach, which is based on the concept of abstract services, allows the business expert to separate the process of describing his business problem and finding the right providers for the fulfillment. From a customer's perspective, a concrete

[13] http://www.sat4j.org/

Fig. 15.9: Service Plan proposal.

instantiation of a business scenario can be seen as a new building block through loose coupling of cross-organizational services.

Concluding, we can say that our prototypical implementation of a service marketplace benefits from concepts introduced by USDL, especially the notion of abstract services allowed us to develop a goal-driven solution for more complex business scenarios. Future work could empirically evaluate the influence on the acceptance rate for business service marketplaces that make full use of USDL's capabilities.

15.5 Deployment Scenarios

As the tools supporting USDL are only loosely coupled, they can be combined for different deployment scenarios and business scenarios, supporting various use cases, outlined in this section. This allows the involved parties to create individual value chains and support new business models across organizational boundaries in an Internet of Services.

15.5.1 Simple USDL Tool-Chain

A first variant of a minimal tool chain is depicted in Figure 15.10 on the left hand side. It is operated within a single unit or organization and merely consists of an editor accompanied by a USDL repository. This simple combination of two USDL-based tools enables the organization to collect all of their relevant services and capture descriptions of the services in a standard format and using standard tools. Establishing such a service inventory is a prerequisite for the reuse of services within the organization.

A second variant of a minimal value chain is shown in the right hand side of Figure 15.10. Within a single unit or organization, a service marketplace can be added to the deployment scenario. With a service marketplace, the services can be exposed for consumption. This simple combination of USDL-based tools would normally be deployed in addition to existing service delivery channels, not depicted in the figure. It generates additional value to be created by exposing the same services into the service marketplace for combination purposes or for consumption by other internal or external stakeholders.

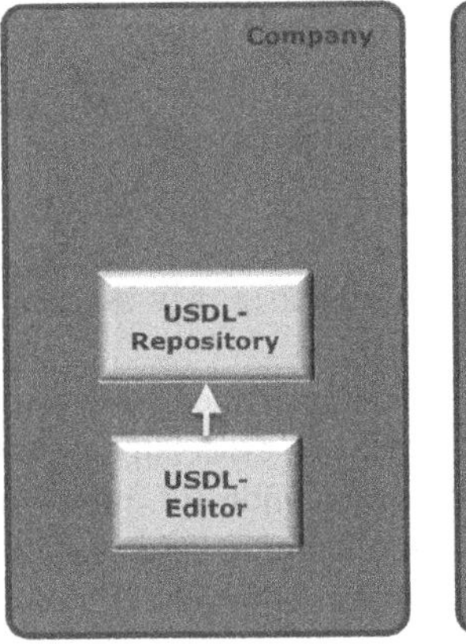

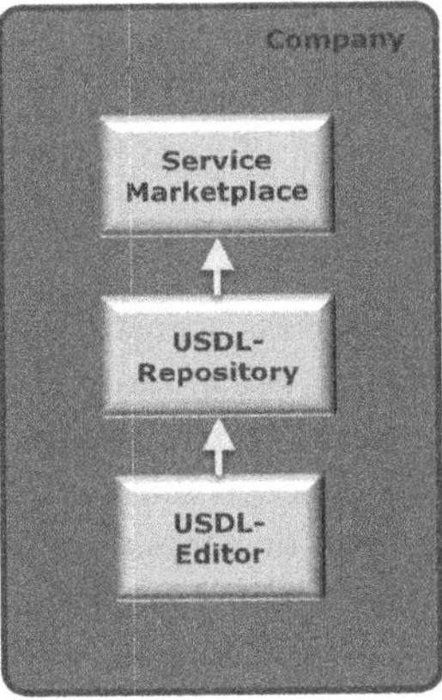

Fig. 15.10: USDL value-chain within a single entity (right: with a service marketplace).

15.5.2 Value Chains for Multiple Stakeholders

The previous section discussed how the USDL tools are used within the boundaries of a single organization. If multiple organizations (or departments, units and the like) are involved, the tools can be deployed in an inter-organizational setting. The exchange of USDL service descriptions (or maybe even fragments, such as price model fragments) between the tools is the enabler for such advanced scenarios.

An example for a possible inter-organizational usage scenario is shown in Figure 15.11. The setting shown can be realized with the goal of establishing a distributed

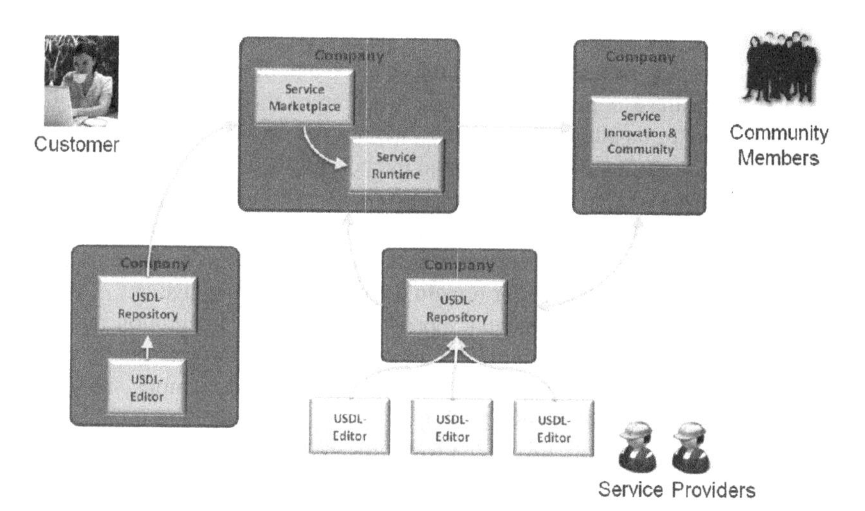

Fig. 15.11: Example deployment of USDL tools in a cross-organizational scenario

service value chain that is spread across multiple organizations. Different companies can install central USDL repositories within their organization as a focal point for storing and maintaining their services (as seen at the bottom of the figure). In order to feed service descriptions into the repository, service providers in the organization use USDL editors to create service description documents and to store them into the USDL repository. These organizational USDL repositories need not (and often should not) be connected directly. Instead, a set of tools is available that can consume and process USDL services, for example, service execution runtime environments that can support the enactment of services that are described in USDL or tools to support the special area of service innovation (right top part of Figure 15.11). These tools can be interconnected with each other, even across organizational borders, and exchange USDL-related information with each other. A service marketplace can either be located within a single organization (top left part of Figure 15.11) which acts as a broker for services from other companies. Multiple service marketplaces can be located in various sites.

15.6 Service Integration Framework

Besides the fundamental tools discussed so far, we also present an application of USDL in the context of service engineering that leverages the potential of USDL. In the remainder, a framework for the integration of services into service-based enterprise systems is described.

In the Internet of Services (cf. Chapter 1), the easy and simple consumption of services constitutes a central challenge. Within service ecosystems, organizations dynamically interact as service providers or service consumers and potentially use a service platform with a service marketplace to create, design, trade and consume services [7]. Business applications running in enterprise systems (e.g., ERP, CRM or SCM systems) constitute a promising channel for service consumption. Users of enterprise systems increasingly demand the possibility to integrate new business functionality from service marketplaces into their core business applications. Enterprise system vendors and service providers for business application functionality extensions (e.g., partners or independent software vendors (ISVs)) have to address this demand in order to establish or remain a competitive advantage.

We can separate two fundamental approaches to service integration into enterprise systems: (i) Service integration in foreseen system interfaces that have been provided at the time of shipping the enterprise system. For example, standard interfaces for Business-to-Business (B2B), cf. [12], or Application-to-Application (A2A) integration scenarios are known from the area of Enterprise Application Integration (EAI), cf. [16]. (ii) The integration of services in unforeseen service interfaces which is the topic of this section.

Before being able to use the integrated service within the context of the enterprise system, the corresponding business application has to be adapted or extended to different affected application layers, e.g., by adding new UI elements (presentation layer), adding a new process step (process layer) or extending a business object with a new field to persist data from the external service (business object layer). Please note, that in both scenarios the structural- and/or behavioral mismatches between the service interfaces of the enterprise system and the service provider are traditionally addressed by service mediation (data and/or process mediation) components (cf. [13]).

The scope of this section is on the adaptation or extension of an enterprise system and business applications in order to consume a given service. We provide a short overview of our service integration framework, i.e., a modeling environment and runtime engine for service integration into business applications. Finally, a brief description of a prototype is given to demonstrate the feasibility of the approach. For a more detailed description of the approach and a discussion of related work we refer the reader to our previous work [4], [5] and [15].

15.6.1 Example: Consumption of a Business Service "Eco Value Calculator" within a PLM Application

In the following, we briefly describe an example for the integration of a service into a business application. After legal changes in the export guidelines, a car seat manufacturer wants to adapt his products to make sure that the changed guidelines are met with all products. In our example, the material of all parts of a car seat has to comply with new environmental ecological regulations.

The manufacturer uses a Product Lifecycle Management System (PLM) that supports its core business processes as one of its core enterprise systems (cf. Fig. 15.12).

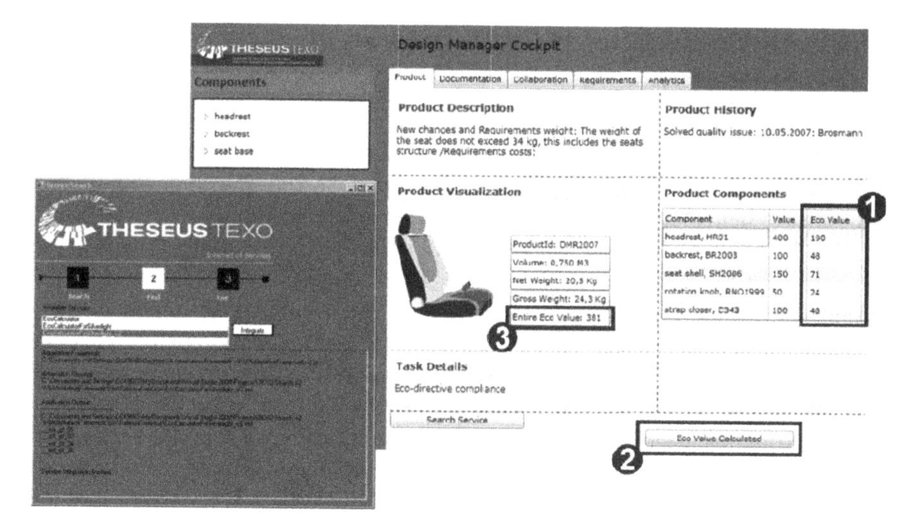

Fig. 15.12: PLM Business Application extended with the complementary service "Eco Calculator."

Originally, the system does not provide the required calculation of eco values for the bill of materials used for a given car seat. Therefore, the manufacturer's system needs to be extended.

A service provider offers a business service, possibly described with a USDL description, on the service marketplace for the required functionality that allows the calculation of eco values for products including certification. A product designer of the manufacturer accesses a service marketplace directly from within his enterprise system and searches for services that provide the missing functionality. From a list of recommended services, which are certified for the enterprise system, the product designer selects the service which best matches his needs. The designer selects a service called *Eco Calculator* and purchases it on the marketplace.

The service is integrated into the business application without running a manual integration project: the user interface of the core business application is extended with (i) an *additional table column* ("Eco Value"), (ii) an *additional button* ("Calculate Eco Value") and (iii) an *additional field* indicating the total eco value for the car seat ("Entire Eco Value"). When the service has been installed automatically, it can subsequently be used: If the total eco value meets the legal requirements, a certificate is generated and passed to the consumer application.

15.6.2 Architecture and Main Components of a Service Integration Framework

The developed service integration framework comprises a set of different components (cf. Figure 15.13). An *enterprise system provider* (left) develops and offers a enterprise system that is consists of several business applications.

A dedicated *adaptation/extension execution environment (runtime)* (depicted at the bottom of Figure 15.13) is responsible for the extension and adaptation of the system after it has been shipped (post-mortem extensibility). An *application extensibility description* document is a model of the core business application's extensibility capabilities and is comprised of the possible extension points where the core business applications can be extended or adapted. On the other hand, a *service provider* (right hand side in the figure) creates a new business service and publishes a *service description* (described, for example, in USDL) for the offered service for future consumption. For this purpose, the USDL tool chain is used: The service provider uses the USDL editor to create and edit the service description. The service description is stored in the USDL repository. Additionally, the service description can be uploaded to service marketplaces (not shown in the figure) in order to offer the service to many potential consumers.

A *service integrator* (in the middle of Figure 15.13) provides a solution that integrates a selected enterprise system with a selected business service. He uses an integration modeling environment to capture all aspects of the service integration, e.g., the integration or adaptation steps on presentation layer, business process layer, or service layer, or other affected application layers. In addition, a dedicated recommendation system [3] is part of such an environment. Finally, an *integration description* is generated that contains all necessary information for parameterizing the adaptation/extension environment.

In order to increase the abstraction level for unforeseen service integration, a new *pattern-driven modeling approach* is used. The set of integration steps is described with instances of a set of (predefined) adaptation patterns in order to avoid describing the changes with lower-level codes, such as program code fragments or other low-level configuration data. An *adaptation pattern* links elements from the application extensibility model (using an *application reference port*) with elements from the service model (using a *service reference port*). Each adaptation pattern describes a relationship between one or more extension points of the application extensibility model and one or more elements of the service model. This approach follows the principle of moderately extending core business applications by none-highly specialized integration experts.

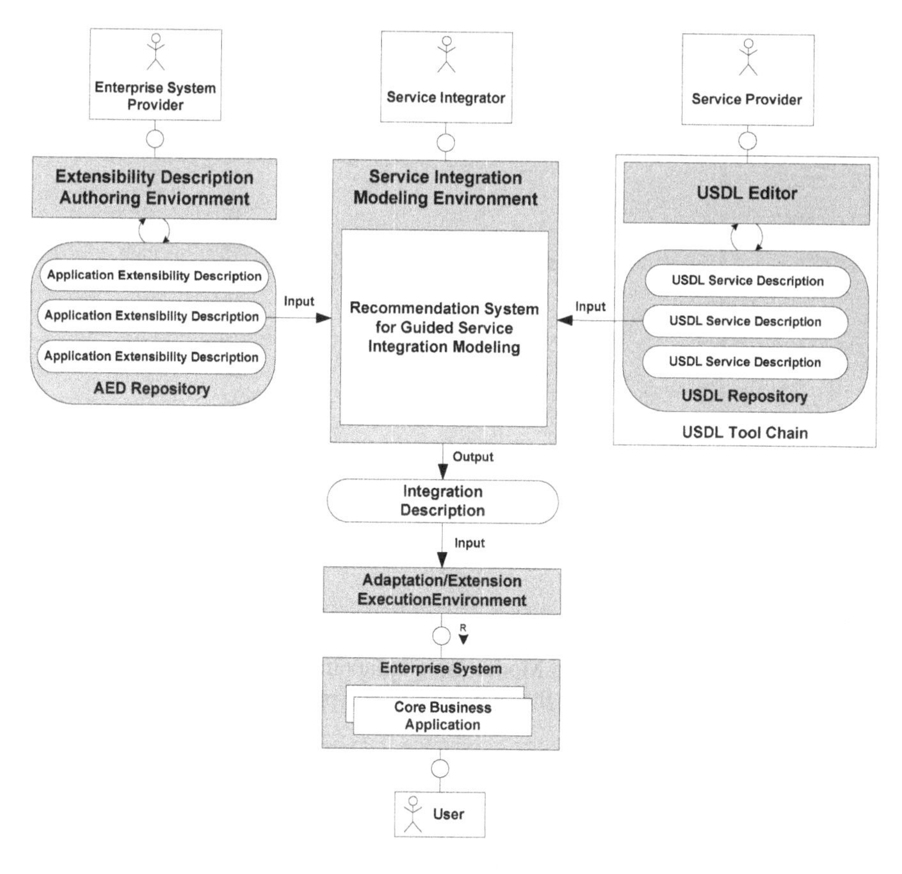

Fig. 15.13: Service Integration Framework (with Recommendation System).

15.6.3 Prototype of a Service Integration Modeling Environment

In this section the modeling approach with the eco calculator service integration scenario is illustrated (cf. Figure 15.14). On the left hand side, the application extensibility model for the enterprise system is shown while on the right hand side, the service description (derived from the USDL service description) is listed. In the middle part the integration model is shown including the visual representation of adaptation patterns.

Four requirements are given to the service integrator: (i) the service should be integrated into the Product-Lifecycle-Management (PLM) part of the enterprise system. (ii) The service should be invoked before the car seat is shipped. (iii) The eco values returned from the service should only be displayed on the user interface in an existing table. (iv) The returned eco values should not be persisted. The integration solution is modeled using one complex and one atomic adaptation pattern. The complex adaptation pattern itself consists of four atomic adaptation patterns.

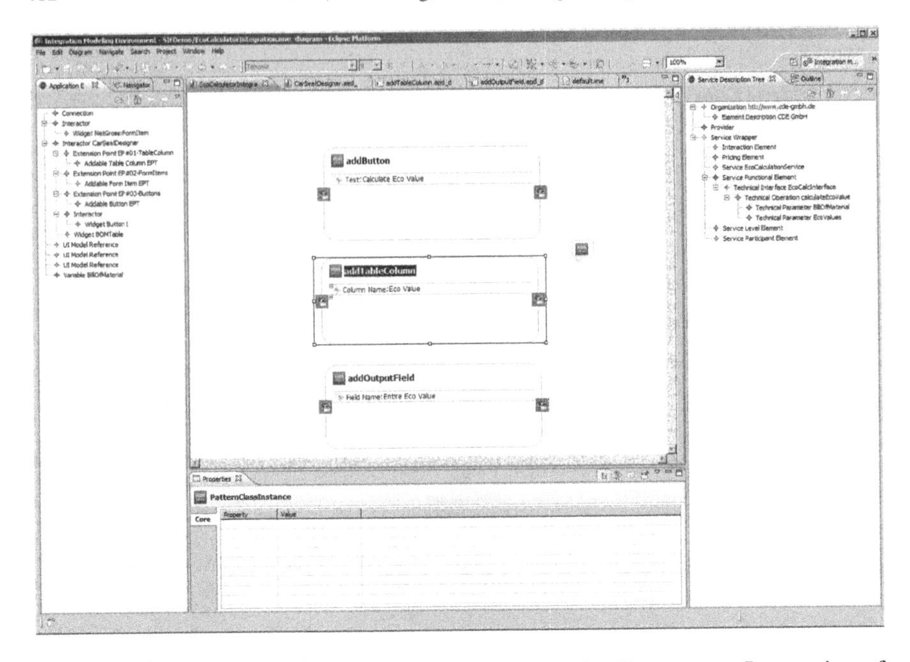

Fig. 15.14: Example Integration Model using Adaptation Patterns — Integration of the Eco Calculator Service into the PLM Business Application.

The complex pattern *"Stateless Service Integration without Data Persistency"* was chosen because it seems to fit to the given requirements. It consists of four atomic adaptation patterns: (i) *addButton*, (ii) *addTableColumn*, (iii) *addDataMeditator* and (iv) *addDataMediator*.

The complex adaptation pattern exposes four application reference ports (*A1 to A4*) and six service reference ports (*A5 to A10*) that are internally linked to the ports of the contained atomic adaptation patterns. All ports are connected to appropriate elements of the application extensibility or service model.

For example *"addButton"* is connected via the application reference port *B1* with the port type Extension Point Type — Button Panel. This port is parameterized with the value *BP-EP#1*. The text for the button is taken from the *Default User Interface* section of the service description (service reference port *B2*). The information for the event handler of this button (service operation that is called when the button is pressed) is taken from the *Operations* section of the service description (service reference port *B3*). The other adaptation patterns are parameterized based on the same principles.

Both adaptation patterns of type *"addDataMediator"* are used to model data mediation problems which are resolved using externally executed data mediation tools. Finally, a new integration description is generated from the modeling environment.

To demonstrate the feasibility of the modeling and the runtime adaptation approach of the framework, a functional prototype has been implemented. A PLM business application is implemented based on Microsoft Silverlight, the business service "Eco Calculator" is implemented as a Web service based on the Axis framework.[14] The prototype addresses the presentation layer. After the integration description has been modeled with the integration modeling environment, it is loaded into the Adaptation/Extension Execution Environment (implemented in Java). It is forwarded to the UI Layer Adaptation Manager that actually adapts the PLM business application by reusing its native extensibility features (Microsoft Silverlight APIs).

15.7 Conclusion

The chapter introduced concrete prototypes of fundamental tools and presented the findings in applying them. Regarding USDL editors, we have learned that the complexity of the language requires different types of editors for expert and casual users. Building USDL repositories requires a careful design of the interface in order to provide a scalable and efficient solution. Service providers can benefit from USDL marketplaces since they allow them to describe their services once and potentially trade them anywhere. All of the tools can be flexibly combined in different deployment scenarios (both single- and cross-organizational). We have also seen that USDL benefits service engineering by a framework for the integration of services into service-based enterprise systems.

References

1. MOF 2.0/XMI Mapping, v2.1.1. OMG Document formal/2007-12-01, OMG, 2007.
2. Meta Object Facility (MOF) Core Specification Version 2.4. OMG Document formal/2008-12-10, Object Management Group, 2008.
3. M. Allgaier. Requirements for a recommendation system supporting guided service integration modelling in extensible enterprise systems. In W. Esswein, M. Juhrisch, M. Nüttgens, and K. Turowski, editors, *MobIS 2010 Modellierung betrieblicher Informationssysteme 15. - 17. September, Dresden. 2010*. CEUR Online Proceedings, 2010.
4. M. Allgaier and M. Heller. Research challenges for seamless service integration in extensible enterprise systems. In B. Pernici, editor, *Proceedings of Industrial Experiences for Service Oriented Computing (IE4SOC), Stockholm 2009, Sweden*, pages 20–24, 2009.
5. M. Allgaier, M. Heller, and M. Weidner. Towards a model-based service integration framework for extensible enterprise systems. In M. Schumann, L. Kolbe, M. Breitner, and A. Frerichs, editors, *Multikonferenz Wirtschaftsinformatik, Göttingen 2010*, pages 1523–1534, 2010.
6. M. Alpuente, D. Ballis, M. Baggi, and M. Falaschi. A fold/unfold transformation framework for rewrite theories extended to cct. In *Proceedings of the 2010 ACM SIGPLAN workshop on Partial evaluation and program manipulation*, PEPM '10, pages 43–52, New York, NY, USA, 2010. ACM.

[14] `http://axis.apache.org/axis/`

7. A. P. Barros and M. Dumas. The rise of web service ecosystems. *IT Professional*, 8(5):31–37, 2006.
8. C. Bartelt, G. Molter, and T. Schumann. A Model Repository for Collaborative Modeling with the Jazz Development Platform. In *42st Hawaii International International Conference on Systems Science (HICSS-42 2009), Proceedings (CD-ROM and online), 5-8 January 2009, Waikoloa, Big Island, HI, USA*, pages 1–10. IEEE Computer Society, 2009.
9. M. Belaunde. A Pragmatic Approach for Building a Flexible UML Model Repository. In R. B. France and B. Rumpe, editors, *UML'99: The Unified Modeling Language - Beyond the Standard, Second International Conference, Fort Collins, CO, USA, October 28-30, 1999, Proceedings*, volume 1723 of *Lecture Notes in Computer Science*, pages 188–203. Springer, 1999.
10. G. Booch, J. Rumbaugh, and I. Jacobson. *Unified Modeling Language User Guide*. Addison-Wesley Object Technology Series. Addison-Wesley Professional, 2nd edition, 2005.
11. F. Budinsky, S. A. Brodsky, and E. Merks. *Eclipse Modeling Framework*. Addison-Wesley, 2003.
12. C. Bussler. *B2B Integration: Concepts and Architecture*. Springer, 2003.
13. O. Corcho, S. Losada, and R. Benjamins. Mediation — bridging between heterogeneous web service systems. In R. Studer, S. Grimm, and A. Abecker, editors, *Semantic Web Services: Concepts, Technology and Applications*. Springer, New York, 2007.
14. K. Duddy, M. Henderson, A. Metke-Jimenez, and J. Steel. Design of a model-generated repository as a service for USDL. In G. Kotsis, D. Taniar, E. Pardede, I. Saleh, and I. Khalil, editors, *iiWAS'2010 — The 12th International Conference on Information Integration and Web-based Applications and Services, 8-10 November 2010, Paris, France*, pages 707–713. ACM, 2010.
15. M. Heller and M. Allgaier. Model-based service integration for extensible enterprise systems with adaptation patterns. In D. A. Marca, B. Shishkov, and M. van Sinderen, editors, *ICE-B 2010 - Proceedings of the International Conference on e-Business, Athens, Greece, July 26 - 28, 2010, ICE-B is part of ICETE - The International Joint Conference on e-Business and Telecommunications*, pages 163–168. SciTePress, 2010.
16. G. Hohpe and B. Woolf. *Enterprise Integration Patterns — Designing, Building, and Deploying Messaging Solutions*. Addison-Wesley Professional, Boston, 2003.
17. M. Koegel and J. Helming. EMFStore: a model repository for EMF models. In J. Kramer, J. Bishop, P. T. Devanbu, and S. Uchitel, editors, *Proceedings of the 32nd ACM/IEEE International Conference on Software Engineering - Volume 2, ICSE 2010, Cape Town, South Africa, 1-8 May 2010*, pages 307–308. ACM, 2010.
18. B. Laurel. *Computers as Theatre*. Addison Wesley, Reading, MA, USA, 1993.
19. M. Lawley and J. Steel. Practical Declarative Model Transformation with Tefkat. In J.-M. Bruel, editor, *Satellite Events at the MoDELS 2005 Conference, MoDELS 2005 International Workshops, Doctoral Symposium, Educators Symposium, Montego Bay, Jamaica, October 2-7, 2005, Revised Selected Papers*, volume 3844 of *Lecture Notes in Computer Science*, pages 139–150. Springer, 2005.
20. J. Nielsen. Progressive disclosure. Alertbox, December 2006.
21. T. Sheard and L. Fegaras. A fold for all seasons. In *Proceedings of the conference on Functional programming languages and computer architecture*, FPCA '93, pages 233–242, New York, NY, USA, 1993. ACM.
22. Talin. A summary of principles for user-interface design, August 1998.
23. A. Wingo. Applications of Fold to XML Transformation. In *Proceedings of the 2007 Workshop on Scheme and Functional Programming*, pages 69–78, 2007.

Chapter 16
Supporting USDL by a Governance Framework

Christian Janiesch and Michael Niemann

Abstract The previous chapter introduced service marketplaces as fundamental tool that enables and benefits service ecosystems. The application of service marketplaces for enterprise resource planning is a growing market. The operation of such an online marketplace requires a governance approach that lies adjacent to the requirements of a SOA and the more general governance of IT. It also has requirements of its own, especially when it comes to the description of services with languages such as USDL. In this chapter, we propose four building blocks as a basis for a governance framework that is capable of supporting the operation of a service marketplace. The research is based on existing frameworks and also takes into consideration the particularities of emerging SOA Governance approaches. We emphasize the processes required for the management of service descriptions.

16.1 Introduction

Service marketplaces in the Internet of Services are an approach to enable and facilitate the trading of services. In the case of trading pure software services, the goal is to make software ubiquitously available as services which can be licensed for use. The aim is to reduce hardware cost and maintenance at the customer side and make (complex) software a commodity. The application of marketplaces for enterprise resource planning is a growing market.

Christian Janiesch
Institute of Applied Informatics and Formal Description Methods (AIFB), Karlsruhe Institute of Technology, Englerstr. 11, Geb 11.40, 76131 Karlsruhe, Germany,
e-mail: christian.janiesch@kit.edu

Michael Niemann
KOM - Multimedia Communications Lab, Technische Universität Darmstadt, Rundeturmstr. 10, 64283 Darmstadt, Germany, e-mail: michael.niemann@kom.tu-darmstadt.de

The operation of an online service marketplace requires a governance approach that lies between the requirements of an SOA and the more general governance of IT. IT Governance aims to reduce the risk of fraud, data inconsistencies, and resulting damages for stakeholders by defining regulations concerning the organizational model (roles and responsibilities) and general procedures (cf. Section 16.2 for an elaborated overview of related work). We use the IT Governance frameworks of COBIT [6] and ITIL [7] as a basis for research not only because they provide insights from unbiased organizations rather than individual enterprises but because they are at both ends of the governance spectrum: COBIT focuses on strategically important tasks (main processes) and ITIL focuses on management tasks (support processes), which are often subject to outsourcing and, thus, the ideal blueprint for managed third-party processes.

Based on an analysis of related work, we propose four building blocks in Section 16.3 to instantiate a governance framework that is capable of supporting the operation of a service marketplace. We base our conceptual considerations on existing frameworks and also take into consideration the particularities of emerging SOA Governance frameworks. We highlight its usefulness and applicability to USDL in Section 16.4. We conclude with a summary and outlook (Section 16.5).

16.2 Related Work

According to a survey conducted among companies that use SOA as enterprise architecture, 79 % of the respondents stated that they feel a large negative risk by taking services into production that are not effectively "governed." On top of that, 88 % of the companies consider their current SOA Governance approach insufficient — only 12% implemented a sufficient approach according to their own estimation [5]. Although companies are aware of the high risk of a governance lack, they have not installed sufficient mechanisms to address it. The need for appropriate governance approaches is high.

In recent years, a number of models and frameworks for SOA Governance have been proposed. While proceeding from diverging challenges and definitions, most of them address similar goals. They propose varying techniques and differing combinations of them to reach these goals. As the awareness of the need for SOA Governance is quite young, only few accepted standard procedures, goals and techniques exist. One reason might be the fact that there is no common definition of SOA Governance that could form a foundation for the different approaches.

We investigated and compared 22 SOA Governance approaches, developed in 35 publications at companies and research institutions. We divided them into three groups: Scientifically published approaches cover reviewed publications such as journal articles, conference papers, as well as book chapters and books. Many governance approaches have been made available by software vendors, published as company whitepapers which target governance for the SOA system aligned with proprietary software products (e.g., SOA infrastructure). The third group is formed by authors from the consulting industry that published their expertise in whitepapers based on achieved experience. During the examination, ten major aspects have been identified.

The approaches, however, show different quality. Approaches which formulate a clear view and opinion backed with arguments concerning a criteria, i.e., whose recommendation of the integration of a corresponding criterion is backed with arguments, are considered a *founded recommendation* in this analysis. Suggested building blocks of SOA Governance are fully integrated, concrete suggestions are made and even examples are given (marked with • in Table 16.1). In contrast to this, some approaches are characterized by a more narrow view of the topic. They show a lack of clear instantiation, explanation, level of detail, or specification. They point out an aspect considered important, however, lack the required level of detail, or precision. These cases are considered as *proposal* of the integration of the given criteria, equally to "partially integrated, mentioned" (marked with ◦ in Table 16.1).

Governance Policies

In almost all approaches, governance policies are informally defined as "means to define what's *right*." Generally, governance policies represent general guidelines, conventions, rules, and best practices that support the controllable and efficient operation of the SOA system. They are often applied in the administration of a service lifecycle, or during a SOA procedure model.

Generally, governance policies are considered distinct from *service performance-related policies* as described by standards such as WS-Policy [47, 24]. Main aspects of governance policies are their application to *roles*, *service design and operation*, and *service documentation*. Some approaches, however, leave the specified policies unclassified. Concerning policy handling, procedures for policy exception handling, as well as recognition of too restrictive policies are suggested.

As a consensus of all authors, policies are considered mighty instruments that combine various application aspects. They represent the most important and quite complex aspect of SOA Governance. Application aspects are roles-related, service design and operation-related, and, explicitly, service documentation-related policies. The latter are to ensure useful retrieval processes that are performed by, e.g., service requesters. Important aspects of policy handling are policy lifecycle management, policy exception regulation, and recognition of inappropriate (too restrictive) regulations.

Organizational Structure

Due to the changed conditions of SOA systems compared to other IT systems, the majority of authors considers to adjust organizational structures. The approaches outline and introduce new boards, councils, and institutions for special accountability around SOA.

Table 16.1: Detailed survey results.

	Governance Policy Catalogue	Organizational Structure	Artefacts Management	Role and Responsibilities	Service Lifecycle	Strategic Aspects	SOA Procedure Model	Governance Processes	Metrics	SOA Maturity Measurement
Books and journal articles										
Schepers et al. [44]	●	○	○	○	○	●	●	-	-	●
Bernhard and Seese [17]	●	●	○	-	●	-	●	●	●	-
Derler and Weinreich [23]	○	-	●	○	●	-	-	-	-	-
Kohnke et al. [31]	●	●	-	○	○	●	-	○	○	-
Bieberstein et al. [18, 19]	○	●	●	●	-	●	●	●	○	○
Marks and Bell [32]	●	●	●	○	○	●	●	●	●	-
Brown et al. [22]	●	●	●	●	●	●	●	●	●	-
Schelp and Stutz [43]	○	●	-	○	-	○	○	-	○	-
Rieger and Bruns [42]	●	●	●	●	○	○	-	-	-	-
Josuttis [29]	○	●	○	○	●	○	○	○	-	-
Software Manufacturers										
Brauer and Kline [21]	●	●	●	-	●	○	○	-	-	-
Hewlett Packard [8]	●	●	●	-	●	○	○	-	-	-
Systinet [3]	●	●	●	-	●	○	○	-	-	-
WebMethods [4]	○	-	●	○	○	-	○	○	-	-
Matsumura [33]	○	-	●	○	○	-	○	○	-	-
Software AG [1, 9]	●	○	●	○	●	○	●	-	○	●
BEA Systems Inc. [2]	○	-	●	-	●	-	○	-	-	-
Afshar [13]	●	●	●	○	○	●	●	-	-	●
Holley et al. [26]	-	●	●	○	●	○	●	-	○	-
McBride [34]	-	●	●	○	●	○	●	-	○	-
Mitra [35]	-	●	●	○	●	○	●	-	○	-
Muriankara [37]	-	●	●	○	●	○	●	-	○	-
Woolf [52]	-	●	●	○	●	○	●	-	○	-
The Open Group [11]	○	●	-	●	●	●	●	●	○	-
Consulting Industry										
Everware-CBDI: Allen [14]	●	●	●	○	-	●	-	●	-	●
BearingPoint: Rane and Lomow[41]	●	○	○	○	○	○	●	-	-	-
ZapThink: Bloomberg [20]	●	○	●	●	-	●	-	○	-	-
Windley [50, 49, 51]	●	●	○	○	●	-	●	-	-	-
Berlecon: Quantz [40]	○	●	●	-	-	-	-	-	-	-

All approaches that give *founded recommendations* concerning organizational changes (15 out of 22), recommend setting up a *SOA Centet of Excellence* (SOA CoE). This institution has convinced in theory (7 mentions) as well as in practice (8 mentions). It can be considered a crucial organizational institution for the operation of an SOA system.

Summarizing, the presented organizational entities (SOA CoE, SOA Board, and SOA Governance Board) are the three most frequently integrated ones. Competencies, however, are not clearly attributable. The approaches give different recommendations especially concerning the question of how decision and consulting competencies are to be distributed among the entities. The majority, however, agrees on the SOA CoE bundling many of the discussed competencies — in some cases even all of them.

In contrast to organizational entities that could also be named *group roles*, the precise definition of (single) roles and responsibilities has been a major aspect of SOA Governance approaches.

Roles and Responsibilities

Almost 80 % of the approaches mention the adjustment of roles and responsibilities for the operation of a SOA system. All these authors consider implementing and operating an enterprise architecture as an SOA to have impact on the organizational structure of the entire company. Besides the introduction of new organizational entities, this covers the definition of new roles and accountability. In order to assign clear and non-overlapping definitions of competencies, a solid concept for roles and accountability is commonly considered to be advantageous for all involved persons and the operation of the SOA system.

In the context of SOA Governance, an important aspect is the targeted *impact on behavior*. The IT Governance goal to "achieve desirable behavior in the use of IT" [48] is an important goal for SOA Governance as well.

Methods such as *RACI charts*, *impact on behavior* [19, 32], *SOA Education Plan* [18], and the *Capability Assessment Method* [22] are considered central components of the discipline SOA Governance.

Artifact Management and Software Support

Clearly more than half of the approaches (14 out of 22) name software support or artifact management a central building block of SOA Governance. Most of them come from the software industry.

During the development process of a SOA, many artifacts are created, e.g., *services*, *meta data*, *service descriptions*, *interface descriptions*, and *message format specifications*. Services in operation are bounded by *policies* and *service contracts*. Further meta data are SOA Governance artifacts such as *roadmaps*, *process descriptions*, and *reference architectures* (cf., e.g., [11]). The approaches suggest the operation of a service registry or service repository, and a Web service management system. They recommend structuring all kinds of artifacts as a meta-model to clarify relationships as well as the establishment of additional data and artifact related roles and responsibilities.

As main function of a service registry, most authors refer to publishing and discovery of services, while the service repository is considered to serve meta data storage. However, none of the approaches recommends operating both of these. Nevertheless, the understanding of service registries and repositories in terms of functionalities astonishingly diverge among the authors (cf., e.g., definitions by [41, 50]).

Service Lifecycle

Service lifecycle management (SLM) is a central aspect of SOA Governance. More than 75% of the approaches mention a service lifecycle to be an integral part of SOA Governance. The majority of approaches emphasizing the service lifecycle are from the author group software vendors.

Lifecycle models, in general, are widely used additives for design, development, operation, and maintenance of software (e.g., in [46]). As a purpose of SOA Governance, the design, implementation, operation, and version management of services can be improved by comprehensive and reasonable regulations in service lifecycles [21, 33, 50]. Their planning and implementation is part of SOA governance. However, the notions of definition and distribution of activities in lifecycle phases vary in wide ranges.

Using lifecycles, many artifacts beyond services can be controlled. Additionally to guidelines, applications composed from services, as well as business processes can be controlled by lifecycles (as proposed by [13]). Also, readjustments of SOA goals to changed business requirements, or frequent transposition of the SOA Governance model are performed using lifecycles (cf., e.g., [22, 26]). Using lifecycles is a powerful instrument of control, i.e., a powerful instrument of governance.

Strategic Alignment

The conception of a strategic plan as well as business-IT alignment, are both considered a further central element of SOA Governance by the experts. 15 out of 22 approaches refer to strategic alignment, the majority with concrete suggestions. Especially authors from the practitioner's domains (consulting industry and software vendors) consider this point crucial. Four aspects of strategic alignment considered most important are *formalization of SOA goals*, *identification and prioritization of services*, *adequate financing of service development*, and *SOA commitment of the management*.

SOA Procedure Model

Besides the management and effective administration of governance methods, the strategy and procedure of adopting and introducing a SOA as enterprise architecture — a procedure model — is considered a crucial part of a SOA Governance approach. 16 of 22 approaches point out the importance of a procedure model, most of them from the software industry and academia.

What is referred to as *SOA procedure model*, are many different variations of procedures for regulated SOA introduction and operation that are called, e.g., *SOA Lifecycle*, *SOA Governance Roadmap*, or *SOA Adoption Model* by the respective approaches. Generally, SOA procedure models act as a global guideline for the future development of the SOA system. They designate and communicate planned future developments of a SOA system and describe the phases from plan to realization.

Governance Processes and Policy Enforcement

Nine out of 22 approaches formulate governance processes and policy enforcement to be crucial aspects of SOA Governance, four of them are founded recommendations from academia. *Governance processes* are the actual implementation of governance. They define the business and IT-internal processes that are required to operate an IT system from the perspective of governance. They provide the activities and accountability for the operation of a SOA on a meta level. Mechanisms for automated policy conformance checks are summarized by the term *policy enforcement* used by the approaches. They target the monitoring of adherence to policies and their operational enactment and are integrated in processes.

Many approaches mention the category *Processes* as a central point of their approach. However, the classification types vary from governing vs. governed processes [18, 22, 12], runtime vs. design time governance [32, 4], policy-related vs. review-related processes [17], organizational structures vs. employees [31], processes vs. organization, infrastructure vs. maturity [14], and architecture review processes [20]. The classification that is mentioned most frequently is *governing vs. governed processes*. *Governing* processes cope with performing and realizing governance methods and structures. They serve as a means for the governance approach. *Governed* processes are subject to governance. They represent activities such as service development, process management, and service operation. Further, all authors agree that concise definition and structuring of governance processes is crucial to the successful operation of a SOA system.

Concerning policy enforcement (as part of *governing* processes), all approaches propose control points that reside in (cyclic) governance processes. Techniques or concrete examples for automated policy enforcement (other than manual revision of artifacts) are provided by none of the approaches.

SOA Maturity Measurement

According to Windley [49], implementations of governance that are not adjusted to the scope and maturity of a SOA system cannot display its full effect: either they exercise too few control, or they limit the involved persons by an overdose of regulation in their freedom of action and have a demotivating effect [50, 49]. Governance methods and procedures are to be planned proactively, in order to keep up with the development of the SOA system and enable controlled growth. Documentation of the planned development as a roadmap is an often proposed method to keep track of the state and development direction of the SOA system. In order to assess the current maturity of a SOA system, SOA maturity models have proven useful [44, 1, 13, 14, 28].

Overall, SOA Maturity Models are explicitly considered in four out of 22 approaches (ca. 23%), where three mentions come from the practitioner's domain, and one from scientific work. As SOA Maturity Models are already widespread and well-known instruments of SOA Governance, it seems astonishing that the integration of maturity models into SOA Governance is proposed by a minority of authors. Obviously, only few authors recognize the benefits of maturity measurement in the context of SOA Governance. However, several additional authors proclaim SOA Maturity Models-related methods. So it might be a lack of awareness which causes the little assignment of maturity models to governance.

SOA Metrics

Almost half of the approaches mention a *metrics system* as an important building block for SOA Governance. Metrics, in general, are defined along with goals and make processes and parameters of the SOA system more transparent. The measurement of goals, combined with a supporting management structure, supports the judgement on the effectiveness of the adoption of an IT system such as SOA. For the implementation of SOA Governance in a company, a set of goals is usually defined that are to be achieved. Metrics, in general, report on the performance of the SOA system as a whole, by measuring the goals set by the governance initiative (cf., e.g., [17, 22]).

Improving the assessment of achievement of SOA goals by the definition of a metrics system is considered an important aspect of SOA Governance by all authors mentioning this issue. Most of the authors especially emphasize the management of service operation, service statistics, project performance, and the relationship to employee behavior to be important in the context of metrics for SOA Governance. Further, the measurement of service reuse is an important aspect.

Summary

We compared the structure and core aspects of several approaches that first structured SOA Governance. As a result, ten components have been identified, that most of the authors make use of to compose their approaches.

The approaches do not usually adhere to consistent criteria, as done by the presented analysis. Most approaches use either organizational means, SOA goals or governance guidelines as a main criterion. In most cases, one important aspect is selected, and other (equally important) ones are presented in a cross-sectional way. The inherently multi-dimensional nature of this area is simplified and reduced to a few structuring criteria in most cases. However, there seems to be no reason for the selected and presented outlines — the choices of main criteria seem arbitrary.

In unison, the authors agree on the necessity of SOA. Based on common characteristics of SOA systems and the emerging challenges, the installation and operation of governance approaches for SOA is considered essential, on the one hand, regarding the management and unification of SOA-inherent heterogeneity and complexity, and, on the other hand, on the regulation and exploitation of new capabilities such as cross-organizational service deployment. SOA Governance turned out to be an area that is structured in various dimensions, e.g., goals and strategy, organizational structures, roles and employee behavior, software support.

Only few proposals [32, 22, 11] present holistic approaches that tackle all or most of the identified components. The overall comparison shows that most approaches are characterized by a *tunnel perspective*, limiting the focus on selected issues. However, the majority of authors agree that a holistic governance approach is crucial for SOA Governance. In the remainder of this chapter, we specify building blocks for a governance approach tailored to the needs of the Internet of Services. In particular, we adopted the results from *Organizational Structure* and *Role and Responsibilities* in our stakeholder map, while the process framework reflects insights from the *SOA Procedure Model* as well as *Governance Processes*. The component *Metrics* is adopted in the Measurement Framework, and *SOA Maturity Measurement* in the Maturity Model and Capability Profile. The service description management for USDL has been influenced by the insights of *Governance Policies* and *Governance Processes*.

16.3 Building Blocks of a Service Governance Framework

Based on an analysis of related work, we propose four building blocks to instantiate a governance framework that is capable of supporting the operation of such a platform. We base our conceptual considerations on existing frameworks and also take into consideration the particularities of emerging SOA Governance frameworks.

First, the *Process Framework* defines tasks and activities required to manage the ISM and its lifecycle. Especially, the areas of "service portfolio management," "service lifecycle management" as well as "broker operations" are not adequately represented by current frameworks and are developed in this component. Roles and responsibilities of the processes, tasks are focused on in the second building block, namely, the *Stakeholder Map*. The third building block, viz., the *Measurement Framework*, describes corresponding key performance indicators and other result measures, which are used to evaluate process quality as well as the compliance with internal, normative, and legal regulations. The fourth building block is a *Maturity Model*. The application to a service-oriented IT system allows the evaluation concerning system maturity and identification of potential gaps, which need to be covered by additional governance processes.

16.3.1 Stakeholder Map

Generally, in the Internet of Services, which we consider as the basis for (future) service marketplaces, several main stakeholders have been identified: *service provider*, *service broker* or *intermediary*, and *service consumer* [16]. While the service consumer and the service provider are actual persons acting as a specific stakeholder, the service broker is a virtual entity, a marketplace, or a piece of software. Nevertheless, it is operated by actual persons who act as a certain stakeholder.

With the emphasis on the complete service lifecycle, including the inception of a service and its after-sales, i.e., the community around the platform, these roles need to be extended. As outlined above, the service broker itself is not a stakeholder in that sense. That means its role cannot be taken by any person but is, instead, a piece of software. However, a supporting stakeholder, such as the platform host, needs to be established. Fig. 16.1 shows all roles. Note that multiple instances of each role but the operating platform host communicate via the service marketplace.

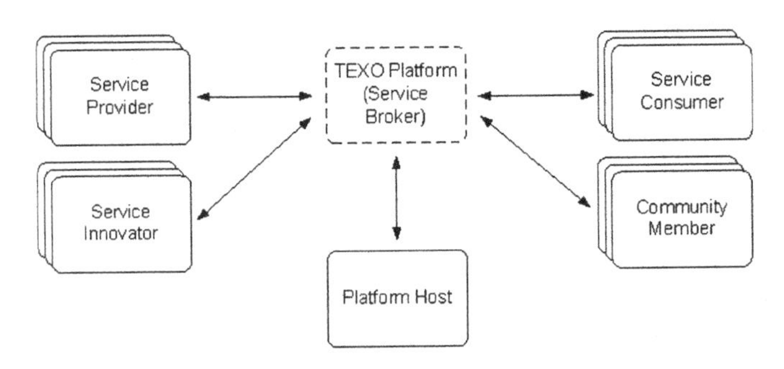

Fig. 16.1: General Service Marketplace Roles [16].

In the following we describe and detail the five *stakeholders*. For each stakeholder, associated activities are also outlined. A *role* is a subordinated entity of a stakeholder. Thus, each stakeholder may have several roles. Primarily a governance framework for marketplace platforms does focus on the platform host's activities.

But the framework also needs to take into consideration that services are neither produced nor consumed by the platform host but a provider and consumer. As such these roles have to be taken into consideration.

16.3.1.1 Service Provider

The service provider supports agencies that hold governance and operational responsibility for a service, including organizational structures and other business aspects, as well as systems and other implementation artifacts. The service provider represents the role of a development party, producing and publishing services ready for execution. Largely, they are the service owners, responsible for the service implementation as well as maintenance. Unlike traditional software producers, service providers develop services that remain in the same organization, rather than being delivered to software clients (what is also possible). Therefore, during requirements engineering, it is the duty of service providers to not only analyze the objective, functionality, interface, and quality of service, but to also consider accessibility, retrievability, how to manage service level agreements (SLA), and define policies, etc. At runtime the provider may have to provide second level support and appropriate change management. Possible roles include the service manager, service clerk, service producer, content provider, service aggregator, service integrator, business expert, service engineer, service designer, and service programmer (for more detail on roles cf. [27]).

16.3.1.2 Platform Host

The platform host administrates the marketplace platform, including the user management, and the maintenance of all running management services, as well as platform governance, risk, and compliance. For the platform host several roles are distinguished. The platform host has to support agencies that specialize in taking services out of markets and driving up their consumption through competitive pricing models. These agencies provide a further intermediation, managing the front-desk delivery of services to customers without encroaching on back-office responsibility. The platform host may have to certify service providers and their offers, since not all offers are acceptable. A role certifier is needed (perhaps even an additional stakeholder). Another point is billing and payment tracking, which also is a role often held by additional stakeholders.

With an increasing number of services, registries are becoming more and more important. They serve as a central location for tracking and managing services. The reusability of services depends on these registries, as these provide a way to share services across organizational borders. The platform host has to keep his registry and search index current as the central information database and its timeliness is crucial to the success of the whole system. Additionally, registry maintenance is of importance. For instance, a service that is updated while being in use should not be interrupted in execution, the removal of services that were never or seldom invoked should be considered, a rating system could be included, etc. During service delivery, a controlled service provisioning has to be ensured. In order to guarantee that clients can be charged by the providers and to ensure security, services are only accessible by authorized users. SLAs are used as contracts and authorization. These usually define costs, assured availability, performance, etc. As soon as services start being executed, the platform host begins the service monitoring process in order to ensure SLAs as well as policies and keeps track of the behavior of published services.

The role platform support gives support (at least second level support) to all marketplace platform processes, assisting various other stakeholders that are interacting with the platform including the service consumer, service innovator, service provider, or community member. Platform support is, e.g., a call center agent or a support consultant.

Further roles can include a business manager, governance officer, ontology and standards engineer/ expert, host architect, hardware admin, and software admin.

16.3.1.3 Service Consumer

The service consumer finds services, based on his functional and non-functional requirements and selects from offered variants (e.g., SLA variants) via the marketplace of the SaaS platform, buys or licences them and then may request and invoke them. For the service consumer several roles are distinguished such as business user, expert user, or manager, and perhaps administrators (each with approval rights). Additionally, a guest (of the platform) may only browse the offers. He has to register and login for ordering.

16.3.1.4 Community Member

Community members are registered and non-registered users of the marketplace. Most other stakeholders can act as community member stakeholders: Roles attached to the platform host are excluded, as they do not take part in the community in this sense but only from an administrative side. Community members — in addition to their possible other stakeholder roles — provide feedback for tradable services (e.g., problem reports) and use wikis, web logs, and forums provided by the marketplace platform to discuss tradable services.

16.3.1.5 Service Innovator

Service innovators use the marketplace platform to innovate on tradable services. The innovator derives new ideas from direct feedback from service consumers, query logs, or other data (e.g., wikis, blogs, etc.) or creates new ideas for services from scratch. Service innovators collect, aggregate, store service ideas in an idea repository, and rank these collections of ideas. Service innovators need to be registered in the marketplace platform, for use of the service browser and service discovery, as well as the community portal to browse consumer feedback.

16.3.2 Process Framework

All relevant governance processes have been grouped in five phases to increase the accessibility of the framework: design, deployment, delivery, monitoring, and change. In each of these phases, several processes constitute the process framework. Figure 16.2 provides an overview. As the framework has been compiled on the basis of existing frameworks, some processes have already been considered in existing frameworks. Most of the time, the existing processes will need to be extended to cater for the specific needs of marketplace platform governance.

The design phase contains all sorts of strategic aspects of the use or operating of the marketplace platform and its traded services. The development and deployment of services, as well as the selection of third-party services are components of the deployment phase. The delivery phase contains all aspects of service and infrastructure operations. It is closely coupled with the monitoring phase as they are executed concurrently. The monitoring phase contains all aspects of service and infrastructure monitoring. It is closely coupled with the delivery phase as they are also executed concurrently. The change phase contains all processes and tasks needed to adjust and change the infrastructure and software traded as service.

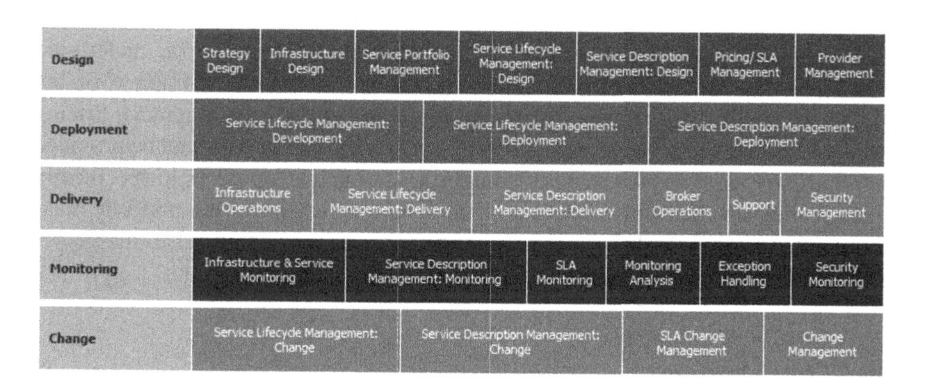

Fig. 16.2: Process Framework.

16.3.2.1 Design

The design phase includes organizational and infrastructure aspects as well as the service portfolio planning and the alignment of business requirements with the IT. It also comprises legal issues concerning general terms and conditions as well as SLAs. Furthermore, a process of provider management has to be introduced (including certification) because a marketplace platform is essentially a supplier/service provider enabling approach.

Strategy design covers all activities that are related to the creating and reviewing a strategic IT plan. This mainly involves strategic alignment, dependency analysis as well as the specific consideration of SOA capabilities. Strategy design also involves financial management tasks. *Infrastructure design* processes comprise activities which ensure a sound architecture specification. Accordingly, general processes on standards and development plans, as well as service marketplace-specific processes on running processes, organizational units, and their relationships are covered. This involves, e.g., the identification of system owners, data owners, and services owners. *Service portfolio management* encompasses all available processes on the infrastructure. Their management is of high importance to smooth operations. Thus, the governance of the composition of services, their granularity, their description as well as portfolio development are the key tasks. In addition to that, capacity planning is of essence as is the management of service continuity. While *service lifecycle management* as such covers all phases of the framework, there are specific tasks which have to be conducted in every of its phases. The design phase is the first. In order to properly deploy and deliver a service, the service has to be configured and validated before a transition strategy for service operations can be designed. Similarly, *service description management* covers all phases of the framework as services have to be described properly according to a certain schema in order to be discovered and used. Since services are traded, the governance of the underlying *pricing models and SLA* becomes important. The design of an SLA framework, the design of standard terms and conditions as well as the design of payment/ pricing models has to precede the service offering. As most of the services can be provided by third-party service providers, it is important to include a process to specifically manage their involvement. The process of *provider management* is a portfolio management process to evaluate and engage or disengage providers. Also, the governance of third-party design processes is part of provider management.

16.3.2.2 Deployment

The deployment phase comprises all processes that surround the deployment of services. This covers the service catalogue management, service continuity management, service validation and testing, and the definition/ negotiation of SLA. *Service development* has been consciously separated from this framework as we only cater for run-time governance. Service development and engineering is a broad topic of its own.

Service lifecycle and *service description management* for *deployment* cover all governance processes that are executed when a service is to be deployed. This involves service catalogue management, service continuity management as well as the execution of validation and testing. In order to create a service offering, the corresponding SLA, operation level agreements (OLA), and general terms and conditions have to be chosen, too.

16.3.2.3 Delivery

The delivery phase contains all aspects which guarantee the delivery of services. Thus, it includes service and infrastructure maintenance. Here, both the *operation* of mere infrastructure software and hardware as well as the assurance of service performance is managed. In addition to that all business functions concerning the brokerage of services have to be governed properly. Thus, the process *broker operations* has to be introduced which contains sub-processes which stem from the common phases of business transactions to reflect all phases of a purchase: Initiation, Agreement, Settlement, After Sales. Furthermore, all support operations (such as help-desk) and the management of the organization are part of this phase. Finally, security and compliance issues are also addressed here. Processes that deal with monitoring have been grouped in a separate phase.

Managing the marketplace platform involves several governance processes which are tightly related to IT management operations. Data as well as the physical environment, i.e., servers, data storage, network, need to be properly protected and/or backed up. In addition, the operation of the infrastructure needs to be safeguarded, i.e., regular maintenance has to be conducted and software updates have to be applied. Besides infrastructure operations the service operation, too, needs to be governed but during *service lifecycle management*. All services which were designed, described, and deployed need to be operated so that performance and capacity requirements can be met. A continuous service must be ensured according to SLAs. The central component of a marketplace platform is the service broker which facilitates the communication between service provider and service consumer. It ensures that a service can be searched for and discovered, contracts can be negotiated, and services can be bought and payed for. Also after-sales routines need to be governed. The structuring of this new component is based on the common phases of transactions for e-commerce purchases on a marketplace [45, 38]. Both, the operation of the infrastructure and of all services, needs to be supported by a proper (multi-level) *support* help-desk. It provides support services and incident management to customers and providers alike. All interaction on the marketplace platform needs to be *secure* as it involves business transactions. In order for the infrastructure to run secure, measures have to be taken to ensure authorization and authentication. Measures for security breaches and constant vulnerability assessments have to be in place. This includes data privacy issues.

Execution is both, the delivery and the monitoring of services and infrastructure. As both topics are closely related but focus on different parts of the execution they are separated logically in the framework. Monitoring covers the observation of services and infrastructure concerning performance, (future) capacity, and fulfilment of SLAs. Data analysis, exception handling, and security specific tasks such as logging are also within the scope of this phase. Error logs have to be analyzed for preparing error corrections.

16.3.2.4 Monitoring

Monitoring is an important governance process to ensure that the infrastructure and services are delivered according to plan. In order to do so, both, the *infrastructure* as well as the *services* including their *descriptions*, need to be closely *monitored*. Governance tasks involve the actual setup of a monitoring organization which specifies the approach, measures, objects to be measured etc. Areas to be monitored are, e.g., third-party services from service providers, overall service performance as well as infrastructure capacity and thresholds. Besides monitoring the infrastructure and service functionality, the contracted service levels need to be safeguarded, otherwise compensation routines have to be executed. This entails that governance processes for the specific monitoring of *SLAs* have to be in place to constantly monitor and review the execution and compensate for violations. Monitoring is no end in itself. The as-is data has to be correlated with planned/ predicted performance, so that weaknesses can be identified and proposals for improvement can be derived. This *monitoring analysis* process usually involves the generation of reports and the use of descriptive data analysis techniques (e.g., Online Analytical Processing). However, in order to allow not only ex post analysis, more intelligent ex ante analyzes are desirable to enable the broker to predict the impact change will have on future operations (architecture management), e.g., when replacing services. While monitoring data and their analysis usually means to compile reports and analyze aggregated data, the *exception handling* processes deals with singular events and is designed to single out irregularities and provide patterns so that events can be correlated and appropriate responses can be selected. This in turn allows the categorization and prioritization of incidents to escalate and recover. Similar to security in the delivery phase, *security monitoring* is necessary to ensure a consistent behavior. Also, in order to comply with legal requirements, certain tasks may need to be monitored and logged.

16.3.2.5 Change

The change phase contains all processes and tasks needed to adjust the infrastructure and services traded on the platform in order to ensure compliance and quality of service. That comprises processes which deal with the change and retirement of the actual services, change management processes for SLAs as well as change management from an organizational perspective.

Within *service lifecycle management*, the deployment of each new release of a service needs to undergo a specified change management governance process in order to ensure the continuous service provisioning. These activities involve, e.g., test plans and deployment verification. In case of a service retirement, contracted warranties have to be enforced, contracts may have to be terminated or changed, and equal functionality may have to be offered as a replacement. Also maintenance requires the removal of defunct services. Service functionality may change, business models may change over time. This needs to be reflected in the contracts which were signed on the platform. Changes within *service descriptions*, *SLAs*, pricing models, and general terms and conditions are governed by these processes to ensure due diligence and traceability. This can be conducted either in conjunction with functionality change or without. All activities in the change phase need to be managed by a proper *change management* organization. All change requests and all managed change need to be documented, prioritized, and evaluated before the change ticket is closed. The change itself — at least in complex cases — will be followed by design, development, deployment, i.e., here the overall process may start again.

16.3.3 Measurement Framework

We propose a multi-stage measurement framework composed of three layers to assess the performance of the governance framework processes. Fig. 16.3 depicts the different stages: Company Scorecard, IT Balanced Scorecard, ITIL/COBIT-based Processes and KPI (key performance indicators).

The first layer describes the company scorecard including the vision, mission, and strategy of the company. The most generic stage is the vision of a company describing the mission statement. It explains the reason for a company to exist. The second stage is the mission of the company describing specific goals in terms of performance, costs, ROI or market goals. The strategy follows as stage three and defines the specific way to achieve the company goals. These three stages are defined by the top management and do not have a standardized way of measuring the achievement of the goals. Instead they are discussed by the responsible managers in person.

The second layer refers to the IT Balanced Scorecard pointing out the strategy of the business area. The IT Balanced Scorecard is composed of six perspectives including financial management, process management, provider management, employee management, innovation management, and product management. Referring to Kaplan and Norton, the four perspectives of the classical Balanced Scorecard (financial, customer, internal process, and innovation & learning perspective) are extensible and should be adjusted to the own needs [30]. Objectives are deduced by concentrating on the main characteristics of the strategy defined on the third stage. These more clearly outlined objectives serve as a basis for the critical success factors which are determined for specific perspectives of the IT Balanced Scorecard.

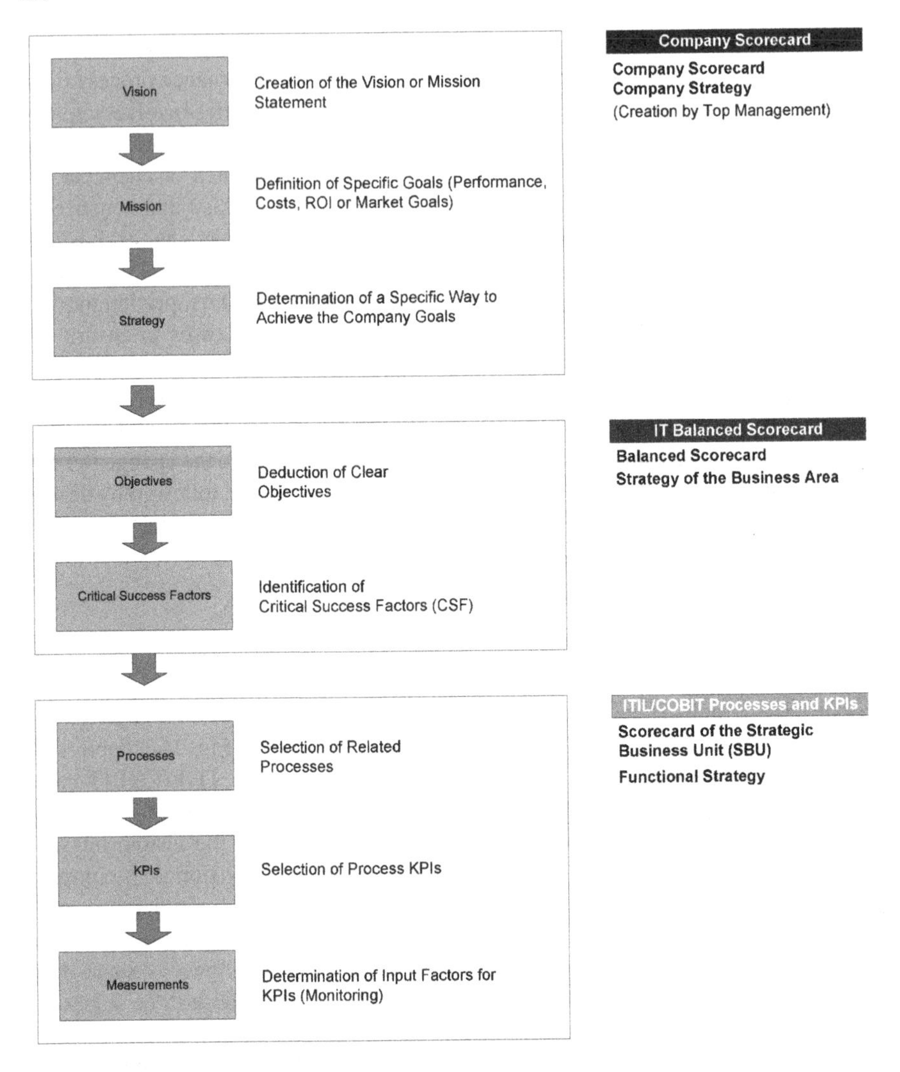

Fig. 16.3: Measurement Framework.

Each critical success factor is measured by few KPIs belonging to one of the six perspectives.

The third layer comprises processes. A suitable selection of processes from the process framework makes up the sixth stage. For each process there exist several KPIs measuring the performance of the specific process, related to the KPIs of the CSF. A proper selection has to be made for the specific case. The final part of the framework comprises monitoring measures serving as input factors for each KPI on the eighth and last stage.

While the distribution and transformation of governance requirements to lower levels follows a top-down approach, the requirements and goals can be measured and controlled on a bottom-up basis by a KPI system. Therefore, the monitoring data on the lowest stage serves as input for KPIs on the next higher stage. The KPIs feed measures on the stage above and so on.

16.3.4 Maturity Model and Capability Profile

A maturity model is a method for evaluating and measuring the current state of service adoption of an organization. Each organization can class in one level based on different characteristics. The maturity model reveals the organization's weaknesses and helps to develop transition plans to achieve the next maturity level [10]. Our maturity model is presented in Table 16.2.

We base our considerations on the following characteristics (cf., e.g., [15, 36, 39]):

Technology/ architecture Level of the underlying architecture, level of integration of for example databases or legacy systems, or the implementation of monitoring and optimization tools

People/ organization The employee's knowledge, characteristics of enterprise culture, and employees' motivation

Adoption scope Organizational focus of SOA, inter-departmental adoption of SOA, supply chain scope of SOA

Process Orchestration of services, business processes

Standards Technical standards, eBusiness standards, standardized approaches

SOA development Maturity of the SOA development process, existence, acceptance, documentation, and communication of an organization-wide standardized SOA approach

Besides these aspects, further dimensions are suggested for a maturity model, for example, the level of tool support, information management as well as lifecycle maturity or governance maturity. Most SOA maturity models consist of five levels such as the Capability Maturity Model (CMMI) [10]. However, their labels are different due to their different focus. The level labels are named according to the CMMI, as various dimensions are considered and the CMMI labels can cover all described aspects.

Table 16.2: Maturity Model.

Dimension	Level 1 – Initial	Level 2 – Repeatable but Intuitive	Level 3 – Defined	Level 4 – Managed and Measurable	Level 5 – Optimized
Technology/ Architecture	Platform-dependent point-to-point services, no standardized or centralized SOA technology, small number of services	Distributed systems, integration of applications, databases, and legacy, planned architecture vision, versioning and security	Reusable and discoverable services, long-running transactions, service versioning	SOA monitoring, event-driven dash-boards and alerts	Business process tools, easy assimilation of new technologies, composite applications, highly flexible architecture
People/ Or-ganization/ Stakeholder	Some self-taught SOA-skills, disconnected SOA project teams, no SOA specific organizational occurrence	SOA leadership and sponsorship through CIO, SOA Competence Centre, SOA skills developed	Incentives established for to encourage SOA adoption, executive commitment for SOA, strong SOA skills	Reuse and measurement culture, CFO sponsorship for SOA	Agile and continuous improvement culture, all responsibilities assigned and defined
Adoption Scope	Intra-departmental adoption	Business unit level adoption	Cross-business unit level adoption	Enterprise-level adoption	Value net respectively supply chain adoption
Process	SOA knowledge available via individual competence	Modeling of business processes with service components, first reusable processes implemented on a project basis	Modeling, documentation, and implementation of business processes based on SOA components across business areas and organizational units	SOA framework and service components are systematically and proactively managed	Systematic approach established for identifying new requirements and detecting gaps, continuous improvement
Standards (sample)	SOAP/REST, XML, WSDL, J2EE	UDDI, WS-Security	WS-BPEL, ebXML	Project management standards, business activity monitoring	Business process modeling standards
SOA Development	Minimal documentation of architecture, no formal service development process, no communication across project teams	Some level of architectural documentation, reusable architecture within project teams, as hoc communication across project teams	Standardized architecture defined, project teams are encouraged to use architecture, support levels are established	Standard approach for SOA, including processes, technologies, and components	Architectural framework in place for each team to expose and consume services including external partners, continuous architecture improvement
SOA Governance	No formal SOA Governance concept in place	Regular SOA Governance practices take place, identified problems are tackled by project teams that are formed when necessary	An organizational and process framework is defined as a basis for SOA Governance, specific procedures for SOA management in place	Target-setting has developed, integration of business BSC	SOA Governance is sophisticated approach using effective and efficient techniques

In addition, capability profiles represent the application of the maturity model on a SOA system and outline the overall abilities of the system compared with the planned targets. Commonly, the purpose of capability profiles is to provide a blueprint of a system's current respective abilities related to specific domains [25]. In the case of IT Governance, a capability profile is created by the assessment of an IT system using a maturity model — it illustrates the situation as-is [6]. Along with a governance framework, adoption models or best practices are often provided, that, in some cases, are part of the framework itself [13]. One kind of adoption support or recommendation is to provide an assessment of reference processes concerning their importance in implementation. This provides an order as well as a benchmark that can be applied later on.

We built a process maturity matrix that includes recommendations for which processes to address with high priority, depending on the targeted maturity level. Our method visualizes capability profiles by emphasizing the importance of specific processes. Thus, it allows a weighting concerning the ordering of process adaptation in order to achieve given maturity levels when implementing the governance approach. It represents an adaptation tool for planning support, especially concerning the implementation details of reference processes, using a *percentage completion-assessment*.

Once a process matrix is defined, generic capability profiles, one for each maturity level, are generated. These capability profiles are aligned along the five phases: design, deployment, delivery, monitoring, and change.

During the generation process, each of the process adoption steps preparation, implementation, and consolidation is weighted concerning expected effort, as well as the respective processes are weighted inside the process domain. This way, the expected percentage value of implemented processes per governance phase is computed. For the following configuration, the resulting radar chart is outlined in Fig. 16.4: process domains: uniformly weighted. Each axis represents one governance phase of the governance framework: Design, Deployment, Delivery, Monitoring, and Change.

Each of the five axes indicates the implementation progress achieved per governance phase on a percentage scale. The diagram shows that the capability profile for maturity level 1, *initial*, poses no requirements concerning any monitoring processes, touches design, change, and delivery-related processes, and demands almost 25% of deployment-related processes to be implemented. This is due to the fact that processes of the deployment phase are considered important for the second level and need to be considered in the first instance when adopting reference processes.

Maturity level 2, *repeatable but intuitive*, demands a solid basis of implemented processes in each of the five phases. Deployment is once more considered far more important than monitoring. The diagram outlines the importance of level 3, *defined*, that covers over 50% of implementation progress of all five phases. In particular, it requires the complete realization (including consolidation) of all deployment processes.

Levels 4 and 5 perform the optimization of processes. For Level 4, *managed and measurable*, the realization of all processes of the phases *Delivery* and *Change* is re-

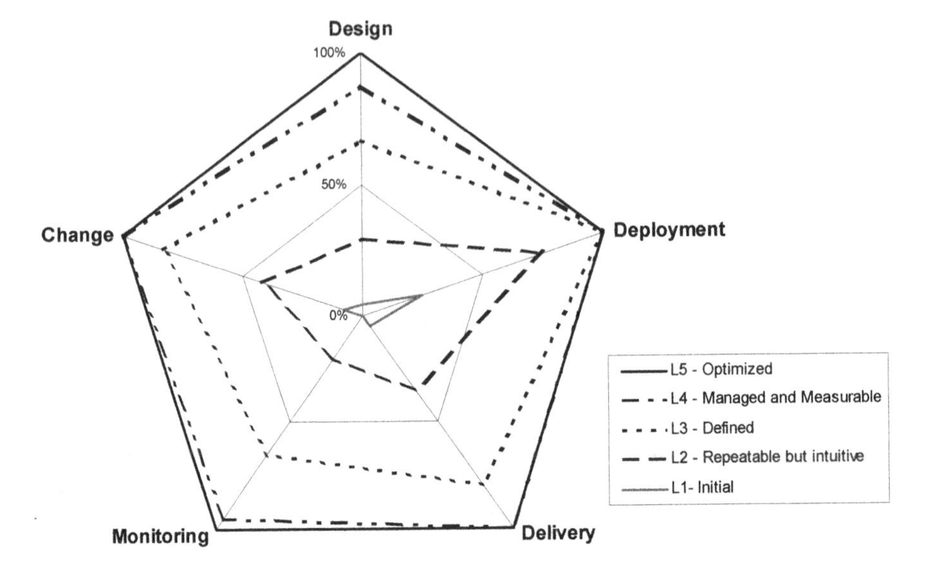

Fig. 16.4: Capability Profiles for Maturity Levels.

quired. Level 5, *optimized*, basically consists of process improvement in *Monitoring* (Monitoring analysis) and *Design* (Strategy design).

This approach provides an overview and visualization of the overall governance process in connection with system maturity. Vice versa, it provides an easy means to estimate the maturity level of the system based on the measured or estimated implementation progress of reference processes by the comparison of respective capability profiles. This analysis allows the combination of the governance framework and the maturity model and hence supports management decision-making.

16.4 Service Description Management for USDL

16.4.1 Service Description Management Processes

In addition to the creation of the initial service description, descriptions are constantly subject to change due to service upgrades, changes in service functionality, changes of the used technical terms, changes in word usage, and many more and lead to different service description variants or configurations (see Chapter 17). In order to reduce the risk of fraud, data inconsistencies, and resulting damages for stakeholders, it is important to define regulations concerning the organizational model (roles and responsibilities) and general procedures. The goal is to organize the handling of descriptions by assuring consistency of the used service description meta-model.

From a software engineering point of view, service descriptions can be seen as *traditional* development objects. Change processes are common and for specific cases (create, update, remove, distribute, ...), there is a number of appropriate (technical) ways to address these issues. However, for the traceable and controllable operation of a service marketplace, it is essential (in the sense of adhering to a general guideline or law, e.g., the Sarbanes Oxley Act) to assure guideline-consistent behavior in the general processes (here: handling service description). In addition to the automation of concrete data handling activities, the according processes need to be observed in order to globally assure efficient control and, in the end, compliance. Governance approaches introduce structures and guidelines aiming at achieving these goals.

We incorporate processes to complement the governance framework (throughout the governance phases) for the regulation and standardization of service description processes, as well as the setup and maintenance processes for service description meta-models (here: USDL). Overall goals are data consistency, sustainability, traceability, reliability and transparency of IT processes. Processes that define the extension of the framework concerning governance of service description are twofold as you need to manage instances, i.e., concrete service descriptions, and the meta-model. Consequently, for all phases of the Service Description Management, we distinguish the processes of *Setup and Maintenance of Service Description Meta-model* and *Service Instance Description.*

The framework defines the phases as a control cycle from an IT system operation perspective. The activities are organized along the phases in Table 16.3.

In the design phase, the host architect (HA) and the business manager (BM) define the standards that will be used to design the actual service description meta-model for the platform. USDL could be one of the standards. The HA and software administrator (SA) define the concrete repository as well as the maintenance and versioning procedures. Also, the BM and the HA define the concrete organizational model to support these maintenance processes for their platform. The processes may be adapted from other governance frameworks such as the TEXO Governance Framework [27]. Similarly, the supporting tools for meta-model design and maintenance have to be chosen. The BM and HA define description guidelines in order to achieve uniformity. This way, they make sure that the same elements in different meta-models are named in a uniform way. Finally in the design phase, the BM then designs the concrete meta-model and the HA checks and implements it from a technical perspective.

The concrete deployment of the meta-model management tools as well as the meta-model itself, takes place in the deployment phase and is executed by the SA.

The delivery of the description is automated, all related governance processes are related to its monitoring.

Accordingly, in the monitoring phase, the BM verifies and certifies that all meta-models conform to the guidelines. He is supported by the meta-model management tools. But ultimately, he is accountable for the data. The SA performs periodic consistency checks to eliminate technical inconsistencies, e.g., after deletions. Advanced monitoring tasks are the monitoring of word changes over time,

Table 16.3: Service Description meta-model Setup and Maintenance Processes. BM — business manager (host), HA — host architect (host), SA — software admin (host), SM — service manager (provider), R stands for responsible, A for accountable.

Phase	Processes	Roles
Design	- Define description standards and meta-modeling language	HA, BM
	- Define repository maintenance and versioning	HA, SA
	- Define organizational model	BM, HA
	- Determine tooling support	BM, HA, SA
	- Define general description guidelines	BM, HA
	- Create meta-model	BM, HA
Deployment	- Meta-model deployment (Establish repository and load meta-model in repository)	BM
Delivery		
Monitoring	- Verify and certify adherence to description meta-model guidelines	BM
	- Perform frequent consistency check	SA
	- Monitor word usage and meaning development over time	BM (A)
	- Monitor changes and perform development trends analysis	BM (A)
	- Analyse feedback from monitoring of semantic applications	BM (R)
Change	- Change description standards	BM (A), HA (R)
	- Change maintenance processes	BM (A), HA (R)
	- Change maintenance roles and responsibilities	BM (A), HA (R)
	- Change peripheral (non-core) modules of description meta-model	BM (A)
	- Change core modules in description meta-model	BM (A)

trends, and semantic checks. Terms change over shorter periods of time, so that meta-model changes may be necessary. Although this more apparent on the instance level, changes of the meta-model may appear. Sometimes, a trend can be calculated from these changes to anticipate necessary future modification. Semantic applications can also provide feedback on the use of technical terms or tags in the meta-models. Again, we assume the latter two are more important on an instance level but still, in a large deployment the may be enough data to perform this kind of monitoring.

In the change phase, standards as well as the associated governance may be adapted if the monitoring of the marketplace platform suggests this. Change may impact the description standard, e.g., the addition of core data types from CCTS to USDL, the roles and responsibilities, e.g., the reassignment of BM tasks to and ontology engineer, the change of processes, e.g., the addition of a community liaison process to bring the service community up to speed about changes in the meta-model, as well as implementing changes in the actual core meta-models or their modules.

For more details on the roles cf. [27]. We also included accountability (A) and responsibility (R) information in the table according to RACI matrices.

Table 16.4 lists essential reference guidelines that are required for successfully designing, setting up, and maintaining a repository used for service description in an environment such as the Internet of Services. Guidelines described in Table 16.4 focus on the usage of the service description repository, the service ontology, throughout the governance lifecycle.

Table 16.4: Service Description Setup and Maintenance Processes. BM – business manager (host), HA – host architect (host), SA – software admin (host), SM – service manager (provider), R stands for responsible, A for accountable.

Phase	Processes	Roles
Design	- Determine tooling support	HA, SA
	- Define general description guidelines	BM
	- Set up repository for service descriptions	HA, SA
Deployment	- Verify description instances	BM
Delivery		
Monitoring	- Verify and certify adherence to description meta-model and description guidelines	SA, BM
	- Monitor word usage and meaning development over time	BM (A)
	- Monitor changes and perform development trends analysis	BM (A)
	- Analysis of semantic applications and usage feed-back	BM, SA
Change	- Change description guidelines	BM, HA
	- Change description process	BM, HA
	- Change service description	BM, HA, SM

Similarly, to the governance of meta-models, the instances of service descriptions have to be governed. In the design phase, approved tooling for actual descriptions needs to be chosen and modeling guidelines have to be published so that the description files are uniform and comparable for the service consumers.

When providers deploy their services so that they can be sold via the marketplace platform, their descriptions have to be checked for compliance to the guidelines set forth in the design phase. This entails a semantic as well as a syntactical evaluation.

Naturally, the actual delivery takes place automatically, similar to the meta-model.

Just like in the case of the meta-model, word changes are tracked, trends are estimated and semantic checks are performed. In addition user feedback is evaluated by the BM and factored into the verification of service descriptions to their meta-mdoel.

In the change phase, the BM and HA adapt processes and guidelines for service descriptions as deemed necessary through monitoring and communicate the change to the users. If necessary, the concrete description of services has to be changed here as well.

16.4.2 Exemplary Application

When designing the service description meta-model, one of the first decisions will have to be on the used standards. If we assume USDL is taken as a basis, then (with a look at the running example) a meta-model for an agent's address might look as follows (rendered as XML for the sake of readability in Listing 16.1):

Listing 16.1: Agent address meta-model.

```
1  <PhysicalAddress xmi:id="">
2    <Street></Street>
3    <StreetNumber></StreetNumber>
4    <City></City>
5    <Postcode></Postcode>
6    <State></State>
7    <Country></Country>
8    <GeographicalPoint xmi:id=" ">
9      <Latitude></Latitude>
10     <Longitude></Longitude>
11   </GeographicalPoint>
12 </PhysicalAddress>
```

It is a design decision to adopt this meta-model or to add or delete information (e.g., remove the state for an exclusively German platform). Also, the meta-model designers need to decide on code lists for the attributes such as country codes or a format for geo-spatial coordinates. USDL is quite comprehensive but changes will be necessary as there may be new forms of communication for virtual addresses or if a country decides to introduce a different system to specify physical addresses. Also, the platform host could decide to implement CCTS and use core data types for addresses which would look significantly different as they carry more detailed data.

On an instance level, the agent's address might look like Listing 16.2.

Listing 16.2: Service description showing a concrete agent's address.

```
1  <PhysicalAddress xmi:id="PhysicalAddress_27437204">
2    <Street>Hacks Cross Road</Street>
3    <StreetNumber>3620</StreetNumber>
4    <City>Memphis</City>
5    <Postcode>38125-8800</Postcode>
6    <State>Tennessee</State>
7    <Country>USA</Country>
8    <GeographicalPoint xmi:id="GeographicalPoint_26101118">
9      <Latitude>35.05107</Latitude>
10     <Longitude>-89.792703</Longitude>
11   </GeographicalPoint>
12 </PhysicalAddress>
```

Here, the change of address is probably the most striking use case. In order to keep addresses (and service interfaces up to date) monitoring and feedback information has to be evaluated to make sure all data is current. Also, legal requirements on reporting the accountability for services might change and require different data which used to be optional. Here a change in the meta-model will cascade to changes in the actual service descriptions.

16.5 Conclusion

Based on evidence from existing governance frameworks and research from academia and practice we derived processes, stakeholders, measurements, and maturity levels as the four building blocks for a governance framework that can support operations of a service marketplace.

In our understanding transactions on a service marketplace involve three entities: a service provider, a service broker and a service consumer. Innovators and communities support this process indirectly. The platform host is the key stakeholder to enable this interaction via the broker. However, the provider and the consumer are also involved directly or indirectly in governance processes in the different phases of the service lifecycle and the generation of KPIs about these processes. We distinguish five phases that cover design, deployment, delivery, monitoring, and change of services. The KPIs are broken down from a company-level to an IT-level to these processes. Furthermore, in order to grow and evaluate the overall architecture we propose a maturity model and capability profile. These four building blocks are necessary to form the core of a Governance framework for the Internet of Services.

Furthermore, this kind of Governance has to pay attention to cross-company legal aspects, e.g., data protection/ security. It must comprise contract management over country borders, country-specific laws for data transmission and protection, and laws concerning the fulfilment of online contracts. It also must cover different service monitoring aspects and includes the interests of multiple stakeholders. Operating a service marketplace platform involves much more stakeholders than common SOA approaches. Being a cross-company setup, a framework must consider the interests of all stakeholders of the marketplace platform.

We acknowledge that there may be more aspects to include as related work points out. However, they can most likely be related to the above four building blocks. In this chapter we provided an initial framework that needs to be instantiated according to the application and its process, roles, maturity levels, and metrics need to be detailed for application. At this stage it is also intended as an overview of the challenges governance faces in the Internet of Services such as brokerage as well as an inspiration for further research in this emerging area.

References

1. SOA Governance: Beherrschen Sie Ihre SOA. Business white paper, Software AG, 2005. http://www.softwareag.com/de/Images/WP_SOA_Governance_D_tcm17-22130.pdf.
2. Service Lifecycle Governance: Timely Policies and Enforcement Help Companies Reap the Full Benefits of SOA. BEA white paper, BEA Systems, Inc., 2006. http://www.itworldcanada.com/Admin/Pages/Assets/DisplayAsset.aspx?id=e0a24263-a10a-4887-8d45-582261587176.
3. SOA Governance: Balancing Flexibility and Control Within an SOA. A systinet white paper, Systinet, 2006. http://www.webservices.org/content/download/83830/

1225383/file/SOA_Gov906.pdf.
4. SOA Governance: Enabling Sustainable Success with SOA. Technical report, webMethods, Inc., 2006. http://www.cioindex.com/nm/articlefiles/44428-SOA_Governance.pdf.
5. 2007 Survey on SOA Governance. Technical report, WebLayers SOA Forum, 2007. http://www.bitpipe.com/detail/RES/1209665671_580.html.
6. CObIT 4.1: Control Objectives for Information and Related Technology. Technical report, IT Governance Institute, Rolling Meadows, IL, USA, 2007.
7. ITIL v3: Information Technology Infrastructure Library Version 3. Core OGC titles, Office of Governance Commerce, The Stationery Office, London, 2007.
8. SOA Governance: Balancing Flexibility and Control within an SOA. Technical report, Hewlett-Packard, 2007. http://hp.com/go/soa.
9. Five Steps for Building the Business Case for SOA Governance. Business white paper, Software AG, 2008. http://www.softwareag.com/de/images/BusCase_SOA_WP_Nov08-web_tcm17-45795.pdf.
10. CMMI Models and Reports. Technical report, Software Engineering Institute, 2009. http://www.sei.cmu.edu/cmmi/models/.
11. SOA Governance Framework. Technical standard, The Open Group, 2009.
12. TOGAF 9: The Open Group Architecture Framework. Technical report, The Open Group, 2009. http://www.opengroup.org/architecture/togaf9-doc/arch/index.html.
13. M. Afshar. SOA Governance: Framework and Best Practices. An Oracle White Paper. An Oracle white paper, 2007. http://www.oracle.com/technologies/soa/docs/oracle-soa-governance-best-practices.pdf.
14. P. Allen. SOA Governance: Challenge or Opportunity? *CBDI Journal*, 7:20–31, 2008.
15. A. Arsanjani and K. Holley. Increase Flexibility with the Service Integration Maturity Model (SIMM). Technical report, IBM developerWorks, 2005.
16. A. P. Barros and M. Dumas. The rise of web service ecosystems. *IT Professional*, 8(5):31–37, 2006.
17. J. Bernhardt and D. Seese. A conceptual framework for the governance of service-oriented architectures. In G. Feuerlicht and W. Lamersdorf, editors, *Service-Oriented Computing - ICSOC 2008 Workshops, ICSOC 2008 International Workshops, Sydney, Australia, December 1st, 2008, Revised Selected Papers*, volume 5472 of *Lecture Notes in Computer Science*, pages 327–338. Springer, 2008.
18. N. Bieberstein, S. Bose, M. Fiammante, K. Jones, and R. Shah. *Service-oriented Architecture Compass: Business Value, Planning, and Enterprise Roadmap.* IBM Press, Upper Saddle River, NJ, USA, 2005.
19. N. Bieberstein, S. Bose, L. Walker, and A. Lynch. Impact of service-oriented architecture on enterprise systems, organizational structures, and individuals. *IBM Systems Journal*, 44(4):691–708, 2005.
20. J. Bloomberg. *SOA Governance: IT Governance in the Context of Service Orientation.* ZapThink, LLC, Baltimore, MD, USA, 2004.
21. B. Brauer and S. Kline. SOA governance: a key ingredient of the adaptive enterprise. Whitepaper, HP, 2005.
22. W. Brown, R. Laird, C. Gee, and T. Mitra. *SOA Governance: Achieving and Sustaining Business and IT Agility.* IBM Press, Upper Saddle River, NJ, USA, 2008.
23. P. Derler and R. Weinreich. Models and Tools for SOA Governance. In D. Draheim and G. Weber, editors, *Trends in Enterprise Application Architecture, 2nd International Conference, TEAA 2006, Berlin, Germany, November 29 - December 1, 2006, Revised Selected Papers*, volume 4473 of *Lecture Notes in Computer Science*, pages 112–126. Springer, 2006.
24. T. Erl. *Service-oriented Architecture: Concepts, Technology, and Design.* Prentice Hall, Englewood Cliffs, NJ, 2005.
25. D. Herron and D. Garmus. Identifying your organization's best practices. *CrossTalk: Journal of Defense Software Engineering*, 19:22–25, 2005.

26. K. Holley, J. Palistrant, and S. Graham. Effective SOA Governance. IBM White Paper, 2006. ftp://ftp.software.ibm.com/software/uk/flexible/wp/effective_soa_governance.pdf.
27. C. Janiesch, M. Niemann, and R. Steinmetz. The TEXO Governance Framework. Technical report, SAP Research White Paper, 2011. http://www.internet-of-services.com.
28. W. Johannsen and M. Goeken. *Referenzmodelle für IT-Governance: Strategische Effektivität und Effizienz mit COBIT, ITIL & Co.* dpunkt, Heidelberg, 2007.
29. N. Josuttis. *SOA in Practice: The Art of Distributed System Design.* O'Reilly Media, Sebastopol, CA, USA, 2007.
30. R. Kaplan and D. Norton. *Translating Strategy into Action: The Balanced Scorecard.* Harvard Business School, Boston, MA, 1996.
31. O. Kohnke, T. Scheffler, and C. Hock. SOA-Governance — Ein Ansatz zum Management serviceorientierter Architekturen. *Wirtschaftsinformatik*, 50(5):408–412, 2008.
32. E. Marks. *Service-Oriented Architecture Governance for the Services Driven Enterprise.* John Wiley & Sons, Inc., Hoboken, NJ, USA, 2008.
33. M. Matsumara. The Definitive Guide to SOA Governance and Lifecycle Management. Technical report, 2007. http://www.scribd.com/doc/7056416/Guide-to-SOA-Governance.
34. G. McBride. The Role of SOA Quality Management in SOA Service Lifecycle Management. IBM developerWorks, 2007. ftp://ftp.software.ibm.com/software/rational/web/articles/soa_quality.pdf.
35. T. Mitra. A Case for SOA Governance. IBM developerWorks, 2005. http://www.ibm.com/developerworks/webservices/library/ws-soa-govern/.
36. B. Moreland and M. Afshar. The Path to SOA (Part I of III). http://www.ebizq.net/topics/systems_management/features/7193.html, 2006.
37. M. Muriankara. SOA Governance Framework and Solution Architecture. IBM developerWorks, 2008. http://www.ibm.com/developerworks/webservices/library/ws-soa-govframe/.
38. A. Picot, R. Reichwald, and R. Wigand. *Die grenzenlose Unternehmung: Information, Organisation und Management.* Gabler, Wiesbaden, 4th edition, 2001.
39. A. Pugsley. Assessing your SOA Program. HP White Paper, Hewlett Packard, 2006.
40. J. Quantz. SOA in der Praxis: Wie Unternehmen SOA erfolgreich einsetzen. Berlecon report, BerleCon Research, Mar 2006. http://www.berlecon.de/studien/downloads/Berlecon_SOA.pdf.
41. D. Rane and G. Lomow. SOA Governance: More than just Registries and Repositories. Technical report, BearingPoint, 2008. http://www.digitalgovernment.com/media/Knowledge-Centers/asset_upload_file77_1753.pdf.
42. I. Rieger and R. Bruns. SOA-Governance und -Rollen: Sichern des Mehrwerts einer serviceorientierten Architektur. pages 20–24, 2007.
43. J. Schelp, O. Schmitz, J. Schulz, and M. Stutz. Governance des IT-Sourcing bei einem Finanzdienstleister. *HMD - Praxis Wirtschaftsinform.*, 250, 2006.
44. T. G. J. Schepers, M.-E. Iacob, and P. van Eck. A lifecycle approach to SOA governance. In R. L. Wainwright and H. Haddad, editors, *Proceedings of the 2008 ACM Symposium on Applied Computing (SAC), Fortaleza, Ceara, Brazil, March 16-20, 2008*, pages 1055–1061. ACM, 2008.
45. B. Schmid. Elektronische Märkte: Merkmale, Organisation und Potentiale. In A. Hermanns and M. Sauter, editors, *Management-Handbuch Electronic Commerce: Grundlagen, Strategien, Praxisbeispiele.* Vahlen, Munich, 1999.
46. I. Sommerville. *Software Engineering.* Addison-Wesley, Harlow, 8th edition, 2006.
47. A. S. Vedamuthu, D. Orchard, F. Hirsch, M. Hondo, P. Yendluri, T. Boubez, and U. Yalcinalp. Web services policy 1.5 - framework. W3C Recommendation, http://www.w3.org/TR/ws-policy/ 2007.
48. P. Weill and J. Ross. *IT Governance: How Top Performers Manage IT Decision Rights for Superior Results.* Harvard Business School Press, Boston, MA, USA, 2004.

49. P. Windley. Governing SOA. InfoWorld.com, Jan 2006.
50. P. Windley. SOA Governance: Rules of the Game. InfoWorld.com, 2006.
51. P. Windley. Teaming up for SOA. InfoWorld.com, 2007.
52. B. Woolf. Introduction to SOA Governance. IBM developerWorks, 2007. http://www.ibm.com/developerworks/library/ar-servgov/.

Chapter 17
Managing Variants of USDL

Gunther Stuhec, Daniel Oberle, Christian Baumann, Christian Janiesch, Michael Dietrich, Jens Lemcke, Jörg Rech, and Wolfgang Karl Rainer Schwach

Abstract Different variants of USDL are required for different contexts. This is already shown by the Legal Module which requires different contents depending on the jurisdiction of a country. The issue aggravates if more and more parameters are relevant to determine the correct variant. The chapter presents one possible solution for variant management consisting of a canonical data model, a context driver mechanism, governance processes, and appropriate tooling. Although the solution for variant management is targeted at existing business documents, such as a purchase order, it provides a powerful and adequate means for dealing with USDL variants as well.

17.1 Introduction

Chapter 14 discussed that the contents of USDL are required in different representations. Different contexts of use, e.g., industries or countries, even require variants of USDL. As shown in Chapter 10, formalizing licenses according to German law and

Gunther Stuhec
SAP Global Partner & Ecosystems Group, Dietmar-Hopp-Allee 16, 69190 Walldorf, Germany, e-mail: `gunther.stuhec@sap.com`

Daniel Oberle, Michael Dietrich, Jens Lemcke, Jörg Rech, Wolfgang Karl Rainer Schwach
SAP Research Karlsruhe, Vincenz-Priessnitz-Str. 1, 76131 Karlsruhe, Germany, e-mail: `firstname.lastname@sap.com`

Christian Baumann
Berkeley Center for Law & Technology, University of California, Berkeley, School of Law, 376 Boalt Hall Berkeley, 94720 CA, USA, e-mail: `baumann@berkeley.edu`

Christian Janiesch
Institute of Applied Informatics and Formal Description Methods (AIFB), Karlsruhe Institute of Technology, Englerstr. 11, Geb 11.40, 76131 Karlsruhe, Germany, e-mail: `christian.janiesch@kit.edu`

US common law requires different Legal Modules. However, both modules are not completely disjoint. There are overlaps in that specific classes are identical in both modules. Along the same lines, one can conjecture that other contexts, e.g., industries, business processes, or business roles, have specific needs on the content level that lead to further variants of USDL. First, extensions or alterations might be required with different granularity (attributes, classes or relations, or whole modules). Second, different representation terms might be applied for the same entity (both because of internationalization and also within the same language). Third, identical representation terms might have different meanings in different contexts. For example, *Reward* has many different meanings in different contexts and requires disambiguation. Reward could be either a bestow of honor to actors in the music industry or it could be a payment return for a service in the context of legal aspects. Fourth, the ownership of an alteration might be proprietary, i.e., only required for a specific user, or should be made normative and channeled back as an official alteration to the standard.

If not managed properly, the situation will lead to a proliferation of USDL variants much like what has happened with business documents in the area of B2B interoperability which requires cost and time intensive mappings between variants. B2B interoperability requires a technical realization of connectivity between systems. This is readily addressed through the use of existing technical standards and supporting middleware such as Web services. However, the bigger challenge, which still remains for achieving B2B interoperability, is the lack of a common understanding at the collaborative business process and data level.

This lack of common understanding is caused by different representations, different purposes of use, different contexts of use, and different syntax-dependent approaches in solving the problem. For more than 30 years, different B2B standardization approaches have only considered their own specific requirements, at a syntax-dependent level, without consideration of adopting a common approach at the semantic level. This lack of a common approach has led to significant interoperability issues when attempting to use the resultant solutions outside of the narrow scope for which they were developed [30].

As an example business document, consider a purchase order, which is being exchanged between a buyer and a retailer, and eventually also between the retailer and a supplier. The communication happens via different e-Business standards. Buyer and retailer might exchange the purchase order in the OAGIS v8.0 syntax; retailer and supplier communicate via RosettaNet [13]. Both the OAGIS and RosettaNet representations use significantly different structures [9] and element names to represent the purchase order document. Therefore, the parties have to negotiate their different meanings and align before the communication takes place and establish cost and time intensive mappings from and to their internal interfaces in ERP systems, respectively.

Besides the RosettaNet and OAGIS standards, there is a proliferation of standardized syntaxes for representing business documents [24]. Typically, different industries in combination with geopolitical settings introduce their own standard. An example set is the family of EDIFACT variants [5].[1] Several industries developed variants, in this particular case subsets of the EDIFACT umbrella standard, to reflect their specific representation needs. Examples are ODETTE for the European Automotive Industry, EDIFICE for the high-tech industry, CEFIC for the chemicals industry, EDILIBE for book retails, and many more. Although an estimated 60% of the represented information in the (competing) standards is similar, most of them established their own terminology, structure, and way of representation.

A possible solution for managing variants of USDL therefore stems from the area of B2B interoperability. The solution consists of the following four pillars:

1. A *Canonical Data Model* to ensure a controlled vocabulary
2. A *Context Driver* mechanism to manage the different contexts of use
3. *Governance Processes* to manage the canonical data model
4. *Tooling* to realize and support 1-3

The chapter will introduce each of the four pillars in more detail in Sections 17.3 to 17.6. Before, Section 17.2 will detail some examples for variants in USDL. Finally, we discuss related works and conclude in Section 17.7 and 17.8, respectively.

17.2 Motivation

The fact that there will be several variants of USDL is already apparent in the Legal Module. Modeling of legal aspects such as licenses depends on the legislation of the country and, consequently, different modules for the German and the US common law are required (cf. Chapter 10). However, it is also apparent that there are overlaps between both modules. For instance, the class Work appears in both modules although with different attributes. In addition, there are also differences on the level of classes and relations. The class of CopyrightTransfer is apparent in US but not in German law, the relation authors points to the foundational classes Person (German variant) and Agent (US variant), respectively. Figure 17.1 provides an overview of the overlaps and differences related to the class Work and will serve as a running example in the remainder of the chapter. The following enumeration provides a detailed description of the running example. More differences for legal aspects are discussed in Chapter 10.

Class Work: Difference with respect to attributes The German and US variants of the class Work differ in individual attributes. According to US common law, a work can have a registration status. The registration of a work is not possible in German law. In contrast, the registration serves as *prima facie evidence* in the case of a copyright lawsuit in US common law.

Class CopyrightTransfer: Only in US Copyrights can be *transferred* in US common law. This option is not available in German law, but you can only *grant* usage

[1] The acronym stands for Electronic Data Interchange For Administration, Commerce and Transport developed under the United Nations.

rights. The copyright always remains with the original author[2] in German law. Therefore, the US Legal Module contains a class CopyrightTransfer to specify whether the copyright is *assigned* or *transferred*. The type of conveyance is specified by the enumeration type:TransferType. The following types are possible: Assignment (usage right is assigned) and Authorization (copyright is transferred).

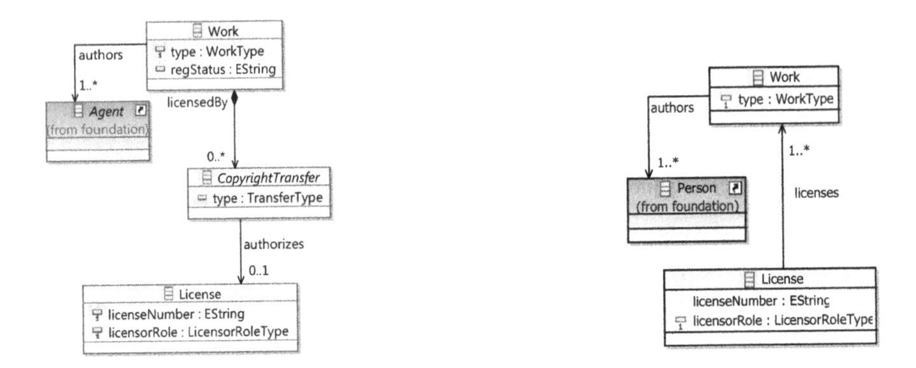

Fig. 17.1: Differences between Work in the US (left) and the German (right) Legal Modules.

Relation authors: Difference with respect to relations German law only considers natural persons as authors of a work. Hence, the German Legal Modules introduces a relation authors which points to the class Person of the Foundation Module. US common law considers both natural and legal persons as authors of a work. Consequently, authors points to the more generic class Agent (of the Foundation Module).

Relation licensedBy: Difference with respect to relations In the US module the relation LicensedBy links the classes Work and CopyrightTransfer. A work is licensed by a specific type of copyright transfer.[3] The CopyrightTransfer then authorizes the License.
The German module represents this via the relation licenses, since it is not an option to transfer the complete copyright but only to grant usage rights.

Putting both the US and German variants in one module is not an option since further countries might even require additional conceptualizations and equally named classes might differ in their conception. Yet, there is overlap, e.g., the notion of Work is indeed equal in both countries, so introducing two classes of the same name in different modules is also not an option.

[2] The copyrights can be transferred only in the case of the death of an author in the German Copyright Law.

[3] Cf. description for class CopyrightTransfer.

Our experience so far also gives reason to expect several industry-specific extensions or variants of USDL. We expect a situation similar to EDIFACT which is governed by the UN and additionally several sub-committees that govern industry-specific derivations of EDIFACT. For example, ODETTE for the European automotive industry or EDIFICE for the electronics industry. Besides, many countries have their own EDIFACT derivation. For example, there is a DIN (German norm) in Germany which is derived from EDIFACT.

In addition, there is the need for proprietary extensions to USDL. Consider a platform provider who might want to integrate a class about Feedback specifically for his usage of USDL. The following sections introduce a possible solution to cope with all the aforementioned challenges by means of four pillars, viz., a canonical data model, a context driver mechanism, governance processes, as well as corresponding tooling.

17.3 Canonical Data Model

Even if the terminology of the two Legal Modules follows a strict grammar, it is unlikely that separately introduced variants of USDL apply the same grammar and arrive at compatible naming schemes. Therefore, the first remedy for countering the problems outlined in the previous section is a Canonical Data Model (CDM) based on a standard naming convention to ensure a common vocabulary.

The *Core Components Technical Specification (CCTS)* [8] by the United Nations Center for Trade Facilitation and Electronic Business (UN/CEFACT) is such a grammar and provides the platform independent definition of data types for document exchange and, in addition to that, a concept on how to organize, i.e., aggregate and associate, data on a higher level to create business documents (e.g., a purchase order). CCTS provides a way of standardizing business semantics, i.e., the document data's meaning.

Core components are building blocks for semantically correct and meaningful business documents. Core components can be instantiated to assemble a business document via core component types (CCT), basic business information entities (BBIE), association business information entities (ASBIE), and aggregate business information entities (ABIE).[4] CCT are of a data type and consist of a content component and one or more supplementary components. Content components carry the actual value while supplementary components further define the content. According to the specification, CCT do not have business semantics. BBIE have business semantics and provide a singular element of an ABIE (e.g., `Type` in `Work`). An ABIE (e.g., `Work`) is a collection of BBIEs and can be part of other ABIEs via an ASBIE. Figure 17.2 shows the class `Work` of the US Legal module rendered in CCTS.

CCTS prescribes naming conventions for specifying the dictionary entry names (DEN) for supplementary components, content components, BDTs, BBIEs, AS-

[4] In terms of object-orientation, ABIEs are roughly equivalent with classes, BBIEs with attributes, and ASBIEs with associations.

<<ABIE>>

Work Details

Type	OCQ	OCT	PQ	PT	RT	Oc
<<BBIE>>		Work	Registration	Status	Text	[0..1]
<<BBIE>>		Work		Type	Identifier	[0..1]
<<ASBIE>>		Work		License	Transfer	[0..*]
<<ASBIE>>		Work		Author	Agent	[1..*]
...						

Fig. 17.2: `Work` of the US Legal Module rendered in CCTS as ABIE.

BIEs, and ABIEs. The basis of the naming conventions is ISO 11179 Part 5 and can be compared with a grammar of a natural language for the consistent and grammatically correct building of element or attribute names. The DEN consists of an Object Class Term (OCT), a property term (PT), and a representation term (RT). OCTs and PTs can be assigned additional qualifiers (OCQ and PQ). The OCT can be compared with the object of a sentence, the PT is usually the predicate and the RT can be regarded the subject of a sentence. The optional qualifiers can be used for the semantic refinement of OCTs and PTs. All of them can only be represented by nouns and adjectives (verbs of the USDL class models are nominalized accordingly). An example of a complete Dictionary Entry Name would be `Work.Registration.Status.Text`. The Object Class Term represents the ABIE for which the BBIE or ASBIE is defined, e.g., `Work`. The Property Term is the first part of the name of an administered item. The Property Term (e.g., `Status`) expresses a property of an ABIE, and defines the characteristics that belong to an ABIE. The PT can be further refined by providing a Property Qualifier (PQ), in our case `Registration`. The Representation Term is the second part of the property. It defines the subject that is used to express the kind of representation of the property and reflects its datatype. For example, `Status` gives the apparent proposition and `Text` specifies how the `Status` will be represented. A BBIE always gets its characteristics for the representation of the values by an appropriate Core Data Type. Type awareness of the core components is easily determined by examining the Representation Term of the BBIE, as a Representation Term is always the same term as the Object Class Term of its corresponding Core Data Type. ASBIEs link to other ABIEs. In our example, the ASBIE `License.Transfer` links to another ABIE identified by `Copyright.Transfer.Details` (not shown in Figure 17.2).

17.4 Context Driver

CCTS specifies how to assemble USDL variants and their data types as well as a methodology on how to create *context-based* instantiations of these general USDL variants. For example, the aforementioned `Work` might feature different BBIEs and ASBIEs in different business contexts. CCTS defines business context as the formal description of a specific business circumstance as identified by the values of a set of context categories, allowing different business circumstances to be uniquely distinguished. A context category is a group of one or more unordered values used to express a characteristic of a business circumstance. Examples for context categories are industry, business role, or country. Each context category consists of a set of related context values. Each value of the context category is defined by a classification scheme which is expressed as code list or identifier scheme.

This is already depicted in Fig. 17.4 by the leftmost columns. Wildcards (*) mean that the BBIE or ASBIE is valid in every industry (Ind), business role (Rol), or country (Co), respectively. Specific values, such as `US` for the country (column Co) for `Registration.Status.Text`, mean that this BBIE is only valid for the United States but in all industries and countries. Thus, both the German and the US version of Work can be captured in one ABIE.

Context			<<ABIE>>						
Ind	Rol	Co	Work Details						
			Type	OCQ	OCT	PQ	PT	RT	Oc
*	*	US	<<BBIE>>		Work	Registration	Status	Text	[0..1]
*	*	DE,US	<<BBIE>>		Work		Type	Identifier	[0..1]
*	*	US	<<ASBIE>>		Work		License	Transfer	[0..*]
*	*	US	<<ASBIE>>		Work		Author	Agent	[1..*]
*	*	DE	<<ASBIE>>		Work		Author	Person	[1..*]

Fig. 17.3: Context categories and CCTS.

In [8], CCTS foresees eight context categories[5] but does not specify how to apply and realize such context drivers. This is where the UN/CEFACT Unified Context Methodology (UCM) [16] comes into play as a methodology for managing the representation and use of business context information, especially for UN/CEFACT technologies. This includes representing the business contexts of business documents, using business context during message assembly to produce messages for specific business contexts, and using business context to (semi-) automatically customize generic business processes. UCM uses graph theory and set theory to define and understand the concept of context, context categories, context values, and context operands [21]. An alternative to the UCM is proposed in [17].

[5] Business process, business process role, supporting role, industry classification, product classification, country, official constraints, and system capabilities.

17.5 Governance Processes

The canonical data model and context driver mechanism are the prerequisites for a common vocabulary and the prevention of the proliferation of incompatible USDL variants. However, with the modification and adding of USDL variants by users, the canonical data model might lose structure, consistency, conciseness, and even correctness. Therefore, it has to be normalized, harmonized, optimized, corrected, and consolidated by a schema engineer or by an automatic analyzer both with the backup of a governance team and corresponding governance processes.

Current business document standards, such as OAGIS or RosettaNet, are typically managed centrally by such a governance board consisting of several schema engineers. Users can inquire changes to the schema and the governance board eventually decides if and how the change will be reflected in future versions. This task is non-trivial and time-consuming since a multitude of users might request similar extensions and alterations. The processes for the inquiry and changes are more or less explicit depending on the standards body and include (cf. also [25]):

Harmonization The semantic analysis of the canonical data model in order to determine equivalence, similarity, or dissimilarity in dictionary entry names. Names and identifiers within the canonical data model should use identical names and a similar structure. The aim of harmonization is to improve the canonical data model's interoperability, reusability, and maintainability. For example, elements have synonyms, related business terms or terms in different languages such as author, auth, writer, creator, etc. These should be harmonized by finding a meaningful common denominator (independent of different languages). The governance board needs to find these names and select a standardized reference term.

Normalization The structural analysis of the canonical data model in order to determine duplication, flatness, or over-nesting of elements. The structure of schema elements (e.g., addresses) should be balanced. The aim of normalization is to improve the readability, interoperability, portability, reusability, accessibility, and maintainability of the canonical data model. For example, an element containing information on a person in a flat structure can contain (academic) titles, the middle name, job titles, zip code, etc. and can be subdivided into reusable sub-schema such as name, address, or job information. The governance board needs to identify flat or otherwise malformed substructures and split or normalize them.

Consolidation The structural and semantic analysis of the canonical data model in order to determine overlap, similarity, or dissimilarity of elements or structural parts of a schema. Schema elements similar to each other but spread over different schemas should be integrated. The aim of consolidation is to improve the canonical data model's understandability, reusability, and maintainability. For example, elements containing the address of a person, company, or field should be integrated and linked — potentially in few different scenarios (e.g., long and short addresses). The governance board needs to identify distributed but similar elements and consolidate them.

Optimization The structural, semantic, or pragmatic analysis of static or dynamic characteristics of the canonical data model in order to determine the usability/ applicability, distribution, usage, etc. of dictionary entry names. A schema needs to be optimized regarding various quality attributes. The aim is to improve a quality attribute such as understandability, reusability, interoperability, portability, reusability, accessibility, maintainability, consistency, conciseness, and even correctness.

The governance board has to process several activities to collaboratively make a decision on the identified problem propositions. These activities are:

Rating of diagnosed problems by each individual in order to prioritize them.

Commenting of diagnosed problems by each individual in order to persist one owns thought.

Discussion forum to argue about the problems, potential treatments (solutions), and side effects.

Comparison mechanism for comparing semantically same or similar elements or structures

Context driver mechanism for refining the unambiguous definition in which business context is an element relevant.

Voting mechanism for the proposed treatments (schema refactorings) or further process (e.g., annotate for deprecation).

Refactoring mechanisms for the semi-automated changes on the CDM. This might include descriptions of (manual) refactorings as well as semi-automated schema refactorings [10] such as Rename Element or Extract Sub-schema.

Annotation mechanism for the annotation or tagging of elements within the CDM. Annotations such as `@deprecated`, `@ambiguous`, `@see`, or `@plannedForVersion` are aimed at supporting governance activities by persisting information, decisions, or background knowledge.

Mapping mechanism for identifying, if an element is not a part of the standard CDM anymore but this should further considered by users. This means a mapping mechanism to the new standard CDM element should be provided.

Furthermore, the changes to the canonical data model should be versioned to be able to rollback erroneous, premature, or inappropriate changes.

17.6 Tooling

The CCTS-based naming conventions, the canonical data model and the context driver are just specifications which need to be realized by appropriate tooling. In addition, the governance processes are usually implicit and realized without any tool support. A community of experts, i.e., the governance board, analyzes, discusses, and treats problems within the canonical data model. They have to manually identify these problems by analyzing the canonical data model on their own. Therefore, we propose a solution, as presented in [29], whose key features are presented in the following subsections.

17.6.1 Common Repository

First and foremost, the solution consists of a repository containing the canonical data model discussed in Section 17.3. The canonical data model is structured according to CCTS and supports the context driver principle. That means every element of the canonical data model can be marked to be valid in a specific number of context categories. The common repository can be accessed remotely and collaboratively. Users can formulate queries to obtain the business document in the corresponding context.

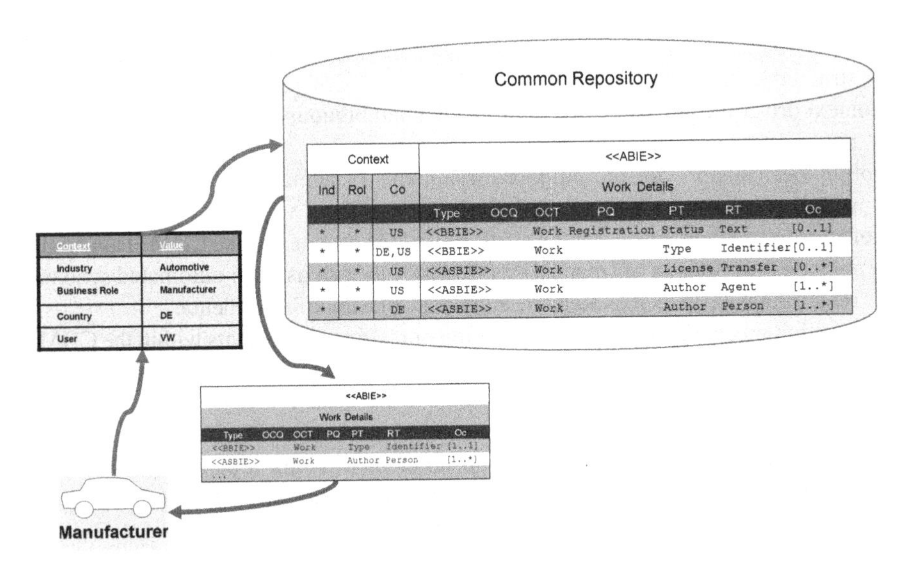

Fig. 17.4: Obtaining context-specific variants of USDL.

In order to illustrate this let us have a look at Fig. 17.4 and consider the following example. Assume an automotive manufacturer in Germany who is in need of obtaining USDL in its context. The manufacturer accesses the common repository through its web browser. The Web interface allows formulating a context specific query for the USDL variant that fits the user's context. Rather than responding with a one-size-fits-all USDL where most fields will remain unused, the common repository provides a customized variant of USDL that most closely fits to the unique requirements of the manufacturer. The solution achieves this by creating a context specific assembly of the most appropriate and reusable building blocks. The result is a tailor made suggestion designed specifically for the user's context that might be very close to the final user's requirements. This suggestion can be modified or customized by the user afterwards and by using the same tool. Coming back to the running example of the Legal Module, the manufacturer would retrieve the `Work` ABIE with ASBIEs and BBIEs only valid in his context.

17.6.2 Collaborative Governance

In the running example, the obtained variant is the recommended structure for the manufacturer. The manufacturer can now determine if the elements identified are appropriate for its context, or if some elements are missing. An online modification in the manufacturer's context according to its business needs can be performed. For the elements that are not required, the solution is simple. A delete action not only removes the item from the USDL variant in the corresponding context, but also provides valuable information that is used to improve future results (cf. Section 17.6.3). For the missing elements, there exists the possibility to add a data element (ABIE, BBIE, or ASBIE) to the variant in the given context. However, one typically struggles to determine the correct name, structure, position and so on. In order to remedy the struggles, the solution contains a natural language processor to provide a CCTS conformant name and definition, identify where else it is being used and in what context, and the most appropriate position in an ABIE. The tool can also provide a list of similar ABIEs, which come close to the required semantics of the users. The tool can additionally provide some statistical results to the users such as *how often used, in which contexts used, what is the substructure* etc. This statistical information should support the user to prefer an already defined ABIE instead of defining a new ABIE.

Although deleting and adding new elements is like a self service for user, governance is required spanning all contexts and variants of the common repository when new elements are created. In this case the governance body is notified and has to decide on the change inquiry. The solution allows the governance body to configure both the notification level and priority of inquiries made by the users. With respect to the level, the governance body is able to choose at which granularity it shall be notified (creation of a new ABIE, BBIE or ASBIE). The governance body is provided tool support to inquire further with the user and eventually decides whether a new element is created or whether the suggested element is harmonized with an existing one. This governance support is a strict workflow to guarantee the consistency of the canonical data model. This workflow observes all activities made by the users and analyzes if the modifications can be harmonized, normalized, consolidated or even optimized. Based on this information, the following actions can be performed:

- The user will just accept the proposal of the system
- The user and governance group will perform a mediation, if there any doubtful questions on both sides
- The governance group will urge the user to take the recommendation, otherwise the canonical data model will lose the consistency, or

- The governance group will consider the user's extension as a mapping, if the user is not willing to change his model according the recommendation.

Therefore, the second key feature of the solution is collaborative governance whose fundamental idea is to include the wisdom of the crowd and abstract from the view of few [31]. In order to support the schema engineers and members of the governance board in their quality assurance activities, the solution provides an intelligent assistance system which informs about potential problems within the canonical data model and recommends potential treatments (e.g., schema refactorings). The assistance functionality has to identify patterns regarding the structure, usage, syntax, or semantics of elements within the canonical data model with a negative impact on the canonical data model's quality. The collaborative governance should enable and assist the community and a subgroup of experts — the governance board — to keep the canonical data model in high quality.

17.6.3 Evolutionary Optimization

Suppose a multitude of users are independently adding similar elements to the common repository. A proliferation of such elements is avoided through governance, yet there is further potential for optimizing the contents of the common repository over time. Suppose more than 50% of the users in different contexts require the same element as the manufacturer. If this is the case, the element might as well become valid in every possible context. Vice versa, if elements are never used, they might be deleted completely from the repository.

Therefore, the third key feature of the solution is the tracking of statistics of the frequency of usage of data in the common repository for optimizing its contents. This comprises a trend analysis of the elements in the common repository as well as information about non-usage in different periods. Data of the trend analysis can be viewed to the users to facilitate maintenance of their business documents. In addition, the information can improve the abilities for semi-automatic consolidation and harmonization.

17.6.4 Semi-automatic Mapping

The fourth key feature of the solution is a Semi-automatic Mapping component facilitating the import from and export to existing industry standards (e.g., RosettaNet, CIDX, EDIFACT) as well as proprietary schemas with different syntax representations (e.g., XML Schema, CSV, or XMI). For the data import, several matching algorithms ([11], [22] in particular) are applied to identify corresponding entities within the repository. The data is mapped to the CCTS-based elements of the common repository. New elements will be generated if no similar element can be identified.

The mapping is realized by a set of matching algorithms implementing various comparative methods. In this manner, comparisons of definitions, descriptions, types, dictionary entry names, context categories and values, attributes and properties are carried out by the various matchers. If the match similarity value of the considered B2B entity and some CCTS entity is disproportionately high comparing to the match similarity values of the other CCTS entities analyzed and above a predefined threshold, then a conclusion can be drawn that this B2B entity can be mapped onto the correspondent CCTS entity with high probability. Each matching algorithm can be assigned a weight so that it contributes more or less to the cumulated total similarity value accordant to its importance. The result of this process is a list of highly probable correspondences that will be verified by a user.

The proprietary or B2B schemas are also stored in the common repository. When a new schema is imported by a user, the user already has the possibility to indicate preferred mappings of elements between his schema and the canonical data model as well as to propose new element names. Then the new schema is stored in the common repository along with the mapping and naming proposals from the user. This triggers the iterative governance process: An automatic analyzer starts to compare the schema elements of the new schema including the information provided by the user with the canonical data model and the other schemas in the repository. The results of that analysis are presented to the governance board: first, how the different elements of the newly imported schema can be mapped to the elements of the canonical data model or if new elements should be inserted. Second, how a newly created or extended element can be named. The proposed names follow the CCTS DEN standard [21].

17.7 Related Work

There are numerous related works concerning canonical data models, context driver, governance processes, and appropriate tooling. However, we argue that the approach for variant management discussed in this chapter supersedes the related work since none of the latter covers all four areas.

17.7.1 Canonical Data Model

There is some related work to the canonical data model. As Janiesch and Thomas point out in [20], the introduction of a semantic layer is a new concept for business documents which older standards do not encompass. As of now, there is only limited transfer of this concept to other applications. For business rules definition, there is the Semantics of Business Vocabulary and Business Rules (SBVR) [1]. While SBVR contains a section on vocabulary, it is focused on the definition of statements rather than the structure of entities such as WSDL specifications or business documents.

In conceptual modeling, domain specific modeling methods (e.g., [2]) use domain concepts to clearly define (and name) certain tasks. Alternatively, domain ontologies or technical term models can be added to the original models [18]. This facilitates comparability and reuse of the models. Most other approaches to data models are of syntactical nature and specify the notation rather than the semantics of the content (e.g., camelback notation).

17.7.2 Context Driver Principle

Related work to the context driver principle is twofold. Numerous approaches to configuration exist [3], [12], [14], [19], [26], [28]. Common modeling frameworks, however, only provide few options for variant handling [10], [27], [33].

CCTS used to provide only indirect support for the configuration of its core components. Rather than providing a structured means to project the variants of the models, it provided a constraint language to trace back the change that has been made to a variant; it is of inverse nature so to speak. This has certain drawbacks, the most striking ones being: No guidance on new projection is offered and each variant has to be stored redundantly [8]. The approach followed in this document operates in the opposite direction, a method is needed that provides a means to derive business document variants from master models in a structured way. In that, it is similar to the approach of Janiesch [17], who applied the mechanisms of configurative reference modeling to business documents.

Concerning the definition of context parameters, there is no unified list if there is one at all. Often, the definition of the attributes, which define a context, are at the liberty of the modeler. The context methodology project aims at providing a solution in the long term.

17.7.3 Governance Processes

The governance (i.e., maintenance, management, or administration) and optimization (i.e., evolution, harmonization, normalization, consolidation) of schemas are relatively new fields and only little research focuses on canonical models fusing other schema into one. Today, approaches on the optimization [23], [15], harmonization [6], normalization [23] or consolidation of business schemas are non-collaborative and unsupported.

17.7.4 Tooling

Collaborative tools for the construction and governance of ontologies are emerging. A survey is given in [7] starting with knowledge bases such as Ontolingua and Co4 to collaborative systems such as BibSonomy (collaborative tagging), DBin (collaborative sharing), Hozo (collaborative ontology construction), OntoWiki (collaborative semantic authoring), Collaborative Protégé (collaborative ontology construction), or SOBOLEO (collaborative taxonomy development). But these tools are, typically, based on RDF or OWL and not easy to migrate to a proprietary XML schema as used for the CDM.

The iSURF eDoCreator (Electronic Document Design and Customization Tool) enables the creation and customization of CCTS-based document schemas. It does not reuse a repository of context driver values for the (semi)automatic projection of variants but merely facilitates the tedious process of defining the structure and serializing it. Therefore, it does not allow for comprehensive variant management to limit document proliferation [32].

H2 is a meta-modeling tool developed by the European Research Center for Information Systems (ERCIS) which enables the dynamic modeling of nested structures such as XML structures. It also allows for the configuration of these to project context specific variants. In many ways, it is a very promising tool for the configuration of documents. It is, however, still in development and lacks necessary import and export functionality as its focus is on being a generic meta-modeling tool [4].

17.8 Discussion

The chapter introduced a *possible* solution for variant management of USDL based on findings and experience in the area of business document modeling, configuration, and variant management. The chapter exemplified how the four pillars of this solution would be applied to USDL by means of the Legal Module. In general, applying the proposed solution would require the following disruptive steps.

First, USDL would have to be represented in CCTS. On the one hand, this comprises the consistent application of the CCTS naming scheme to classes, associations and attributes. On the other hand, the main challenge is the difference in expressiveness of CCTS and USDL's current meta-modeling language. USDL is originally modeled by UML class models and makes extensive use of interfaces and inheritance. Both modeling primitives are not provided by CCTS and workarounds as well as extensions would have to be found.

Second, in order to put the context driver mechanism into action, a first step is to agree on a valid set of context categories. Consequently, the current version of USDL could be defined as valid in all contexts (apart from the Legal Modules which could already be split according to the geopolitical context).

Third, the solution would require one or more centrally governed repositories for USDL variants based on USDL master templates. This repository should be under the auspices of an independent governing body. This could be a standardization committee such as the W3C. Another option is to deploy repositories in specific settings, such as large enterprises. Chapter 16 already discussed additional aspects of governance which are a prerequisite to this solution.

Fourth, with regards to tooling available nowadays, there is only one tool that supports all required pillars as discussed in Section 17.6. However, this tool remains a prototype for the time being [29].

References

1. Semantics of business vocabulary and business rules (SBVR). Available specification, OMG.
2. J. Becker, L. Algermissen, D. Pfeiffer, and M. Räckers. Bausteinbasierte Modellierung von Prozesslandschaften mit der PICTURE-Methode am Beispiel der Universitätsverwaltung Münster. *Wirtschaftsinformatik*, 49(4):267–279, 2007.
3. J. Becker, P. Delfmann, and R. Knackstedt. Adaptive reference modeling: Integrating configurative and generic adaptation techniques for information models. In *Proceedings of the Reference Modeling Conference (RefMod2006), Passau*, 2006.
4. J. Becker, C. Janiesch, and S. Kramer. Modellierung und Konfiguration elektronischer Geschäftsdokumente mit dem H2-Toolset. Arbeitsberichte des Instituts für Wirtschaftsinformatik 116, University of Münster, Germany, 2007.
5. J. Berge. *The EDIFACT Standards.* Blackwell Publishers, 2nd edition, 1994.
6. Y.-F. Chen, X. Sun, and C.-C. J. Kuo. XML schema harmonization: design methodology and examples. In *Proc. SPIE*, volume 5241 of *Multimedia Systems and Applications VI*, pages 221 – 232, 2003.
7. G. Correndo and H. Alani. Survey of tools for collaborative knowledge construction and sharing. In *Proceedings of the 2007 IEEE/WIC/ACM International Conference on Web Intelligence and International Conference on Intelligent Agent Technology - Workshops, 2-5 November 2007, Silicon Valley, CA, USA*, pages 7–10. IEEE, 2007.
8. M. Crawford. Core Components Technical Specification - Part 8 of the ebXML Framework. Technical report, UN/CEFACT, 2003.
9. S. Damodaran. B2B integration over the Internet with XML: RosettaNet successes and challenges. In *Proceedings of the 13th international World Wide Web conference on Alternate track papers & posters*, WWW Alt. '04, pages 188–195, New York, NY, USA, 2004. ACM.
10. D. Delen, N. P. Dalal, and P. C. Benjamin. Integrated modeling: the key to holistic understanding of the enterprise. *Commun. ACM*, 48(4):107–112, 2005.
11. H.-H. Do. *Schema Matching and Mapping-based Data Integration: Architecture, Approaches and Evaluation.* PhD thesis, University of Leipzig, Germany, 2007.
12. A. Dreiling, M. Rosemann, W. M. P. van der Aalst, L. Heuser, and K. Schulz. Model-based software configuration: patterns and languages. *EJIS*, 15(6):583–600, 2006.
13. M. Flebowitz. OAGIS 8.0: Practical Integration meets XML Schema. *XML Journal*, 3(9), 2002.
14. U. Frank and C. Lange. E-MEMO: a method to support the development of customized electronic commerce systems. *Inf. Syst. E-Business Management*, 5(2):93–116, 2007.
15. F. Gessner. CCTS modeler concept and development: Semantic-driven optimization and harmonization of data models. Master's thesis, University of Applied Sciences Würzburg, Germany, 2008.
16. S. Hinkelmann. UN/CEFACT Unified Context Methodology (UCM) Direction And Concepts. presentation slides, Mar 2009.
17. C. Janiesch. Implementing views on business semantics. In *15th European Conference on Information Systems (ECIS 2007), St. Gallen*, 2007.

18. C. Janiesch. Enhancing the accessibility of enterprise system documentation with domain ontologies. *AIS Transaction on Enterprise Systems*, pages 27 – 35, 2009.
19. C. Janiesch, A. Dreiling, U. Greiner, and S. Lippe. Configuring processes and business documents - an integrated approach to enterprise systems collaboration. In *2006 IEEE International Conference on e-Business Engineering (ICEBE 2006), 24-26 October 2006, Shanghai, China*, pages 516–521. IEEE Computer Society, 2006.
20. C. Janiesch and S. M. Thomas. Business document taxonomy - comparison of the state-of-the-art and recommendations for future applications. *IBIS*, 2(2):59–78, 2006.
21. P. O'Connor, A. Coates, and M. Crawford. Unified context methodology technical specification. Technical report, UN/CEFACT, Apr 2010.
22. E. Peukert, J. Eberius, and E. Rahm. AMC - a framework for modelling and comparing matching systems as matching processes. In S. Abiteboul, K. Böhm, C. Koch, and K.-L. Tan, editors, *Proceedings of the 27th International Conference on Data Engineering, ICDE 2011, April 11-16, 2011, Hannover, Germany*, pages 1304–1307, 2011.
23. H. Proper and T. Halpin. Conceptual schema optimisation — database optimisation before sliding down the waterfall. Technical Report 341, Department of Computer Science, University of Queensland, Jul 1995.
24. J. Quantz and T. Wichmann. E-Business-Standards in Deutschland Bestandsaufnahme, Probleme, Perspektiven. Technical report, Berlecom Research, 2003.
25. J. Rech, W. Schwach, M. Dietrich, and G. Stuhec. Intelligent assistance for collaborative schema governance in the german agricultural ebusiness sector. In G. Kotsis, D. Taniar, E. Pardede, I. Saleh, and I. Khalil, editors, *iiWAS'2010 - The 12th International Conference on Information Integration and Web-based Applications and Services, 8-10 November 2010, Paris, France*, pages 867–870. ACM, 2010.
26. M. Rosemann and W. M. P. van der Aalst. A configurable reference modelling language. *Inf. Syst.*, 32(1):1–23, 2007.
27. A.-W. Scheer. *ARIS: Business Process Modeling*. Springer Verlag, Berlin, 3rd edition, 2000.
28. P. Soffer, B. Golany, and D. Dori. ERP modeling: a comprehensive approach. *Inf. Syst.*, 28(6):673–690, 2003.
29. G. Stuhec. Using CCTS modeler Warp 10 to customize business information interfaces. Technical report, SAP Developer Network (SDN), Nov 2007.
30. G. Stuhec and M. Crawford. How to solve the business standards dilemma — the CCTS standards stack. Technical report, SAP Developer Network (SDN), Nov 2007.
31. J. Surowiecki. *The wisdom of crowds: Why the many are smarter than the few*. Abacus, 2005.
32. F. Tuncer, A. Dogac, S. Postaci, S. Gonul, and E. Alpay. iSURF edocreator.
33. J. A. Zachman. A framework for information systems architecture. *IBM Systems Journal*, 26(3):276–292, 1987.

Part IV
USDL — Evaluation

The USDL has been designed as a means to describe services so that they can be traded via the Internet. The previous parts outlined the status-quo of service description research and presented detail of USDL's model and methods. For evaluating the actual worthiness of a modeling language such as USDL, potential users will consider the fit of the language with the contingent influences their organizations have to deal with. Therefore, Part IV is concerned with multiple evaluations of the Unified Service Description Language. The part consists of four chapters that move from empirical to more theoretical evaluations.

Chapter 18 starts off with the findings of four different case studies: services in the energy domain, services for mobile users, manual services for insurances, and services for B2B integration. Each case study originates from a different company and provides its own conclusion. Case studies are the first step to a comprehensive validation of USDL, because they can create feedback regarding usability, feasibility, completeness etc. of certain aspects of USDL. In contrast to the specific case studies on selected business projects, Chapter 19 presents a multiple-enterprise evaluation study that follows an iterative methodology with a survey part which delivers several quantitative results. The purpose of Chapter 20 is to identify requirements for a service description language from potential USDL users. The presented research takes a semiotic theory perspective to the design of modeling languages. Through a Delphi study approach, i.e., an anonymous, written multi-stage survey process, the chapter elaborates a set of requirements. The requirements can be used to ex-post test if the features of the USDL actually address the users' needs and to recheck the underlying assumptions of the USDL design and development process. Finally, Chapter 21 focuses on evaluating the expressive power of the language to specify *software* services from a theoretical point of view. In particular, it discusses which information should be provided in order to support the discovery and combination of software services.

Chapter 18
Case Studies

Martin Schäffler, Anke Thede, Bastian Leferink, Kay Kadner, Andrea Horch, Maximilien Kintz, Monika Weidmann, and Moritz Weiten

Abstract Case studies play an important role for validations in general. In particular, they are the first step to a comprehensive validation of USDL, because they can create feedback regarding usability, feasibility, completeness etc. of certain aspects of USDL. This chapter contains the findings of executing four different case studies: services in the energy domain, services for mobile users, manual services for insurances, and services for B2B integration. Each case study originates from a different company and provides its own conclusion. In general, the case studies show that USDL can be used to realize the described scenarios. However, they identify also room for improvement: the scope of USDL should be limited (don't try to be "universal"), semantic technologies should be used, and a simple API is necessary.

Martin Schäffler
Siemens AG, Otto-Hahn-Ring 6, 81739 Munich, Germany,
e-mail: `martin.schaeffler@siemens.com`

Anke Thede, Bastian Leferink
B2M Software AG, Emmy-Noether-Str. 17, 76131 Karlsruhe, Germany,
e-mail: `a.thede@b2m-software.de`, e-mail: `b.leferink@b2m-software.de`

Kay Kadner
SAP Research Dresden, Chemnitzer Strasse 48, 01187 Dresden, Germany,
e-mail: `kay.kadner@sap.com`

Andrea Horch, Maximilien Kintz
University of Stuttgart, Institute IAT, Allmandring 35, 70569 Stuttgart, Germany,
e-mail: `firstname.lastname@iat.uni-stuttgart.de`

Monika Weidmann
Fraunhofer Institute for Industrial Engineering IAO, Nobelstr. 12, 70569 Stuttgart, Germany,
e-mail: `monika.weidmann@iao.fraunhofer.de`

Moritz Weiten
SEEBURGER AG, Edisonstrasse 1, 75015 Bretten, Germany,
e-mail: `m.weiten@seeburger.de`

18.1 Introduction

Although case studies are rather a research method in social science, the concept can be applied to other sciences as well. Especially in the beginning of any research endeavor, it may make sense to describe and explain the universe of discourse (or "case") with some case studies. They help to understand causes of particular phenomenons, which creates or fosters research questions or hypotheses. A nice side effect is that case studies are typically easy to understand by stakeholders due to their descriptive nature. Case studies typically have a high conceptual validity because they have an in-depth understanding of the case itself and its context and processes [2]. Therefore, case studies are often used to investigate a certain area, which is very specific and can hardly be put into formal methods. Nevertheless, the case study results can support the derivation of general principles and generic answers to the case.

Besides the initial clarification of a case, case studies are also a quite pragmatic approach of evaluating concepts because the same advantages apply here (depth, conceptual valid, context understanding among others). Although this kind of evaluation can hardly be seen as validation in the strict sense, it can be used to detect problems and reveal issues of the discussed matter with limited effort. Therefore, we decided to evaluate USDL by specifying various case studies. Since case studies cannot be the only way of validation for a concept, a more formal way of validation can be found in the subsequent chapters.

The case studies described in this chapter originate from various companies and institutions. They originate in various projects with customers or research endeavors. This heterogeneity of sources is a perfect situation for proofing the "U" in USDL, being "unified." USDL is meant to be useful for nearly all kinds of services, independent of their nature. Consequently, the case studies involve electronic services (Web services) as well as manual services from different domains. Is has to be noted that the case studies where carried out using USDL3 milestone 4. The book is based on the USDL3 milestone M5 which already includes some of the feedback. Both versions of the specification are available online.

The first case study is about services in the energy domain. Energy gets more important in our everyday life and an increasing number of people produce energy on their own premises. Such people are in need of new services, e.g., a weather forecast service. Therefore, services for energy prediction, energy metering, billing, etc., will be described with USDL and offered via a marketplace.

The second case study investigates how USDL supports the use of services in mobile environments. This is of interest because mobile devices such as smart phones allow the user to access the internet almost everywhere and at anytime. The Mobility Mediation Layer (MML) is a mediator between mobile devices and the actual services, which can easily be added to the MML because they are described with USDL.

The third case study is carried out in the area of insurance organization and management. Insurance companies need to have close contact with craftsmen and other people that handle the insurance claim. The case study investigates the suitability of USDL for describing services of craftsmen, putting these on a marketplace and making this marketplace searchable and accessible for insurance companies.

Finally, the fourth case study describes the use of USDL in the area of B2B integration. Small companies face major problems if they want to integrate their IT landscape with the systems, protocols, and data structures of larger companies. Services for mapping and adaptation tasks are required to facilitate the integration. They are described with USDL so that it's easy to find the appropriate services for the given requirements.

It can be seen from these short descriptions that USDL can be used in diverse contexts for different purposes. The following sections contain a more detailed and comprehensive description of the case studies, including background information of projects, system architecture and — most importantly — conclusions that the authors draw from practical experience.

18.2 Case Study: Energy

18.2.1 Changes in the Energy Domain

The energy market is changing right now. Today, big energy companies are responsible for the production, transmission, and billing of energy. In contrast, the energy network of the close future will be dominated by new players on the energy market (cf. [13], [15]). The ability to store energy in devices such as electric cars or hydrogen tanks enables the energy consumer to buy and use the energy at different points in time. The consumer is then able to buy at a cheap price at times of low demand but is not forced to use the energy immediately. Instead, he could decide to use the energy later or sell it at times of high demand for a better price.

Consequently, the ordinary consumer of energy evolves to a kind of broker in addition to his role as consumer. And if this consumer produces energy with solar panels, wind generators or other facilities of his own he could be referred to as "prosumer."

The Siemens AG is part of this change right now. The company participates in public research projects on the subject "smart grid" (cf., e.g., E-DeMa [8]) and develops hard and software to support the change (cf. [17]).

18.2.2 Services as Emerging Technology

The change in the energy markets goes hand in hand with a change in the network topology and the network devices. This "smart grid" is able to provide the prosumer with intelligent cabling for energy supply, delivery, and smart devices to meter and supply energy (cf. [16]).

These smart devices enable a new market, viz., the market for services around the production, storage, and supply of energy of prosumers. These kinds of services are value-added services on top of the core businesses of energy and correspond directly with the needs of the emerging prosumer.

As an example, the prosumers could use services to optimize their use and productivity of energy, e.g., by calculating the optimal point in time for consuming or producing energy. Combined with a forecast of the overall demand and the own possible energy production and consumption the prosumers are then able to optimize their behavior (cf. [3]). Other services could be selling, delivery, maintenance, and service of the devices and energy production systems.

18.2.3 Standards Pave the Way to User Acceptance

Currently, many service marketplaces are being developed in diverse contexts and domains (e.g., E-DeMa [8], Microsoft Pinpoint [10]). Therefore, similar effort is spent in many different companies on common building blocks for service marketplaces. One such building block is a comprehensive description of a service and how consumers or systems can decide for the similarity of services. A standard would help to foster synergies because the standards development process in standardization organizations (e.g., W3C) combines forces of some of the main players of the new market. Also, a well defined and accepted standard could facilitate the development, provision, and usage of services in the near future for all roles involved on different service marketplaces. From the point of view of a developer of marketplaces, a standard could be the basis for his software development. Also from a service provider's point of view, a standard is the freedom to develop and describe a service for many platforms. In addition, a standard is a clear definition basis for the service consumer.

Siemens could benefit from the USDL standard as well. As a global player in the B2B market for industry, energy and healthcare products, the offering and consuming of services could become a significant part of its business in the near future. As the USDL standard could be a basis of cross-sector initiatives for service development, it would provide considerable synergy even inside the company.

18.2.4 The Use Case: Services for Energy Markets

For the scenario, let us assume that a prosumer of energy wants to benefit from his investments in an electric car, solar panels, and intelligent devices to manage the energy production. Consequently, he would like to optimize his energy production and sell the energy when the price for energy supply is at its peak. For this, he needs a detailed forecast of the productivity of his devices to negotiate the price of his produced energy on the energy market.

The productivity of his device is directly related to the sun's radiation on the solar panels mounted on his house. To calculate the productivity of the near future the prosumer needs a detailed weather forecast. He could then use his calculation to decide, e.g., to load the battery of his car a bit later but sell the production of the solar panels to the network in the next minutes.

This weather service could be investigated via a search engine on the internet, analyzed "by hand" if it copes with the requirements and connected via a proprietary Web service implementation. Or it can be offered on a service marketplace for services related to prosuming of energy, compared with other offers by the system, recommended to the customer and then subscribed to via a standard Web service interface — all of that on the basis of the USDL description.

18.2.5 Example: Services for Weather Forecast

In the implementation of a prototype at Siemens Corporate Technology, we prepared a corresponding weather service to provide the basis for the calculation of the device productivity. We conducted research and found that no standard weather service provides this kind of prediction we needed. Hence, we took a standard weather service with a fine granularity in sun probability and calculated a sun radiation factor for the near future. This service was implemented and then described by means of USDL 3M4 to evaluate the suitability of the language at this stage.

18.2.6 Implementation of the Service Marketplace

At the time of the case study, no service marketplace was available which could process USDL service descriptions. To execute the use case and evaluate the usability of marketplaces at this stage, we simply implemented a service marketplace on our own.

A screenshot of the end-user area of this marketplace is shown in Figure 18.1. The screenshot depicts the marketplace with the navigation area on the left side, the possibility to filter services for "Private Customers" and "Business Customers" on the top left, on the top right a "Service Search" box to search services by keywords and in the center the detail area for the specifics of a service. The buttons on the bottom right side can be used to "Subscribe" the service shown in the detail area or to "Subscribe and Activate" the service by one click.

By using the navigation area the user could navigate to the subscribed services of the his own ("My Services"), to navigate through the service catalogue by help of service categories ("Service Catalogue"), to get the top 25 services ("Top25") and the newest services ("New Services") on the marketplace. If the user selects one service listed in the navigation area or the search result, he could see the details of this service in the center area of the browser window. The information is distributed over several tabs to organize the user interface.

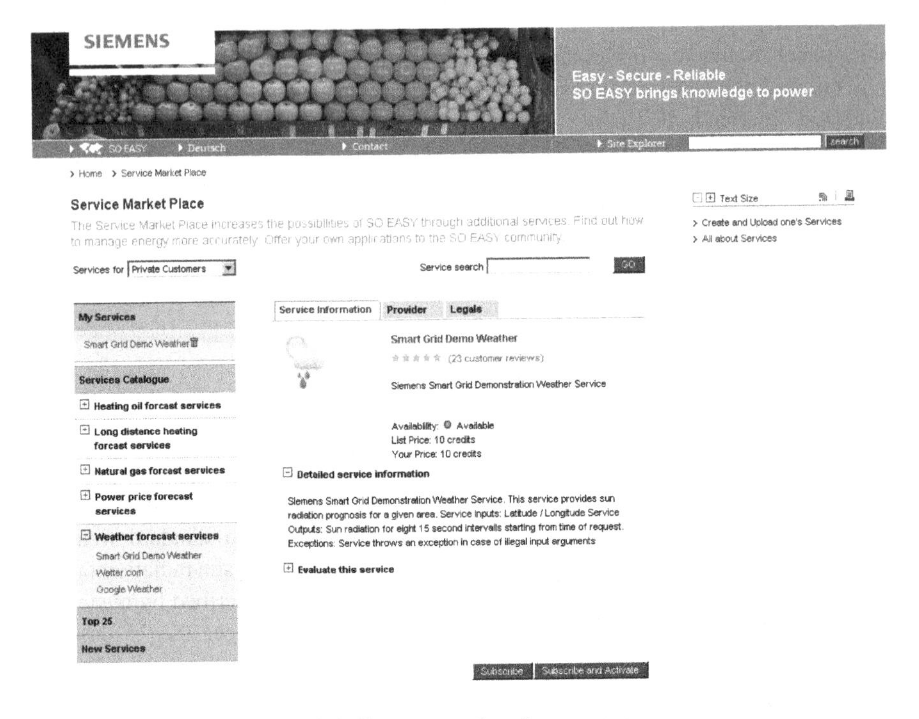

Fig. 18.1: Energy marketplace prototype.

The implementation was based on the Google Web Toolkit (GWT, cf. [4]) and tailored to the needs of a service marketplace for a Siemens energy marketplace. The user interface was adapted to the Siemens web appearance and an import mechanism for services described in USDL was added. The marketplace prototype is used for demonstration purposes.

18.2.7 Marketplace Functionalities and Roles

As not all details of the UDSL description are needed, the import functionality only selects some details, e.g., the price model, provider details, and legal information. After the import of the USDL file, this information is stored in the service repository and is accessible for the marketplace customer via the Service Catalogue in the user interface. With the import, the provider could choose or define a topic cluster in which the services are organized.

Details of the service are displayed when the customer navigates to a service of interest in the catalog. The overall and coarse-granular description, availability and the price model are displayed in the "Service Information" tab of the detail side and the user can navigate to other details by selecting the other tabs "Provider" and "Legal."

The roles "administrator," "provider," and "customer" are implemented to use the marketplace. On top of that, brokering and clustering of services will be realized in the near future.

18.2.8 Lessons Learned

The implementation of the marketplace and the use of USDL 3M4 as its basis provided a deep insight in the mechanisms and pitfalls of the adoption of the standard. In our scenario, different roles are implemented and therefore different stakeholders have to overcome some hurdles.

The standard benefits the provider since the service must be described only once but can be published on several marketplace platforms. Nevertheless, a well-regulated process has to be defined for the editing and provisioning to the marketplaces. An additional editor with versioning and deployment mechanisms could assist here. The editor also has to assist during the design phase of the service — the provider has to describe complex correlations, e.g., price models or service deployment interfaces. Therefore, clear and supportive user interfaces and user interaction guidance is needed here at all costs.

The service consumer needs guidance as well, e.g., to select the right services for his own purposes. Hence, a clear and supportive implementation of the marketplace interface is needed. On top of that, an intuitive filtering functionality to navigate through the offered services would be helpful here. Additional functionalities such as "compare services," "related services," and use of customer preferences for pre-selection of services would complete the picture of a marketplace, which satisfies the customers demand.

One special overall finding for all roles is that the USDL should not provide a standard for all possible services in every domain. Instead, it should provide a framework sufficient to describe the basics of services and mechanisms for domain specific extension. Otherwise, it is too complex for all parties involved to use the standard for service description.

18.2.9 Demand of a USDL Standard in the Energy Domain

The energy smart grid movement has started and will influence many other parts of our lives as well, e.g., personal mobility (e-car) and intelligent buildings (cf. [11]). These new areas could be accepted faster by business and consumer if these areas are accessible by clear and well-defined interfaces. Hence a standard for service description could contribute to simplify the provision and delivery of value-added energy services and to establish these new technologies in the society fast and stable.

18.3 Case Study: Mobility

Today, many business processes can be interacted with through mobile interfaces. Mobile applications often involve connections to backend services without which they would not properly function. Mobile applications are developing at a quick pace. Operating systems are evolving, new devices are developed and old ones are abandoned. A standardized interface to external services is very important to keep the chances and the effort to maintain and port mobile applications at an acceptable level. Furthermore, mobile applications have specific requirements on the amount of data provided to the user. Due to small displays and a mostly small bandwidth it is important to deliver as little information as necessary. The limited possibilities of user interactions on mobile devices raise the need for pre-populated data fields with, e.g., context information as location data. There are basically two distinct types of mobile applications:

- The *mobile synchronization* is a backend process made available for mobile usage. The basic process is executed on a backend system and can mostly run without any mobile support. Most importantly, the current process status is persisted in the backend. Some (even all) parts of the process can be performed on the mobile device. Before a process step is executed on a mobile device the current process status is copied to the device. Once the step is finished the result is copied back to the backend system. Here, of course, synchronization problems may occur and have to be dealt with (synchronous reading and writing).
- the *genuine mobile application* is an application of which the basic process is executed on the mobile device. Such an application may use external services to gather information needed to execute the service. But the process itself persists on the mobile device, that is the current process state is kept on the device. Such a process may still be invoked by an external event and the last step may be an external call.

We will concentrate on genuine mobile applications because the range of service types is larger. Mobile applications may use services as

- information source: large sets of information cannot be kept on a mobile device
- external calculation: complex calculations often require much data and computing power
- centralization: when mobile users interact, the backend provides the centralized data store

We show how USDL can facilitate the integration of backend services as well as the flexibility to design new backend services to suit specific mobile needs. The use of a mediation layer especially for mobile access is shortly described. Such a layer interacts with USDL service on the one hand and with the mobile devices and their particularities on the other hand. Combining services on the server-side, providing only the required content for the actual situation, merged together in a comfortable interface is the aim of our approach to mobile, service-based applications. The composition of services, described in the next section, is required to achieve this goal as well as the ability to augment the technical service description with further information as described in Section 18.3.2.

18.3.1 Composition of Services

Many services are reusable in different contexts. As an example, consider a routing service that calculates a route from one geographical point to another, using different means of transport. Such a service can be used in a mobile device to display routing information to a contact. It can be included in a value-added mobility service that calculates shortest times to get from a point to another, using a combination of public transportation, taxi, pedestrian routes and car sharing. It can be used to find alternative routes when confronted with a closed motorway section. It may be the basis for proposing circular jogging routes that fit your personal sports level and distance. This list shows that the composition of basic and very general services is essential for building new value-added services that are useful for more specific needs.

The composition of services requires a standardized language for service and interface description. If services shall be composed or recomposed automatically and on-the-fly then the descriptions have to be extensive and have to include semantic information in order to match correct services. Technical description standards such as WSDL describe the technical interface of a service comprehensively but nothing can be described about the service's content nor is the interface composed of necessarily standardized data types. Either, the service providers have to adhere to a standardized set of data types or a semantic description of services has to be created in order to gain the remaining knowledge about services. USDL provides the basis for the latter as it contains additional information about the task of a service and others. The most important parts of USDL in this aspect are described in detail in the next section.

18.3.2 Technical Components

A mediation layer is required to optimize and unify the communication between the mobile devices and the Internet of Services. Accordingly, our Mobility Mediation Layer (MML, see Figure 18.2) offers functionality for unified mobile security, for optimization of service calls and contents with respect to the restricted bandwidth, for transformation and adaption of service contents to mobile needs and service composition directly in the mediation layer for adding mobile features such as context and localization to services.

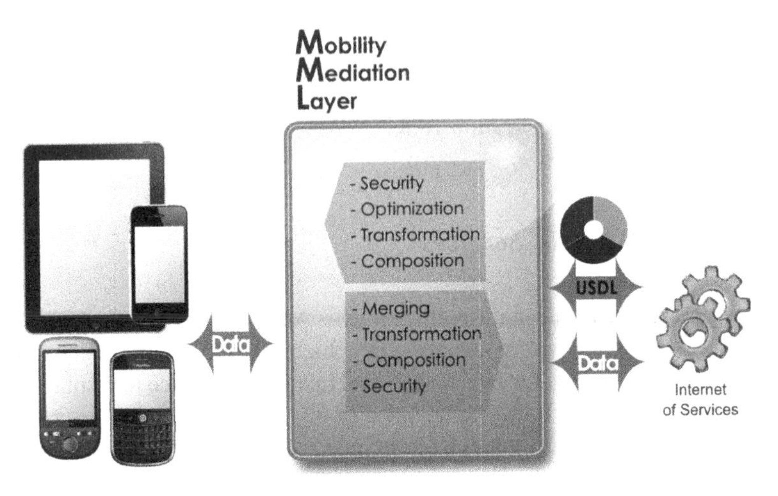

Fig. 18.2: Mobile Architecture including Mobility Mediation Layer.

Mobile devices issue service calls to the MML that translates the calls into backend service calls and transforms the request with respect to the service's interface and protocol. This means that making services available to mobile devices requires them to be known in the MML. An import of the USDL description is usually enough to make the service usable. The standardized and comprehensive description through USDL facilitates the automated processing of services by mediation layers in general. The MML makes specific use of many of the elements of USDL. Among the most important parts are

- Technical interface description: the description is used for automatically translating between the mobile communication and the service interface. In particular, by semantic means data types are automatically mapped against another so different services with the same semantic interface become exchangeable.
- Security: services may have distinct security requirements and mechanisms. Implementing all these on different mobile infrastructures is tedious. By a comprehensive description of the security specifications of the services, the MML can offer unified authentication and authorization to the mobile devices and translate this information to the service specific requirements.
- Categorization and description of the service itself: USDL offers many ways to describe what a service actually does. This information can be semantically enriched and then used for matching. If the description already contains semantic annotations then this step is further facilitated.
- Price models for automatic accounting: Ad-hoc composition of services also requires to specify a composed price plan. A detailed and machine readable de-

scription of a service's price model allows to automatically derive such a composed price plan.
- Legal description: when composing services the legal conditions of all underlying services have to be considered.

To sum it up, USDL can be imported into the MML. The MML then generally has sufficient information to make the service available to mobile devices and to offer different options for the mobile usage. Such options comprise automatic service selection among different substitutable services (e.g., with respect to performance, availability or price), automatic inclusion of services into an application (plug-and-play) or calculation of combined price models for composed services. USDL delivers the information that is needed by the MML in order to understand the service and what it does and in order to translate the service results to the unique protocol and data structure delivered to the mobile devices.

18.3.3 Lessons Learned

Overall, the usage of USDL as a description language for our services proved to be convenient. However, the deployment itself turned out to be non-trivial. Due to the missing of a machine-readable technical semantic description, an automated parameter-matching was not possible. Therefore, a fully automated deployment process was not realizable. To annotate the mobile interface parameter to external (web-) service parameters our USDL import only estimates a parameters nature. A service administrator still needs to manually apply or create an ontology for each service parameter corresponding to the idea of annotation in the field of Semantic Web Services (cf. Chapter 7).

Despite this minor challenge — which can be solved in future versions — USDL meets our requirements. We are able to automatically search for services, to combine and compose them. Security and legal information help us to provide our users only with services they have registered for including a single-sign on and which they are allowed to use. Lessons learned are that a comprehensive description such as USDL is very useful, especially for software as the MML that needs to interpret these data. But it is essential to have an editor that is easy to use and to understand as well as possibilities to import information from other standardized sources such as WSDL specifications and so forth.

In the future, utilizing the price model data will allow us to offer additional, interest services based on context and service branch to our users, taking care of the billing process for them as the price models allow for automatic price calculation of composed services.

18.4 Case Study: Craft Services

18.4.1 The openXchange Project

The openXchange[1] project deals with the development and improvement of IT support in active property damage claim management. The project's goal is to simplify the associated processes by offering better and more integrated IT support.

One important task in this field is to improve the communication between the insurance companies and their partners, such as craftsmen, regulators, or technical experts.

As of today, there is no adequate standard for the communication between the insurance companies and the mentioned partners. Therefore, the openXchange project extended existing standards and, when needed created new ones.

One special problem is the description of the craft service offered by crafts companies, which insurance companies can look for when searching craftsmen for repairing specific damage. For this purpose the openXchange team identified USDL as an adequate and easy to use tool for describing the crafts services and using the USDL description of crafts services in a search engine [1].

18.4.2 The Roles of Services and Service Description in Craft Services

In Germany, craft services are regulated by the German Crafts Code,[2] which lists all craft service categories, their admission requirements and defines other craft service related rules. In the state of the art analysis conducted at the beginning of the openXchange project we identified several problems specific to craft services classification and search:

- craft services are a large field consisting of many subcategories, some of which are overlapping (for example in that they seek to repair the same damaged objects or depend on each other);
- when customers are looking for craftsmen, they in many cases do not know which categories or keywords they should give into the search engine to find the appropriate answer, and cannot ensure that they found all the services they actually need;
- additionally, craftsmen are confronted with the problem to decide on which internet platforms and search engines they should register their services. Adding and updating their data in the heterogeneous environment of today's internet platforms is a highly time consuming task. Craftsmen usually do not know which

[1] http://www.openxchange-projekt.de/

[2] Gesetz zur Ordnung des Handwerks (Handwerksordnung), cf. `http://bundesrecht.juris.de/hwo/`

platforms offer the higher benefits in terms of possible new customer contacts and contracts.

The solution for these problems is standardizing the description of crafts services. A standard crafts service description ensures the consistent description of crafts services and the recoverability of a once created service description. As part of our project, we built up a new ontology describing crafts services, possible damaged objects and the relationships between each craft service category and object type in a standardized way, as other current classification schemes, such as, [6], appeared insufficient to meet our requirements.

Because we needed a description language for craft services, we chose to use the USDL for describing the services and using these descriptions in a crafts search engine. USDL had the advantage of being well specified service description language, which made our task easier, and appeared better suited than other description languages restricted to Web services [9]. The following sections will present the approach of the team and its experience by using USDL for these purposes.

18.4.3 Use Case: openXchange Crafts Search

One of the supporting IT tools created in openXchange is a crafts service search engine for insurance companies (see Figure 18.3). The search engine is an ontology-based Web service with the ability to understand semantic input of the user: the user does not need to distinguish between search criteria and can simply enter, for example, a specific service he needs, an object that has been damaged and should be repaired or a ZIP code in the same text field. The query is sent to the Web service in a specific XML format. The output of the Web service uses another specific XML format to encapsulate the descriptions of the matching craft services. Each craft service returned is described using the XML serialization of USDL.

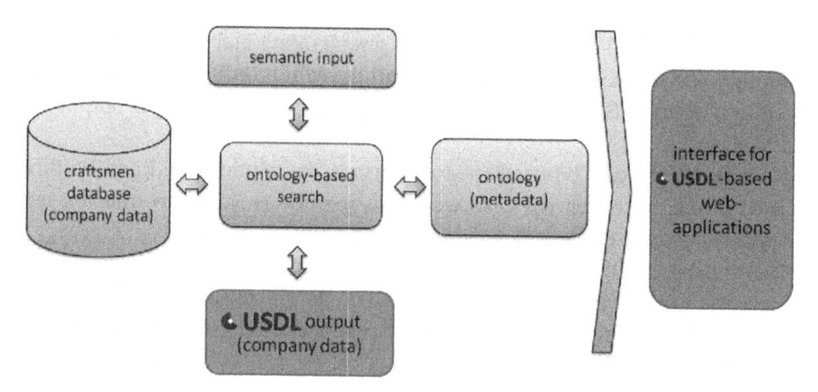

Fig. 18.3: Current structure of the openXchange crafts search.

The crafts search engine uses an ontology for understanding the semantic input of the user [14]. “Understanding” in this case means identifying what kind of input the user sent (address, company name, crafts category, damaged item, etc.) or if there are synonymous items for the query keywords the engine can search for in the crafts database. After finding the matches in the database, the search engine returns the results as USDL descriptions of the matching craftsmen companies. The ontology describes the metadata (classifications of and relations between craftsmen, craft services and damaged items) and the USDL result shows the specific craftsman service data found by the search.

The automatic conversion of the database results into USDL descriptions is done by the search engine by filling the data of each result dataset into a previously manually generated general USDL crafts service description (an example is shown in Listing 18.1) [7].

Listing 18.1: Part of the XML serialization of craft service USDL description.

```
1 <Service xmi:id="ServiceWrapper_8732241">
2     <ServiceElements xmi:id="Service_29245520">
3       <Guid>bdcb0674-9d90-479f-8fb7-92b6e6142fe0</Guid>
4       <Version>1.0</Version>
5       <Name>Sanitaer Rataplan GmbH</Name>
6       <PublicationTime xsi:type="foundation:AbsolutePointInTime"
7         xmi:id="AbsolutePointInTime_32657640" />
8 </Service>
```

18.4.4 Experiences

Integrating USDL in our service turned out to be quite easy. The most difficult part of the work consisted in the description of a dummy craft service using all attributes that were needed in our use case (capabilities, certificates, basic pricing information and of courses general information such as contact data and postal address). Once this dummy service had been described with the USDL editor, we used the XML serialization of this USDL description and integrated it as an example output format in our Web service.

Although USDL can be used to store service descriptions and can be directly searched, in the crafts service case we use it mainly as a way to transfer service descriptions between the search service and the users.

We also used USDL to describe our crafts services search, as our search engine itself consists in a service that could be traded on an online service marketplace [5]. In this case, USDL provided a way to describe search engine capabilities and its technical specifications.

18.4.5 Lessons Learned

Our work has shown that creating an interface for a search engine to USDL-based web applications is simple. When USDL becomes a better recognized standard for the description of services, for reasons of consistency and interaction it will become necessary to be able to convert the description of services like craft services to USDL. The project openXchange found an easy way for doing this conversion by using a general description for a special type of service.

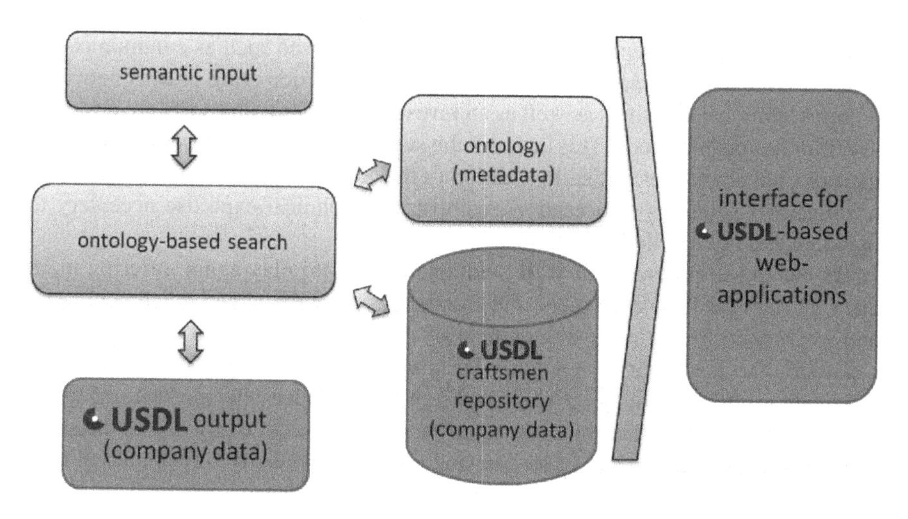

Fig. 18.4: Future structure of the openXchange crafts search.

A further improvement of our craft services search engine would be storing the craftsmen data directly in a USDL repository, as shown on Figure 18.4. Creating a USDL repository as base of a search engine is an interesting task for future work and further evaluation of USDL.

18.5 Case Study: Business Integration

The use case described in this section illustrates how the next generation of B2B- ("Business-to-Business"-) solution benefits from USDL. It shows how USDL is applied in two different contexts: as a building component for a platform-data model (in terms of a service engineering approach) and as an enabler for external channels (in terms of a service description approach).

For many inter-company relations, the B2B-integration has become indispensable nowadays. According to a survey conducted by Forrester Research, 75% of the participating enterprises exchange EDI/B2B-documents with more than 40% of all their trading partners [12]. B2B-solutions significantly improve the overall efficiency for procurement, production, and distribution based on the integration of internal operations as well as the extension of processes to trading partners. The automation of business processes significantly reduces production and transportation costs. The partners can react much faster on changing demands and achieve a faster time-to-market. In addition, time consuming and error-prone manual data entries can be avoided.

Despite the obvious advantages, small companies typically face barriers when trying to integrate into a B2B network due to the complexity of the field. Among the reasons are the heterogeneous requirements in different countries and industries as well as the number of technologies and standards involved such as communication protocols, message formats, process monitoring etc. Participating in B2B networks requires technical decisions as well as an investment in hard- and software designed to fulfill the specific needs. This includes the supported business processes as well as the expected amount of traffic. In addition to this monetary investment, participants hosting B2B solutions also need to establish the technical expertise necessary to maintain and operate their solution.

At this point, providers of B2B solutions come into play again, offering managed services for their customers. This model has recently gained much popularity and works well for an increasing number of B2B-participants who do not have the capabilities to host their own solution or simply consider managed services to be more attractive for a couple of reasons. From the provider's perspective, the question remains how to operate the managed services cost-effectively and at the same time keep the service attractive for the customers — especially for smaller companies with their limited technical expertise. It soon turned out that this goal cannot be achieved by just providing "software as a service" but only by offering a complete solution to the customers. This was the starting point for research and development activities conducted by the SEEBURGER AG as part of the TEXO project [19].

18.5.1 The Next Generation of Hosted B2B Solutions

Several questions have been driving this project:

- How does the next generation of hosted B2B-solutions look like?
- How can entry barriers (primarily costs and technical expertise) for potential B2B participants be lowered?
- How can the "on-boarding" of new B2B-participants and -relations be improved?
- How can "pay per use" be made more attractive for both parties: provider and customer?

From the perspective of the underlying infrastructure, cloud computing has been identified as a key component for next generation hosted solutions. The application of cloud technology increases the flexibility for hosted B2B solutions; cloud-based solutions can be adapted more easily to changes in the market. In addition, cloud technology supports pay-per-use models that base directly on the correlation between customer costs and (solution-) provider costs.

This section concentrates on the answers that have been found to the previous questions on application level. The core idea is to establish a hosted B2B platform that significantly reduces operational costs on the provider's and the customer's side. A significant cost reduction can merely be achieved by relying onto configurable standard B2B-solutions and by delegating major parts of the necessary configuration effort to the customer, in form of a self-service. As an example, B2B participants can register themselves on their own at the platform and can on their own establish business relations to other partners. A "B2B Directory" supports the B2B participants in setting up and maintaining their partner profiles and business relations in a centralized manner. This B2B Directory servers two main tasks: (i) reduce the technical complexity of configuring a B2B solution to configuring a simple business task using a business terminology the B2B participant is able to understand and (ii) establish a central data repository free of redundancies across all trading partners.

The "B2B Directory" provides data and metadata for services offered by the platform, which include:

- the automation of B2B relations based on industry-specific standard processes;
- the management of business partner profiles;
- data (message) conversion based on standard mappings;
- profiling and tracking services;
- etc.

Thus, the "B2B Directory" is the base for the configuration of executable processes for B2B-relations. It also maintains all master data necessary to run those processes.

18.5.2 Integrating USDL into the "B2B Directory" Data Model

The "B2B Directory" relies on a central data model. Apart from B2B-specific functional requirements, important goals of model-design have been to establish a flexible but extensible model, to avoid conceptual gaps (between application and end user) and to support the exchangeability of service metadata. For these reasons, the "B2B Directory" data model integrates several parts of the Unified Service Description Language to describe different aspects of platform services. This includes core concepts, such as the descriptions of participants, technical protocols, or service levels.

There are several advantages of integrating USDL modules or parts of them into the platform data model: From a service engineering perspective, USDL provides a well-founded and generic base for managing the kind of data that the B2B-platform requires — especially when the set of services offered shall be extended in the future. A typical use case concerns the description of partner profiles. Partner profiles provide information about the "B2B capability" of a trading or business partner, such as the technical communication channels and the message formats supported. The information provided in these partner profiles is restricted to the services that the platform supports, e.g., with respect to the communication protocols and message standards that can be used.

However, for certain industries and customers, there might be additional interfaces that are of interest for participants that have been registered — in terms of an emerging de facto standard. E.g., one might think of a new interface beyond the typical ordering-, invoicing etc. standard processes. Generally, the platform can handle such interfaces on two levels: (i) registering/describing them and (ii) supporting them technically (through the processes running on it). The advantage of registering interfaces that are not yet supported technically is that the provider can keep track of possible extensions that have a high potential for a number of participants more easily. However, the description of this kind of additional ("non-standard") interface needs to be provided on a rather generic level. Based on USDL, a generic but still formal description can easily be made on the base of a common model.

In the more general case, the platform might be extended by more general services beyond typical B2B connections, e.g., infrastructure services such as the ones that the platform uses itself (in order to exploit free capacities) or community services in terms of business networks. The metadata needed for those additional services overlaps with the general B2B case: aspects such as "participants" or "service level" need to be described.

18.5.3 USDL for Platform Services: Marketplaces as Channels

Once USDL is used for the data model of the "B2B Directory," USDL-based service metadata can be extracted easily — without any conceptual gaps. USDL metadata is available for different kinds of services provided by the B2B platform: (i) a general B2B connection service, (ii) partner-specific connection services and platform-internal services that are exposed as (iii) "standalone" service for external use (e.g., data conversion).

The first option represents a generic description of what the platform has to offer: a service to establish B2B connections between business partners registered against the "B2B Directory." This service can be offered in any context where the automatic processing of orders, invoices, etc. is relevant. Important aspects that are part of the USDL-instance include:

- general information (textual, internationalized description of the service);
- roles (service-provider, B2B-partner);
- supported technical interfaces ("communication channels," message standards, etc.);
- service level.

The second option concerns the connection to specific business partners. To motivate this use case we introduce the concept of a "hub." A hub in the B2B context is typically a large enterprise that maintains business relations to a number of partners, such as, suppliers in a production scenario for example. Business partners, which have to exchange data with a hub, need to meet several technical requirements, starting with the communication protocol, supported message standards, etc. From the business partner's perspective, the possibility to establish B2B connections to certain hubs is of strategic value. For the hub, on the other hand, automated B2B-connections influence the decisions regarding the maintenance of a network of partners.

Each possible connection to specific hubs that are already registered against the B2B-directory is considered a service the platform has to offer. Consequently, USDL descriptions can be provided for each such connection. This description contains the same information as for the general B2B connection service, but also provides additional information about the individual requirements in terms of technical constraints, etc.

With the USDL description, the information about possible B2B-relations to certain hubs and partners can be made available on service marketplaces — in addition to other service-descriptions that are partially provided by the (potential) B2B participants themselves. This is illustrated in Figure 18.5.

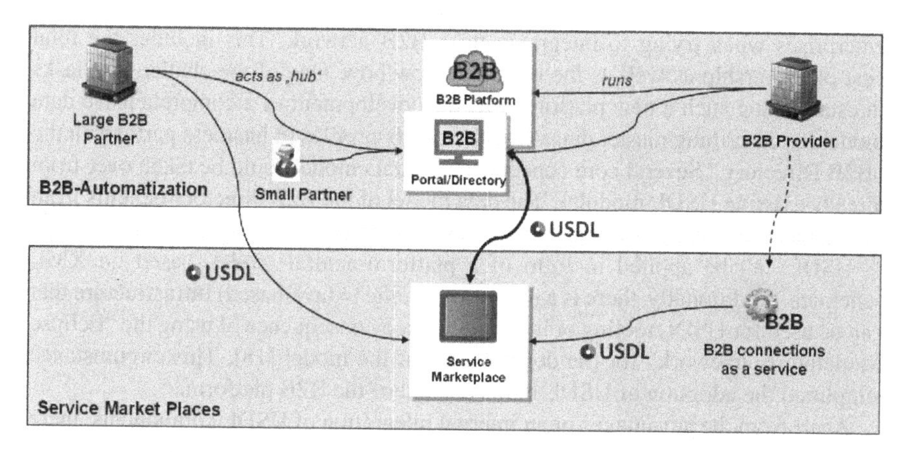

Fig. 18.5: Scenario for the interoperability of a B2B platform and service marketplaces based on USDL.

Figure 18.5 shows a possible scenario that illustrates the role of USDL in the different approaches that can be used to combine B2B hosting with service marketplaces. Services offered by the B2B platform are made visible on one or more service marketplaces via USDL along with other services. The following scenarios illustrate the synergies that are a consequence of this approach:

- Companies/institutions acting as a small partner search on the marketplace for new hubs offering certain services; next to potential matches the marketplace offers automated B2B-connections (as a generic or as a partner-specific service);

- Companies/institutions acting as a small partner (typically suppliers) search new customers (hubs) that they can address

From the perspective of the B2B platform (or its provider), the service marketplace acts as a channel maker: by publishing USDL instances on marketplaces, platform customers can be acquired. The effort for the platform provider to make B2B services visible is comparatively low - independent of the number of marketplaces involved and their degree of specialization on a certain domain or industry.

Finally, USDL can be used to describe standalone services that the platform has to offer - the third of the options mentioned above. This could be a data conversion or transformation service bridging different standard formats. Such a service is a typical component of B2B-solutions and is implicitly part of the services described above (options (i) and (ii)). Offered as a standalone variant, transformation services support additional use cases, which do not require full-fledged B2B connections.

18.5.4 Lessons Learned

The next generation of B2B platform will significantly reduce the barriers for small enterprises when trying to integrate into a B2B network. This includes the total cost of ownership as well as the required know how. One of the challenging tasks in establishing such a new platform was the development of a comprehensive data model for describing master data — such as B2B-profiles of business partners in the "B2B Directory." Several core concepts of the data model could be taken over from already existing USDL modules. The data model of the B2B directory benefits from the modular, extensible design of the USDL components integrated.

USDL can be applied in form of a platform-neutral model, based on XML schemata. Additionally, there is a platform-specific (=Java-based) infrastructure that can be used for USDL tooling or integration - as a consequence of using the "Eclipse Modeling Framework" for the development of the model [18]. This circumstance simplified the adoption of USDL in the context of the B2B platform.

Apart from the advantages of an internal integration of USDL components, there is a strategic value of supporting USDL in terms of an external metadata interface. One of the main goals establishing the next generation B2B platform is to gain customers that currently do not participate in the B2B business - especially small enterprises. Having such an effective and efficient channel in form of service marketplaces will help to accelerate the adoption by those customers. The channel will be effective because of the awareness that can be achieved by placing B2B connection services as an additional offer to the domain-specific services provided by marketplace participants. The channel will be efficient because of the low effort that is being required to generate USDL for any number or marketplaces.

18.6 Conclusion

This chapter contains various use cases that describe how USDL is useful in different application domains. In the first use case, USDL was used to realize trading energy services via some marketplace. The marketplace was developed and service providers were able to upload service descriptions. The second use case demonstrated how USDL supports the adaptation of services to mobile environments and their requirements. Basic information provided through the MML is extracted from the service's USDL description. Third, a search engine for craftsmen returns the service descriptions of their offering also in USDL and the search engine itself has been described in USDL. Finally, USDL was used to describe services for B2B integration that were published by service providers on a marketplace where they can be found by customers. Based on the conducted case studies, the following conclusions can be drawn:

- USDL should not try to be a service description for all possible services. It should rather provide a framework and allow for extensions. Otherwise, USDL will become even more complex and hard to understand, which will result in a high entry barrier.
- USDL could use a semantic description for less manual effort in deployment. If USDL is semantically annotated, the meaning of particular elements is predefined. This facilitates the mapping to internal data structures and use by custom applications.
- The API of USDL's data model is too complicated. The model makes sense from the model point of view. However, it lacks some object orientation aspects. Since the code is automatically generated from the model files, some features are too complex.
- Using Eclipse Modeling Framework facilitates integration in existing landscapes. Since Eclipse is a common tool for development environments, it is easy to extend current systems with USDL capability.

These valuable comments and feedback help to improve USDL and tailor it to its users' needs. The next version of USDL will tackle them if possible.

References

1. J. Cardoso, A. Barros, N. May, and U. Kylau. Towards a Unified Service Description Language for the Internet of Services: Requirements and First Developments. In *Proceedings of the 2010 IEEE International Conference on Services Computing*, SCC '10, pages 602–609, Washington, DC, USA, 2010. IEEE Computer Society.
2. N. K. Denzin and Y. S. Lincoln. *The SAGE handbook of qualitative research.* Sage Publications, 2011.
3. C. Gellings. *The Smart Grid: Enabling Energy Efficiency and Demand Response*. Taylor and Francis, 2009.
4. Google Inc. Google Web Toolkit. Online available, http://code.google.com/webtoolkit/, 2011. Accessed 15 July 2011.

5. S. Heinzl, B. Schmeling, and B. Freisleben. Using temporal policies for managing changing meta-data of web services. *International Journal of Web and Grid Services*, 6:331–356, November 2010.
6. M. Hepp, J. Leukel, and V. Schmitz. A Quantitative Analysis of eClass, UNSPSC, eOTD, and RNTD: Content, Coverage, and Maintenance. In *Proceedings of the IEEE International Conference on e-Business Engineering*, pages 572–581, Washington, DC, USA, 2005. IEEE Computer Society.
7. R. Kursawe. Evaluierung der Modellierung realer Dienstleistungen in der Dienstbeschreibungssprache USDL. Technical report, Dresden University of Technology, Institute of Systems Architecture, Chair of Computer Networks, 2010.
8. M. Laskowski. E-DeMa: Entwicklung und Demonstration dezentraler vernetzter Energiesysteme hin zum E-Energy-Marktplatz der Zukunft. In *E-Energy Wandel und Chance durch das Internet der Energie*. Springer, 2009.
9. S. A. McIlraith, T. C. Son, and H. Zeng. Semantic Web Services. *IEEE Intelligent Systems*, 16:46–53, March 2001.
10. Microsoft Corp. Microsoft Pinpoint. Online available, http://pinpoint.microsoft.com, 2011. Accessed 1st June 2011.
11. OECD. Smart sensor networks for green growth. *OECD Information Technology Outlook 2010*, 2010.
12. K. Perillo and K. Vollmer. Joint Industry EDI/B2B Survey. Technical report, Forrester Consulting, 2010.
13. D. T. Roy. *Smart Metering und Smart Grid im Jahre 2025: Szenarioanalyse "ber die Entwicklung der Energiewirtschaft in Deutschland.* Grin Verlag, 2011.
14. M. E. Saleh. Semantic-based query in relational database using ontology. *Canadian Journal on Data, Information and Knowledge Engineering*, 2011.
15. Siemens AG. Energy for Everyone - Scenario 2020. In *Pictures of the Future Magazine Spring 2008*, 2008. Available online, Accessed 7 July 2011, http://www.siemens.com/innovation/en/publikationen/publications_pof/pof_spring_2008/energy/scenario.htm.
16. Siemens AG. City of Wind - Scenario 2023. In *Pictures of the Future Magazine Fall 2010*, 2010. http://www.siemens.com/innovation/apps/pof_microsite/_pof-fall-2010/_html_en/scenario-2023.html, Accessed 11 July 2011.
17. Siemens AG. Smart Metering Solutions. Online available, http://www.energy.siemens.com/hq/en/energy-topics/smart-grid/smart-consumption/smart-metering-solution.htm, Accessed 21 June 2011, 2011.
18. D. Steinberg, F. Budinsky, M. Paternostro, and E. Merks. *EMF: Eclipse Modeling Framework*. Addison-Wesley, Boston, MA, 2. edition, 2009.
19. O. Terzidis, A. Fasse, B. Flügge, M. Heller, K. Kadner, D. Oberle, and T. Sandfuchs. Texo: Wie THESEUS das Internet der Dienste gestaltet - Perspektiven der Verwertung. In W. W. Lutz Heuser, editor, *Internet der Dienste*, acatech diskutiert, pages 141–161. Springer, 2011.

Chapter 19
Experience Report on Real-World Manual Service Modeling in USDL

Josef Spillner, Ronny Kursawe, and Alexander Schill

Abstract The Internet of Services promotes distributable, composable and tradeable services as first-class entities. Such services are assumed to encompass the full range from technical Web services to conventional business services delivered by humans. However, research and development of service models and platforms to realize the Internet of Services vision has largely been concentrating on pure technical services. With the introduction of the Unified Service Description Language (USDL), this is about to change. In a multiple-enterprise evaluation study performed in parallel to the ongoing USDL specification process, we have applied modeling and registration techniques to existing services with none or few technical components provided by businesses in a locality-constrained area with the purpose of determining suitability and acceptance aspects. We outline the results of our study and include an evaluation of USDL language facilities in the context of real-world service representation. This chapter connects evaluation results with early USDL drafts [16] with a discussion on recent USDL capability changes up to the specification milestone M5.

19.1 Introduction

Businesses involved with e-business activities struggle to find a balance between keeping trade secrets and exposing information about their products and services on the untrusted Internet [5]. On the contrary, any participation in value chains assisted by distributed business processes requires a public, detailed and accurate description of goods and services. The Internet of Services (IoS) idea suggests that these services become tangible entities which can be traded and composed on marketplaces on the basis of declarative descriptions to facilitate the creation of value chains (cf.

Josef Spillner, Ronny Kursawe and Alexander Schill
Technische Universität Dresden, Faculty of Computer Science, Nöthnitzer Str. 46, 01187 Dresden, Germany, e-mail: `firstname.lastname@tu-dresden.de`

Chapter 1). However, techniques from this area still differ in how they apply to technical services compared to manual services. Today, most publicly available research prototypes for service trading and composition focus on technical Web services with purely syntactical interface descriptions and little semantic information about their behaviour. For the future, it is expected that the service continuum handled by IoS service platforms be extended to conventional business services manually delivered or performed by humans over time, possibly as part of solutions related to product exchange, aligned with the general trend towards servitisation as explained in Chapter 3. Another weakness of existing prototypes especially around USDL is that, even when considering the full spectrum of service definitions, service marketplace concepts have so far only been analyzed with artificial scenarios, for instance in the THESEUS/TEXO research project, and not with real-world services. We define real-world services as the slice of the service continuum which is offered in the real world, under legal jurisdiction of their respective location, and in a certain economic and publicity context. This definition includes manual business services as well as technical services offered for free or for compensation by businesses, as shown in Figure 19.1. We omit the definition of the complementary slice which likely includes the mentioned artificial offerings for research purposes and test instances of services without a required notion of providers and consumers.

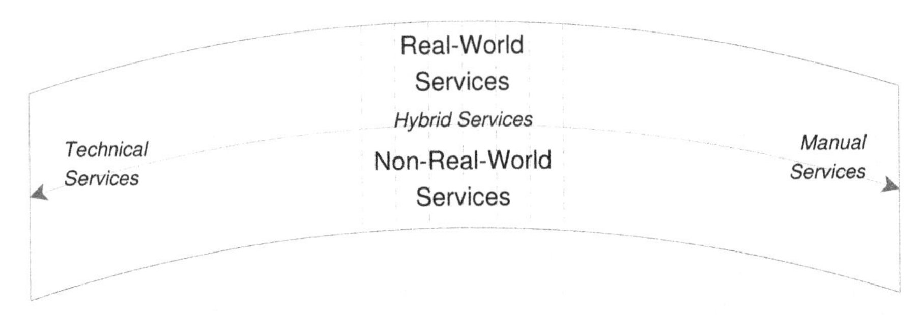

Fig. 19.1: Real-world services in the service continuum.

One indicator backing the expectation of the continuum expansion is that a surge in human-provided services has recently been acknowledged, for instance in their variety as composable human-as-a-service in crowdsourcing environments [14]. A second indicator is the progress of generic models in service sciences [12] and the General Service Model presented in Chapter 4. According to the typical service lifecycle in the IoS (cf. Section 4.4), which is divided into the five phases offering, matchmaking, usage, feedback and innovation, unavoidable differences would most likely occur during the usage phase due to automated technical service execution only being present in electronic Web services. Further differences would be expected for the development and deployment of the service implementation prior to its announcement to potential consumers (e.g., programming + installation vs. training + service center opening), while the development and deployment of the declarative service artifacts such as descriptions (modeling, registration) would not differ.

The other phases would also mostly remain comparable, thus hiding the service implementation (manual or technical) as suggested by implementation neutrality [7]. Therefore, we conducted an evaluation study on the extent of reusability and fungibility of Internet of Services ideas and techniques especially from the modeling, registration and portfolio review processes to real-world services with an expectedly high degree of manual services among them. Our central goal has been the practical evaluation of using USDL as information carrier throughout these processes. The superordinated offering phase scope of the work is hachured in the IoS service lifecycle in Figure 19.2, which follows the lifecycle model given in the TEXO Service Ontology presentation in Chapter 4 (Fig. 4.3 on page 86).

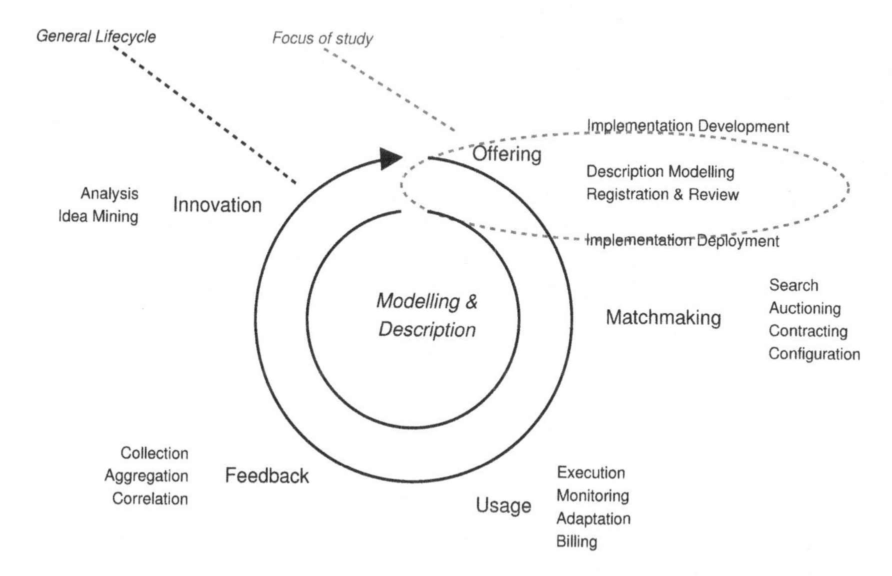

Fig. 19.2: Scope of evaluating real-world service trading in the IoS lifecycle.

Two specific measurable aims were extracted from the goal which influenced the methodology in conjunction with the process selection within the offering phase.

The first aim was to determine the technical suitability of semantically rich declarative description languages for conventional business services or hybrid IT-supported services. Languages previously designed purely for Web services such as WSDL, WADL and DAML-S have been criticized for not conveying enough information about the value and the content of manual services [1]. On the other hand, computationally complete semantic languages such as WSML or OWL (in its application as OWL-S) have not been received well beyond academic projects due to high complexity and missing standardized base ontologies. Therefore, we wanted to find out if USDL offers a suitable formalized spectral view from conventional to electronic services, thus making it possible for service providers to change the service implementation and delivery without having to change the description. Of particular interest has been a practical readiness assessment of the language which has been specifically designed to describe dynamically tradeable and contractually protectable business services as opposed to services with just informal business aspect descriptions.

Our second aim was to see to which extent companies accept the offering processes on service marketplaces, independent of the actual description language, but aligned with the marketplace and registry architecture envisioned to be used for large-scale use of USDL. It implies a consideration of the willingness to publicly expose descriptive service and provider information in registries. To reach both aims, viz., the determination of suitability and acceptance, we performed an evaluation study called Real World Services in Dresden (RWSDD) with the additional goal of extrapolating its results to a general discussion on IoS concepts.

In contrast to the previous chapter about specific case studies on selected business projects, our multiple-enterprise evaluation study follows an iterative methodology with a survey part which delivers several quantitative results [13].

The remainder of this chapter is structured as follows: First, we present our applied multi-phase evaluation methodology including the selection of target companies and services and the steps required to bring them into the IoS. Then, we explain our findings regarding the suitability of selected service description technologies and acceptance of our chosen approach concerning the survey, modeling and registration of commercial business services. The methodology and the findings are compared to similar studies. Finally, the chapter is concluded with remarks on the prospective value of the IoS for real-world services.

19.2 Evaluation Methodology

The starting point for the evaluation study is a local setting of companies offering services within the city of Dresden in Saxony, Germany, in the year 2010. The economical-structural characteristic of having mostly small, often tiny, and only some medium-sized companies poses a challenge regarding the ease of use of Internet of Services techniques. The criteria for the evaluation of suitability and acceptance aspects in applying IoS techniques to real-world services have been chosen accordingly. They encompass *service identification* to isolate services from the overall offering of the companies, *adequate tooling* to express the service concepts and constraints appropriately without overhead or under-specification, and *self-empowerment* to let companies model, manage and register the descriptions by themselves.

The methodology was designed to lead to increased precise knowledge about our evaluation aims and criteria through an iterative process with each iteration introducing a new phase, which are not to be confused with the service lifecycle phases. Consequently, we prepared a five-phase methodology which included an identification of a decent number of various service companies, and subsequent reduction of quantity, hence gaining quality, by filtering with a questionnaire, active modeling of service descriptions and registration of the services at an internal and a public service portal. We also planned a selective implementation of Web services leading to hybrid conventional-electronic service offerings although eventually decided to postpone this activity for reasons to be explained later. Instead, we added a re-evaluation phase after some delay to determine the suitability of the most recent USDL specification milestone. The resulting methodology is visualized in Figure 19.3. The time frame for the evaluation (the first four phases) was limited to three months from February to May 2010, and the delay for the fifth phase lasted until April 2011. Hence, the evaluation was overlapping with the USDL milestone phases towards M3 and M4 whereas the recent re-evaluation of the results performed in the context of M5 allows us to include an analysis of the specification re-engineering progress partially based on our own feedback from the first four phases.

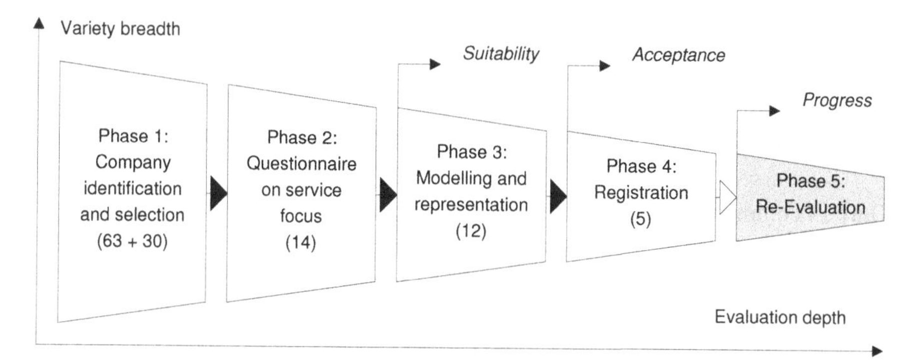

Fig. 19.3: Overall methodology of real-world service selection and integration into the IoS in multiple phases, including number of participating companies in each phase.

The iterative evaluation was mainly conducted by a student, first by creating two sets of diverse companies, followed by a survey by e-mail in multiple feedback rounds, and finally by interacting with the most relevant companies by e-mail based on the modeling and registration work with the USDL editor and a registration portal. The following paragraphs describe all evaluation phases in detail. A summary and excerpt of the questionnaires will then be given at the end of the section in Listing 19.1.

19.2.1 Identification of Service-Centric Companies

We limited our study to 63 individually identified local companies from all industry sectors and in addition to 30 companies from the tourism sector known to us through a local industry association (chamber of commerce, IHK).

The first set of companies included mostly small and tiny companies which, according to their own estimations, offer 95% services and 5% products, matching our expectation regarding the initial identification. The industry representation consisted of a mix of event agencies, design studios, translation and correction of documents, consulting and delivery services. Target customers were a mix of private consumers, businesses and the public sector.

The second set of companies included mostly branch offices of larger tourism companies which are characterized by a higher level of IT-supported operations. In contrast, decisions about IoS integration are less likely taken directly by the local companies. The chosen tourism companies offer travel planning and booking services, sightseeing tours as well as lodging and catering for tourists.

All selected companies were sent background information on the Internet of Services and a questionnaire about their service offering and procurement habits. The tourism companies received an industry association statement about the importance of Internet innovations as additional incentive.

19.2.2 Inventory of Existing Service Offering and Usage

Only 14 companies completed the questionnaire, while some others filled the forms only partially. Missing key data such as company names even made it impossible to identify duplicates, therefore we dropped all incomplete questionnaires. The response rate of smaller companies was higher than the rate of medium-sized ones.

Of note are some statistics about the companies' previous usage of Internet and World Wide Web facilities according to the results of the questionnaire. 100% are advertising on a dedicated web site. 71% make use of online business registries, while only 36% are paying for inclusion in printed industry registries. Regarding the descriptions of offered services, 100% found the company name and web site required attributes, compared to 64% for both the offering details and telephone contacts. Physical address information was deemed important in 29% of all cases, either as text or as geographical location on a map. Electronic services are offered by half of the companies. The statistical information served as input for the service identification criterion.

19.2.3 Service Modeling

Among all companies which completed the questionnaire, 12 indicated that they be interested in having their service offering modeled in a declarative service description language, and the remaining 2 were undecided. We started modeling with USDL while in parallel evaluating concepts from the Web Services Modeling Ontology (WSMO) for a comparison which is not presented in this chapter but can be found in [16].

The results of this phase determine the suitability of IoS service description languages and are thus presented in detail in the first half of the next section.

19.2.4 Service Registration

After creating the service descriptions, we submitted them for a brief review to the companies and offered them to have them registered publicly on a service marketplace, not registered at all, or registered with pseudonymization techniques applied so that the real-world business could not easily be found out from the service description. For publicly registered services, we submitted the resulting information pages again for a public service portfolio appearance review.

The results of this phase determine the acceptance of IoS marketplace concepts and are thus presented in detail in the second half of the next section.

19.2.5 Further Phases

The questionnaire results indicated that similar to the modeling needs, 9 companies would be interested in a Web service implementation regarding parts of their business services, with the remaining 5 being undecided. Our study did not cover the actual realization of Web services or any post-registration activities such as search for offered services, service usage contract establishment or feedback submission. Therefore, we omit a discussion of using USDL in practice beyond the offering phase.

19.2.6 Methodology Summary and Questionnaire

The applied methodology helped us in the reduction of efforts to gain experience with the most relevant companies and services. It accompanied the whole offering phase for 5 providers while still leaving room for considering implementation aspects beyond our study. The structural excerpt of the used questionnaire given in Listing 19.1 reflects the survey part (phase 2) of the methodology. In total, 24 questions were to be answered by the participants.

Listing 19.1: Excerpt from the questionnaires

```
1 Company data {name, contact person, title, web site, e-mail
     address}
2 Service overview and identification {service centrism, portfolio,
      advertisement forms, advertisement information}
3 (Technical) Services - export {applicability, target groups,
     technologies, operator, quality assurance}
4 (Technical) Services - import {applicability, contact
     establishment}
5 --- Answers determined further questions
6 Realisation of formalized service offering {modeling,
     registration, implementation}
```

```
7 --- Answers determined participation in phase 3 (modeling) and 4
      (registration)
```

19.3 Findings

Our study concentrated on two main aspects: Suitability of current service description and registration techniques, as well as acceptance of offering services at IoS marketplaces among the participating companies. We give a detailed explanation about possibilities and weaknesses of modeling manual services with USDL to determine the suitability, and we analyze the responses of companies to our realization to determine the acceptance. We draw conclusions mostly from the 3rd and 4th phase of our evaluation. It should be noted that due to the low number of services considered in the late phases, the results are not representative. However, given the lack of similar studies, they help to understand the challenges of bringing real-world services into the IoS and to identify problems early in the process.

19.3.1 Suitability of USDL

USDL is a recent and still evolving service description language which has so far not yet been used on public service marketplaces [4]. As outlined in Chapter 8, it is being constructed under a design science approach. Major modules in the latest milestone M5 encompass a foundation and core description, service level objectives, pricing, functional descriptions and interaction.

The USDL syntax is kept synchronized with an Ecore model so that Ecore-based transformations can be used in addition to XML transformations, which is popular with Java tool developers [6] and hence stimulates the refinement and extension through domain-specific modules and vocabulary. At the time of the evaluation study, around the introduction of M4, not all modules of version 3 (USDLv3) were completed, as shown in Figure 19.4. Especially the Legal Module has only been created as a stub, and the Service Level module was work in progress. This limitation, combined with an exclusion of syntax intended to represent features of technical services, restricted our study to human-descriptive, functional and pricing aspects. We deliberately left out interaction descriptions as we focused on isolated request-reply style offerings.

For pragmatic reasons, the suitability of the language was analyzed in conjunction with the syntax and semantic expressivity generated by the USDL editor, since we assume that this description creation method will be the dominant one in the near future. The complementary theoretical analysis will be given in the chapters 20 and 21. The Eclipse-based USDL editor (cf. Section 15.2.1) available to us could be considered adequate and complete for expressing all language features needed for the evaluation. However, it required special user skills and training, hence it seemed to be unsuitable to let companies model their services on their own. A remedy would have been the USDL light editor (cf. Section 15.2.2) which is however still in a prototypical state. A further limitation, restraining the self-empowerment criterion, was the missing unrestricted availability of the editor to companies at evaluation time. We worked closely together with some of the USDL authors and tooling creators to ensure that our feedback could help to avoid some of the early issues we found. Some results can be verified with the now widely-available USDL editor for M5 [8]. The progress on USDL development in general can be tracked on an Internet of Services commercial community site.[1]

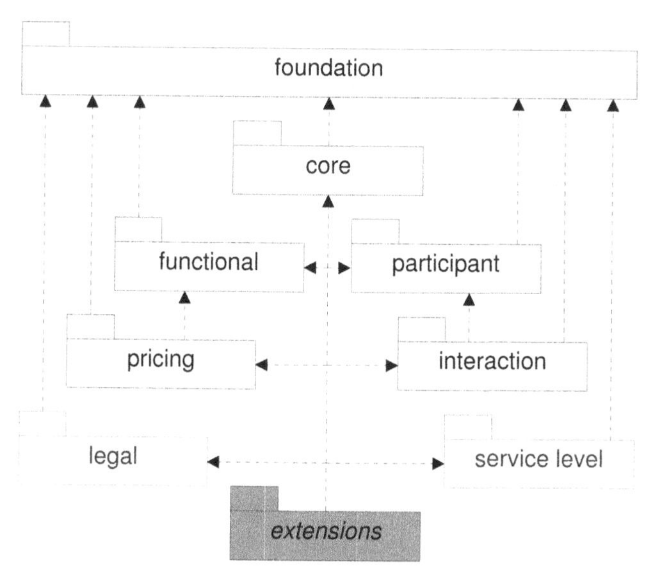

Fig. 19.4: Modular structure of USDL version 3, as presented to the participating companies.

In general, USDL can be seen as one of the first accessible formalized description languages which is suitable for real-world manual and Web services alike. The example in Listing 19.2, which is an excerpt from one of the services modeled by us during the study, demonstrates how we represented an actual certified consulting service without explicit pricing information. The example is given in compact USDL pseudo-notation which omits syntax elements for better readability.

Listing 19.2: USDL representation of consulting service

```
1 core:Service {
2  guid "urn:Consulting Mueller Services"
3  nature "Human"
```

[1] USDL community: `http://internet-of-services.com`

```
4  naturalPersons {
5   firstName "Gustav"
6   lastName "Mueller"
7   contactProfiles {
8    virtualAddress "+49 (351) 0000000"
9  }}
10 organizations {
11  guid "urn:Consulting Mueller Company"
12  representatives "//@naturalPersons.0"
13  certifications {
14   certificate "DIN ISO 9001"
15  }
16  contactProfiles {
17   physicalAddress {
18    street "Dienstboulevard"
19    city "Dresden"
20 }}}
21 serviceLevelElements {
22  availabilityAttribute {
23   type "Telefonische Erreichbarkeit"
24   timeZone "CEST"
25   timeRanges {
26    Mo-Fr: 9:30 - 16:00
27  }}
28 }
29 paymentTerms {}
30 pricePlans {
31  name "Standardstundensatz"
32  currency "EUR"
33  planComponents {
34   componentFloor "70"
35  }
36  taxes {}
37 }
38 capabilities {
39  name "Unternehmensberatung"
40 }}
```

According to the feedback gathered from the participants, several interesting USDL features map real-world situations to service description fragments in a much better way than with conventional description approaches. For example, the modeling of roles can map a restaurant, an associated delivery service and a training with the chef de cuisine as services with corresponding opening or operating times to the same provider structure. Compared to printed business registries, these structures already present an immense advantage when looking for a service to cover specific needs under selection constraints due to the higher expressivity and transparency of the information.

Yet, a number of weaknesses result from precisely this broad scope. Additional weaknesses were found in the realization of certain XML language constructs and in the presentation within the Eclipse-based standalone USDL editor of version 0.5. A subset of ostensive weaknesses found by us will be briefly mentioned here, ranging from general language characteristics to module-specific remarks. A full unedited (but anonymized) report on the suitability of modeling particular real-world services in USDL can be found in an online report [10].

- While industry-specific perspectives are planned, they are currently not part of the public specification. Therefore, modeling always exposes the entire structure which is often not desired, especially for services in the creative industries such as marketing or event management.
- The modeling characteristics, especially the extensibility entry points, impede an axiomatic approach which lets providers start with a minimal description first and extend towards more advanced concepts as needed. The unique identifiers (GUIDs) of elements and the term expression languages contribute to this issue.
- Composite services in the real world can be extremely complex. For instance, a food delivery service may have separate product shopping cart and delivery options. USDL offers limited support for composite services.
- The Pricing Module of USDL is very powerful by allowing a great degree of granular price decomposition. However, this leads to difficult price evaluation at runtime. A possible solution would be to add non-normative simple pricing information to each full specification, e.g., a service *starting at 99 EUR*. This enhancement would reflect imprecise price statements often seen in manual service advertisements.
- A related issue is the combination of services with products. If a service delivery price depends on chosen products as the service configuration, the dynamic price calculation is hard to realize. A proposed meta-model found in related literature [2] primarily considers the four service system dimensions component, resource, product and process and could serve as valuable input for overcoming this limitation.
- Availability as a temporal and spatial concept is not used to its full potential. This is a key metric to distinguish functionally identical or similar services and filter out services which are out of the question because they cannot be reached.
- USDL service descriptions are likely to be processed in user-friendly portals. However, USDL lacks unified graphical elements such as provider and service logos, banners and similar visual hints. In M5, description elements are translatable which makes it possible to create single-source documents for multiple languages. However, there are still weaknesses in the XML serialization given that instead of using the xml:lang attribute, an under-specified custom language tag is introduced.
- The Service Level Module is only basic. It does not support ranges or dependency specifications between objectives. Legally valid contractual rights and obligations of real-world service providers and consumers require an expressive syntax if the aim is to replace currently used paper contracts with machine-readable documents.
- Legally required information such as tax numbers and court locations are not yet part of the language. For the completion of the Legal Module, such real-world requirements need to be taken into account.

- The editor refuses to work properly with Unicode data and mandates to use XML encoding for non-ASCII characters, resulting in larger and harder to read documents.

We expect that the upcoming standardization attempt will reduce the number and impact of the identified weaknesses and therefore increase the suitability of describing tradeable services in USDL.

19.3.2 Acceptance of IoS Concepts among Companies

Service marketplaces represent an intuitive concept to exchange USDL files between providers and consumers. Acceptance problems found by previous studies center around a mismatch between supply and demand in certain configurations, for instance, consumers expecting better quality in one domain and more choice in another [9]. Solving the mismatch requires thorough specification and processing of non-functional properties during matchmaking, which is provided by most advanced service description concepts in USDL. It certainly also requires means to discover under-supply and subsequently techniques to describe and offer competitive offerings to close the gaps, which is outside the scope of service descriptions and needs to be provided by service platforms and marketplaces. Acceptance is an important competitive metric which directly affects the quantity of turning potential into actual users [9]. This necessity led us to include the registration process at a marketplace into our evaluation.

Out of the 12 companies whose services we modeled, 5 indicated to us that they would like to immediately see the service descriptions being used on a public service marketplace independently from whether or not the companies would get direct access to modify the registry entries if needed. The remaining 7 were unsure. Therefore, we measured the combined acceptance of modeling and registration techniques in the IoS with 5 real-world service providers.

Service descriptions approved for public registration were registered into an instance of ConQo, a freely available semantic service description registry software which ships with domain-independent base ontologies for common service concepts such as quality and price [17]. Its service model mandates a domain-dependent (and possibly automatically generated) Web Service Modeling Language (WSML) file of arbitrary complexity and usage of the base ontologies, with an arbitrary number of attached documents and hyperlinks, including USDL or WSDL files. For a preference-driven search and discovery over the service offering, USDL information would have to be transformed into WSML constructs to become visible to the ConQo reasoner. However, since search was not a goal of this study, we only considered few attributes such as service name, description and ownership specification, and confined all others within the USDL file. A service catalogue within a social network of users, communities and companies has been produced which reflects the contents of the service registry. We requested extended company information such as the logo from the 5 participants and invited them them to take a look and tell us what they think about the registration process and their public appearance own on the network. The resulting service catalogue can be found online.[2]

The general feedback from companies about the catalogue was a positive one, although the resulting service information pages were lacking some information due to concerns over commitment and privacy.

We perceived a general reluctance to specify all prices and conditions upfront. Therefore, we suggest to include negotiation capabilities into the service description languages. Service providers would specify which input parameters affect the price on a mandatory or optional level. While SLA negotiation including QoS and pricing conditions is an established topic among SOA researchers, specifying negotiation metadata for a prior meta-negotiation about which properties are negotiable in which situation is a relatively new topic [3].

USDL in particular offers specification fields for employee roles and birth dates. The usage of these fields will depend on trading employee privacy and company protection versus customer needs and curiosity. If all service description fields are public, crawlers could be created to exploit this information in non-intended ways. We expect this to become a major area of interest for privacy research along with the establishment and rising popularity of the IoS, and suggest to reflect the issue as privacy control mechanism in the USDL editor.

By closely observing the progress of some participants and the modeling and registration work performed by the student, we also experienced that companies would need much easier-to-use guided tools and editors to bring their services into the IoS. The USDL editor (Eclipse version) and comparable tools from the Semantic Web Services domain such as WSMO Studio are exclusively targeting technical users. The newly prototyped USDL light editor (cf. Section 15.2.2) is certainly a step in the right direction to enable also casual users. Easy applications or web-based wizards could help to parallelize the collection of registration data from companies. After all, the presence of a large number of service descriptions is one of the key success factors for service marketplaces due to the network effect [15].

Regarding the registration, our plan to let companies register additional descriptions by themselves after our initial registration had to be put on hold despite the availability of a step-by-step documentation. We perceived both a problem of understanding the resulting declarative service descriptions and an incentive problem of what the benefits of registering the files would be. Therefore, we as evaluation marketplace operators performed the registration entirely on behalf of the companies, which will obviously not scale. We recommend the standardization of service package publishing interfaces and the inclusion of publishing functionality into modeling tools to streamline this process. In the M5 version of the editor, a lightweight publication function for automated registration at marketplaces has been added, but the standardization issue remains. We also suggest that service description ownership and dynamic content adaptation (e.g., provider address updates) within service platforms be treated explicitly to convey the advantage of roaming between different hosters without losing information, and to keep the maintenance efforts low. Our

[2] RWSDD service catalogue: `http://crowdserving.com/real-world-services-in-dresden`

discussion with the affected companies strengthened this view, but a general statement regarding the influence of effort on the acceptance needs future research.

Finally, the acceptance of post-registration processes such as service search, usage and rating needs to be taken into account. While we don't cover these aspects in our current work, we intend to explore it later through complex service scenarios. For example, a manual service provider needs to offer a composed service which includes a Web service interaction, and the web service shall be dynamically replaceable under the constraint that it won't ever cost more than a certain limited amount of money.

19.4 Conclusion

In the spectrum from fully electronic services to IT-independent business services, current Internet of Services approaches and platforms mostly concentrate on electronic or at least IT-supported services. However, our experience from the evaluation study RWSDD suggests that central service offering tasks such as modeling and registration of service descriptions have become more feasible and inclusive than with previous conventional SOA techniques. While the study criteria are only partially fulfilled, especially concerning the adequate tooling, the expanded view on practical service trading helps to refine research results in the IoS.

Major barriers to wide adoption of the techniques can still be found in service description languages such as USDL. They require a further evolution, especially in terms of standardized base vocabulary for expressing service properties, and improved user-friendly tooling for entering into an economy of scale by reaching a critical mass of self-empowered IoS participants.

Further crucial factors for durable acceptance of IoS ideas will be the availability of permanent service marketplace installations with guided service provisioning processes, the long-term manageability of service portfolios and the inclusion of user feedback into the service evolution. Future studies about real-world services are expected to yield more concrete results about the post-registration usage phases. We expect such studies to be performed once the IoS infrastructures are completely developed and actively used.

References

1. Z. Baida, H. Akkermans, and J. Gordijn. Serviguration: Towards Online Configurability of Real-World Services. In *Proceedings of the 5th International Conference on Electronic Commerce*, volume 50 of *ACM International Conference Proceeding Series*, pages 111–118, 2003. Pittsburgh, Pennsylvania, USA.
2. M. Böttcher and K.-P. Fähnrich. Service Systems Modeling. In *Proceedings First International Symposium on Services Science (ISSS)*, March 2009. Leipzig, Germany.

3. I. Brandic, S. Venugopal, M. Mattess, and R. Buyya. Towards a Meta-Negotiation Architecture for SLA-Aware Grid Services. In *Workshop on Service-Oriented Engineering and Optimizations*, December 2008. Bangalore, India.
4. J. Cardoso, M. Winkler, and K. Voigt. A Service Description Language for the Internet of Services. In *Proceedings First International Symposium on Services Science (ISSS)*, March 2009. Leipzig, Germany.
5. T. Deelmann and P. Loos. Trust Economy: Aaspects of Reputation and Trust Building for SMEs in E-Business. In *Eighth Americas Conference on Information Systems*, pages 2213–2221, 2002. Dallas, Texas, USA.
6. K. Ehrig, G. Taentzer, and D. Varro. Tool Integration by Model Transformations based on the Eclipse Modeling Framework. *EASST Newsletter*, 12, June 2006.
7. M. N. Huhns and M. P. Singh. Service-Oriented Computing: Key Concepts and Principles. *IEEE Internet Computing*, 1(9):75–81, 2005.
8. K. Kadner and A. Kümpel. USDL-Editor 1.0.0.3M4. Available online: http://sourceforge.net/projects/usdleditor/, April 2011.
9. T. Kollmann. Measuring the Acceptance of Electronic Marketplaces: A Study Based on a Used-car Trading Site. *Journal of Computer-Mediated Communication (JCMC)*, 6(2), 2001.
10. R. Kursawe. Evaluierung der Modellierung realer Dienstleistungen in der Dienstbeschreibungssprache USDL. Evaluation Study Report, June 2010. Available online: http://texo.inf.tu-dresden.de/publications/reports/usdl-rwsdd.pdf.
11. D. Oberle, N. Bhatti, S. Brockmans, M. Niemann, and C. Janiesch. Countering Service Information Challenges in the Internet of Services. *Business & Information Systems Engineering (BISE)*, 5, 2009.
12. E. Ostasius, Z. Petraviciute, and G. Kulvietis. Constructing a Generic E-Service Model in Public Sector. In *Proceedings of the 16th International Conference on Information and Software Technologies*, April 2010. Kaunas, Lithuania.
13. A. Pinsonneault and K. L. Kraemer. Survey research methodology in management information systems: an assessment. *Journal of Management Information Systems - Special section: Strategic and competitive information systems*, 10(2):75–105, September 1993.
14. D. Schall. A Human-Centric Runtime Framework for Mixed Service-Oriented Systems. *Distributed Parallel Databases*, March 2011.
15. C. Shapiro and H. R. Varian. *Information Rules: A Strategic Guide to the Network Economy*. Harvard Business School Press, 1998.
16. J. Spillner, R. Kursawe, and A. Schill. Case Study on Extending Internet of Services Techniques to Real-World Services. In *International Symposium on Services Science (ISSS)*, September 2010. Leipzig, German.
17. G. Stoyanova, B. Buder, A. Strunk, and I. Braun. ConQo – A Context- and QoS-Aware Service Discovery. In *Proceedings of IADIS Intl. Conference WWW/Internet*, October 2008. Freiburg, Germany.

Chapter 20
Requirements for a Service Description Language — Findings from a Delphi Study

Martin Matzner and Jörg Becker

Abstract The USDL has been designed as a means to describe services so that they can be traded via the Internet. The previous parts outlined the status-quo of service description research and practice and highlighted by feature comparison that USDL outstands related approaches in various concerns. However, for evaluating the actual worthiness of a modeling language such as USDL, potential users will consider the fit of the language with the contingent influences their organizations have to deal with. To fill this gap, the purpose of this chapter is to identify requirements for a service description language from potential USDL users. The presented research takes a semiotic theory perspective to the design of modeling languages. Through a Delphi study approach, i.e., an anonymous, written multi-stage survey process, the chapter elaborates a set of requirements. The requirements can be used to ex-post test if the features of the USDL actually address the users' needs and to recheck the underlying assumptions of the USDL design and development process. While finding broad consent with most requirements, we also observed differentiated needs related to the intended use of the USDL.

20.1 Introduction

The emerging research discipline of service science and its design research oriented "normative" [12] part — service sciences management and engineering (SSME) — focus on the design and delivery of services in service systems. In particular, SSME strives to build and evaluate IT artifacts of utility for the service economy.

Service systems are dynamic value co-creation configurations of resources, including people, organizations, shared information, and technology, all connected internally and externally to other service systems by value propositions [23].

Martin Matzner and Jörg Becker
University of Münster, European Research Center for Information Systems, Leonardo-Campus 3, 48149 Münster, Germany, e-mail: `firstname.lastname@ercis.uni-muenster.de`

In order to get the relational processes of value creation between (multiple) providers of a service, the customer as co-creator of value and all potential intermediaries established, various information about a service needs to be (digitally) exchanged between them. However, a specific description might focus on particular aspects of a service or address multiple aspects of a service within one representation.

Information models are a means to represent a service. Information models are computer-based symbol structures which capture the meaning of information and organize it in ways that make it understandable and useful to people [16, p. 127]. Information models are represented in a modeling language. Accordingly, to provide a proper service description as a service representation for a specific purpose there is a need for a purposeful language to represent the services.

In recent years a plethora of languages that are intended to describe specific aspects of a service have been proposed (cf. Part I of this book). Now, the Unified Service Description Language (USDL) has been proposed as a platform-independent language that is intended to facilitate web-based services offered via the Internet. Particularly, it shall address a variety of services, from "manual services" to rather technical services such as infrastructure services (e.g., CPU and file storage).

Against this background of a "blooming production of modeling methods" [22], any new proposed modeling language has to demonstrate its "worthiness." Such demonstration could be done by applying evaluation techniques that fall in one of these categories: feature comparison, theoretical and conceptual investigation, and empirical evaluation [22].

In this context, the purpose of this work is to identify requirements for a service description language. We therefore employ an empirical Delphi study approach to potential USDL users and pilot users of the USDL draft specification. The elaborated list of requirements can be used to ex-post test if the features of the USDL language cover these requirements. Furthermore, the requirements are used to check the underlying assumptions of the USDL design process as they reflect the intended modeling scope of (potential) USDL users and thereby shed light on a sufficient pool of concepts to describe services.

This chapter remains as follows: Section 20.2 discusses information models, modeling languages and methods to construct models, as well as method engineering as the process of respectively constructing the methods as research background of this work. Section 20.4 introduces the Delphi study method and outlines the specific setting of our survey. Section 20.5 then informs on findings from our study and outlines requirements for a service description language. The chapter ends with conclusions and a research outlook.

20.2 Fundamentals of Modeling Language Design

20.2.1 Models and Information Models

Models are abstract representations of an original (real-world or thought-of). Models are artifacts that are made of constructs. *Constructs* form the vocabulary of a domain. Constructs build the basis for defining problems and specifying their solutions [13]. Models are sets of statements expressing relationships between constructs. Building on constructs, models are meant to represent certain situations of problems or solutions [13].

Information models in particular, serve the "construction of computer-based symbol structures which capture the meaning of information and organize it in ways that make it understandable and useful to people" [16, p. 127].

Typically, several modeling techniques are applied in different phases of information systems (IS) development, leading to an increasing degree of formalization and an decreasing level of abstraction — from conceptual modeling over design specification to implementation [1]. *Conceptual models* beneath their representational aspect — i.e., to capture knowledge about the application domain of an IS — have at least two additional properties [24]. Conceptual models do not represent all characteristics of the original. Based on the intentions of the modelers and the objectives of the modeling project certain aspects of the domain are simplified or omitted (reduction feature). A conceptual model is also always embedded into a context consisting of a group of subjects, an intended purpose, and a certain time (pragmatic feature).

To better handle the complexity arising from adequately representing an IS, models may be divided into partial models with respectively focus on the description of a specific aspect. Further, complexity might be addressed by multi-perspective modeling that allows for projections on the overall model by assigning certain model types to different perspectives [11].

20.2.2 Characteristics of Modeling Languages

Conceptual models are expressed in a conceptual modeling language. *Modeling languages* in general might be described in the semiotic perspectives of syntax, semantics and pragmatics. Modeling languages are embedded into modeling methods.

A Semiotic Theory Perspective to Modeling Languages

Semiotics is the theory of signs and designation processes [15]. It considers language as a system of signs and distinguishes three perspectives on sign relations. The syntax describes the set of signs and their interrelations. The semantics denotes the relationships of signs to the concept they refer to (i.e., their meaning). The pragmatics is related to the relationship between the sign system and a sign-user, taking into account how signs are interpreted and what the intention of the language use is [17].

IS research regards modeling languages in the context of Modeling Techniques. Such modeling techniques feature two aspects (cf. Fig. 20.1): the modeling language itself and related instructions.

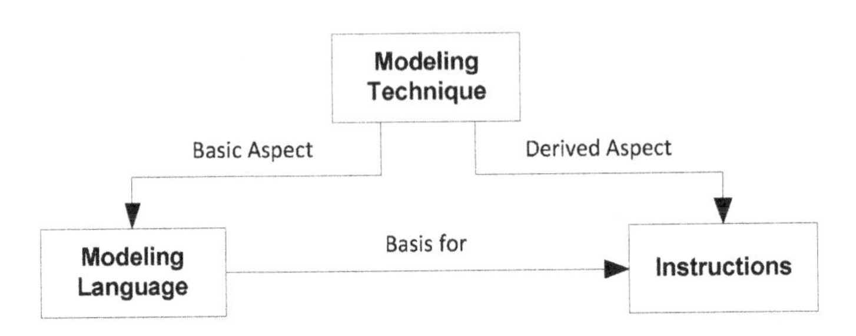

Fig. 20.1: Modeling methods and modeling languages.

With respect to the semiotic perspectives, a *modeling language's syntax* is described by the conceptual and the representational aspect (cf. Fig. 20.2).

The *conceptual aspect* refers to the language elements and syntactical relationships among them. It thereby describes the "abstract syntax" of the language. It thus provides syntactical rules that describe how different language constructs may be properly combined to build well-formed models [17]. Thereby, the set of all grammatically correct models that are constructible with that specific language is determined [6]. Typically, this language aspect is formally described in form of a (language-based) meta-model [7].

The representational aspect of a language refers to the representation of the language elements and their relationships. In diagrammatic modeling languages these symbols typically occur as pictorial or geometrical shapes [2]. It thereby describes the "particular syntax" of a language. There might exist multiple forms of representation for one conceptualization. A new representation does not constitute a new language. In general, the design of the concrete syntax depends on the intended model use and should hence adhere to quality measures such as utility and comprehensibility [10].

The *semantics* of a language is also defined by the conceptual aspect of a language. It explicates the meaning of the language elements (the used constructs) and their relationships in the context of a modeled domain [25]. The semantics is typically defined in a natural language description, a technical term model or by ontologies. In any case, a clear definition of the semantics is important, since different users of a modeling language should develop a compatible understanding of the language's elements in order to discuss and exchange models with that specific language [7].

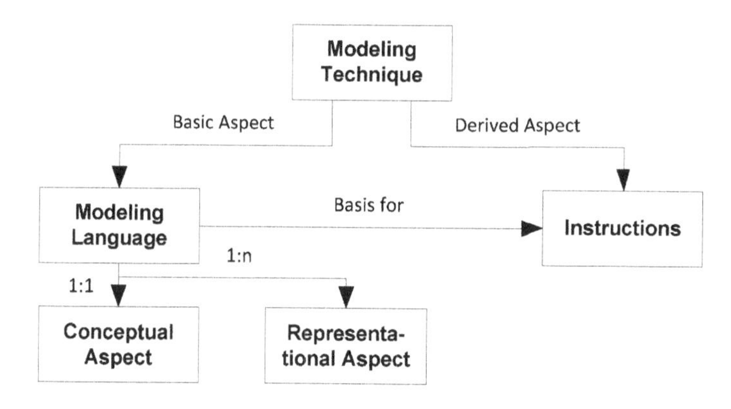

Fig. 20.2: Aspects of modeling languages.

The *pragmatics* of a modeling language refers to its intended scope and objectives with respect to a given modeling need in a particular area of application. The intended scope of a modeling language especially determines which domain concepts are to be incorporated into the language. Hence, the pragmatics has a significant influence on the syntactic dimension of a language and should therefore be specified before designing the language constructs [10].

Modeling Methods

A *Modeling Technique* combines a modeling language with *instructions*. The instructions describe how to construct a model with a modeling language. Instructions are therefore derived from the modeling language. The *Modeling Method* (cf. Fig. 20.3) defines the essential tasks of one or more model development phase(s). There might be several modeling techniques assigned to these phases. Modeling techniques might particularly be assigned to certain modeling views (e.g., organization, data, control, functions) or development layers (concept, data processing concept, implementation).

20.2.3 Method Engineering and Language Design Process

Modeling languages and methods are designed to address the modeling needs of new business and research areas [18]. The design of a modeling method is commonly considered to be an engineering task; leading to the notion of *method engineering* as the discipline that comprises all activities dealing with the creation of methods in general and modeling methods in particular. Method engineering has

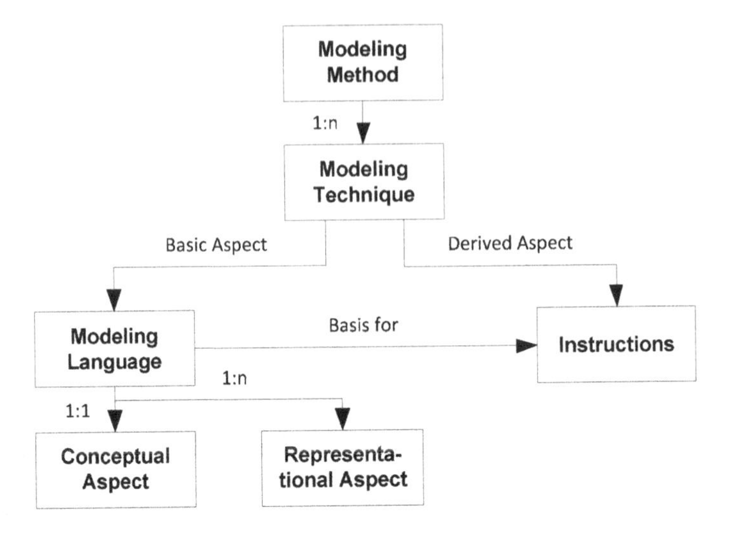

Fig. 20.3: Modeling method.

been explained as a process that comprises of three phases: requirements engineering, method design and method implementation [8].

Requirements engineering (RE) comprises discovering, prioritizing, documenting, representing and maintaining a set of requirements for a specific artifact, which is in case the modeling method. Further it is concerned with analyzing and documenting requirements, by processes including needs analysis, requirements analysis and requirements specification [9]. Some authors also emphasize an organizational aspect in RE, outlining the involved political, sensemaking, and communicative processes [4].

Method design comprises all activities of the actual modeling method construction. Amongst them are the identification of basic language concepts, their supplementation with attributes, and the definition of relationships between attributes, e.g., by a meta-model. To make the design choices traceable and to facilitate a better understanding of the method's capabilities, the reasons for building the meta-model in a specific way should be documented in form of a method rationale [19]. The specification of the semantics completes the specification of the conceptual aspect. Then, the representational elements have to be defined in a way that every language concept must have an according representation. Finally, modeling instructions are specified.

Method implementation and *evaluation* finally subsumes the activities of implementing the method in an information system and testing it in real conditions.

20.2.4 User Requirements for the Unified Service Description Language (USDL)

The Unified Service Description Language (USDL) is designed as a platform-independent service description language. USDL is intended to facilitate web-based services offered via the Internet. It shall address a variety of services, amongst them "manual services" (e.g., project management and consulting), transactional services (e.g., purchasing, order), information services (e.g., request geo information), software services (e.g., software widgets), digital media services (e.g., video and audio clips), platform services (e.g., storage and forwarding of messages) or infrastructure services (e.g., CPU and file storage). Thus, USDL obviously breaks new grounds. It is an endeavor to develop a modeling method that addresses business needs for a unified digital representation of services in a new fashion.

The USDL method was and still is subject to a method engineering (ME) process (cf. Section 20.2.3). At the time of this study, a well advanced draft of the USDL method's specification — namely the USDL 3.0 M4 — was proposed by the developers after several times passing through the ME phases of requirements engineering, method design and method implementation.

Accordingly, this research is an effort to explore the requirements of (potential) USDL users. Thereby, we want to justify the underlying assumptions of the USDL ME process. Further, a precise analysis of the requirements the users raise to USDL is a vital means for evaluating characteristics of the specifications against the users' needs [22].

Particularly, an elaboration of the user requirements for a service description language would help to review (see above):

- The reduction feature of conceptual modeling: As a service description cannot represent all characteristics of a service — which aspects of the domain can be simplified or omitted?
- The pragmatic feature of conceptual modeling: As a service description is always embedded into a context consisting of a group of subjects, an intended purpose, and a certain time — what is the intended use of the USDL specification?

20.3 A Delphi Method Process for Exploring Requirements for a Service Description Language

20.3.1 Delphi Method

The Delphi study is a method for structuring group communication and decision processes [5]. It is a means to systematically capture expert opinions on a complex problem. Typically, the Delphi process is applied if there is only doubtful or fragmentary knowledge about the area under investigation. The Delphi method has been applied in various areas, such as scenario development in business forecasting and public policy-making [21]. In the IS literature, the Delphi method has previously been applied — amongst others — to reach consensus between researchers and practitioners on a common quality framework for evaluating data model quality [14].

The Delphi study has three characteristic features [3]:

- Anonymous response — opinions are obtained by questionnaire. Individual assessment is not passed to the other group members.
- Iteration and controlled feedback — interaction is effected by conducting several iterations of the survey, with carefully controlled feedback between the rounds.
- Statistical group response — the group opinion is aggregated from the individual opinions. Thus, biasing effects of dominant individuals and group pressure towards conformity are minimized.

The Delphi process follows an anonymous, written, multi-stage survey process [5]. In the run-up to each stage, each member of the expert group is given feedback on the results of the previous stage.

Delphi is also subject to some (early described) limitations [20]. Amongst these limitations are esp. constraints regarding the reliability and validity of Delphi results for predictive models. However, Delphi studies are therefore used to serve rather exploratory purposes in an uncharted domain as it is in our case.

20.3.2 Specific Setting and Procedure

A two-stage survey layout was chosen to serve our purpose — i.e., to identify requirements for a service description language. Members of the Delphi expert panel were invited via email and were contacted by telephone in advance. The survey was run from November 2010 to February 2011. Potential participants were provided with a cover letter that informed them about the goals and overall procedure of the survey and with a questionnaire. The questionnaire informed about the research question addressed by the survey, i.e., to assess criteria that shall be "applied for assessing the quality of a service description language."

In the first round of the Delphi study, the experts were asked about their experiences with service description in general and USDL in particular. The panel has been provided with a list of quality criteria towards a service description that we identified from the literature. The criteria were grouped according to the USDL modular structure into the sections "discovery and matchmaking," "ordering, negotiation and access," "composition and aggregation," "multi-channeling, TCO, and outsourcing," and "resources." The panelists then were asked to rate the relevance of several criteria in the described sections on a 6-point Likert scale, ranging from "completely disagree" to "completely agree." For every section an open-ended question was added to enable the panelists to add additional items.

Subsequently, the results of the first round were analyzed and compiled into a document that was sent back to the panelists for revision. For each item of the list, median, arithmetic mean ($\bar{x}$), inter-quartile range (IQR), and standard deviation (σ) were communicated. In the second round, besides again rating the criteria, the panelists were asked to add a reason if choosing a rating that was outside the IQR in order to reach an overall consensus. Further rounds were not conducted because the results from the second round reached a satisfactory degree of saturation, and because we also felt that the respondents would be unwilling to participate in a third round.

20.4 Results from the Study: Requirements for a Service Description

In total, 20 experts participated in the first round, 17 of which again answered in the second round. The list below shows the institutions the respondents are affiliated to.

- Attensity Europe GmbH
- Carl von Ossietzky Universität Oldenburg
- Deutsches Zentrum für Künstliche Intelligenz GmbH, Saarbrücken
- FOKUS - Fraunhofer Institut für Offene Kommunikationssysteme
- Fraunhofer-Institut für Arbeitswissenschaft und Organisation IAO
- imc - information multimedia communication AG
- Intelligent views GmbH
- Metasonic AG
- Metris GmbH
- Public Research Centre Henri Tudor, Luxemburg
- Queensland University of Technology, Brisbane, QLD, Australia
- SAP AG
- Seeburger AG
- Siemens AG
- Smart Services CRC, Eveleigh, NSW, Australia
- The Open University, Knowledge Media Institute, Milton Keynes, UK
- University of Helsinki, Finland
- Westfälische Wilhelms-Universität Münster - ERCIS
- YellowMap AG

Most of the participants by now use description techniques for their service portfolio. Widely-used approaches comprise enterprise architecture and resource models (e.g., ArchiMate, ARIS) and business process models (e.g., ARIS, BPMN). Further, formal descriptions for technical services are used at least in an experimental setting. BPEL and Web service description approaches such as WSDL, OWL-S and WSMO have been frequently stated. We also asked the panelists which information systems in their enterprises process service-related information and, thus, might be potential target systems to a service description. Amongst the answers dominate standard enterprise software systems (CRM, ERP, SCM), product configurators, master data management systems and search engines. Answering the questionnaire did not presume USDL modeling experience as we wanted to explore requirements for a service description in general. Thus, nine of the responds said that they did not have prior USDL modeling experience. The remaining panelists stated that they modeled up to 20 services with USDL. Among the modeled services were geo information services, supplier services, point-of-interests search or Web service description as well as the application of the USDL data structure as part of an internal data model.

We present key findings from the study in the subsequent subsections. We also refer to Table 20.1 that summarizes the results from the Delphi study and that also provides data that supports our presentation of the key findings.

20.4.1 Level of Detail

An important decision in the course of designing a service description such as USDL is to choose a certain level of detail. A very broad specification would rather focus on the description of a service's core properties. On the contrary, a very specific specification would provide detailed constructs and structures for various aspects of a service. Thus we wanted to explore the users' needs to the reduction feature of a modeling language (see above). When asked, which granularity of a service description is more likely to meet their needs, the respondents' answers were diverse. The results feature relative high variance (e.g., standard deviation in round 1 $\sigma_1 = 1.69$) and a measure of central tendency nearby "undecided" (arithmetic means in round 2, $\bar{x}_2 = 3.875$) (cf. Figure 20.4 and Table 20.2).

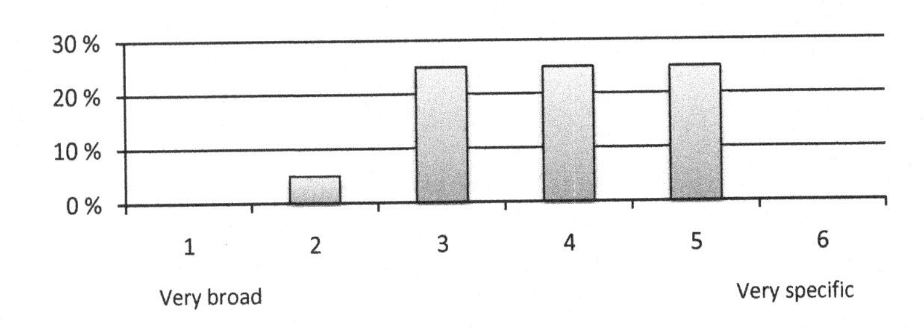

Fig. 20.4: Answer distribution to "level of detail."

In the first round of the Delphi survey the panelist could (optionally) state reasons for their assessments. These statements further attest the heterogeneous assessment. By way of example, one respondent answered: "There is an obvious tradeoff between granularity and comprehensiveness of service description and ease of use."

Table 20.1: Results from the Delphi study. 'Negotiating functional service properties' and 'Configuration rules (exclusion)' were control questions from Delphi round 1.

Requirement	$\bar{x}_1$	$\bar{x}_2$	σ_1	σ_2
General Requirements				
Granularity of service description	3.83	3.875	1.69	0.956
Domain-specific extensions	5.26	5.19	1.05	1.17
Discovery and Matchmaking				
Search based on functional properties	5.55	5.81	0.78	0.4
Search based on non-functional properties	5.3	5.69	0.66	0.48
Search based on cross-references	4.22	4.5	1.17	0.76
Reducing search space through input or output compatibilities	4.3	4.38	1.53	1.15
Ordering, Negotiation and Access				
Understandability of the terms of service-usage	5.1	4.88	0.85	0.96
Service variants based on functional properties	4.1	4.13	1.17	0.89
Service variants based on non-functional properties	4.5	4.56	1.24	0.81
Negotiating non-functional service properties	3.58	3.4	1.68	1.06
Negotiating functional service properties		3.2		1.26
Composition and Aggregation				
Composition 1 (through process descriptions)	4.3	4.31	1.75	1.3
Composition 2 (with intermediary services)	4.05	4.13	1.76	1.36
Aggregation	4.9	4.94	0.97	0.46
Configuration rules (inclusion)	4.89	5.06	1.33	1.06
Configuration rules (exclusion)		4.25		1.18
Substitution relationship	4.1	4.33	1.71	0.82
Provider-sided service integration	5	5	0.87	0.58
Multi-channeling, TCO and Outsourcing				
Multi-channeling 1 (channel-based diversification)	4.13	4.19	1.55	1.05
Multi-channeling 2 (based on properties of devices)	4.28	4.25	1.41	1.18
Total-Costs-of-"Ownership"	3.95	4	1.58	1.46
Reliability	4.75	4.53	1.29	1.3
Availability	4.95	4.93	1.28	1.43
Resources				
Integrating customer resources	4.56	4.69	1.38	0.6
Operant resources	3.72	3.19	1.53	1.11
Operand resources	3.53	3.19	1.39	1.22
Linking services to product life-cycle	3.65	3.6	1.22	0.51

Table 20.2: Delphi rating on "level of detail."

$\bar{x}_1$	3.83	σ_1	1.69	n/a_1	2 (10%)
$\bar{x}_2$	3.875	σ_2	0.956	n/a_2	0

As reasons that would contradict a rather fine-granular description of services was mentioned by the respondents: increased "effort," "future occurrences are not foreseeable," "acceptance," and limited "reuse."

As reasons for a rather fine-granular description was mentioned: increased "usability," "transparency," "combinability," "interoperability," "interpretability by information systems," "comparability in details."

20.4.2 Domain-specific Extensions

Domain-specific extensions to a rather narrowly focus description of the service's core could be a design option that overcame the previously discussed dichotomy of a service description. Such extensions would address the above introduced principles of modularization and multi-perspective modeling. In the Delphi one of the respondents commented correspondingly to the previous question on a desired level-of-detail of the service representation: "From my point of view, a generic service description language with an extension mechanism for domain-specific features would be ideal solution."

Thus, we asked the panel whether — given the existence of such domain-specific extensions — as much aspects of the description as possible should reside with these extensions. The overall feedback states great support by the experts for this thesis. The panel recommends transferring as much aspects of the service description from the core to the extensions ($\bar{x}_1 = 5.26$, $\bar{x}_2 = 5.19$) as can be seen in Figure 20.5 and Table 20.3.

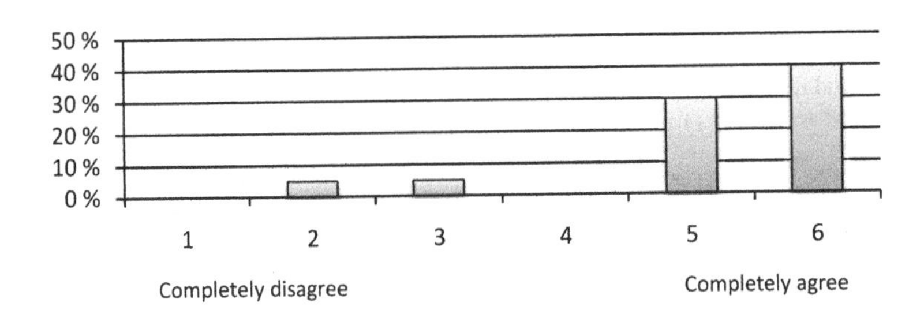

Fig. 20.5: Answer distribution to "domain specific extensions."

Table 20.3: Delphi rating on "domain-specific extensions."

$\bar{x}_1$	5.26	σ_1	1.05	n/a_1	1 (5%)
$\bar{x}_2$	5.19	σ_2	1.17	n/a_2	0

20.4.3 Discovery and Matchmaking

To allow for an effective discovery and matchmaking of services turned out to be a major requirement for a service description. The panel expressed support to four provided categories that might be addressed by a service description language, i.e., to support a search on functional as well as on non-functional properties of a service, to support a search based on cross-references amongst services and to support a reduction of the search space of potential service candidates through an analysis of input and output compatibilities among services. Central tendency indicators for all four aspects expressed "agreement" in both rounds of the survey (all $\bar{x}_1 > 4$ and all $\bar{x}_2 > 4$).

20.4.4 Ordering, Negotiation and Access

In the view of the respondents, a service description should be very good understandable, referring to a very comprehensible representation of expectations, rights, duties and penalties related to a service provision. Here, we observed a high mean affirmation ($\bar{x}_1 = 5.1$, $\bar{x}_2 = 4.88$). As a means that might facilitate understandability, the use of multi-view modeling (see above) has been proposed in an expert's comment. Thus, e.g., contracting issues would reside within a specific view with a specific modeling technique assigned to this view.

Also the need that a service description could represent service variants based on functional as well as on non-functional properties was expressed. In the first type, the service variants differ in the subsets of capabilities they feature. In the latter type, the service variants differ in their non-functional properties, e.g., regarding price or availability.

In contrast, the support of negotiation processes on the actual values of non-functional service properties as well as on the actual customer-specific functional properties of a service was considered to be less important.

20.4.5 Composition and Aggregation

According to the panel's opinion from the given features inclusive configuration rules for services are most important. The mechanism refers to a service description's capability to represent dependencies between services in the form of "component A requires component B" ($\bar{x}_2 = 5.06$). Instead, exclusive rules, i.e., component A is incompatible with component B, was attached less importance ($\bar{x}_2 = 4.25$).

In order to loosely couple several of a provider's services in the form of marketable service bundles the "aggregation" of services seems to be a convenient mechanism. All experts agree in the second round that a service description should facilitate aggregation.

The respondents set a special emphasis on the requirement for a service description to allow for the description of integrated service bundles comprising offers from not only one, but several providers (provider-sided service integration, $\bar{x}_{1/2} = 5$).

A very high variance could be observed in the panelists' need for a service description's ability to support a coupling of services through process descriptions of a service (coupling 1, $\sigma_1 = 1.75$, $\sigma_2 = 1.3$). Here, certain process steps might be realized by (other) services. Accordingly, the ability to support a coupling of services with intermediary services (e.g., different delivery-related services such as Payment Engines and Service Broker (coupling 2, $\sigma_1 = 1.76$, $\sigma_2 = 1.36$) was rated with highest variance among all characteristics.

20.4.6 Multi-Channeling, TCO and Outsourcing

Requirements for a service description in this category seem to depend heavily on the intended use of the service description, e.g., providing a Web service versus offering "conventional" services to customers via the Internet. The responses of the panelists vary intensely (all $\sigma > 1.0$).

However, there is still clear support to all items of the provided requirement list (all $\bar{x}_2 > 4$). Namely these requirements comprise the support of channel-based diversification, e.g., through integrating into several channel partner's offerings (multi-channeling 1), the diversification of services based on properties of devices the service is accessed with (e.g., browser, mobile phone; multi-channeling 2), as well as providing decision support to a customer by means of supporting an easy assessment of the total costs related to service usage for a certain time and based on certain usage assumptions.

The panel set a strong focus on the service description's ability to allow for a comprehensive understanding of availability conditions of service usage ($\bar{x}_1 = 4.95$, $\bar{x}_2 = 4.93$).

20.4.7 Resources

The customer as co-creator of value has a particular important role in the relational processes of service value creation. Accordingly, the resource aspect of a service description might require also representing resources of the customer, if these resources have to be integrated into the providers' service fulfillment processes. In the second round of the survey, no respondent disagrees with that notion ($\bar{x}_2 = 4.69$ with low variance $\sigma_2 = 0.6$).

We also asked for the panelists' need to further distinguish between several types of resources and proposed the concepts of "operant" and "operand" resources that are distinctively advocated in the service science literature. In this notion operant resources refer to competencies such as human resources, knowledge, business processes, culture, business relations, whereas operand resources refer to, e.g., machinery, components, parts, and material. Such a distinction was of less importance in the eyes of the respondents.

20.5 Conclusion and Outlook

The purpose of this chapter was to identify requirements for a service description language from potential USDL users. We introduced information models as a means to serve the construction of computer-based symbol structures. As information models are intended to capture the meaning of information about an original and organize it in ways that make it understandable and useful to people the question occurred what the particular demands of (potential) USDL users on a service description language are. We focused on the representational aspect of a modeling language — that is to capture knowledge about the application domain. Here, we particularly tried to assess how the two main features of (conceptual) information models would be met by USDL. These aspects are, first, the reduction feature of conceptual modeling, i.e., "Are the right aspects of a service selected respectively omitted?," and second, the pragmatic feature, i.e., "Is the users' intended use of the USDL compliant with its addressed scope?"

We addressed this objective by employing a Delphi study approach in order to explore required characteristics of a service description — in particular concepts and relationships among these concepts. The observations headmost suggest a very broad support of the respondents with the provided list of requirements.

Two forces might have added to this convergence and thus might constitute limitations of this work. The study was conducted at a late stage of the method engineering process that comprises of multiple iterations of requirements engineering, method design and method implementation followed by phases of sense-making and discussions. Against this background, first, many panelists actively participated in the development process of the USDL specification. Accordingly, many of the respondents were able to make their wishes and concerns heard at an early stage of method design. Second, also the participants' expectations to a service description might be biased by this participation and prior USDL modeling experiences in a sense that the respondents' perceived requirements for a service description converged with actual outcomes of the participatory USDL design process.

At the time of this study, only a well advanced draft of the USDL method's specification — namely the USDL 3.0 M4 — was proposed by the developers. Particularly, the diagrammatic modeling approach and corresponding modeling tools were then in a premature stage. Some criticism expressed by respondents should be met as final and improved tool support and documentation are available.

Still the desired design of the concrete syntax depends heavily on the intended scope and objectives of the service models, referring again to the pragmatics of the modeling method. Here, we see a further contribution of the work as it outlines differences in the needs of those users that strive for the representation of rather technical services and the users that also consider representing "conventional" services.

The groups' needs correspond to a discussion in the USDL design process about the trade-off between a rather generic or rather specific description of the services. Necessarily such a decision is a tender spot in the design of a modeling method such as the USDL. Further work will need to further elaborate on these oppositional paradigms. Here, modularization — as currently advanced in the W3C Unified Service Description Language Incubator Group — and multi-perspective modeling are promising strategies to bring together the users' interests.

References

1. J. Becker and P. Delfmann. *Reference Modeling. Efficient Information Systems Design Through Reuse of Information Models*. Physica, Berlin, Germany, 2007.
2. A. F. Blackwell. Pictorial representation and metaphor in visual language design. *J. Vis. Lang. Comput.*, 12(3):223–252, 2001.
3. N. Dalkey. An experimental study of group opinion: The delphi method. *Futures*, 1(5):408 – 426, 1969.
4. E. Davidson. Technology frames and framing: A socio-cognitive investigation of requirements determination. *MIS Quarterly*, 26(4):329–358, 2002.
5. A. L. Delbecq, A. H. van De Ven, and D. H. Gustafson. *Group Techniques for Program Planning: A guide to nominal group and Delphi processes*. Scott. Foresman, Gleview, IL, USA, 1975.
6. G. Guizzardi. *Ontological foundations for structural conceptual models*. PhD thesis, CTIT, Centre for Telematics and Information Technology, University of Twente, Enschede, 2005.
7. G. Guizzardi, L. F. Pires, and M. J. van Sinderen. On the role of domain ontologies in the design of domain-specific visual modeling languages. In J.-P. Tolvanen, J. Gray, and M. Rossi, editors, *Proceedings of the 2nd OOPSLA Workshop on Domain-Specific Visual Modeling Language*, pages 25–38, 2002.
8. D. Gupta and N. Prakash. Engineering methods from method requirements specifications. *Requir. Eng.*, 6(3):135–160, 2001.
9. J. Holmström and S. Sawyer. Requirements engineering blinders: exploring information systems developers' black-boxing of the emergent character of requirements. *EJIS*, 20(1):34–47, 2011.
10. M. Karow. *Business Process Documentation in Creative Work Systems*. PhD thesis, University of Münster, Germany, 2011.

11. R. Knackstedt. *Fachkonzeptionelle Referenzmodellierung einer Managementunterstützung mit quantiativen und qualitativen Daten. Methodische Konzepte zur Konstruktion und Anwendung*. PhD thesis, University of Münster, Germany, 2006.
12. P. P. Maglio, S. L. Vargo, N. Caswell, and J. Spohrer. The service system is the basic abstraction of service science. *Inf. Syst. E-Business Management*, 7(4):395–406, 2009.
13. S. T. March and G. F. Smith. Design and natural science research on information technology. *Decision Support Systems*, 15(4):251 – 266, 1995.
14. D. L. Moody. Theoretical and practical issues in evaluating the quality of conceptual models: current state and future directions. *Data Knowl. Eng.*, 55(3):243–276, 2005.
15. C. W. Morris. *Foundations of the theory of signs*. The University of Chicago Press, Chicago, IL, USA, 1938.
16. J. Mylopoulos. Information modeling in the time of the revolution. *Inf. Syst.*, 23(3-4):127–155, 1998.
17. D. Pfeiffer. *Semantic Business Process Analysis — Building Block-based Construction of Automatically Analyzable Business Process Models*. PhD thesis, University of Münster, 2008.
18. J. Ralyté, C. Rolland, and R. Deneckère. Towards a meta-tool for change-centric method engineering: A typology of generic operators. In A. Persson and J. Stirna, editors, *Advanced Information Systems Engineering, 16th International Conference, CAiSE 2004, Riga, Latvia, June 7-11, 2004, Proceedings*, volume 3084 of *Lecture Notes in Computer Science*, pages 202–218. Springer, 2004.
19. M. Rossi, J.-P. Tolvanen, B. Ramesh, K. Lyytinen, and J. Kaipala. Method rationale in method engineering. In *33rd Annual Hawaii International Conference on System Sciences (HICSS-33), 4-7 January, 2000, Maui, Hawaii, Track 2: Decision Technologies for Management. IEEE Computer Society*, 2000.
20. H. Sackman. Delphi assessment: expert opinion, forecasting and group process. Technical Report AD0786878, Rand Corp., Santa Monica, CA, USA, Apr 1974.
21. R. C. Schmidt. Managing delphi surveys using nonparametric statistical techniques. *Decision Sciences*, 28(3):763–774, 1997.
22. K. Siau and M. Rossi. Evaluation techniques for systems analysis and design modelling methods — a review and comparative analysis. *Inf. Syst. J.*, 21(3):249–268, 2011.
23. J. Spohrer, S. L. Vargo, N. Caswell, and P. P. Maglio. The service system is the basic abstraction of service science. In *41st Hawaii International International Conference on Systems Science (HICSS-41 2008), Proceedings, 7-10 January 2008, Waikoloa, Big Island, HI, USA*, pages 1530–1605. IEEE Computer Society, 2008.
24. H. Stachowiak. *Allgemeine Modelltheorie*. Springer, Wien, 1973.
25. Y. Wand, D. E. Monarchi, J. Parsons, and C. C. Woo. Theoretical foundations for conceptual modelling in information systems development. *Decision Support Systems*, 15(4):285 – 304, 1995.

Chapter 21
How Complete is the USDL?
A Theoretical Evaluation of its Capability to Specify Software Services

Dominik Q. Birkmeier, Sven Overhage, Sebastian Schlauderer, and Klaus Turowski

Abstract The USDL aims at providing comprehensive descriptions of business and software services which cover all aspects relevant to support their discovery and combination in the envisioned Internet of Services. In this chapter, we specifically evaluate the expressive power of USDL to specify *software* services. Based on an analysis of literature on software description requirements and related approaches, we derive a theoretically grounded evaluation framework. This framework is used as a benchmark to evaluate the constructs of the USDL. According to the presented evaluation framework, comprehensive descriptions of software services should cover commercial information, implemented business semantics, technical binding information, and service quality. The evaluation shows that the USDL provides the most detailed approach to date to comprehensively describe software services, which nevertheless should be harmonized in some aspects.

21.1 Motivation

The Unified Service Description Language (USDL) has been developed as a vendor-independent approach to facilitate the trading of both business and software services. One of its main goals is to provide a single, complete source of information that supports the evaluation of services by consumers and the combination in the envisioned Internet of Services (cf. Chapter 1). Therefore, the USDL introduces a meta-model which unifies capabilities to specify business, operational, and technical service aspects into a coherent framework. The preceding chapters have shown

Dominik Q. Birkmeier, Sven Overhage, Sebastian Schlauderer
University of Augsburg, Universitätsstrasse 16, 86159 Augsburg, Germany,
e-mail: firstname.lastname@wiwi.uni-augsburg.de

Klaus Turowski
Otto-von-Guericke University Magdeburg, Universitätsplatz 2, 39106 Magdeburg, Germany,
e-mail: klaus.turowski@ovgu.de

that the USDL is applicable to describe services of various domains and that it is suitable to support the trading and reuse of both business and software services in practice. In this chapter, we focus on evaluating the expressive power of the language to specify *software* services from a theoretical point of view. In particular, we will discuss which information should be provided in order to support the discovery, contracting, and combination of software services.

Especially for service-oriented computing approaches, evaluating the expressive power of a service description language is a crucial task since the success of such approaches considerably depends on the ability of consumers to discriminate between offered software services and choose the ones best suited to fulfill their requirements [24]. Without adequate information, consumers are forced to treat software services as "experience goods" [16] whose properties cannot be sufficiently assessed against existing requirements until after buying. If consumers are left unable to discriminate between different service offerings before buying, the corresponding service market is likely to malfunction, though [2]. For this reason, an approach to comprehensively specify software services is a key prerequisite for the prophesied emergence of so-called service ecosystems [4].

The development of the USDL acknowledges the critical importance of providing comprehensive descriptions of software services. Yet, it has to be clarified whether the expressive power of the USDL is sufficient to facilitate the trading of software services in the intended way. To evaluate the expressive power of the USDL, we introduce a theoretical evaluation framework with several description aspects that have been identified as relevant in literature. This framework is used as a benchmark to evaluate the USDL against. The evaluation results document to what extent the expressive power of the USDL meets theoretically motivated requirements. Where differences are identified, the potentially resulting room for improvements is analyzed.

The remaining chapter is organized as follows: in Section 21.2, we discuss approaches that aim at a complete description of reusable software artifacts. This discussion helps identifying relevant properties that have to be described in order to support a successful trading and reuse of software services. Based on the results of the discussion, we define the theoretical evaluation framework in Section 21.3. It is used to evaluate the expressive power of the USDL, milestone version M4, in Section 21.4. We conclude by summarizing and reflecting the evaluation results in Section 21.5.

21.2 Existing Approaches to Describe Reusable Software Artifacts

Various approaches exist in literature which aim at providing a complete specification of software artifacts. Some of these approaches have already been discussed in the chapters belonging to Part I. While that part of the book focused on describing the state of the art in service description, we concentrate on approaches to comprehensively specify software artifacts in the following. Similar to the USDL, most of the approaches build upon specialized techniques to describe specific aspects, which are tied together in order to form a comprehensive approach. The majority of approaches to completely describe software artifacts originated as part of the component-based software engineering (CBSE) discipline, which aims at an efficient trading and reuse of coarse-grained software parts, viz., software components [27]. The CBSE discipline is conceptually close to service-oriented computing [27].

In an approach to describe reusable software components based on the concept of software contracts, [6] combined existing techniques to describe programming interfaces, assertions (pre- and post-conditions as well as invariants), interaction protocols, and quality attributes into a unifying specification framework. The approach, which aims at providing a comprehensive overview of the functionality provided by a software artifact, is depicted in Figure 21.1.

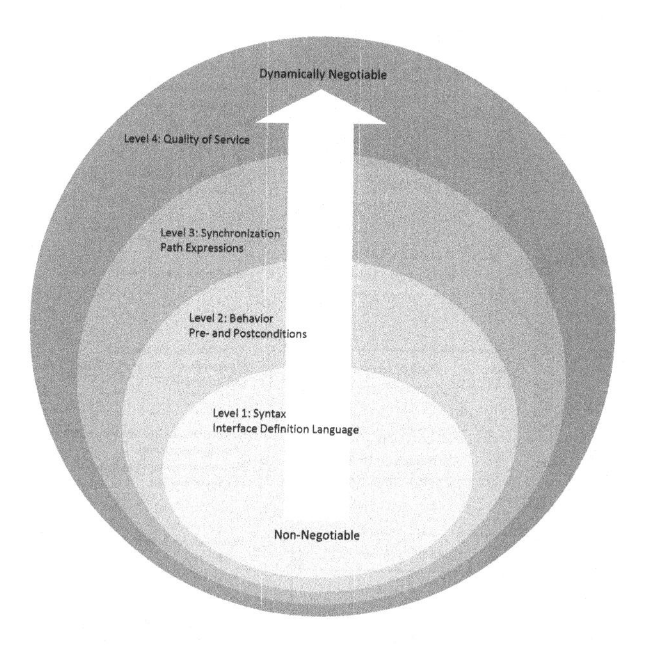

Fig. 21.1: Contract-based approach to comprehensively describe software components [6].

The specification framework introduced by [6] has been criticized since it focuses solely on technical aspects of the provided functionality and ignores the implemented domain semantics as a relevant part of a component description. The domain semantics represents the understanding of the concepts underlying a certain application domain and determines the meaning of domain aspects such as exchanged information, actions of stakeholders, work-flows, incidents etc. [7].

As software components cannot be assessed for reusability without knowing about the domain semantics that they implement, the framework proposed by [6] has been augmented in subsequent approaches. In an interdisciplinary setting, members of the special interest group on component-based business applications of the German Informatics Society jointly created the Standardized Specification of Business Components framework [1]. Its structure is depicted in Figure 21.2. The contributions of this framework are two-fold. Firstly, it complements the description of technical aspects of the provided functionality with a description of the implemented business semantics. That description explains the meaning of business terms from the component providers' point of view. Secondly, the framework introduces additional aspects which describe marketing-related information such as distribution channels, license agreements, or administrative contacts.

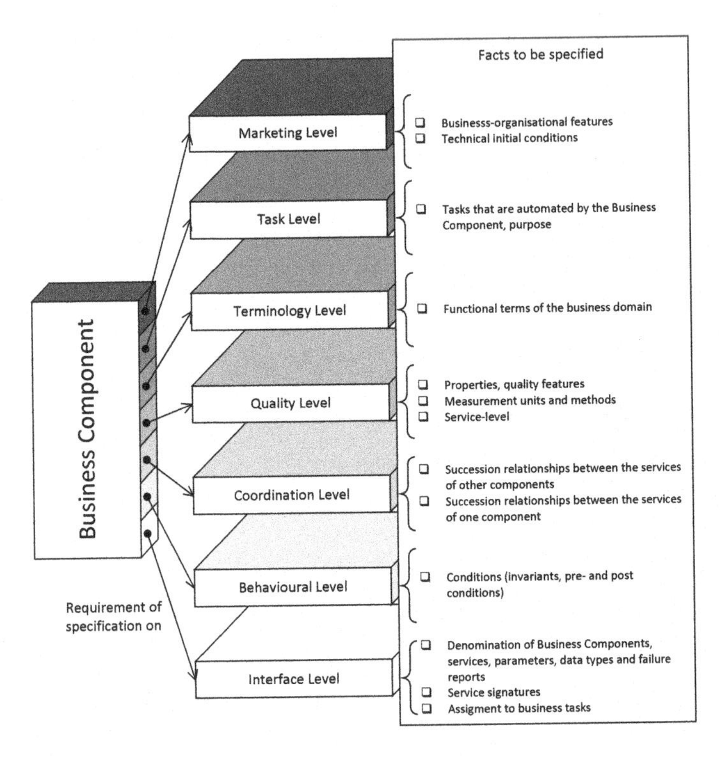

Fig. 21.2: Standardized Specification of Business Components framework [1].

The Standardized Specification of Business Components framework has been used as a conceptual basis to create the WS-Specification framework which aims at a comprehensive description of software services [21, 22, 23]. Like the Universal Description, Discovery, and Integration (UDDI) standard, WS-Specification thematically groups parts of a service description (see Figure 21.3):

- the white pages contain general and commercial information about the service;
- the yellow pages comprise service classifications according to standardized taxonomies;
- the blue pages describe the business semantics of the operations offered by a service;
- the green pages contain technical information to bind and execute the operations of a service;
- the grey pages describe the quality characteristics of services in the form of service levels.

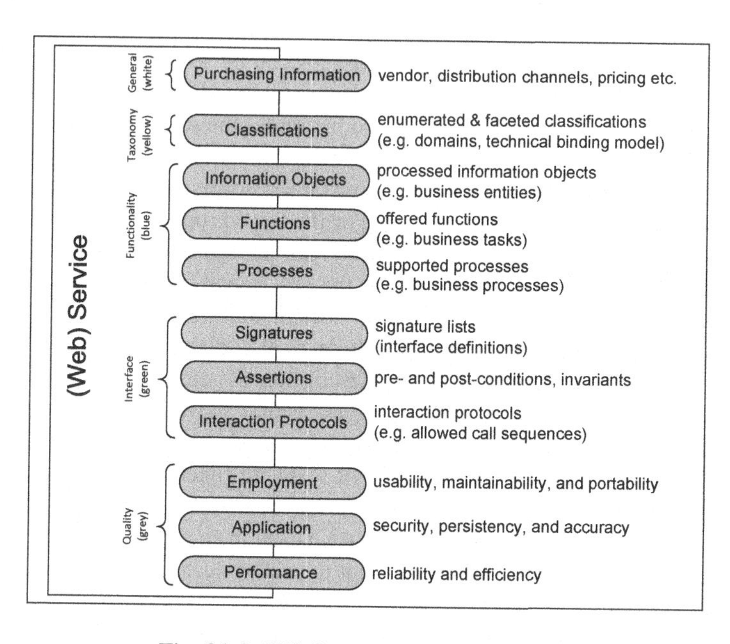

Fig. 21.3: WS-Specification framework.

As we can see, approaches to comprehensively describe the properties of reusable software artifacts have a variety of commonalities. All of them describe the functionality that is provided by a software artifact. The functionality is described as the effects of a software artifact, i.e., the situation which precedes and/or results from its execution. The effects of a software artifact are described in different ways however, as the approaches describe different aspects of the provided functionality.

Approaches, which have been created by interdisciplinary communities, also include marketing-related information into the description of software artifacts. While such information does not characterize the provided functionality per se, it is required to determine whether a software artifact can be reused in a certain project context.

21.3 Theoretical Evaluation Framework

We use the findings presented in Section 21.2 as the starting point for the development of our theoretical evaluation framework. As discussed, approaches to comprehensively describe the characteristics of software services should cover both marketing-related as well as functionality-related information. Marketing-related information should cover legal and pricing aspects as well as the form of distribution. This information usually is required to initiate and complete commercial transactions.

In addition, consumers need to know about the functionality provided by software services in order to discriminate between alternate candidates and choose the ones best suited to fulfill their requirements. As can be seen from the findings in Section 21.2, a service description should document "the intended effect of a piece of software in precise terms" [11]. Doing so facilitates the searching and binding of software services that users have to establish when integrating a service into a composite application. In order to evaluate in how far a description language completely specifies the intended effects of a software artifact, we propose the following evaluation framework. This evaluation framework can particularly be used to highlight the properties which can be specified using the USDL and shows to what extent the USDL is capable of describing software services in a unified way.

The evaluation framework builds upon theoretical findings regarding the description of software functionality. Both the description of reusable software artifacts (program routines, algorithms, classes etc.) in general as well as the description of software services in particular have repeatedly been subject to research. Regardless of the underlying paradigm, a description of software functionality documents the task that a software artifact is meant to perform in detail [13]. It can be employed to assess whether this artifact is suitable to be (re-) used in an application development project. The task a software artifact is meant to perform can generally be described by the following three levels of abstraction [6, 9, 17, 20, 23, 25].

The *business semantics* is defined during the conceptual design of the software artifact. It expresses the business context in which a software artifact is meant to operate and describes the supported tasks from a business-oriented viewpoint (i.e., in business terms). A typical business context could, e.g., be "annual accounting" and a supported task could be "balancing." Thereby, the definition of "balancing" has to be made explicit. A description of the business semantics may for example point out that a service supports a balancing process according to the US-GAAP as opposed to a balancing process according to the IFRS. The information provided on this abstraction level allows users to assess if a software artifact fulfills functional requirements and if it implements the desired business tasks.

The *software architecture* is determined during the technical design phase of the software artifact and describes its programming interface. It provides information about how to technically integrate the artifact into a composite application. A software architecture description could for example document the in- and outputs and the pre- and post-conditions of interface operations [15, 26]. Based on this information, it becomes possible to correctly invoke operations at run-time.

The *quality* of a software artifact is determined by its implementation. It is described in the form of quality characteristics and metrics [12]. Quality characteristics could for example document that a service has a mean response time of less than 0.2 seconds or that it has a throughput of 150 calls per second. Such a description allows users to assess the fulfillment of non-functional requirements such as the efficiency or the reliability of a service.

The information provided on the different abstraction levels targets varying user groups (such as designers, analysts, business people, personnel responsible for procurement etc.). Accordingly, the software description should articulate the information contained in the various abstraction levels in language formats that are understandable for the respective target groups. The business semantics should therefore be documented using an end-user oriented format. The software architecture should be described in a computer-oriented (e.g., a programming) format, and the quality should preferably be represented as a list of characteristics that is grouped into service levels.

To get a more detailed overview what has to be specified on each abstraction level, they can be further structured according to the generic model views of General Systems Theory [5]. These model views are similar to the perspectives that are traditionally distinguished in modeling disciplines [3].

The *static view* describes structural properties of a software artifact. Such properties range from the processed information items (things) and organizational aspects (like the involved stakeholders) on the business semantics level to type declarations on the software architecture level to fixed characteristics like the usability or maintainability on the quality level.

The *functional view* focuses on the capabilities of a software artifact. This view documents the supported business tasks and activities on the business semantics level, the pre- and post-conditions of interface operations on the architectural level, as well as the provided security or persistency on the quality level. Interestingly, the ISO 9126 quality model defines extra-functional attributes such as the security or the persistency as *functionality* that is provided additionally [12].

The *dynamic view* specifies how a software artifact executes at run-time. The execution is documented as business process or work-flow on the business semantics level. On the software architecture level, it is described by the timing constraints and interactions between the operations of the programming interface. Properties like the reliability or the efficiency characterize the execution of a software artifact on the quality level.

To assess whether a software artifact in general or a service in particular is reusable in a certain context, all three abstraction levels have to be analyzed and compared to the requirements by the consumer. To completely specify the functionality of a software artifact, a description language should accordingly be able to document all of the above-mentioned properties. From a theoretical perspective, these properties hence form a framework against which description languages can be evaluated.

The resulting evaluation framework is depicted in Table 21.1. In the following, we use it to evaluate the expressive power of the USDL. Note that a description

Table 21.1: Aspects of a comprehensive specification.

Systems View \ Abstraction level	Business level (conceptual design)	Architectural level (technical design)	Quality level (implementation)
Static view (structure)	Organizations, stakeholders, information items	Signatures (type and interface declarations)	Usability, maintainability, portability
Functional view (capabilities)	Tasks, activities, events	Assertions (pre- and post-conditions, invariants)	Functionality (security, persistency, transactions)
Dynamic view (execution)	Processes, work-flows	Timing constraints (interaction protocols)	Reliability, efficiency

which covers all abstraction levels and model views is still a black-box specification as long as it solely describes what a software artifact does without detailing on how this is achieved algorithmically. It is thus up to the providers to decide whether they use the specification aspects depicted in Table 21.1 to document external or internal properties of a software artifact. While external properties specify what a software artifact does, internal properties document how this is achieved.

21.4 Evaluation of the USDL

Comprehensively specifying software services leads to complex documents that contain contents for different target groups. Like related approaches, the USDL therefore makes use of a structuring mechanism to organize and thematically group the different aspects of service descriptions. To separate specification parts with different contents, the USDL differentiates between a business perspective, an operational perspective, and a technical perspective as core clusters. Thereby, general, legal, and commercial information belong to the business core. The description of the business semantics and the software architecture belongs to the operational core. The technical core has been reduced to only contain interactions and the protocol information.

When analyzing the USDL structuring against the theoretically motivated grouping of specification content as illustrated in Table 21.1, it becomes evident that abstraction levels and system views are currently intermixed. The operational perspective in fact contains information that describes services from a domain-oriented (business semantics) and an architectural (programming interfaces) perspective. Nevertheless, the USDL is among the few approaches to provide a holistic service description which addresses all areas of the defined evaluation framework.

In the remainder of this section, the USDL is analyzed for completeness based on the defined framework. Besides, the analysis reveals how comprehensive the USDL is compared to related approaches regarding the *general and commercial* information (21.4.1), the information about the *business semantics* (21.4.2) and the *software architecture* (21.4.3), as well as the information about the *quality* (21.4.4).

21.4.1 General and Commercial Information

To support the reuse of software services, a service description needs to provide information about the provider as well as the commercial and legal conditions under which a service may be used. Additionally, to uniformly catalog services in marketplaces and to support an efficient browsing for services with a certain property, software services need to be classified according to standard taxonomies. Therefore, various classification systems have been introduced which have repeatedly been applied to catalog reusable software artifacts in general and services in particular.

The specification of general and commercial properties has only been addressed by some related approaches yet. The UDDI standard for example supports a specification of the organization that provides a service. As part of the UDDI white pages, one can describe the company's name, business identifiers (e.g., tax number, D-U-N-S number etc.), addresses, administrative contacts, and communication media. The Standardized Specification of Business Components framework supports the description of the provider name, administrative contacts, prices, license agreements, as well as the scope of supply. WS-Specification combines both approaches and introduces a distribution channel concept to specify pricing, scope of supply, and legal constraints. For each distribution channel, the pricing, scope of supply, and legal constraints can be specified separately. A software service can so be offered with different pricing policies (e.g., pay per use, flat fees) and license agreements that are adapted for each distribution channel.

Compared to these approaches, the USDL offers a similar, in most aspects even superior functionality. Other than the specification of distribution channels in WS-Specification, the USDL Pricing Module offers a more detailed functionality that covers several pricing models. Additionally, the description of the provider and the administrative contacts is more refined in the USDL than in other approaches. With the Participants Module, providing organizations can be specified in detail. In contrast to most other approaches, the USDL makes use of a highly customizable design pattern for many elements of the meta-model such as the VirtualAddress or the AdministrativeArea. For example, instead of hard-coding the different parts of a VirtualAddress, the element is designed to consist of multiple generic address parts. Each address part can be modeled as a structure that contains a string and a type. Address part types can be listed as a taxonomy (enumeration), which can dynamically be updated to cover addresses of different regions in the world.

Apart from this general information, related approaches such as the UDDI or WS-Specification also support to classify various aspects of software services like the offered functionality, the underlying technology, or the provider location. For the classification, standard taxonomies with keywords are used. The following taxonomies are commonly used to classify software services:

- the North American Industry Classification System (NAICS)
- the United Nations Standard Products and Services Code (UNSPSC)
- the Microsoft Geo Web (MSGEO)

The Foundation Module of the USDL contains a Classification element which supports classifying description elements into defined taxonomies. In the Service Module, this element is used to classify the functionality of services. Among others, the NAICS and the UNSPSC are applicable to classify the functionality of a service. USDL hence provides a similar capability for classifying services as it can be found in related approaches. All in all, USDL thus provides the most detailed support for the description of relevant general and commercial information.

21.4.2 Information about the Business Semantics

To efficiently assess whether the functionality of services is suited to fulfill existing requirements, users need to know about the business semantics that a service implements. The business semantics usually is difficult to analyze on the basis of the programming interfaces. For example, it is cumbersome to assess if an interface operation "createBalance" implements a balancing strategy according to the US-GAAP or the IFRS.

The specification of the implemented business semantics has not been addressed by all approaches to comprehensively describe reusable software artifacts. In fact, specifying the business semantics of software parts only became a topic of interest as the complexity of the artifacts began to cause assessment problems. Coarse-grained software units such as components and software services have to be described more comprehensively to make their complex functionality understandable. In literature, there exist approaches to augment programming interfaces with business terms as keywords [28]. More sophisticated approaches such as the Standardized Specification of Business Components framework and WS-Specification support a specification of the business semantics on the basis of business concepts, which can be put into relationship with the programming interfaces. WS-Specification uses a system of interrelated concepts to describe the business semantics as ontology. Its concept types and relationships are deduced from the conceptual modeling domain, which utilizes conceptual models to capture and communicate a formal understanding of business domains [21].

Compared to such approaches, the USDL offers a similar functionality to specify the business semantics of services. The Functional Module makes use of capability modeling approaches to specify the abstract functionality of services. Capabilities express the ability to perform a course of action and consist of actions. Actions, however, are characterized as being service-internal actions that should only be used in a glass-box specification. Both capabilities and actions can have input and output parameters, faults, and conditions. Capabilities, actions, parameters, faults, and conditions can be linked to the technical interface if they have direct counterparts (see Section 21.4.3).

From an analytical point of view, the current way of describing business semantics in the USDL also has to be criticized in some ways though. First of all, the artificial distinction between capabilities and actions only partially corresponds to the way in which business functionality is modeled traditionally. Especially in conceptual modeling approaches, business functionality is specified as a set of business functions. Business functions can be decomposed hierarchically to depict a stepwise refinement from coarse-grained top-level business functions to more elementary business actions which might be automated by software applications. In the USDL, complex business functions would have to be described as capabilities, while more elementary business functions would have to be specified as actions. A more gradual refinement cannot be modeled as the USDL only makes use of these two hierarchy levels.[1]

Furthermore, business information can only be described as input or output parameters to business functions (i.e., capabilities or actions). It is not possible to define complex business information objects which usually consist of several parts and have both inheritance-based as well as compositional relationships to other information objects. As the business information that is processed by software services usually is complex, it is necessary to describe the business semantics of this information comprehensively. Finally, in order to describe a course of actions, schema elements which describe business processes ought to be introduced. Business processes describe temporal relationships between business functions and can be used to show how the operations of a software service have to be used together. Such a description of business processes would furthermore complement the specification of technical interactions, which is already supported in the meta-model.

The business semantics that actually needs to be described hence is in some parts more complex compared to what is currently proposed by the USDL. Furthermore, a service description should only specify externally accessible business functions, exchanged information objects, and visible business processes in order to follow the aspired black-box strategy. In contrast to what is stated in the USDL specifications, we hence recommend avoiding the inclusion of service internal information objects and functions into a service description. All in all, it can be concluded that USDL already covers important aspects regarding the description of the business semantics, but some points still require further improvement which ought to be addressed in future versions.

21.4.3 Information about the Software Architecture

To technically bind and execute software services, the programming interface has to be specified and communicated to the service consumer. From a theoretical point of view, the programming interface defines the software architecture of a service. It consists of interface signatures, assertions (pre-, post-conditions and invariants) and

[1] This has already been addressed in the latest milestone M5.

timing constraints that might exist between the operations of an interface. As the technical binding of software services is a central step to realize a working composite application, there already exist numerous approaches to specify the necessary characteristics of services. These approaches build upon well-researched findings from the distributed systems domain. For software services, the most widespread approach to describe programming interfaces is the Web Service Description Language (WSDL), version 1.1.

The WSDL is a standardized specification language to define signature lists. It therefore uses the following objects: Services, Endpoints, Bindings, Interfaces, Operations, and Types [8]. Services and Endpoints generally describe where to find a service. Bindings specify the concrete protocol as well as the data formats for a certain port. Types and Operations define the corresponding data types and interface methods independently of a concrete communication technology. Related data types and interface methods are grouped together by the Interface construct. The WSDL does not provide mechanisms to specify pre- and post-conditions or interaction protocols, but only documents structural properties of the software architecture. Other approaches, such as the Semantic Annotations for WSDL and XML Schema (SAWSDL), have upgraded the WSDL accordingly. This approach allows users to annotate WSDL files with additional semantics [10] (cf. also Chapter 7). The SAWSDL therefore specifies how semantic annotations can be made within the WSDL in order to make a reference to semantic models such as ontologies. In contrast to the WSDL, the SAWSDL approach hence also allows for the description of functional aspects of the software architecture. Yet, the SAWSDL does not specify a language to describe these semantic models, but only shows how they can be referenced to when using WSDL.

The Web Ontology Language for Services (OWL-S) goes beyond the scope of the WSDL. Regarding the technical aspects, it for instance furthermore describes pre- and post-conditions [14]. It also allows specifying timing constraints and therefore describes all three aspects of the software architecture of a service as depicted in Table 21.1. As well as the WSDL, the OWL-S uses an XML-based representation format. Other approaches such as the Service-Oriented Architecture Modeling Language (SOAML) or the Systems Modeling Language (SysML) build upon the UML to describe the software architecture of a service. These languages have a broad scope and can be used to specify signature lists, assertions, and timing constraints amongst others [18, 19]. Consequently, all of the three views of the software architecture can also be described using graphical languages.

To describe the software architecture of services, the Functional Module of the USDL supports the specification of technical interfaces, operations, parameters, and faults. To describe interface signatures, the TechnicalOperation element has been introduced. It covers the specification of an interface method. Besides the method name, parameter names, types, and faults can be specified. However, in contrast to most related approaches and programming languages, the USDL does not support the definition of complex data types. Although the TechnicalOperation includes inputs/outputs which are used/created during the execution of an operation, these in- and outputs cannot be further specified. Therefore, in the next version of USDL, a TypeDeclaration element should be introduced along with a connection to the TechnicalInterface. The TypeDeclaration element could be utilized to define the in- and outputs used during the execution of the operation. In particular, composite data types such as enumerations or compositions of primitive data types (structures) could be defined so that they can later on be referred to. Another option would be to refer to corresponding WSDL definitions. However, in such a case the USDL would become dependent on additional languages to specify details.

The USDL also supports the specification of functional software architecture properties such as pre- and post-conditions as well as incoming and outgoing faults. Pre- and post-conditions describe which conditions have to be fulfilled before an operation can start and which conditions will result after the execution of an operation. Faults report unexpected states during the execution of a service, which can either be reported or reacted to. The definition of functional architecture properties is fully comparable to that supported by related approaches.

In addition, dynamic architecture properties such as timing constraints and interactions between interface operations can be specified with the USDL. To define such aspects, the USDL provides a specific Interaction Module which implements a less widespread, block-structured approach to define interactions. However, it seems that this block-structured approach might not be scalable to support the specification of arbitrarily complex interaction protocols, as the USDL allows service providers to leave the ordering of interactions unspecified and attach a formal definition of the protocol in a formalism of their choice. Such a procedure has to be criticized as it does not enforce homogeneous service descriptions which are comparable. We therefore recommend implementing a more widespread graph-based approach to specify interaction protocols which is able to handle intuitive and complex protocols on the same basis. An optimal representation of interaction protocols to service consumers will then have to be reached by choosing an adequate formalism. As the technical interaction protocol is usually assessed by designers and programmers, graphical notations (e.g., the UML state chart) should be adequate notations. To sum up, it can be concluded that the USDL addresses the relevant aspects of software architecture information. However, it still has to be improved in some details.

21.4.4 Information about the Quality

In the theoretical evaluation framework the third abstraction level describes the quality aspects of software artifacts. Quality considerations are an important part when specifying the functionality of software and, thus, are part of most approaches to specify software components and services as well. Quality assurances are commonly realized through the concept of service level agreements.

As a "comprehensive specification and evaluation of software product quality is a key factor in ensuring adequate quality," the International Organization for Standardization (ISO) has defined the ISO 9126 standard [12]. This "Software engineering - Product quality" standard covers a quality model, internal and external metrics, as well as quality-in-use metrics. The quality model itself, as developed by the ISO, categorizes software quality attributes into six different characteristics:

- the *usability* characteristic is defined as the capability of services to be understood, learned and used, as well as being attractive to the user under specified conditions
- the *maintainability* characteristic is defined as the capability of services to be altered for corrections and improvements or be adapted for changes in the environment, requirements or functional specifications
- the *portability* characteristic is defined as the capability of services to be relocated between different environments
- the *functionality* characteristic is defined as the capability of services to provide functions which meet predefined, as well as implicit needs with respect to specified conditions
- the *reliability* characteristic is defined as the capability of services to maintain a predefined level of performance with respect to specified conditions
- the *efficiency* characteristic is defined as the capability of services to provide suitable performance in relation to the amount of resources expended under specified conditions

All of them are further subdivided into sub-characteristics, which can be measured by internal or external metrics. Thus, the result is a strictly hierarchical model as depicted in Figure 21.4.

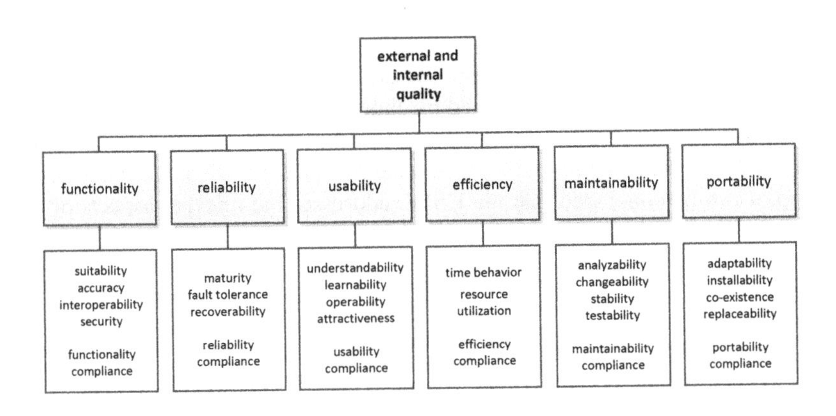

Fig. 21.4: Quality model for external and internal quality [12].

The current USDL version recognizes the importance of quality specifications and, hence, envisions a service level package as one of the central modules. However, in the current USDL specification it is still under investigation and only a rough draft of the corresponding module exists. In this draft, numerous aspects of related approaches (e.g., excerpts of several different WS-* specifications) are included. In its current version, the Service Level Module in the USDL already covers several of the above-mentioned characteristics. Within others, it contains security, transactional behavior, reliability, performance and availability aspects. Furthermore it defines concrete quality attributes to further refine the basic quality aspects. However, it does not cover all of the characteristics described in the ISO quality model (e.g., usability aspects are not yet included). While some of them might be negligible, as the ISO standard was developed for traditional software products, others ought to be examined more closely.

One possible impairment might be the lack of any clear structure in the current USDL Service Level Module. First of all, the contained characteristics are described on one level, so there is no hierarchy as implied by the ISO standard. Moreover, most elements are explicitly related in some way and additional implicit relations are included through a color coding. The upcoming module should therefore build upon the existing ISO standard for software product quality. In so doing, missing characteristics could be included together with a more sophisticated, standardized hierarchy of characteristics.

21.5 Conclusions

Due to the existing plethora of description languages and their different motivations and representation formats, the searching, comparison, and trading of software services today is a complicated task. In order to facilitate the emergence of an Internet of Services, the USDL aims at providing a coherent and complete source of information to select and integrate services. In this chapter, we have analyzed the expressive power of the USDL to specify software services from a theoretical point of view. Therefore, we proposed a theoretically motivated evaluation framework with several description aspects that were found to be relevant in literature. Doing so allowed us to evaluate in how far the USDL in its current form is capable of providing a universal and complete source of information. It moreover revealed areas where the USDL lacks details and future work should concentrate on.

The evaluation of the USDL against a theoretical framework and related approaches has shown that the USDL aims at documenting all relevant specification perspectives. All of the abstraction levels discussed in Section 21.3 have been addressed with the USDL as the language considers the specification of marketing-related information as well as of functionality-related information. Regarding the functionality-related information, it particularly endorses the specification of business semantics next to technical and quality properties. As regards the state of the art of description languages for software services, the business semantics of a service can only be specified by a very few approaches. Consequently, the USDL can indeed be regarded as a comprehensive approach to specify services.

As the evaluation has shown, the concepts provided by the USDL to specify software services are in many aspects superior to the current state of the art or at least comparable (see Table 21.2). This is especially true for the description of general

Table 21.2: Summary of evaluation results

Abstraction Level	USDL description content	Comparison to state of the art
General Information	Vendor information	+
	Legal information	+
	Commercial information (pricing, distribution)	++
Business	Information items	+
	Capabilities	-
	Processes	-
Architecture	Signatures (interface declarations)	+
	Assertions (pre- and post-conditions, invariants)	++
	Timing constraints (interaction protocols)	+
Quality	Usability, maintainability, portability	-
	Functionality (security, persistency, transactions etc.)	+
	Reliability, efficiency	+

Legend: ++: superior, +: comparable, -: inferior

and commercial information and the classification of the functionality of services. With some limitations regarding the description of interaction protocols, this can also be attested for the documentation of the software architecture. Other aspects of the USDL still lack important details or reveal room for improvements, however. Especially the description of quality attributes still seems to be under investigation. Hence, various quality attributes such as the usability of services cannot be described at all yet. Furthermore, the quality module of the USDL is lacking a clear hierarchical structure. We therefore propose building revisions of this module on the basis of a conceptual quality model, e.g., the ISO 9126 model of software quality [12]. Room for improvements was also identified with respect to the specification of the business semantics of services. In particular, specifying stepwise refinements of business functions and describing the execution of business functions according to a business process ought to be better supported in future versions of the USDL.

While the presented results provide several insights regarding the adequacy of the USDL for the description of services, they are nevertheless subject to some limitations. Most notably, we have limited our analysis to software services and left out the examination of business services. The reasons behind this decision are two-fold. Firstly, approaches aiming at the trading and reusing of software services are expected to build a cornerstone for the Internet of Services [4]. Analyzing the suitability of the USDL to describe software services therefore is of particular importance. Secondly, the authors have already gathered several years of expertise in the description of reusable software artifacts which could be incorporated into the analysis. However, as a consequence of the chosen evaluation strategy, additional shortcomings of the USDL regarding the specification of business services might have remained undiscovered during our analysis. Evaluating the completeness of the USDL as to the description of business services accordingly is an important direction for future research. Furthermore, we did not discuss the trade-off between the completeness of a description on the one hand and its usability on the other hand. Usually, more complete descriptions are more difficult to create and cannot be algorithmically processed anymore [20]. As such a trade-off has to be taken into account during the design of a specification language, sub-optimal solutions are sometimes unavoidable.

Based on the results of our evaluation, the USDL nevertheless can be classified as a vendor-independent approach that overall appears to be suited to facilitate the trading of software services. All in all, the USDL seems to be on a promising way to become a complete source of information that supports the assessment of software services by consumers. Fundamentally, it aims at specifying all aspects of a service description that were found to be necessary in the corresponding literature. As the USDL is still in the standardization process, the responsible W3C Incubator Group can address aspects in which details are still missing or requiring improvement. Especially, it should be possible to eliminate identified shortcomings and to effectively complement additional aspects that were found to be missing during our evaluation. With our work, we therefore hope to contribute to the ongoing endeavor of making USDL a comprehensive description language for the Internet of Services.

References

1. J. Ackermann, F. Brinkop, P. Fettke, A. Frick, E. Glistau, H. Jaekel, O. Kotlar, P. Loos, H. Mrech, E. Ortner, S. Overhage, U. Raape, S. Sahm, A. Schmietendorf, T. Teschke, and K. Turowski. Standardized Specification of Business Components. Technical report, German Society of Informatics (GI), 2002.
2. G. A. Akerlof. The Market for "Lemons": Quality Uncertainty and the Market Mechanism. *The Quarterly Journal of Economics*, 84(3):488–500, 1970.
3. I. Arbnor and B. Bjerke. *Methodology for Creating Business Knowledge*. Sage Publications, London, 3 edition, 2009.
4. A. P. Barros and M. Dumas. The Rise of Web Service Ecosystems. *IEEE IT Professional*, 8(5):31–37, 2006.
5. L. v. Bertalanffy. *General System Theory*. George Braziller, New York, NY, 1976.
6. A. Beugnard, J.-M. Jézéquel, N. Plouzeau, and D. Watkins. Making Components Contract Aware. *IEEE Computer*, 32(7):38–45, 1999.
7. M. Bunge. *Treatise on Basic Philosophy, Volume 3: Ontology I: The Furniture of the World*. Reidel, Boston, MA, 1977.
8. R. Chinnici, J.-J. Moreau, A. Ryman, and S. Weerawarana. Web Services Description Language (WSDL) Version 2.0 Part 1: Core Language. Technical report, World Wide Web Consortium, 2007.
9. D. F. D'Souza and A. C. Wills. *Objects, Components, and Frameworks with UML: The Catalysis Approach*. The Object Technology Series. Addison-Wesley, Upper Saddle River, NJ, 1999.
10. J. Farrell and H. Lausen. Semantic annotations for wsdl and xml schema. Recommendation, W3C, 2007.

11. N. Gehani and A. T. McGettrick. *Software Specification Techniques*. Addison-Wesley, Wokingham, 1986.
12. ISO/IEC. Software Engineering - Product Quality - Part 1: Quality Model. Technical Report ISO/IEC Standard 9126-1, International Organization for Standardization, 2001.
13. B. H. Liskov and V. Berzins. *Software Specification Techniques*, chapter An Appraisal of Program Specifications, pages 3–24. Addison-Wesley, Upper Saddle River, NJ, 1986.
14. D. Martin, M. Paolucci, S. McIlraith, M. Burstein, D. McDermott, D. McGuinness, B. Parsia, T. Payne, M. Sabou, M. Solanki, N. Srinivasan, and K. Sycara. Bringing Semantics to Web Services: The OWL-S Approach. In J. Cardoso and A. P. Sheth, editors, *Semantic Web Services and Web Process Composition, First International Workshop, SWSWPC 2004, Revised Selected Papers*, volume 3387, pages 26–42. Springer, Berlin, Heidelberg, 2005.
15. B. Meyer. *Object-Oriented Software Construction*. Prentice Hall, Englewood Cliffs, NJ, 2. edition, 1997.
16. P. Nelson. Information and Consumer Behavior. *The Journal of Political Economy*, 78(2):311–329, 1970.
17. T. W. Olle, J. Hagelstein, I. G. MacDonald, and C. Rolland. *Information Systems Methodologies. A Framework for Understanding*. Addison-Wesley, Wokingham, 1991.
18. OMG. Service oriented architecture Modeling Language (SoaML). Revised Submission ad/2008-11-01, Object Management Group, 2008.
19. OMG. OMG Systems Modeling Language (SysML), Version 1.2. OMG Specification formal/2010-06-02, Object Management Group, 2010.
20. S. Overhage. UnSCom: A Standardized Framework for the Specification of Software Components. In M. Weske and P. Liggesmeyer, editors, *Object-Oriented and Internet-Based Technologies, 5th Annual International Conference on Object-Oriented and Internet-Based Technologies, Concepts, and Applications for a NetworkedWorld, Net.ObjectDays 2004*, volume 3263 of *Lecture Notes in Computer Science*, pages 169–184. Springer, Berlin, Heidelberg, 2004.
21. S. Overhage. Vereinheitlichte Spezifikation von Komponenten: Grundlagen, UnSCom Spezifikationsrahmen und Anwendung. Dissertation, Universität Augsburg, 2006.
22. S. Overhage and P. Thomas. WS-Specification: Specifying Web Services Using UDDI Improvements. In A. B. Chaudri, M. Jeckle, E. Rahm, and R. Unland, editors, *Web, Web-Services, and Database Systems, NODe 2002 Web and Database-Related Workshops, Revised Papers*, volume 2593 of *Lecture Notes in Computer Science*, pages 100–119. Springer, Berlin, Heidelberg, 2002.
23. S. Overhage and P. Thomas. WS-Specification: Ein Spezifikationsrahmen zur Beschreibung von Web-Services auf Basis des UDDI-Standards. In O. Ferstl, E. J. Sinz, S. Eckert, and T. Isselhorst, editors, *Wirtschaftsinformatik 2005: eEconomy, eGovernment, eSociety*, pages 1539–1558. Physica, Heidelberg, 2005.
24. M. P. Papazoglou, P. Traverso, S. Dustdar, and F. Leymann. Service-Oriented Computing: State of the Art and Research Challenges. *IEEE Computer*, 40(11):38–45, 2007.
25. A. Scheer. *ARIS - Business Process Frameworks*. Springer, Berlin, Heidelberg, 3. edition, 2000.
26. M. Shaw and D. Garlan. *Software Architecture: Perspectives on an Emerging Discipline*. Prentice Hall, Englewood Cliffs, NJ, 1996.
27. C. Szyperski, D. Gruntz, and S. Murer. *Component Software. Beyond Object-Oriented Programming*. Addison-Wesley, Harlow, 2. edition, 2002.
28. P. Vitharana, F. Zahedi, and H. Jain. Knowledge-Based Repository Scheme for Storing and Retrieving Business Components: A Theoretical Design and an Empirical Analysis. *IEEE Transactions on Software Engineering*, 29(7):649–664, 2003.

Handbook of Service Description